NUMERICAL METHODS FOR TWO-POINT BOUNDARY-VALUE PROBLEMS

Herbert B. Keller
CALIFORNIA INSTITUTE OF TECHNOLOGY

DOVER PUBLICATIONS, INC.
NEW YORK

To the memory of my parents

Published in Canada by General Publishing Company, Ltd., 30 Lesmill Road, Don Mills, Toronto, Ontario.
Published in the United Kingdom by Constable and Company, Ltd., 3 The Lanchesters, 162–164 Fulham Palace Road, London W6 9ER.

This Dover edition, first published in 1992, is an unabridged, corrected, and expanded republication of the work first published by the Blaisdell Publishing Company, Waltham, Massachusetts, in 1964. Two appendices have been added following the Index to the original work containing additional material by Herbert B. Keller and co-authors. Bibliographical information concerning these appendices will be found in the Preface to the Dover Edition, and in the acknowledgements at the start of each section. The publisher is grateful to the organizations and journals which gave permission for their material to be reprinted here.

Manufactured in the United States of America
Dover Publications, Inc., 31 East 2nd Street, Mineola, New York 11501

Library of Congress Cataloging-in-Publication Data

Keller, Herbert Bishop.
Numerical methods for two-point boundary-value problems / Herbert B. Keller.
p. cm.
Includes bibliographical references and index.
ISBN 0-486-66925-4
1. Boundary value problems—Numerical solutions. I. Title.
QA379.K43 1992
515′.35—dc20 92-31818
CIP

PREFACE

In this monograph we present a brief, elementary yet rigorous account of practical numerical methods for solving very general two-point boundary-value problems. Three techniques are studied in detail: initial value or "shooting" methods in Chapter 2, finite difference methods in Chapter 3, integral equation methods in Chapter 4. Each method is applied to nonlinear second-order problems, and the first two methods are applied to first-order systems of nonlinear equations. Sturm-Liouville eigenvalue problems are treated with all three techniques in Chapter 5. The application of shooting to generalized or nonlinear eigenvalue problems is also discussed in Chapter 5.

Initial-value methods are seldom advocated in the literature, but we find them extremely practical and theoretically powerful. A modification, called parallel shooting, is introduced to treat those "unstable" cases (with rapidly growing solutions) for which ordinary shooting may be inadequate. High-order accurate methods are stressed, and the well-developed theory of numerical methods for initial-value problems is employed to obtain corresponding results for boundary-value problems. Continuity techniques (related to imbedding) are discussed and illustrated in the examples.

To help maintain an elementary level, we include in Chapter 1 brief accounts of some of the basic prerequisites: existence theorems for initial-value and two-point boundary-value problems, numerical methods for initial-value problems and iterative methods for solving nonlinear systems. In addition, several other areas of numerical analysis enter our study, such as numerical quadrature, eigenvalue-eigenvector calculation, and solution of linear algebraic systems. Since so many diverse numerical procedures play an important role, the subject matter of this monograph is useful in motivating, introducing, and/or reviewing more general studies in numerical analysis. The level of presentation is such that the text could be used by well-prepared undergraduate mathematics students. Advanced calculus, basic numerical analysis, some ordinary differential equations, and a smattering of linear algebra are assumed.

Many of the problems contain extensions of the text material and so should at least be read, if not worked out. More difficult problems are starred.

It is a pleasure to acknowledge the helpful advice received from Professor Eugene Isaacson and Mr. Richard Swenson, both of whom read the entire manuscript. The computations for the examples were accomplished by the superior work of W. H. Mitchell at Caltech and J. Steadman at the Courant Institute, N.Y.U. The typing was done by Miss Connie Engle in her usual flawless manner.

Finally, the support of the U.S. Army Research Office (Durham) and the U.S. Atomic Energy Commission is gratefully acknowledged.

H. B. K.

PREFACE TO THE DOVER EDITION

This edition contains numerous minor corrections to the original edition, a new Appendix B containing my CBMS/NSF regional conference monograph: *Numerical Solution of Two-Point Boundary-Value Problems* published by SIAM (1976) and a new Appendix C, *Some Further Results*, containing nine research papers. Of course much recent work, and in particular the development of powerful general purpose codes, could not be included.

Several areas which have been particularly active in the past ten years or so are not well covered. This includes the treatment of boundary layers, adaptive grid generation, collocation methods, path following and bifurcation theory, multiple or parallel shooting and problems posed on infinite intervals. The powerful continuation and bifurcation code AUTO developed by E. Doedel applies and extends the methods of §5.6 (or Appendix C.7) using collocation for solving very general boundary-value problems depending upon parameters. The methods in AUTO include automatic mesh generation and refinement to obtain specified accuracy and to resolve boundary layers. Thus it is a very advanced code and I judge it to be the most powerful such software currently available.

With the rapid development of parallel or concurrent computers we are beginning to see corresponding changes in algorithm development for two-point boundary-value problems. Our choice of the name "parallel shooting" in §2.4 was based on the early stages of such designs and indeed these methods are becoming more important. The recent interest in dynamical systems is also beginning to influence the numerical methods. In particular the study of homoclinic and heteroclinic orbits must employ methods valid for infinite intervals.

I hope that this somewhat updated edition will help those interested in the theory and practice of solving two-point boundary-value problems.

H. B. KELLER

Pasadena, May, 1991.

CONTENTS

CHAPTER 1

INTRODUCTION

1.1 Initial-Value Problems

The theory of boundary-value problems for ordinary differential equations relies rather heavily on initial-value problems. Even more significant for the subject of this monograph is the fact that some of the most generally applicable numerical methods for solving boundary-value problems employ initial-value problems. Thus we must assume that the reader is somewhat familiar with the existence and uniqueness theory of, as well as numerical methods for solving, initial-value problems. We shall review here and in Section 1.3 some of the basic results required.

Since every nth-order ordinary differential equation can be replaced by an equivalent system of n first-order equations, we confine our attention to first-order systems of the form

$$\mathbf{u}' = \mathbf{f}(x; \mathbf{u}).$$

Here $\mathbf{u} \equiv (u_1, u_2, \ldots, u_n)^T$ is an n-dimensional column vector with the dependent variables $u_k(x)$ as components; we say that $\mathbf{u}(x)$ is a vector-valued function; $\mathbf{f}(x; \mathbf{u})$ is vector-valued with components $f_k(x, u_1, u_2, \ldots, u_n)$, which are functions of the $n + 1$ variables $(x; \mathbf{u})$. An initial-value problem for the above system is obtained by prescribing at some point, say $x = a$, the value of $\mathbf{u}$, say

$$\mathbf{u}(a) = \boldsymbol{\alpha}.$$

The existence, uniqueness, and continuity properties of the solutions of such problems depend on the continuity or smoothness properties of the function $\mathbf{f}$ in a neighborhood of the initial point $(a; \boldsymbol{\alpha})$. We shall use as a measure of distance between two points in n-space the *maximum norm*

$$|\mathbf{u} - \mathbf{v}| \equiv \max_{1 \le k \le n} |u_k - v_k|.$$

However, all results are equally valid if we employ the *Euclidean norm*

$$|\mathbf{u} - \mathbf{v}|_2 \equiv [(u_1 - v_1)^2 + (u_2 - v_2)^2 + \cdots + (u_n - v_n)^2]^{1/2}.$$

One of the basic results can now be stated as follows.

THEOREM 1.1.1. *Let the function* $\mathbf{f}(x; \mathbf{u})$ *be continuous on the infinite strip*

$$R: a \leq x \leq b, \qquad |\mathbf{u}| < \infty$$

and satisfy there a Lipschitz condition in $\mathbf{u}$ *with constant K, uniformly in x; that is,*

$$|\mathbf{f}(x; \mathbf{u}) - \mathbf{f}(x; \mathbf{v})| \leq K|\mathbf{u} - \mathbf{v}| \quad \textit{for all} \quad (x; \mathbf{u}) \quad \textit{and} \quad (x; \mathbf{v}) \in R.$$

Then

(a) *the initial-value problem*

$$\mathbf{u}' = \mathbf{f}(x; \mathbf{u}), \qquad \mathbf{u}(a) = \boldsymbol{\alpha} \tag{1.1.1}$$

has a unique solution $\mathbf{u} = \mathbf{u}(x; \boldsymbol{\alpha})$ *defined on the interval*

$$[a, b] \equiv \{x \mid a \leq x \leq b\};$$

(*b*) *this solution is Lipschitz-continuous in* $\boldsymbol{\alpha}$, *uniformly in x; in fact we have*

$$|\mathbf{u}(x; \boldsymbol{\alpha}) - \mathbf{u}(x; \boldsymbol{\beta})| \leq e^{K(x-a)}|\boldsymbol{\alpha} - \boldsymbol{\beta}| \quad \textit{for all} \quad (x; \boldsymbol{\alpha}) \quad \textit{and} \quad (x; \boldsymbol{\beta}) \in R.$$

Proof. We merely sketch the proof; more details can be found in Ince (1944), pp. 62–72. If a solution exists, we obtain from (1.1.1) by integration

$$\mathbf{u}(x) = \boldsymbol{\alpha} + \int_a^x \mathbf{f}(\xi; \mathbf{u}(\xi))\, d\xi.$$

Conversely, if $\mathbf{u}(x)$ is continuous and satisfies this integral equation, it is differentiable and hence, by differentiation, satisfies (1.1.1). We now construct a solution of this integral equation by means of the *Picard iteration procedure*

$$\mathbf{u}^{(0)}(x) \equiv \boldsymbol{\alpha},$$

$$\mathbf{u}^{(\nu+1)}(x) = \boldsymbol{\alpha} + \int_a^x \mathbf{f}(\xi; \mathbf{u}^{(\nu)}(\xi))\, d\xi, \qquad \nu = 0, 1, \ldots .$$

From the Lipschitz continuity of $\mathbf{f}(x; \mathbf{u})$ we have

$$|\mathbf{u}^{(\nu+1)}(x) - \mathbf{u}^{(\nu)}(x)| \leq \int_a^x K|\mathbf{u}^{(\nu)}(\xi) - \mathbf{u}^{(\nu-1)}(\xi)|\, d\xi, \qquad \nu = 1, 2, \ldots .$$

Then, by induction, with $|f(x; \boldsymbol{\alpha})| \leq M$ on $[a, b]$, it follows that

$$|\mathbf{u}^{(\nu+1)}(x) - \mathbf{u}^{(\nu)}(x)| \leq \frac{M}{K} \frac{(K[x - a])^{\nu+1}}{(\nu + 1)!}.$$

Now the sequence of continuous functions $\{\mathbf{u}^{(\nu)}(x)\}$ can be shown to converge uniformly on $[a, b]$ since

$$\mathbf{u}^{(\nu+1)}(x) = \mathbf{u}^0 + \sum_{\mu=0}^{\nu} [\mathbf{u}^{(\mu+1)}(x) - \mathbf{u}^{(\mu)}(x)].$$

The limiting function clearly satisfies the integral equation and hence (1.1.1); thus, existence is established for any $\boldsymbol{\alpha}$.

The uniqueness of this solution will follow from the continuity of $\mathbf{u}(x, \boldsymbol{\beta})$ by letting $\boldsymbol{\beta} \to \boldsymbol{\alpha}$.

To demonstrate (b) we form

$$[\mathbf{u}(x; \boldsymbol{\alpha}) - \mathbf{u}(x; \boldsymbol{\beta})] = [\boldsymbol{\alpha} - \boldsymbol{\beta}] + \int_a^x [\mathbf{f}(\xi; \mathbf{u}(\xi; \boldsymbol{\alpha})) - \mathbf{f}(\xi; \mathbf{u}(\xi; \boldsymbol{\beta}))]\, d\xi.$$

Then, by the Lipschitz continuity of $\mathbf{f}$, it follows that

$$|\mathbf{u}(x; \boldsymbol{\alpha}) - \mathbf{u}(x; \boldsymbol{\beta})| \leq |\boldsymbol{\alpha} - \boldsymbol{\beta}| + K \int_a^x |\mathbf{u}(\xi; \boldsymbol{\alpha}) - \mathbf{u}(\xi; \boldsymbol{\beta})|\, d\xi.$$

Calling

$$E(x) \equiv \int_a^x |\mathbf{u}(\xi; \boldsymbol{\alpha}) - \mathbf{u}(\xi; \boldsymbol{\beta})|\, d\xi,$$

we may write the above inequality as

$$E'(x) - KE(x) \leq |\boldsymbol{\alpha} - \boldsymbol{\beta}|.$$

This differential inequality may be "solved" by multiplying by the integrating factor $e^{-K(x-a)}$ and integrating over $[a, x]$ to get

$$E(x) \leq \frac{|\boldsymbol{\alpha} - \boldsymbol{\beta}|}{K} [e^{K(x-a)} - 1].$$

The original inequality now yields (b). ∎

In many problems of interest the function $\mathbf{f}(x; \mathbf{u})$ does not have the required continuity properties in the infinite strip R but rather in a finite domain $R_B: a \leq x \leq b$, $|\mathbf{u} - \boldsymbol{\alpha}| \leq B$. In this case a similar theorem holds but the solution may exist only in some smaller interval $a \leq x \leq b_0 \equiv \min(b, B/M)$, *ibid.* (See Problems 1.1.4 and 1.1.5 for other, more powerful, generalizations.)

With little additional effort the result in Theorem 1.1.1 can be strengthened to show that in fact the solution $\mathbf{u}(x; \boldsymbol{\alpha})$ is uniformly differentiable with respect to the initial values α_k, $k = 1, 2, \ldots, n$. However, for some of our applications we require a still stronger result which shows that each derivative $\partial\mathbf{u}(x; \boldsymbol{\alpha})/\partial\alpha_k$ is the solution of a specific initial-value problem for a linear system of ordinary differential equations. This is the content of the following theorem.

THEOREM 1.1.2. *In addition to the hypothesis of Theorem* 1.1.1 *let the Jacobian of* $\mathbf{f}$ *with respect to* $\mathbf{u}$ *have continuous elements on R; that is, the nth-order matrix*

$$F(x; \mathbf{u}) \equiv \frac{\partial \mathbf{f}(x; \mathbf{u})}{\partial \mathbf{u}} = \left(\frac{\partial f_i(x; \mathbf{u})}{\partial u_j}\right)$$

is continuous on R. Then for any $\boldsymbol{\alpha}$ *the solution* $\mathbf{u}(x; \boldsymbol{\alpha})$ *of* (1.1.1) *is continuously differentiable with respect to* α_k, $k = 1, 2, \ldots, n$. *In fact, the derivative* $\partial \mathbf{u}(x; \boldsymbol{\alpha})/\partial \alpha_k \equiv \boldsymbol{\xi}^{(k)}(x)$ *is the solution, on* $[a, b]$, *of the linear system*

$$\frac{d}{dx}\boldsymbol{\xi}(x) = F(x; \mathbf{u}(x; \boldsymbol{\alpha}))\boldsymbol{\xi}(x), \tag{1.1.2a}$$

subject to the initial condition

$$\boldsymbol{\xi}(a) = \mathbf{e}^{(k)}. \tag{1.1.2b}$$

(Here $\mathbf{e}^{(k)} \equiv (0, \ldots, 0, 1, 0, \ldots, 0)^T$ is the kth unit vector in n-space.)

Proof. Again we sketch the proof; details can be found in Birkhoff and Rota (1962), pp. 123–124. For arbitrarily small $|h| > 0$ we define the difference quotient

$$\boldsymbol{\eta}(x, h) \equiv \frac{\mathbf{u}(x; \boldsymbol{\alpha} + h\mathbf{e}^{(k)}) - \mathbf{u}(x; \boldsymbol{\alpha})}{h}$$

in terms of the indicated solutions of the initial-value problems (1.1.1) (with initial data $\boldsymbol{\alpha}$ and $\boldsymbol{\alpha} + h\mathbf{e}^{(k)}$). Then it follows that

$$\begin{aligned}\frac{d}{dx}\boldsymbol{\eta}(x, h) &= \frac{1}{h}[\mathbf{f}(x; \mathbf{u} + h\boldsymbol{\eta}) - \mathbf{f}(x; \mathbf{u})] \\ &= F(x; \mathbf{u}(x; \boldsymbol{\alpha}))\boldsymbol{\eta}(x, h) + \boldsymbol{\delta}(x, h)|\boldsymbol{\eta}(x, h)|,\end{aligned}$$

where we have used Taylor's theorem and $|\boldsymbol{\delta}| \to 0$ as $h \to 0$. But by (b) of Theorem 1.1.1, $|\boldsymbol{\eta}|$ must be bounded as $h \to 0$ and so the above equation formally tends to Equation (1.1.2a). We also note that $\boldsymbol{\eta}(a, h) = \mathbf{e}^{(k)}$ for all $|h| \geq 0$. Since the system (1.1.2a) is linear, we can show that the solution $\boldsymbol{\eta}(x, h)$ of the above initial-value problem converges, as $h \to 0$, to the solution of (1.1.2). That the solution $\boldsymbol{\xi}(x) \equiv \boldsymbol{\xi}(x; \boldsymbol{\alpha})$ of (1.1.2) is continuous in $\boldsymbol{\alpha}$ follows easily as in the proof of Theorem 1.1.1. ∎

The system (1.1.2a) is known as the *variational equation* for the system $\mathbf{u}' = \mathbf{f}(x; \mathbf{u})$. We note that it can be obtained formally by assuming $\mathbf{u}$ to depend upon some parameter, say c, differentiating with respect to this parameter and setting $\partial \mathbf{u}/\partial c \equiv \boldsymbol{\xi}$. If this is done to both the system and the initial condition in (1.1.1), using $c = \alpha_k$ as a parameter, we obtain the *variational problem* (1.1.2).

For some of our later studies of eigenvalue problems, we shall require additional results for initial-value problems which contain a parameter in the equations, that is, problems of the form

$$\mathbf{u}' = \mathbf{f}(\lambda, x; \mathbf{u}), \qquad \mathbf{u}(a) = \boldsymbol{\alpha}, \tag{1.1.3}$$

where $\mathbf{f}$ is now a vector-valued function of the $n + 2$ variables $(\lambda, x; \mathbf{u})$. If for all λ in $\lambda_* \leq \lambda \leq \lambda^*$ the hypothesis of Theorem 1.1.1 is satisfied, and in addition $\mathbf{f}(\lambda, x; \mathbf{u})$ depends Lipschitz-continuously on λ, uniformly for $(x; \mathbf{u}) \in R$, then the initial-value problem (1.1.3) has a unique solution, $\mathbf{u}(\lambda, x)$, which is Lipschitz-continuous in λ. Again, this can be strengthened to continuous differentiability of the solution with respect to λ. But if $\mathbf{g}(\lambda, x; \mathbf{u}) \equiv \partial\mathbf{f}(\lambda, x; \mathbf{u})/\partial\lambda$ and $F(\lambda, x; \mathbf{u}) \equiv \partial\mathbf{f}(\lambda, x; \mathbf{u})/\partial\mathbf{u}$ are continuous, we can show that $\boldsymbol{\xi}(x) \equiv \partial\mathbf{u}(\lambda, x)/\partial\lambda$ is a solution of the linear variational problem

$$\frac{d}{dx}\boldsymbol{\xi} = F(\lambda, x; \mathbf{u}(\lambda, x))\boldsymbol{\xi} + \mathbf{g}(\lambda, x; \mathbf{u}(\lambda, x)), \qquad \boldsymbol{\xi}(a) = 0. \tag{1.1.4}$$

Note that this problem is obtained by formal differentiation of the initial-value problem (1.1.3). The initial condition in the variational problem (1.1.4) is homogeneous, while the variational equation is not; this is just the reverse of the situation in the variational problem (1.1.2).

THEOREM 1.1.3. *Let* $\mathbf{f}(\lambda, x; \mathbf{u})$ *satisfy the hypothesis of Theorem* 1.1.1 *for all* λ *in* $\lambda_* \leq \lambda \leq \lambda^*$. *In addition let* $\mathbf{g}(\lambda, x; \mathbf{u}) \equiv \partial\mathbf{f}(\lambda, x; \mathbf{u})/\partial\lambda$ *and the Jacobian matrix*

$$F(\lambda, x; \mathbf{u}) \equiv \frac{\partial\mathbf{f}(\lambda, x; \mathbf{u})}{\partial\mathbf{u}} \equiv \left(\frac{\partial f_i(\lambda, x; \mathbf{u})}{\partial u_j}\right)$$

be continuous. Then for any λ *in* (λ_*, λ^*) *the solution* $\mathbf{u}(\lambda; x)$ *of* (1.1.3) *is continuously differentiable with respect to* λ. *The derivative* $\partial\mathbf{u}(\lambda; \mathbf{x})/\partial\lambda \equiv \boldsymbol{\xi}(\lambda; x)$ *is the solution on* $[a, b]$ *of the variational problem* (1.1.4).

Proof. The proof is almost identical to that of Theorem 1.1.2. [See Coddington and Levinson (1955), pp. 29–31.] ▮

Problems

1.1.1 Show that, in the Euclidean norm, with $b > a$, we have

$$\left|\int_a^b \mathbf{f}(x)\,dx\right|_2 \leq \int_a^b |\mathbf{f}(x)|_2\,dx.$$

[HINT: Use the triangle inequality, that is, $|\mathbf{f} + \mathbf{g}|_2 \leq |\mathbf{f}|_2 + |\mathbf{g}|_2$ (which follows from the Schwarz inequality $|\mathbf{f}\cdot\mathbf{g}| \leq |\mathbf{f}|_2|\mathbf{g}|_2$) in the Riemann sums defining the integral.]

Note that in the maximum norm this result follows trivially from its scalar form.

1.1.2 For each $i = 1, 2, \ldots, n$, let the scalar component $f_i(x; \mathbf{u})$ of $\mathbf{f}(x; \mathbf{u})$ satisfy a uniform Lipschitz condition in each component u_j of $\mathbf{u}$ with constant K_{ij}; that is, let

$$|f_i(x; u_1, \ldots, u_j, \ldots, u_n) - f_i(x; u_1, \ldots, v_j, \ldots, u_n)| < K_{ij}|u_j - v_j|.$$

Then show that $\mathbf{f}(x; \mathbf{u})$ is Lipschitz in $\mathbf{u}$ with constant

$$K = \left(\sum_{i=1}^{n} \sum_{j=1}^{n} K_{ij}^2\right)^{1/2}$$

if we use the Euclidean norm. With the maximum norm, show that the Lipschitz constant for $\mathbf{f}$ may be taken as

$$K = n \max_{i,j} K_{ij}.$$

1.1.3 Show that the solution $\mathbf{u}(\lambda, x)$ of (1.1.3) satisfies

$$|\mathbf{u}(\lambda, x) - \mathbf{u}(\mu, x)| \leq K_1(b - a)e^{K_2(b-a)}|\lambda - \mu|, \qquad a \leq x \leq b,$$

where K_1 and K_2 are the Lipschitz constants for $\mathbf{f}(\lambda, x; \mathbf{u})$ in λ and $\mathbf{u}$, respectively.

1.1.4* Prove the following.

THEOREM 1.1.4. *Let an n-vector* $\boldsymbol{\alpha}_0$ *and positive numbers* ρ, K, *and* M *be given such that, with* $R_{\rho,M}(\boldsymbol{\alpha}_0) \equiv \{(x, \mathbf{u}) \mid |\mathbf{u} - \boldsymbol{\alpha}_0| \leq \rho + M(x - a), a \leq x \leq b\}$,

(a) $\mathbf{f}(x, \mathbf{u})$ *is continuous in* $R_{\rho,M}(\boldsymbol{\alpha}_0)$,
(b) $|\mathbf{f}(x, \mathbf{u})| \leq M$ *for all* $(x, \mathbf{u}) \in R_{\rho,M}(\boldsymbol{\alpha}_0)$,
(c) $|\mathbf{f}(x, \mathbf{u}) - f(x, \mathbf{v})| \leq K|\mathbf{u} - \mathbf{v}|$ *for all* $(x, \mathbf{u})$ *and* $(x, \mathbf{v}) \in R_{\rho,M}(\boldsymbol{\alpha}_0)$.

Then the initial-value problem

$$\mathbf{u}' = \mathbf{f}(x, \mathbf{u}), \qquad \mathbf{u}(a) = \boldsymbol{\alpha} \tag{$*$}$$

has a unique solution $\mathbf{u} = \mathbf{u}(\boldsymbol{\alpha}, x)$ *on* $[a, b]$ *for each*

$$\boldsymbol{\alpha} \in N_\rho(\boldsymbol{\alpha}_0) \equiv \{\boldsymbol{\alpha} \mid |\boldsymbol{\alpha} - \boldsymbol{\alpha}_0| \leq \rho\}.$$

The solution is Lipschitz-continuous in $\boldsymbol{\alpha}$; *more precisely,*

$$|u(\boldsymbol{\alpha}, x) - u(\boldsymbol{\beta}, x)| \leq e^{K|x-a|}|\boldsymbol{\alpha} - \boldsymbol{\beta}|$$

for all $x \in [a, b]$ *and all* $\boldsymbol{\alpha}$ *and* $\boldsymbol{\beta} \in N_\rho(\boldsymbol{\alpha}_0)$.

[HINT: Use the Picard method, showing by induction that the iterates $\mathbf{u}_n(x)$ are all such that $(x, \mathbf{u}_n(x)) \in R_{\rho,M}(\boldsymbol{\alpha}_0)$ for $x \in [a, b]$. The proof then follows as in Theorem 1.1.1.]

1.1.5* Replace condition (c) above by the requirement that $\partial\mathbf{f}(x, \mathbf{u})/\partial\mathbf{u}$ be continuous on $R_{\rho,M}(\boldsymbol{\alpha}_0)$ and then show that the solutions $\mathbf{u}(\boldsymbol{\alpha}, x)$ of ($*$) are continuously differentiable with respect to $\boldsymbol{\alpha}$ in the *open* sphere $|\boldsymbol{\alpha} - \boldsymbol{\alpha}_0| < \rho$; that is, $\partial \mathbf{u}(\boldsymbol{\alpha}, x)/\partial \boldsymbol{\alpha}$ is continuous in $\boldsymbol{\alpha}$ for $\boldsymbol{\alpha}$ in the open sphere and $x \in [a, b]$.

[HINT: The proof of differentiability follows that of Theorems 1.1.2 and 1.1.4. Using the variational problem $V' = [\partial \mathbf{f}(x, \mathbf{u}(\boldsymbol{\alpha}, x))/\partial \mathbf{u}]V$, $V(a) = I$ satisfied by $V(\boldsymbol{\alpha}, x) \equiv \partial \mathbf{u}(\boldsymbol{\alpha}, x)/\partial \boldsymbol{\alpha}$ and the continuity of $\partial \mathbf{f}/\partial \mathbf{u}$, then show that $V(\boldsymbol{\alpha}, x)$ is continuous in $\boldsymbol{\alpha}$.]

1.2 Two-Point Boundary-Value Problems

A boundary-value problem for an ordinary differential equation (or system of equations) is obtained by requiring that the dependent variable (or variables) satisfy subsidiary conditions at two or more distinct points. By means of Theorem 1.1.1 we know that a unique solution of an nth-order equation is determined (for a very large class of equations) by specifying n conditions at one point (that is, for initial-value problems). However, with a total of n boundary conditions imposed at more than one point it is possible that a very smooth nth-order equation has many solutions or even no solution. Thus, as we may expect, the existence and uniqueness theory for boundary-value problems is considerably more complicated and less thoroughly developed than that for initial-value problems. When the boundary conditions are imposed at only two points, which is the usual case in applications, a simple theory can be developed for many special classes of equations and systems of equations. This existence and uniqueness theory plays a role in devising and analyzing numerical methods for solving boundary-value problems, and therefore we study some of its aspects here.

Let us consider first an important class of boundary-value problems in which the solution, $y(x)$, of a second-order equation

$$y'' = f(x, y, y') \tag{1.2.1a}$$

is required to satisfy at two distinct points relations of the form

$$\begin{aligned} a_0 y(a) - a_1 y'(a) &= \alpha, \qquad |a_0| + |a_1| \neq 0; \\ b_0 y(b) + b_1 y'(b) &= \beta, \qquad |b_0| + |b_1| \neq 0. \end{aligned} \tag{1.2.1b}$$

The solution is sought on the interval $[a, b] \equiv \{x \mid a \leq x \leq b\}$.

A formal approach to the exact solution of this problem is obtained by considering a related *initial*-value problem, say

$$u'' = f(x, u, u'), \tag{1.2.2a}$$

$$a_0 u(a) - a_1 u'(a) = \alpha, \qquad c_0 u(a) - c_1 u'(a) = s. \tag{1.2.2b}$$

The second initial condition is to be independent of the first. This is assured if $a_1c_0 - a_0c_1 \neq 0$. Without loss in generality we require that c_0 and c_1 be chosen such that

$$a_1c_0 - a_0c_1 = 1. \tag{1.2.2c}$$

With c_0 and c_1 fixed in this manner, we denote the solution of (1.2.2) by

$$u = u(x; s),$$

to focus attention on its dependence on s. Evaluating the solution at $x = b$, we seek a value of s for which

$$\phi(s) \equiv b_0u(b; s) + b_1u'(b; s) - \beta = 0. \tag{1.2.3}$$

With b and β fixed Equation (1.2.3) is, in general, a transcendental equation in s. If $s = s^*$ is a root of this equation, we then expect the function

$$y(x) \equiv u(x; s^*)$$

to be a solution of the boundary-value problem (1.2.1). This is true in many cases, and in fact all solutions of (1.2.1) can frequently be determined in this way. To be precise, we have the following.

THEOREM 1.2.1. *Let the function $f(x, u_1, u_2)$ be continuous on*

$$R: a \le x \le b, \qquad u_1^2 + u_2^2 < \infty,$$

and satisfy there a uniform Lipschitz condition in u_1 and u_2. Then the boundary-value problem (1.2.1) *has as many solutions as there are distinct roots, $s = s^{(\nu)}$, of Equation* (1.2.3). *The solutions of* (1.2.1) *are*

$$y(x) = y^{(\nu)}(x) \equiv u(x; s^{(\nu)});$$

that is, the solutions of the initial-value problem (1.2.2) *with initial data $s = s^{(\nu)}$.*

Proof. Introducing the new dependent variables $u_1(x) \equiv u(x)$ and $u_2(x) \equiv u'(x)$, the initial-value problem (1.2.2) can be written as

$$\begin{aligned} u_1' &= u_2, & u_1(a) &= a_1s - c_1\alpha, \\ u_2' &= f(x, u_1, u_2), & u_2(a) &= a_0s - c_0\alpha. \end{aligned} \tag{1.2.4}$$

Now with the notation

$$\mathbf{u} \equiv \begin{pmatrix} u_1 \\ u_2 \end{pmatrix}, \qquad \mathbf{f}(x; \mathbf{u}) \equiv \begin{pmatrix} u_2 \\ f(x, u_1, u_2) \end{pmatrix}, \qquad \boldsymbol{\alpha} \equiv \begin{pmatrix} a_1s - c_1\alpha \\ a_0s - c_0\alpha \end{pmatrix},$$

we can apply Theorem 1.1.1, since each component of **f**, and hence **f** itself, is Lipschitz-continuous on R (see Problem 1.1.2). Thus the initial-value problem (1.2.2), has a unique solution, $u_1(x) = u(x; s)$, which exists on $a \le x \le b$.

But clearly, if $\phi(s) = 0$ for some s, then this solution is also a solution of the boundary-value problem (1.2.1). If $s^{(i)}$ and $s^{(j)}$ are distinct roots of Equation (1.2.3), then $u(x; s^{(i)}) \not\equiv u(x; s^{(j)})$, by the uniqueness, so that each distinct root of Equation (1.2.3) yields a distinct solution of (1.2.1).

Now suppose $y(x)$ is a solution of (1.2.1). Then it is also a solution of the initial-value problem (1.2.2) with the parameter value $s = c_0 y(a) - c_1 y'(a)$. But this value of s must satisfy Equation (1.2.3). Thus every solution of the boundary-value problem yields a root of Equation (1.2.3). ∎

By means of this theorem the problem of solving a boundary-value problem is "reduced" to that of finding the root, or roots, of an (in general, transcendental) equation. A very effective class of numerical methods, which we call *initial-value or shooting methods*, is based on this equivalence. As we shall see, more general boundary-value problems than (1.2.1) can be reduced in this way to solving systems of (transcendental) equations.

There is an important class of problems for which we can be assured that Equation (1.2.3) has a unique root. The existence and uniqueness theory for the corresponding boundary-value problems is then settled.

THEOREM 1.2.2. *Let the function* $f(x, u_1, u_2)$ *in* (1.2.1a) *satisfy the hypothesis of Theorem* 1.2.1 *and have continuous derivatives on R which satisfy, for some positive constant M,*

$$\frac{\partial f}{\partial u_1} > 0, \qquad \left|\frac{\partial f}{\partial u_2}\right| \le M.$$

Let the coefficients in (1.2.1b) *satisfy*

$$a_0 a_1 \ge 0, \qquad b_0 b_1 \ge 0, \qquad |a_0| + |b_0| \ne 0.$$

Then the boundary-value problem (1.2.1) *has a unique solution.*

Proof. Since Theorem 1.2.1 is applicable, we need only show that Equation (1.2.3) has a unique root. By the assumed continuity of the derivatives of f it easily follows, from the formulation (1.2.4), that Theorem 1.1.2 is also applicable. Thus let $u(x; s)$ be the solution of the initial-value problem (1.2.2) and define $\xi(x) \equiv \partial u(x; s)/\partial s$. Then $\xi(x)$ is the solution, on $[a, b]$, of the variational equation

$$\xi'' = p(x)\xi' + q(x)\xi,$$

subject to the initial conditions

$$\xi(a) = a_1, \qquad \xi'(a) = a_0.$$

Here we have introduced

$$q(x) \equiv \frac{\partial f(x, u(x; s), u'(x; s))}{\partial u_1}, \qquad p(x) \equiv \frac{\partial f(x, u(x; s), u'(x; s))}{\partial u_2}.$$

The solution $\xi(x)$ of the above variational problem has a continuous second derivative and is nonzero in some arbitrarily small interval $a < x \leq a + \varepsilon$, by virtue of the initial conditions (we recall that $|a_0| + |a_1| \neq 0$). We shall show that $\xi(x)$ is also nonzero in $a < x \leq b$. With no loss in generality, we may assume that $a_0 \geq 0$ and $a_1 \geq 0$ since $a_0 a_1 \geq 0$ and the variational problem is linear. Then we will show that $\xi(x)$ is positive in $a < x \leq b$. If this is not true, say $\xi(x) \leq 0$ for some x_* in $a < x \leq b$, then $\xi(x)$ must have a positive maximum at some point x_0 in $a \leq x_0 < x_*$. However, the maximum cannot be at $x_0 = a$ if $a_0 \neq 0$, since then $\xi'(a) > 0$. For $a_0 = 0$, the variational equation and $\xi(a) = a_1$ imply $\xi''(a) = q(a)a_1 > 0$ and so the maximum is not at $x_0 = a$ in this case either. Hence we have $a < x_0 < x^*$ and at such an interior maximum by the continuity properties of $\xi(x)$,

$$\xi(x_0) > 0, \qquad \xi'(x_0) = 0, \qquad \xi''(x_0) \leq 0.$$

But the variational equation at this point yields, since $q > 0$,

$$\xi''(x_0) = q(x_0)\xi(x_0) > 0.$$

The contradiction implies that $\xi(x) > 0$ on $a < x \leq b$.

From this result it follows that $q(x)\xi(x) > 0$ on $a < x \leq b$, and the variational equation yields the differential inequality

$$\xi''(x) > p(x)\xi'(x), \qquad a < x \leq b.$$

We may "solve" this inequality by the same procedure used in the proof of Theorem 1.1.1. Thus we multiply through by the integrating factor

$$\exp\left[-\int_a^x p(t)\,dt\right]$$

(which, since $|p| \leq M$, must be positive), to get

$$\left\{\exp\left[-\int_a^x p(t)\,dt\right]\xi'(x)\right\}' > 0.$$

Now integrating the above inequality over $[a, x]$ we get, on recalling that $\xi'(a) = a_0$,

$$\xi'(x) > a_0 \exp\left[\int_a^x p(t)\,dt\right].$$

Another integration and application of $\xi(a) = a_1$ gives

$$\xi(x) > a_1 + a_0 \int_a^x \exp\left[\int_a^t p(t')\,dt'\right] dt, \qquad a < x \leq b.$$

But, since $p \geq -M$, it follows that

$$\exp\left[\int_a^t p(t')\,dt'\right] > e^{-M(t-a)}$$

and, using this, we obtain finally

$$\xi'(x) = \frac{\partial u'(x;s)}{\partial s} > a_0 e^{-M(x-a)} \geq 0,$$

$$a < x \leq b. \quad (1.2.5)$$

$$\xi(x) = \frac{\partial u(x;s)}{\partial s} > a_1 + a_0\left(\frac{1 - e^{-M(x-a)}}{M}\right) > 0,$$

With no difficulty it can be seen that for the case in which $a_0 \leq 0$ and $a_1 \leq 0$, the inequality signs in Equation (1.2.5) need only be reversed.

In particular, then, setting $x = b$, the function $u(b; s)$ is a monotone function of s with derivative bounded away from zero for any value of a_0. The same is true of $u'(b; s)$ if $a_0 \neq 0$; if $a_0 = 0$, it is not bounded away from zero. But since b_0 and b_1 do not both vanish or have opposite signs, and $b_0 \neq 0$ if $a_0 = 0$, the function $\phi(s)$ in Equation (1.2.3) must have a derivative of one sign which *is* bounded away from zero for all a_0. Such a function takes on each real value once and only once, and hence $\phi(s) = 0$ has a unique root. ▮

We have not sought the weakest possible conditions in the above theorems. It is not difficult to replace $|p| \equiv |\partial f/\partial u_2| \leq M$ by a less-stringent condition requiring only boundedness from above of integrals of p or to relax $q \equiv \partial f/\partial u_1 > 0$ by allowing $q \geq 0$ in some cases (see Problem 1.2.3). The proof can also be extended to apply to more general boundary conditions (see Problems 1.2.5 and 1.2.6). M. Lees (1961) proves existence and uniqueness for the special case of problem (1.2.1) with $a_1 = b_1 = 0$ under much weaker conditions on $f(x, y, y')$.

A very important special case of Theorem 1.2.2 occurs when the function $f(x, u_1, u_2)$ is linear in u_1 and u_2, that is, for linear second-order boundary-value problems. We state this result as follows.

COROLLARY. *Let the functions $p(x)$, $q(x)$, and $r(x)$ be continuous on $[a, b]$ with*

$$q(x) > 0, \qquad a \leq x \leq b.$$

Let the constants a_0, a_1, b_0, and b_1 satisfy

$$a_0 a_1 \geq 0, \quad |a_0| + |a_1| \neq 0,$$
$$|a_0| + |b_0| \neq 0.$$
$$b_0 b_1 \geq 0, \quad |b_0| + |b_1| \neq 0,$$

Then the boundary-value problem

$$Ly \equiv -y'' + p(x)y' + q(x)y = r(x), \qquad a < x < b, \tag{1.2.6a}$$

$$a_0 y(a) - a_1 y'(a) = \alpha, \qquad b_0 y(b) + b_1 y'(b) = \beta \tag{1.2.6b}$$

has a unique solution for each α, β.

Proof. Writing Equation (1.2.6a) in the form of Equation (1.2.1a) we see that $\partial f/\partial u_1 = q(x)$ and $\partial f/\partial u_2 = p(x)$. Then, since a function continuous on a closed interval is bounded there, $|p(x)| \leq M$ for some $M > 0$, and Theorem 1.2.2 applies. ∎

There is a more general result concerning the solution of linear boundary-value problems which we state as follows.

THEOREM 1.2.3. *Let $p(x)$, $q(x)$, and $r(x)$ be continuous on $[a, b]$. Then for any α and β the following mutually-exclusive alternatives hold: either the boundary-value problem*

$$Ly \equiv -y'' + p(x)y' + q(x)y = r(x), \qquad y(a) = \alpha, y(b) = \beta, \quad (1.2.7)$$

has a unique solution, or else the corresponding homogeneous boundary-value problem,

$$Ly = 0; \qquad y(a) = y(b) = 0, \quad (1.2.8)$$

has a nontrivial solution [*that is, problem* (1.2.7) *has a unique solution if and only if problem* (1.2.8) *has only the trivial solution, $y \equiv 0$*].

Proof. We define two functions $y^{(1)}(x)$ and $y^{(2)}(x)$ as solutions of the respective initial value problems

$$Ly^{(1)}(x) = r(x), \qquad y^{(1)}(a) = \alpha, \qquad y^{(1)\prime}(a) = 0;$$
$$Ly^{(2)}(x) = 0, \qquad y^{(2)}(a) = 0, \qquad y^{(2)\prime}(a) = 1.$$

From the continuity properties of p, q, and r we are assured by Theorem 1.1.1 that unique solutions of these initial-value problems exist on $[a, b]$. Now form the function

$$y(x; s) \equiv y^{(1)}(x) + sy^{(2)}(x)$$

and seek s such that

$$y(b; s) \equiv y^{(1)}(b) + sy^{(2)}(b) = \beta.$$

This equation, which corresponds to Equation (1.2.3), is linear in s and has a unique solution if and only if $y^{(2)}(b) \neq 0$. But if $y^{(2)}(b) = 0$, then $y^{(2)}(x)$ is a nontrivial solution of the homogeneous problem (1.2.8), and the result follows on applying Theorem 1.2.1. ∎

There is no difficulty in extending the above theorem to include more general boundary conditions of the form of Equation (1.2.6b). We pose this as Problem 1.2.4.

This theorem, or more properly a slight extension of it, is called the *Alternative Theorem.* (The extension gives conditions for the existence of solutions of problem (1.2.7) which are not unique when problem (1.2.8) *does* have nontrivial solutions; a very general treatment can be found in Ince (1944) pp. 204–214.) Clearly the case in which the homogeneous system has a nontrivial solution is important and exceptional. In fact, linear *eigenvalue problems*, which we consider later, are concerned with adjusting a parameter in the coefficients so that this exceptional case occurs (see Chapter 5).

1.2.1 Boundary-Value Problems for General Systems

It is not difficult to extend most of the previous results, valid for single second-order equations, to much more general boundary-value problems. To illustrate the possibilities, we shall consider the general system of n first-order differential equations

$$\mathbf{y}' = \mathbf{f}(x, \mathbf{y}), \qquad a < x < b, \tag{1.2.9a}$$

subject to the most general linear two-point boundary conditions

$$A\mathbf{y}(a) + B\mathbf{y}(b) = \boldsymbol{\alpha}. \tag{1.2.9b}$$

Here $\mathbf{y}(x)$ is an n-dimensional vector with components $y_j(x)$; $\mathbf{f}(x, \mathbf{y})$ is an n-vector with components $f_k(x, \mathbf{y})$ which are functions of the $n + 1$ variables $x, y_j, j = 1, 2, \ldots, n$; A and B are nth-order matrices with constant elements and $\boldsymbol{\alpha}$ is a fixed n-vector. It would seem that most two-point boundary-value problems with linear boundary conditions can be written in the above form. Note that the n conditions in Equation (1.2.9b) are linearly independent if and only if the $n \times 2n$ order matrix $[A, B]$ has rank $= n$.

We may reduce the study of existence and uniqueness of solutions to the boundary-value problem (1.2.9) to the study of the roots of some systems of (transcendental) equations by means of initial-value problems, just as we did for the boundary-value problem (1.2.1). This can be done in many ways, but perhaps the most obvious is as follows. We consider the initial-value problem

$$\text{(a)} \quad \mathbf{u}' = \mathbf{f}(x, \mathbf{u}), \qquad \text{(b)} \quad \mathbf{u}(a) = \mathbf{s}, \tag{1.2.10}$$

where $\mathbf{s}$ is an n-vector to be determined. In terms of the solution $\mathbf{u} = \mathbf{u}(\mathbf{s}; x)$ of the problem (1.2.10) we define the system of n equations

$$\boldsymbol{\phi}(\mathbf{s}) \equiv [A\mathbf{s} + B\mathbf{u}(\mathbf{s}; b)] - \boldsymbol{\alpha} = 0. \tag{1.2.11}$$

Clearly, if $\mathbf{s} = \mathbf{s}^*$ is a root of this equation, we now expect that

$$\mathbf{y}(x) = \mathbf{u}(\mathbf{s}^*; x)$$

is a solution of the boundary-value problem (1.2.9). In fact, just as in Theorem 1.2.1, we now have the following.

THEOREM 1.2.4. *Let* $\mathbf{f}(x, \mathbf{u})$ *be continuous on*

$$R: a \le x \le b, \qquad |\mathbf{u}| < \infty,$$

and satisfy there a uniform Lipschitz condition in $\mathbf{u}$. *Then the boundary-value problem* (1.2.9) *has as many solutions as there are distinct roots* $\mathbf{s} = \mathbf{s}^{(\nu)}$ *of Equation* (1.2.11). *These solutions are*

$$\mathbf{y}(x) = \mathbf{u}(\mathbf{s}^{(\nu)}; x),$$

the solutions of the initial-value problems (1.2.10) *with* $\mathbf{s} = \mathbf{s}^{(\nu)}$.

Proof. The proof is almost identical with that of Theorem 1.2.1 and is left to the reader. ∎

We have thus reduced the problem of solving the boundary-value problem (1.2.9) to that of finding the roots of a system of n transcendental equations. It is generally quite difficult to prove the existence of roots of such systems. Of course if the vector-valued function $\mathbf{f}(x, \mathbf{u})$ is linear in $\mathbf{u}$, then it is easy to show that the system of equations (1.2.11) reduces to a linear algebraic system. In this case existence and uniqueness would follow from the nonsingularity of the appropriate coefficient matrix. The detailed study of a particular class of linear problems is presented in Theorem 1.2.5 below. We present this case in part to develop some results required in our analysis of the more general case of nonlinear boundary-value problems.

THEOREM 1.2.5. *The linear boundary-value problem*

$$\text{(a)}\quad \mathbf{y}' = K(x)\mathbf{y}, \qquad \text{(b)}\quad A\mathbf{y}(a) + B\mathbf{y}(b) = \boldsymbol{\alpha} \tag{1.2.12}$$

has a unique solution for each $\boldsymbol{\alpha}$, *provided that*

$$\begin{aligned} &\text{(a)}\quad K(x) \quad \textit{is continuous on } [a, b]; \\ &\text{(b)}\quad (A + B) \quad \textit{is nonsingular}; \\ &\text{(c)}\dagger \quad \int_a^b \|K(x)\|_\infty \, dx < \ln\left(1 + \frac{1}{m}\right), \end{aligned} \tag{1.2.13}$$

where

$$m \equiv \|(A + B)^{-1}B\|_\infty .$$

Proof. By virtue of condition (1.2.13a) we may apply Theorem 1.2.4 to the linear boundary-value problem (1.2.12). Thus we consider the appropriate initial-value problem

$$\mathbf{u}' = K(x)\mathbf{u}, \qquad \mathbf{u}(a) = \mathbf{s}. \tag{1.2.14a}$$

† For any matrix $\Gamma \equiv (\gamma_{ij})$ the maximum norm is $\|\Gamma\|_\infty = \max_i \sum_j |\gamma_{ij}|$, and is compatible with the maximum vector norm, $|\mathbf{u}|$, in the sense that $|\Gamma\mathbf{u}| \le \|\Gamma\|_\infty \cdot |\mathbf{u}|$.

The unique solution of this linear problem can be represented by means of the matrizant, $\Omega_a^x \{K\}$, as

$$\mathbf{u}(\mathbf{s}, x) = \overset{x}{\underset{a}{\Omega}} \{K\}\mathbf{s}. \tag{1.2.14b}$$

The matrizant is defined [see Ince (1944), pp. 408–411] by the absolutely and uniformly convergent matrix series

$$\overset{x}{\underset{a}{\Omega}} \{K\} \equiv I + \int_a^x K(\xi_1)\, d\xi_1 + \int_a^x \int_a^{\xi_1} K(\xi_1)K(\xi_2)\, d\xi_1\, d\xi_2 + \cdots. \tag{1.2.15}$$

A derivation of this result is indicated in Problem 1.2.7 and we note that it is the fundamental solution of $U' = K(x)U$ which satisfies $U(a) = I$.

The solution (1.2.14b) of the initial-value problem (1.2.14a) will be a solution of the boundary-value problem (1.2.12) *if and only if* $\mathbf{s}$ satisfies

$$\left[A + B\overset{b}{\underset{a}{\Omega}} \{K\}\right]\mathbf{s} = \boldsymbol{\alpha}. \tag{1.2.16}$$

We note that this is a *linear algebraic system* for the determination of $\mathbf{s}$, as anticipated. It is exactly the system of Equations (1.2.11) in a more explicit notation. The coefficient matrix in Equation (1.2.16) can be written, recalling (1.2.13b), as

$$\begin{aligned} A + B\overset{b}{\underset{a}{\Omega}} \{K\} &= (A + B) + B\left(\overset{b}{\underset{a}{\Omega}} \{K\} - I\right) \\ &= (A + B)\left[I + (A + B)^{-1}B\left(\overset{b}{\underset{a}{\Omega}} \{K\} - I\right)\right]. \end{aligned}$$

However, it is known that a matrix of the form $[I + H]$ is nonsingular if $\|H\|_\infty < 1$ [see Isaacson and Keller (1966), Theorem 1.5 in Chapter 1]. But using the fact that for $x \geq a$

$$\left\|\int_a^x K(\xi)\, d\xi\right\|_\infty \leq \int_a^x \|K(\xi)\|_\infty\, d\xi,$$

it follows with little effort (see Problem 1.2.8) that

$$\left\|\overset{b}{\underset{a}{\Omega}} \{K\} - I\right\|_\infty \leq \exp\left(\int_a^b \|K(\xi)\|_\infty\, d\xi\right) - 1. \tag{1.2.17}$$

Thus (1.2.13b) and (1.2.13c) imply that the coefficient matrix in Equation (1.2.16) is nonsingular since

$$\begin{aligned} \left\|(A + B)^{-1}B\left(\overset{b}{\underset{a}{\Omega}} \{K\} - I\right)\right\|_\infty &\leq \|(A + B)^{-1}B\|_\infty \left\|\overset{b}{\underset{a}{\Omega}} \{K\} - I\right\|_\infty \\ &\leq m\left[\exp\left(\int_a^b \|K(\xi)\|_\infty\, d\xi\right) - 1\right] < 1. \quad \blacksquare \end{aligned}$$

The conditions (1.2.13) are only sufficient for the existence of a unique solution to the linear boundary-value problem (1.2.12). Given (1.2.13a), the *necessary and sufficient* conditions are clearly that the coefficient matrix in Equation (1.2.16) be nonsingular. Our condition (1.2.13b) is not necessary for rank $[A,B] = n$, and thus many boundary conditions of interest are not covered by this result. A slight almost trivial weakening of condition (1.2.13c) is indicated in Problem 1.2.10. It should be observed that when conditions (1.2.13a) and (1.2.13b) are satisfied, condition (1.2.13c) is automatically satisfied if the interval $[a, b]$ is sufficiently small. In particular, condition (1.2.13c) is satisfied if the length of the interval is so small that

$$|b - a| < \frac{\ln (1 + 1/m)}{\max\limits_{a \le x \le b} \|K(x)\|_\infty}.$$

We now return to the study of the general nonlinear boundary-value problem (1.2.9). Under appropriate conditions, resembling those in (1.2.13), we shall prove the existence of a unique solution. This will be done by showing that the system of Equations (1.2.11) has a unique root. Since the function $\boldsymbol{\phi}(\mathbf{s})$ is no longer linear in the general case, we can try to employ some type of fixed-point theorems to show that it has a zero. The simplest such result is the Contracting Mapping principle which, when applicable, also yields uniqueness. We shall use this principle, and all the theory required is developed in Section 1.4 (in particular, Theorem 1.4.1). Contracting maps will be employed throughout this work and can be studied independently of the preceding sections. We have delayed their study simply in order to get into the basic subject matter more rapidly.

Our existence and uniqueness theorem for the problem (1.2.9) is as follows.

THEOREM 1.2.6. *Let* $\mathbf{f}(x, \mathbf{u})$ *satisfy on* R: $a \le x \le b$, $|\mathbf{u}| < \infty$

(a) $\mathbf{f}(x, \mathbf{u})$ *continuous*;

(b) $\dfrac{\partial f_i(x, \mathbf{u})}{\partial u_j}$ *continuous*, $\quad i, j = 1, 2, \ldots, n$; (1.2.18)

(c) $\left\| \dfrac{\partial \mathbf{f}(x, \mathbf{u})}{\partial \mathbf{u}} \right\|_\infty \le k(x)$.

Furthermore, let the matrices A and B and the scalar function k(x) satisfy

(d) $(A + B)$ *nonsingular*;

(e) $\displaystyle\int_a^b k(x)\, dx \le \ln \left(1 + \frac{\lambda}{m}\right)$, (1.2.18) *for some* λ *in* $0 < \lambda < 1$, *where*

$$m \equiv \|(A + B)^{-1}B\|_\infty .$$

Then the boundary-value problem (1.2.9) *has a unique solution for each* $\boldsymbol{\alpha}$.

Proof. From conditions (1.2.18b, c) it follows that $\mathbf{f}(x, \mathbf{u})$ satisfies a uniform Lipschitz condition in $\mathbf{u}$ (see Problem 1.2.11). But then Theorem 1.2.4 is applicable and implies that the boundary-value problem (1.2.9) has a unique solution if and only if $\boldsymbol{\phi}(\mathbf{s})$ in Equation (1.2.11) has a unique zero.

Let Q be a nonsingular matrix; then it follows that the system

$$\mathbf{s} = \mathbf{g}(\mathbf{s}) \equiv \mathbf{s} - Q\boldsymbol{\phi}(\mathbf{s}) \tag{1.2.19}$$

has roots that are identical with those of Equation (1.2.11). However, if

$$\left\| \frac{\partial \mathbf{g}(\mathbf{s})}{\partial \mathbf{s}} \right\|_\infty \leq \lambda < 1, \tag{1.2.20}$$

we can then apply the Contracting Mapping principle (see Problem 1.2.11 and Theorem 1.4.1) to conclude that Equation (1.2.19) and hence Equation (1.2.11) has a root which is unique. Our proof is thus reduced to verifying the inequality (1.2.20) for some nonsingular matrix Q.

With the definition of the matrix W as

$$\partial \mathbf{u}(\mathbf{s}; x)/\partial \mathbf{s} \equiv W(\mathbf{s}; x),$$

it follows from Equations (1.2.11) and (1.2.19) that

$$\partial \mathbf{g}(\mathbf{s})/\partial \mathbf{s} = I - Q[A + BW(\mathbf{s}; b)]. \tag{1.2.21}$$

An application of Theorem 1.1.2 to the initial-value problem (1.2.10) reveals that $W(\mathbf{s}; x)$, or $W(x)$ for brevity, satisfies the variational system

$$W' = J(\mathbf{s}; x)W, \qquad W(a) = I. \tag{1.2.22}$$

Here we have introduced the Jacobian matrix

$$J(\mathbf{s}; x) \equiv \partial \mathbf{f}(x, \mathbf{u}(\mathbf{s}; x))/\partial \mathbf{u},$$

and by condition (1.2.18c),

$$\|J(\mathbf{s}; x)\|_\infty \leq k(x).$$

The solution of the linear system (1.2.22) is the matrizant of J; that is, recalling the definition (1.2.15),

$$W(x) = \overset{x}{\underset{a}{\Omega}} \{J\}.$$

Now we can write Equation (1.2.21) as

$$\partial \mathbf{g}(\mathbf{s})/\partial \mathbf{s} = I - Q\left[(A + B) + B\left(\mathop{\Omega}_a^b \{J\} - I\right)\right].$$

Since $(A + B)$ is nonsingular by condition (1.2.18d) we take $Q = (A + B)^{-1}$ to get, using the inequalities (1.2.17) and (1.2.18e)

$$\begin{aligned}\left\|\frac{\partial \mathbf{g}(\mathbf{s})}{\partial \mathbf{s}}\right\|_\infty &= \left\|(A + B)^{-1}B\left(\mathop{\Omega}_a^b \{J\} - I\right)\right\|_\infty \\ &\leq m\left(\exp\left(\int_a^b k(x)\,dx\right) - 1\right) \\ &\leq \lambda < 1. \quad \blacksquare\end{aligned}$$

We observe again that Condition (1.2.18e) will certainly be satisfied, given conditions (1.2.18a–d), if $|b - a|$ is sufficiently small. (Of course, one of the objectives in theoretical studies is to prove existence for relatively larger intervals; see Problem 1.2.13.) The relaxation to possibly smaller m as in Problem 1.2.10 is also valid now. If $(A + B)$ is singular, a different choice for Q in (1.2.21) may still result in a contraction. In fact, we shall later advocate the use of Newton's method, which corresponds to the choice $Q = [A + BW(\mathrm{s}, b)]^{-1}$. Existence proofs based on this choice or related choices are indicated in Section 5.5.

These results are far from sufficient to cover most problems that arise in practice. Additional existence proofs (some under much weaker conditions) are contained in Problems 1.2.12, 1.2.13, 1.2.14, and the results of Theorem 4.1.3 covering systems of second-order equations. (See also the discussion in Section 5.5.) However, solutions of boundary-value problems and roots of transcendental systems can exist without formal proofs of these facts. Thus in many of the difficult and important applied problems leading to boundary-value problems we may use the techniques to be developed here without the benefit of existence theorems.

Problems

1.2.1 (a) Determine the solution of the initial-value problem

$$Ly \equiv y'' - k^2y = 0, \qquad y(a) = \alpha, \qquad y'(a) = s.$$

(b) Use this solution to solve the boundary-value problem

$$Ly = 0, \qquad y(a) = \alpha, \qquad y(b) = \beta.$$

Find conditions on $k(b - a)$ so that a unique solution exists. Does a solution always exist? (Note that Theorem 1.2.2 applies here.)

1.2.2 Repeat Problem 1.2.1 for the equation $Ly \equiv y'' + k^2 y = 0$. Under what conditions do nonunique solutions exist?

1.2.3 Prove the analog of Theorem 1.2.2 under the hypothesis

$$\frac{\partial f}{\partial u_1} \geq 0, \qquad \frac{\partial f}{\partial u_2} = 0, \qquad a_0 = b_0 = 1, \qquad a_1 = b_1 = 0,$$

that is, with $f(x, u_1, u_2)$ independent of u_2 and the solution specified at the boundary points [see Henrici (1962), pp. 347–348, for a proof]. Note that the inequalities (1.2.5) give the bounds $\partial u(x; s)/\partial s > (x - a)$ and $\partial u'(x; s)/\partial s > 1$ upon letting $M \to 0$.

1.2.4 Replace the boundary conditions in the boundary-value problems (1.2.7) and (1.2.8) by those in Equations (1.2.1b) and prove the corresponding Alternative Theorem.

1.2.5 Replace the boundary-value problem (1.2.1) by $y'' = f(x, y, y')$ subject to

(a) $a_0 y(a) - a_1 y'(a) = \alpha, \qquad |a_0| + |a_1| \neq 0,$

(b) $b_0 y(b) + b_1 y'(b) + b_2 y(a) + b_3 y'(a) = \beta, \qquad |b_0| + |b_1| \neq 0.$

Show that a unique solution to this problem exists if $f(x, y, y')$ satisfies

$$N > \frac{\partial f}{\partial y} > 0, \qquad \left|\frac{\partial f}{\partial y'}\right| \leq M,$$

and the coefficients a_i, b_i satisfy

$$a_i \geq 0, \qquad b_i \geq 0, \qquad i = 0, 1; \qquad a_0 + b_0 > 0; \qquad a_0 b_2 + a_1 b_3 \geq 0.$$

1.2.6 Replace the boundary conditions in Problem 1.2.5 by

$$g_1(y(a), -y'(a)) = 0,$$
$$g_2(y(a), y'(a), y(b), y'(b)) = 0.$$

Under the same conditions on f show that a unique solution to the boundary-value problem exists if the partial derivatives of $g_1(x_1, x_2)$ and $g_2(x_1, x_2, x_3, x_4)$ denoted by $g_{ij} \equiv \partial g_i/\partial x_j$ are continuous and satisfy

$$g_{i,j} \geq 0, \qquad i = 1, \qquad j = 1, 2; \qquad i = 2, \qquad j = 3, 4;$$
$$g_{1,1} + g_{1,2} \geq \varepsilon > 0, \qquad g_{2,3} + g_{2,4} \geq \varepsilon > 0;$$
$$g_{1,1} + g_{2,3} > 0, \qquad g_{2,1} g_{1,2} + g_{2,2} g_{1,1} \geq 0.$$

[HINT: Use the result in Problem 1.2.5.]

1.2.7 Use the Picard iteration scheme (see proof of Theorem 1.1.1) to derive formally the representation of the matrizant given in Equation (1.2.15).

1.2.8 Derive the inequality (1.2.17) from the definition (1.2.15).

[HINT: Use the iterated-integral formulas, for scalar functions,

$$\int_a^x \int_a^{\xi_1} k(\xi_1)k(\xi_2)\,d\xi_2\,d\xi_1 = \frac{1}{2!}\left(\int_a^x k(\xi)\,d\xi\right)^2, \ldots .]$$

1.2.9 If the matrix $K(x)$ is self-commuting, that is, if $K(x)K(x') = K(x')K(x)$, show that

$$\overset{x}{\underset{a}{\Omega}}\{K\} = \exp \int_a^x K(\xi)\,d\xi .$$

[HINT: Use induction with the matrix iterated-integral formula to derive the general term of the exponential series.]

1.2.10 Show that Theorems 1.2.5 and 1.2.6 remain valid if we replace the constant m by

$$m = \min\left(\|(A + B)^{-1}B\|_\infty, \|(A + B)^{-1}A\|_\infty\right).$$

[HINT: Use the appropriate initial-value problems with "initial" point $x = b$ in place of $x = a$.]

1.2.11 If the function $\mathbf{g}(\mathbf{s})$ has continuous first derivatives which satisfy, for all $\mathbf{s}$, $\|\partial\mathbf{g}(\mathbf{s})/\partial\mathbf{s}\|_\infty \le \lambda$, then $\mathbf{g}(\mathbf{s})$ satisfies the Lipschitz condition

$$|\mathbf{g}(\mathbf{s}) - \mathbf{g}(\mathbf{t})| \le \lambda|\mathbf{s} - \mathbf{t}| .$$

1.2.12 Consider the two-point boundary-value problem

$$\mathbf{y}' = \mathbf{f}(x, \mathbf{y}), \qquad A\mathbf{y}(a) = \boldsymbol{\alpha}, \qquad B\mathbf{y}(b) = \boldsymbol{\beta},$$

where $\mathbf{y}$ and $\mathbf{f}$ are n vectors, A is a $p \times n$ matrix with rank $p > 0$, B is a $q \times n$ matrix with rank $q > 0$, $p + q = n$, $\boldsymbol{\alpha}$ is a p-vector and $\boldsymbol{\beta}$ is a q-vector. Show that with no loss in generality we may take A in the form

$$A \equiv (I_p, A_1),$$

where I_p is the p-order identity matrix and A_1 is some $p \times q$ matrix. Show further that if $\mathbf{f}(x, \mathbf{y})$ satisfies Conditions (1.2.18a–c, e), where

$$m = \|Q^{-1}B\|_\infty \cdot \left\|\binom{-A_q}{I_1}\right\|_\infty \qquad \text{and} \qquad Q \equiv B\binom{-A_1}{I_q}$$

is assumed nonsingular, then the above boundary-value problem has a unique solution for each $\boldsymbol{\alpha}$, $\boldsymbol{\beta}$.

[HINT: Use the initial-value problem

$$\mathbf{u}' = f(x, \mathbf{u}), \qquad A\mathbf{u}(a) = \mathbf{x}, \qquad C\mathbf{u}(a) = \mathbf{s},$$

where $\mathbf{s}$ is a q-vector and the $q \times n$ matrix $C \equiv (0, I_q)$. The system to be solved is $\boldsymbol{\phi}(\mathbf{s}) \equiv B\mathbf{u}(\mathbf{s}; b) - \boldsymbol{\beta} = 0$, which is of order q. Show that a contracting map is given by $\mathbf{s} = \mathbf{g}(\mathbf{s}) \equiv \mathbf{s} - Q^{-1}\boldsymbol{\phi}(\mathbf{s})$.]

1.2.13* Prove the following.

THEOREM 1.2.7 *Let an n-vector* $\mathbf{s}_0$ *and positive constants* ρ, K, *and* M *satisfy, for* $R_{\rho,M}(\mathbf{s}_0) \equiv \{(x, \mathbf{u}) \mid |\mathbf{u} - \mathbf{s}_0| \leq \rho + M(x - a),\quad a \leq x \leq b\}$,

(a) $\mathbf{f}(x, \mathbf{u})$ *continuous on* $R_{\rho,M}(\mathbf{s}_0)$,
(b) $|\mathbf{f}(x, \mathbf{u})| \leq M$ *on* $R_{\rho,M}(\mathbf{s}_0)$,
(c) $|\mathbf{f}(x, \mathbf{u}) - \mathbf{f}(x, \mathbf{v})| \leq K|\mathbf{u} - \mathbf{v}|$ *for all* $(x, \mathbf{u})$, *and* $(x, \mathbf{v}) \in R_{\rho,M}(\mathbf{s}_0)$.

Further, let the matrices A *and* B *and interval length,* $|b - a|$, *be such that*

(d) $(A + B)$ *is nonsingular,*
(e) $\|(A + B)^{-1}B\| \cdot M(b - a) + \|(A + B)^{-1}\boldsymbol{\alpha} - \mathbf{s}_0\| \leq \rho$.

Then the boundary-value problem

$$\mathbf{y}' = \mathbf{f}(x, \mathbf{y}), \qquad A\mathbf{y}(a) + B\mathbf{y}(b) = \boldsymbol{\alpha} \tag{1.2.23}$$

has at least one solution with $\mathbf{y}(a) \in N_\rho(\mathbf{s}_0)$.

[HINT: Use the initial-value technique and Theorem 1.1.4 in Problem 1.1.4 to show that the function $\mathbf{g}(\mathbf{s})$, defined in the usual way, is continuous and maps the closed sphere $N_\rho(\mathbf{s}_0)$ into itself. Then by the Brouwer fixed-point theorem $\mathbf{s} = \mathbf{g}(\mathbf{s})$ has at least one root in $N_\rho(\mathbf{s}_0)$. Note that Condition (e) suggests using as the "initial" iterate the value $\mathbf{s}_0 = (A + B)^{-1}\boldsymbol{\alpha}$.]

1.2.14* Demonstrate the following.

COROLLARY. *If in addition to the hypothesis of Theorem* 1.2.7 *we have*

(f) $K|b - a| \leq \log\left(1 + \dfrac{\lambda}{m}\right), \qquad m \equiv \|(A + B)^{-1}B\|$

then the solution of the boundary-value problem (1.2.23) *with* $\mathbf{y}(a) \in N_\rho(\mathbf{s}_0)$ *is unique.*

[HINT: Show that the map $\mathbf{g}(\mathbf{s})$ is now contracting on $N_\rho(\mathbf{s}_0)$.]

1.3 Numerical Methods for Initial-Value Problems

We present here some basic material on numerical methods for solving initial-value problems. The initial-value problems will be assumed of the form

$$\mathscr{L}[\mathbf{u}(x)] \equiv \mathbf{u}' - \mathbf{f}(x, \mathbf{u}) = 0, \qquad a \leq x \leq b \tag{1.3.1a}$$

$$\mathbf{u}(a) = \boldsymbol{\alpha} \tag{1.3.1b}$$

and as indicated the solution is sought on $[a, b]$. The numerical methods for solving the problem (1.3.1), or rather for approximating its solution, all employ some set of discrete points $\{x_j\}$ on $[a, b]$ called a *net* or *lattice*. Although it is not required for all of our procedures, we will take this net to be uniformly spaced, say

$$x_0 = a, \qquad x_j = a + jh; \qquad j = 1, 2, \ldots, J + 1, \qquad h \equiv \frac{b - a}{J + 1}. \quad (1.3.2)$$

The quantity h is called the *mesh size* or *net spacing*. A rule which assigns to each point x_j of the net a corresponding n vector $\mathbf{U}_j$ is called a *net function*, $\{\mathbf{U}_j\}$. Clearly any solution of problem (1.3.1) evaluated on the net determines a net function $\{\mathbf{u}(x_j)\}$. The numerical methods of interest define net functions $\{\mathbf{U}_j\}$ which will be close approximations to $\{\mathbf{u}(x_j)\}$ for sufficiently small net spacing h. The question of close approximations by computed net functions is related to the concepts of consistency, convergence, and stability for which there is a very general and quite complete theory. We shall indicate some of the simpler aspects of this theory to convey its flavor before presenting practical schemes for solving the initial-value problem (1.3.1).

A numerical method assigns to each point x_j of the net an n-vector $\mathbf{U}_j$ by means of some "algebraic" equations involving "neighboring" net points and corresponding values $\mathbf{U}_i$. Specifically, we assume that at most $k + 1$ successive points are involved and write the algebraic system as

$$\mathscr{L}_h[\mathbf{U}_{j+1}, \mathbf{U}_j, \ldots, \mathbf{U}_{j+1-k}] = 0, \qquad j = k - 1, k, \ldots, J, \quad (1.3.3a)$$

$$\mathbf{U}_j = \boldsymbol{\gamma}_j, \qquad j = 0, 1, \ldots, k - 1. \quad (1.3.3b)$$

As this notation implies $\mathbf{U}_{i+1}$ is determined from the quantities $\mathbf{U}_i, \mathbf{U}_{i-1}, \ldots, \mathbf{U}_{i-k+1}$. Thus, as indicated in Equation (1.3.3b), the first k values of $\mathbf{U}_i$ must be specified and such a scheme is termed a *k-step method.* To simplify the notation, we shall write

$$\mathscr{L}_h[\mathbf{U}_j] \equiv \mathscr{L}_h[\mathbf{U}_{j+1}, \mathbf{U}_j, \ldots, \mathbf{U}_{j+1-k}].$$

To relate the numerical scheme to the initial-value problem, we define the *local truncation error*, for any sufficiently smooth n-vector $\mathbf{v}(x)$, by†

$$\boldsymbol{\tau}_j[\mathbf{v}] \equiv \mathscr{L}_h[\mathbf{v}(x_j)] - \mathscr{L}[\mathbf{v}(x_j)], \qquad k - 1 \le j \le J. \quad (1.3.4)$$

We say that $\mathscr{L}_h[\cdot]$ *is consistent with* $\mathscr{L}[\cdot]$ *if and only if*

$$\lim_{J \to \infty} |\boldsymbol{\tau}_j[\mathbf{v}]| = 0, \qquad k - 1 \le j \le J, \quad (1.3.5)$$

for all sufficiently smooth $\mathbf{v}(x)$. (This definition of consistency is frequently weakened to apply only to functions $\mathbf{v}(x)$ which are solutions of $\mathscr{L}[\mathbf{v}] = 0$.)

† In specific cases, the argument of $\mathscr{L}[\cdot]$ may be shifted slightly to obtain a better definition of local truncation error.

A consistent scheme is said to have an *order of accuracy at least p if and only if*†

$$|\boldsymbol{\tau}_j[\mathbf{v}]| = \mathcal{O}(h^p), \qquad k - 1 \leq j \leq J, \tag{1.3.6}$$

for all sufficiently smooth $\mathbf{v}(x)$.

The numerical scheme of Equations (1.3.3) is said to be *convergent* for the initial-value problem (1.3.1) if their solutions satisfy

$$\lim_{J \to \infty} |\mathbf{U}_j - \mathbf{u}(x_j)| = 0, \qquad 0 \leq j \leq J + 1. \tag{1.3.7}$$

A basic notion which relates consistency and convergence is that of *stability*. In contrast to the previous definitions, stability is solely a property of the numerical scheme. We define it as follows: $\mathscr{L}_h[\cdot]$ *is stable (or determines a stable scheme) if there exists a constant M, independent of the mesh size h, such that for all net functions* $\{\mathbf{V}_j\}$ *and* $\{\mathbf{W}_j\}$,

$$|\mathbf{V}_j - \mathbf{W}_j| \leq M\Big\{\max_{i \leq k-1} |\mathbf{V}_i - \mathbf{W}_i| + \max_{i \geq k-1} |\mathscr{L}_h[\mathbf{V}_i] - \mathscr{L}_h[\mathbf{W}_i]|\Big\},$$
$$k \leq j \leq J + 1. \tag{1.3.8}$$

The fundamental theorem relating the above concepts is the following.

THEOREM 1.3.1. *Let* $\mathscr{L}_h[\cdot]$ *be stable and consistent with* $\mathscr{L}[\cdot]$. *Then the numerical solution* $\{\mathbf{U}_j\}$ *of the scheme* (1.3.3) *and the exact solution* $\mathbf{u}(x)$ *of the problem* (1.3.1) *satisfy*

$$|\mathbf{U}_j - \mathbf{u}(x_j)| \leq M\Big\{\max_{0 \leq i \leq k-1} |\mathbf{u}(x_i) - \boldsymbol{\gamma}_i| + \max_{k-1 \leq i \leq J} |\boldsymbol{\tau}_i[\mathbf{u}]|\Big\},$$
$$k \leq j \leq J + 1. \tag{1.3.9}$$

If the initial data in Equation (1.3.3b) *satisfies*

$$\lim_{J \to \infty} |\mathbf{u}(x_i) - \boldsymbol{\gamma}_i| \to 0, \qquad 0 \leq i \leq k - 1, \tag{1.3.10}$$

then the numerical scheme is convergent.

Proof. From Equation (1.3.1a), evaluated at any net point x_j with $k - 1 \leq j \leq J$, we have, using the definition (1.3.4),

$$\begin{aligned} \mathscr{L}_h[\mathbf{u}(x_j)] &= \mathscr{L}_h[\mathbf{u}(x_j)] - \mathscr{L}[\mathbf{u}(x_j)] \\ &= \boldsymbol{\tau}_j[\mathbf{u}]. \end{aligned}$$

But then Equation (1.3.3a) and the above yield

$$\mathscr{L}_h[\mathbf{u}(x_j)] - \mathscr{L}_h[\mathbf{U}_j] = \boldsymbol{\tau}_j[\mathbf{u}].$$

† For any two scalar quantities, say $f(h)$ and $g(h)$, depending upon h, the notation $f(h) = \mathcal{O}(g(h))$ means that there exists some positive constant, c, independent of h, such that

$$\lim_{h \to 0} |f(h)/g(h)| \leq c.$$

However, since $\mathscr{L}_h[\cdot]$ is stable, this implies the inequality (1.3.9), on recalling Equations (1.3.1b) and (1.3.3b) and using $\mathbf{V}_j \equiv \mathbf{u}(x_j)$ and $\mathbf{W}_j \equiv \mathbf{U}_j$ in inequality (1.3.8). By consistency $\boldsymbol{\tau}_j[\mathbf{u}] \to 0$ as $J \to \infty$ and so convergence follows from condition (1.3.10). ∎

Bounds on the rate at which the numerical solution converges to the exact solution as $h \to 0$ are easily obtained from the above theorem. We have the obvious

COROLLARY. *In addition to the hypothesis of Theorem* 1.3.1, *let the order of accuracy of* $\mathscr{L}_h[\cdot]$ *as an approximation to* $\mathscr{L}[\cdot]$ *be at least p. Let the initial data in Equation* (1.3.3b) *satisfy*

$$|\mathbf{u}(x_i) - \boldsymbol{\gamma}_i| = \mathcal{O}(h^q), \qquad 0 \le i \le k-1. \tag{1.3.11}$$

Then if the solution $\mathbf{u}(x)$ *of the initial-value problem* (1.3.1) *is sufficiently smooth,*

$$|\mathbf{U}_j - \mathbf{u}(x_j)| \le \mathcal{O}(h^q) + \mathcal{O}(h^p). \quad ∎$$

To illustrate the above concepts we examine first the simplest and perhaps best known numerical scheme for solving initial-value problems, namely *Euler's method.* Only the most naive computer user would generally employ this method in practice since, as we shall see, there are much more accurate methods requiring about the same effort.† In the notation of Equations (1.3.3), using the net (1.3.2), Euler's method for approximating the solution of (1.3.1) can be written as

$$\mathscr{L}_h[\mathbf{U}_{j+1}, \mathbf{U}_j] \equiv \frac{1}{h}(\mathbf{U}_{j+1} - \mathbf{U}_j) - \mathbf{f}(x_j, \mathbf{U}_j) = 0, \qquad j = 0, 1, \ldots, J; \tag{1.3.12a}$$

$$\mathbf{U}_0 = \boldsymbol{\alpha}^{(h)}. \tag{1.3.12b}$$

The equation for $\mathbf{U}_{j+1}$ only involves $\mathbf{U}_j$ and so this is a one-step method; that is, $k = 1$. The first term on the right-hand side in Equation (1.3.12a) is clearly a difference quotient, intended to approximate $\mathbf{u}'(x_j)$.

To examine consistency for this scheme let $\mathbf{v}(x)$ be any twice continuously differentiable vector function with $|\mathbf{v}''(x)| \le M_2[\mathbf{v}]$ on $[a, b]$. Then from Equations (1.3.4), (1.3.12a), (1.3.1a), and Taylor's theorem,

$$\begin{aligned} |\tau_j[\mathbf{v}]| &= \left|\frac{1}{h}(\mathbf{v}(x_j + h) - \mathbf{v}(x_j)) - \mathbf{v}'(x_j)\right| \\ &\le \frac{h}{2} M_2[\mathbf{v}]. \end{aligned}$$

† On the other hand, *in rare circumstances*, a sophisticated computer user might intentionally use such a low order scheme in the neighborhood of certain types of singularities to reduce the effect of roundoff errors.

From the first line above we see that $\mathscr{L}_h[\cdot]$ is consistent with $\mathscr{L}[\cdot]$ for all functions with one continuous derivative (that is, the solution of (1.3.1) need not have two continuous derivatives). But, from the second line, it follows that Euler's method is consistent and has at least first-order accuracy for twice continuously differentiable functions. If the function $\mathbf{f}(x, \mathbf{u})$ satisfies the conditions of Theorem 1.1.2 and in addition $\mathbf{f}_x$ is continuous on R, then it clearly follows that the problem (1.3.1) has a unique solution with two continuous derivatives, and in fact

$$\mathbf{u}''(x) = \frac{\partial \mathbf{f}(x, \mathbf{u}(x))}{\partial \mathbf{u}} \mathbf{f}(x, \mathbf{u}(x)) + \mathbf{f}_x(x, \mathbf{u}(x)).$$

In this case the constant $M_2[\mathbf{u}]$ can be bounded in terms of bounds for $\|\partial \mathbf{f}/\partial \mathbf{u}\|$, $|\mathbf{f}|$ and $|\mathbf{f}_x|$ on R.

The stability of Euler's method is easily demonstrated provided $\mathbf{f}(x, \mathbf{u})$ satisfies the hypothesis of Theorem 1.1.1. (Since this is our basic existence proof for the initial-value problem, we are not imposing any strong new restrictions.) For any net function $\{\mathbf{V}_j\}$ the definition of $\mathscr{L}_h[\cdot]$ in (1.3.12a) yields the identity

$$\mathbf{V}_{j+1} = \mathbf{V}_j + h\mathbf{f}(x_j, \mathbf{V}_j) + h\mathscr{L}_h[\mathbf{V}_j], \qquad j = 0, 1, \ldots, J.$$

Subtracting the corresponding identity for $\{\mathbf{W}_j\}$ and using the Lipschitz continuity of $\mathbf{f}$, we obtain

$$|\mathbf{V}_{j+1} - \mathbf{W}_{j+1}| \leq (1 + hK)|\mathbf{V}_j - \mathbf{W}_j| + h \max_{0 \leq i \leq J} |\mathscr{L}_h[\mathbf{V}_i] - \mathscr{L}_h[\mathbf{W}_i]|, \qquad j = 0, 1, \ldots, J.$$

Now apply this inequality recursively in j, sum the resulting geometric progression which occurs and reduce the subscript j by unity to get

$$|\mathbf{V}_j - \mathbf{W}_j| \leq (1 + hK)^j|\mathbf{V}_0 - \mathbf{W}_0| + \frac{(1 + hK)^j - 1}{K} \max_{0 \leq i \leq J} |\mathscr{L}_h[\mathbf{V}_i] - \mathscr{L}_h[\mathbf{W}_i]|, \qquad j = 1, 2, \ldots, J + 1.$$

However, with $x_j = a + jh$ and $1 + hK \geq 0$, it is well known (see Problem 1.3.1) that

$$(1 + hK)^j \leq \exp [K(x_j - a)].$$

Stability clearly follows with the constant M in the inequality (1.3.8) chosen as, say,

$$M = \exp [K(b - a)] \max (1, K^{-1}).$$

We can now apply Theorem 1.3.1 to deduce convergence for Euler's method, or the Corollary to this Theorem to deduce that

$$\max_j |\mathbf{U}_j - \mathbf{u}(x_j)| \leq \mathcal{O}(h),$$

provided $|\boldsymbol{\alpha} - \boldsymbol{\alpha}^{(h)}| = \mathcal{O}(h)$ and $\mathbf{f}$ satisfies the hypothesis of Theorem 1.1.2. A simple modification of this scheme which yields a two-step method with possibly second-order accuracy is presented in Problems 1.3.2 and 1.3.3.

We turn now to more practical or rather higher-order-accurate methods. If we attempt to replace $\mathbf{u}'(x_j)$ by higher-order-accurate approximations than those employed in Euler's method or in the modification of Problem 1.3.2, we are led to schemes which are not stable [see Isaacson and Keller (1966), Section 1.4, Chapter 8]. However, by integrating the system of differential equations (1.3.1a) over the interval $[x_{j-k+1}, x_{j+1}]$, we obtain the system of *integral equations*

$$\mathbf{u}(x_{j+1}) = \mathbf{u}(x_{j-k+1}) + \int_{x_{j-k+1}}^{x_{j+1}} \mathbf{f}(x, \mathbf{u}(x))\, dx, \qquad k - 1 \le j \le J. \quad (1.3.13)$$

Now a large class of stable, higher-order-accurate methods can be devised by approximating the integrals in this system.

For $k = 1$ there are no net points interior to the interval of integration. One procedure is to define a set of tentative approximations to $\mathbf{u}(x)$ at a fixed set of points in $[x_j, x_{j+1}]$ and then use them in approximating the integral to define a final approximation to $\mathbf{u}(x_{j+1})$. One of the schemes suggested by this procedure is the well-known *Runge-Kutta method*

$$\mathbf{U}_{j+1} = \mathbf{U}_j + \frac{h}{6}[\mathbf{f}_{j,1} + 2\mathbf{f}_{j,2} + 2\mathbf{f}_{j,3} + \mathbf{f}_{j,4}], \qquad 0 \le j \le J, \quad (1.3.14\text{a})$$

where

$$\begin{aligned} \mathbf{f}_{j,1} &\equiv \mathbf{f}(x_j, \mathbf{U}_j), \\ \mathbf{f}_{j,2} &\equiv \mathbf{f}\left(x_j + \frac{h}{2}, \mathbf{U}_j + \frac{h}{2}\mathbf{f}_{j,1}\right), \\ \mathbf{f}_{j,3} &\equiv \mathbf{f}\left(x_j + \frac{h}{2}, \mathbf{U}_j + \frac{h}{2}\mathbf{f}_{j,2}\right), \\ \mathbf{f}_{j,4} &\equiv \mathbf{f}(x_{j+1}, \mathbf{U}_j + h\mathbf{f}_{j,3}). \end{aligned} \quad (1.3.14\text{b})$$

Under sufficient smoothness conditions on $\mathbf{f}$ (that is, continuous fourth derivatives), a very long calculation [Ince (1944), Appendix B] shows that *the Runge-Kutta method has order of accuracy* 4. Under much weaker conditions (that is, $\mathbf{f}$ Lipschitz-continuous), it can be shown that the *Runge-Kutta method is stable.* This proof is hardly more complicated than that for Euler's method and so we pose it as Problem 1.3.4. In fact, if $\mathbf{f}$ satisfies the hypothesis of Theorem 1.1 and in addition $|\mathbf{f}| \le B_0$, $|\mathbf{f}_x| \le B_1$, then it is also quite easy to show that the scheme in Equations (1.3.14) is consistent with the first-order system in Equation (1.3.1a). Further, if we consider only functions $\mathbf{v}(x)$ with Lipschitz-continuous first derivatives (with Lipschitz-constant

K'), then the truncation error of the Runge-Kutta scheme can be bounded by (see Problem 1.3.5)

$$|\tau_j[\mathbf{v}]| \leq \frac{h}{2}(2K' + KB_0 + B_1).$$

Thus we see that under rather mild smoothness conditions the Runge-Kutta method has at least first-order accuracy. In principle it is not difficult to define other one-step methods which have an order of accuracy greater than 4 under appropriate smoothness conditions. However, for most applications the Runge-Kutta scheme is sufficiently accurate.

The *multistep methods* suggested by Equation (1.3.13) with $k \geq 1$ can be written as

$$\mathbf{U}_{j+1} = \mathbf{U}_{j-k+1} + h \sum_{\nu=0}^{k'} c_k \mathbf{f}(x_{j-\nu+1}, \mathbf{U}_{j-\nu+1}),$$

$$\max(k, k') \leq j + 1 \leq J + 1. \quad (1.3.15)$$

Here $k' > 1$ and we use only data at the net points x_i to approximate the integral. It should be noted that if $c_0 \neq 0$, this scheme gives an explicit formula for $\mathbf{U}_{j+1}$ if and only if $\mathbf{f}(x, \mathbf{u})$ is linear in $\mathbf{u}$. Actually, by taking h sufficiently small we could be assured that this system always yields a unique value for $\mathbf{U}_{j+1}$ whenever $\mathbf{f}$ satisfies the hypotheses of Theorem 1.1.1, and that it could be obtained from a convergent iteration scheme (see Section 1.4). However, as a practical procedure we advocate the *predictor–corrector* methods which circumvent the above difficulty. First a tentative value, say $\mathbf{U}_{j+1}^*$, is predicted by a formula of the form (1.3.15) but with $c_0 = 0$, and then this tentative value is used in a different formula with $c_0 \neq 0$.

One of the many schemes which can be obtained in this way is the *modified Adams method* given by

$$\mathbf{U}_{j+1}^* = \mathbf{U}_j + \frac{h}{24}[55\mathbf{f}_j - 59\mathbf{f}_{j-1} + 37\mathbf{f}_{j-2} - 9\mathbf{f}_{j-3}], \quad (1.3.16a)$$

$$\mathbf{U}_{j+1} = \mathbf{U}_j + \frac{h}{24}[9\mathbf{f}(x_{j+1}, \mathbf{U}_{j+1}^*) + 19\mathbf{f}_j - 5\mathbf{f}_{j-1} + \mathbf{f}_{j-2}], \quad (1.3.16b)$$

$$\text{where} \quad \mathbf{f}_\nu \equiv \mathbf{f}(x_\nu, \mathbf{U}_\nu), \quad j = 3, 4, \ldots, J.$$

Here we have used Equation (1.3.15) with $k = 1$, $k' = 4$ for Equation (1.3.16a) and $k' = 3$ for Equation (1.3.16b). We call Equation (1.3.16a) the *predictor* and Equation (1.3.16b) the *corrector*. [Such a scheme can be viewed as the first step in an attempt to solve Equation (1.3.16b), with $\mathbf{U}_{j+1}^*$ replaced by $\mathbf{U}_{j+1}$, by iterations. But we do not continue the process.] To start the calculations in Equations (1.3.16), values of

$$\mathbf{U}_0, \mathbf{U}_1, \mathbf{U}_2, \mathbf{U}_3$$

must be determined. This can be done using the Runge-Kutta method (1.3.14), with $\mathbf{U}_0 = \mathbf{u}(x_0)$, and then

$$|\mathbf{U}_j - \mathbf{u}(x_j)| = \mathcal{O}(h^4), \qquad j = 0, 1, 2, 3, \tag{1.3.17}$$

if $\mathbf{f}$ is sufficiently smooth. Alternatively Taylor's expansion could be employed using

$$u' = \mathbf{f}(x, \mathbf{u}), \qquad u'' = \mathbf{f}_x(x, \mathbf{u}) + \frac{\partial \mathbf{f}(x, \mathbf{u})}{\partial \mathbf{u}} \mathbf{u}', \qquad \ldots$$

to evaluate the higher derivatives.

The truncation error $\boldsymbol{\tau}_j[\mathbf{u}]$ in the scheme (1.3.16) can be estimated in terms of the errors $h\boldsymbol{\sigma}_j^*[\mathbf{u}]$ and $h\boldsymbol{\sigma}_j[\mathbf{u}]$ by which the exact solution fails to satisfy Equations (1.3.16a) and (1.3.16b), respectively [with $\mathbf{U}_\nu$ replaced by $\mathbf{u}(x_\nu)$ and $\mathbf{U}_{j+1}^*$ by $\mathbf{u}(x_{j+1})$]. It is well known that

$$|\boldsymbol{\sigma}_j^*[\mathbf{u}]| \leq \tfrac{251}{720} M_5[\mathbf{u}]h^4, \qquad |\boldsymbol{\sigma}_j[\mathbf{u}]| \leq \tfrac{19}{720} M_5[\mathbf{u}]h^4$$

[see Isaacson and Keller (1966), Table 1.1, Section 2, Chapter 8], provided $\mathbf{u}$ has a continuous fifth derivative. Then, recalling that Equations (1.3.16) define $h\mathscr{L}_h[\mathbf{U}_j]$ in the notation of Equation (1.3.3a), it follows that

$$\begin{aligned} |\boldsymbol{\tau}_j[\mathbf{u}]| &\leq \frac{9h}{24} \left| \frac{\partial \mathbf{f}}{\partial \mathbf{u}} \boldsymbol{\sigma}_j^*[\mathbf{u}] \right| + |\boldsymbol{\sigma}_j[\mathbf{u}]| \\ &\leq \tfrac{3}{8} \cdot \tfrac{251}{720} M M_5[\mathbf{u}]h^5 + \tfrac{19}{720} M_5[\mathbf{u}]h^4 = \mathcal{O}(h^4), \end{aligned}$$

where M is a bound on the maximum absolute row sum of $\partial \mathbf{f}/\partial \mathbf{u}$. Thus, for sufficiently smooth functions, *the modified Adams method has order of accuracy* 4.

It is not difficult to show that this scheme is also stable, under the mild condition that $\mathbf{f}(x, \mathbf{u})$ be Lipschitz-continuous in $\mathbf{u}$. Thus if conditions (1.3.17) are satisfied by the initial data we can apply the Corollary of Theorem 1.3.1 to the modified Adams method to see that $\mathcal{O}(h^4)$ accuracy can be obtained. Of course we can easily define other stable multistep methods (that is, $k' > 1$) that have accuracy greater than four, but the scheme in (1.3.16) is adequate for many applications.

Comparing the Runge-Kutta scheme (1.3.14) with the scheme (1.3.16), it should be noted that the former requires four evaluations of $\mathbf{f}(x, \mathbf{u})$ at each step while the latter requires only two (since some evaluations can be retained in going from j to $j + 1$). As this is generally the major source of computations it would seem more efficient to use the modified Adams method. However, as indicated, some separate starting procedure is then required to evaluate $\mathbf{U}_\nu$ for $\nu = 1, 2, 3$, and Runge-Kutta is frequently used for this purpose. This implies that both procedures must be included in an appropriate digital computer code, and as a result one is tempted to use only the one-step method. But this choice, dictated by efforts to simplify the computer code, is frequently a false economy unless only a very few initial-value problems are to be solved.

Problems

1.3.1 Show that for all real numbers z,

$$1 + z \le e^z,$$

and equality holds only if $z = 0$. If $1 + z \ge 0$, show that for all integers $j \ge 0$,

$$(1 + z)^j \le e^{jz}.$$

1.3.2 The *midpoint rule* for solving the initial-value problem (1.3.1) is defined by

$$\text{(a)}\quad \mathscr{L}_h[\mathbf{U}_j] \equiv \frac{1}{2h}(\mathbf{U}_{j+1} - \mathbf{U}_{j-1}) - \mathbf{f}(x_j, \mathbf{U}_j) = 0, \qquad j = 1, 2, \ldots, J;$$

$$\text{(b)}\quad \mathbf{U}_0 = \boldsymbol{\gamma}_0, \qquad \mathbf{U}_1 = \boldsymbol{\gamma}_1.$$

Show that $\mathscr{L}_h[\cdot]$ is consistent with $\mathscr{L}[\cdot]$ for continuously differentiable functions, is at least first-order accurate for twice continuously differentiable functions and at least second-order accurate for three times continuously differentiable functions.

1.3.3 Show that $\mathscr{L}_h[\cdot]$, defined in Problem 1.3.2, is stable if $\mathbf{f}(x, \mathbf{u})$ satisfies the hypothesis of Theorem 1.1.1 (note that $k = 2$ in this scheme). The constant in the stability inequality can be taken as

$$M = \exp\,[2K(b - a)]\max\left(1, \frac{1}{2K}\right),$$

where K is the Lipschitz constant for $\mathbf{f}$.

1.3.4 The operator notation for the Runge-Kutta method is

$$h\mathscr{L}_h[\mathbf{U}_j] \equiv \mathbf{U}_{j+1} - \mathbf{U}_j - \frac{h}{6}[\mathbf{f}_{j,1} + 2\mathbf{f}_{j,2} + 2\mathbf{f}_{j,3} + \mathbf{f}_{j,4}], \qquad 0 \le j \le J,$$

where the $\mathbf{f}_{j,\nu}$ are as defined in Equations (1.3.14b). If $\mathbf{f}(x, \mathbf{y})$ is Lipschitz-continuous in $\mathbf{y}$ with constant K, show that for any net functions $\{\mathbf{U}_j\}$ and $\{\mathbf{V}_j\}$

$$|\mathbf{U}_{j+1} - \mathbf{V}_{j+1}| \le K'|\mathbf{U}_j - \mathbf{V}_j| + h|\mathscr{L}_h[\mathbf{U}_j] - \mathscr{L}_h[\mathbf{V}_j]|, \qquad 0 \le j \le J,$$

where

$$K' = 1 + hK + \frac{1}{2!}h^2K^2 + \frac{1}{3!}h^3K^3 + \frac{1}{4!}h^4K^4.$$

Complete the stability proof for the Runge-Kutta method.

1.3.5 Show that the Runge-Kutta method is *consistent* for all $\mathbf{v}(x)$ with continuous $\mathbf{v}'(x)$ if $\mathbf{f}$ satisfies the hypothesis of Theorem 1.1.1 and if in addition $|\mathbf{f}| \leq B_0$, $|\mathbf{f}_x| \leq B_1$. If $\mathbf{v}'(x)$ is Lipschitz-continuous with constant K', show that

$$|\boldsymbol{\tau}_j[\mathbf{v}]| = |\mathscr{L}_h[\mathbf{v}(x_j)] - \mathscr{L}[\mathbf{v}(x_j)]| \leq \frac{h}{2}(2K' + KB_0 + B_1).$$

1.3.6 Prove the stability of the modified Adams method (1.3.16) when $\mathbf{f}(x, \mathbf{u})$ is Lipschitz-continuous in $\mathbf{u}$.

1.4 Iterative Solution of Nonlinear Systems; Contracting Maps

All of the numerical methods that we study for solving nonlinear boundary-value problems lead to the problem of solving a nonlinear system of equations (either algebraic or transcendental). In some cases (see for example Section 2.2) the system is only a single equation while in other cases (see Section 3.2) the system may contain hundreds of equations. The procedures used to solve these systems are all iterative so we examine the basic theory of some of these procedures.

The nonlinear system will be assumed of the form

$$\boldsymbol{\phi}(\mathbf{s}) = 0, \tag{1.4.1}$$

or else of the form

$$\mathbf{s} = \mathbf{g}(\mathbf{s}). \tag{1.4.2}$$

Here $\mathbf{s}$ is an n-vector and $\boldsymbol{\phi}$ and $\mathbf{g}$ are vector-valued functions. It is a simple matter to convert one of these forms into the other without gaining or losing roots. For instance, if $A(\mathbf{s})$ is any nth-order matrix which is bounded and nonsingular for all $\mathbf{s}$, then the system of the form (1.4.2) with

$$\mathbf{g}(\mathbf{s}) \equiv \mathbf{s} - A(\mathbf{s})\boldsymbol{\phi}(\mathbf{s}), \tag{1.4.3}$$

is equivalent to the system (1.4.1) (that is, they have identical roots).

The simplest scheme for approximating a root of Equation (1.4.2) is known as *functional iteration.* It proceeds from some initial guess at the root, say $\mathbf{s} = \mathbf{s}^{(0)}$, and then generates the sequence of iterates $\{\mathbf{s}^{(\nu)}\}$ by

$$\mathbf{s}^{(\nu+1)} = \mathbf{g}(\mathbf{s}^{(\nu)}), \qquad \nu = 0, 1, \ldots. \tag{1.4.4}$$

This scheme is occasionally quite practical and can even be used to prove existence of a solution. We have the so-called *Contracting Mapping* theorem, which follows.

THEOREM 1.4.1. *Let* $\mathbf{g}(\mathbf{s})$ *satisfy the Lipschitz condition*

$$|\mathbf{g}(\mathbf{s}) - \mathbf{g}(\mathbf{t})| \leq \lambda|\mathbf{s} - \mathbf{t}|, \qquad \lambda < 1$$

for all $\mathbf{s} \in N_\rho(\mathbf{s}^{(0)})$ *and* $\mathbf{t} \in N_\rho(\mathbf{s}^{(0)})$, *where*

$$N_\rho(\mathbf{s}^{(0)}) \equiv \{\mathbf{s} \mid |\mathbf{s} - \mathbf{s}^{(0)}| \le \rho\}.$$

Let $\mathbf{s}^{(0)}$ *be such that*

$$|\mathbf{s}^{(0)} - \mathbf{g}(\mathbf{s}^{(0)})| \le (1 - \lambda)\rho.$$

Then the sequence in (1.4.4) *with initial iterate* $\mathbf{s}^{(0)}$ *satisfies*

(a) $\mathbf{s}^{(\nu)} \in N_\rho(\mathbf{s}^{(0)}), \qquad \nu = 0, 1, \ldots;$

(b) $\lim_{\nu\to\infty} \mathbf{s}^{(\nu)} = \mathbf{s}^*;$

(c) $\mathbf{s}^* = \mathbf{g}(\mathbf{s}^*)$ *and is the unique root in* $N_\rho(\mathbf{s}^{(0)});$

(d) $|\mathbf{s}^{(\nu)} - \mathbf{s}^*| \le \lambda^\nu \dfrac{|\mathbf{s}^{(1)} - \mathbf{s}^{(0)}|}{1 - \lambda}.$

Proof. By the hypothesis $\mathbf{s}^{(1)} = \mathbf{g}(\mathbf{s}^{(0)}) \in N_\rho(\mathbf{s}^{(0)})$ and we prove (a) by induction. Thus, if $\mathbf{s}^{(1)}, \ldots, \mathbf{s}^{(\nu)} \in N_\rho(\mathbf{s}^{(0)})$, we have

$$|\mathbf{s}^{(\nu+1)} - \mathbf{s}^{(\nu)}| = |\mathbf{g}(\mathbf{s}^{(\nu)}) - \mathbf{g}(\mathbf{s}^{(\nu-1)})| \le \lambda|\mathbf{s}^{(\nu)} - \mathbf{s}^{(\nu-1)}|.$$

Applying this recursively in ν we get, since $|\mathbf{s}^{(1)} - \mathbf{s}^{(0)}| \le (1 - \lambda)\rho$,

$$|\mathbf{s}^{(\nu+1)} - \mathbf{s}^{(\nu)}| \le \lambda^\nu(1 - \lambda)\rho.$$

But then, using $0 \le \lambda < 1$,

$$\begin{aligned}|\mathbf{s}^{(\nu+1)} - \mathbf{s}^{(0)}| &\le |\mathbf{s}^{(\nu+1)} - \mathbf{s}^{(\nu)}| + |\mathbf{s}^{(\nu)} - \mathbf{s}^{(\nu-1)}| + \cdots + |\mathbf{s}^{(1)} - \mathbf{s}^{(0)}| \\ &\le (\lambda^\nu + \lambda^{\nu-1} + \cdots + 1)(1 - \lambda)\rho \le \rho,\end{aligned}$$

so $\mathbf{s}^{(\nu+1)} \in N_\rho(\mathbf{s}^{(0)})$.

To prove (b) we show that $\{\mathbf{s}^{(\nu)}\}$ is a Cauchy sequence. In fact, just as above, we get for any integers ν and μ

$$|\mathbf{s}^{(\nu+\mu)} - \mathbf{s}^{(\nu)}| \le \lambda^\nu\rho,$$

and since $|\lambda| < 1$ the Cauchy criterion is satisfied. We call the limit vector $\mathbf{s}^*$. Since $\mathbf{g}(\mathbf{s})$ is continuous, by taking the limit $\nu \to \infty$ of Equations (1.4.4) it follows that $\mathbf{s}^*$ is a root of Equation (1.4.2). If there were two roots in $N_\rho(\mathbf{s}^{(0)})$, say $\mathbf{s}^*$ and $\mathbf{t}^*$, then

$$|\mathbf{s}^* - \mathbf{t}^*| = |\mathbf{g}(\mathbf{s}^*) - \mathbf{g}(\mathbf{t}^*)| \le \lambda|\mathbf{s}^* - \mathbf{t}^*|,$$

and since $|\lambda| < 1$ we must have $\mathbf{s}^* = \mathbf{t}^*$. Finally, letting $\mu \to \infty$ in an inequality above, we get $|\mathbf{s}^{(\nu)} - \mathbf{s}^*| \le \lambda^\nu\rho$. However, if, in the recursions leading to this result, we do not use $|\mathbf{s}^{(1)} - \mathbf{s}^{(0)}| < (1 - \lambda)\rho$ but retain the left-hand factor, we obtain (d). ▮

It should be observed that if $\rho = \infty$ in this theorem, then the iterations (1.4.4) converge for any initial estimate $\mathbf{s}^{(0)}$.

As we shall see in many practical applications of contracting maps, the function $\mathbf{g}(\mathbf{s})$ cannot be evaluated exactly. Thus, in place of Equations (1.4.4), the iterates which are actually generated satisfy, say,

$$\mathbf{s}^{(\nu+1)} = \mathbf{g}(\mathbf{s}^{(\nu)}) + \boldsymbol{\delta}^{(\nu)}, \qquad \nu = 0, 1, \ldots, \tag{1.4.5}$$

where the vectors $\boldsymbol{\delta}^{(\nu)}$ represent the errors in evaluating $\mathbf{g}(\mathbf{s}^{(\nu)})$. We cannot expect convergence to a root in this situation, but under proper conditions close approximations can still be obtained. This can be stated as follows.

THEOREM 1.4.2. *Let* $\mathbf{s} = \mathbf{s}^*$ *be a root of Equation* (1.4.2) *where* $\mathbf{g}(\mathbf{s})$ *satisfies the Lipschitz condition*

$$|\mathbf{g}(\mathbf{s}) - \mathbf{g}(\mathbf{t})| \leq \lambda|\mathbf{s} - \mathbf{t}|, \qquad \lambda < 1,$$

for all $\mathbf{s} \in N_\rho(\mathbf{s}^*)$ *and* $\mathbf{t} \in N_\rho(\mathbf{s}^*)$. *Let the errors* $\boldsymbol{\delta}^{(\nu)}$ *occurring in Equations* (1.4.5) *be bounded by*

$$|\boldsymbol{\delta}^{(\nu)}| < \delta, \qquad \nu = 0, 1, \ldots.$$

If the initial iterate $\mathbf{s}^{(0)}$ *satisfies* $\mathbf{s}^{(0)} \in N_{\rho_0}(\mathbf{s}^*)$, *where*

$$0 < \rho_0 \leq \rho - \frac{\delta}{1 - \lambda},$$

then the iterates $\{\mathbf{s}^{(\nu)}\}$ *generated in Equations* (1.4.5) *satisfy*

(a) $\mathbf{s}^{(\nu)} \in N_\rho(\mathbf{s}^*)$,

(b) $|\mathbf{s}^* - \mathbf{s}^{(\nu)}| \leq \lambda^\nu \rho_0 + \dfrac{\delta}{1 - \lambda}, \qquad \nu = 0, 1, \ldots.$

Proof. Since $\mathbf{s}^{(0)} \in N_{\rho_0}(\mathbf{s}^*)$ and $\rho \geq \rho_0$ it is clear that $\mathbf{s}^{(0)} \in N_\rho(\mathbf{s}^*)$. To prove (a) by induction, assume it valid for $\mathbf{s}^{(1)}, \ldots, \mathbf{s}^{(\nu-1)}$ and then, by Equations (1.4.5) and the hypothesis,

$$\begin{aligned}
|\mathbf{s}^* - \mathbf{s}^{(\nu)}| &= |\mathbf{g}(\mathbf{s}^*) - \mathbf{g}(\mathbf{s}^{(\nu-1)}) - \boldsymbol{\delta}^{(\nu-1)}| \\
&\leq \lambda|\mathbf{s}^* - \mathbf{s}^{(\nu-1)}| + \delta \\
&\leq \lambda^2|\mathbf{s}^* - \mathbf{s}^{(\nu-2)}| + \lambda\delta + \delta \\
&\ \ \vdots \\
&\leq \lambda^\nu|\mathbf{s}^* - \mathbf{s}^{(0)}| + \frac{1 - \lambda^\nu}{1 - \lambda}\,\delta.
\end{aligned}$$

But $|\mathbf{s}^* - \mathbf{s}^{(0)}| \leq \rho_0$ and $0 \leq \lambda < 1$ so that $|\mathbf{s}^* - \mathbf{s}^{(\nu)}| \leq \rho_0 + (\delta/1 - \lambda)$ from which (a) follows. Then (b) is obvious from the final inequality above. ∎

It is clear from (b) that the error bound on the computed root can be made arbitrarily close to $\delta/(1 - \lambda)$, but perhaps no smaller.

To "solve" equations of the form (1.4.1) we need only determine a matrix $A(\mathbf{s})$ such that $\mathbf{g}(\mathbf{s})$ as defined in Equation (1.4.3) satisfies the hypothesis of Theorem 1.4.1 (or Theorem 1.4.2 if errors are permitted). A particularly effective procedure of this form is *Newton's method*, in which we take $A(\mathbf{s}) \equiv J^{-1}(\mathbf{s})$, where

$$J(\mathbf{s}) \equiv \frac{\partial \boldsymbol{\phi}(\mathbf{s})}{\partial \mathbf{s}} \equiv \left(\frac{\partial \phi_i(\mathbf{s})}{\partial s_j}\right)$$

is the Jacobian matrix of $\boldsymbol{\phi}$ with respect to $\mathbf{s}$. The corresponding iteration scheme is

$$\mathbf{s}^{(\nu+1)} = \mathbf{s}^{(\nu)} - J^{-1}(\mathbf{s}^{(\nu)})\boldsymbol{\phi}(\mathbf{s}^{(\nu)}), \qquad \nu = 0, 1, \ldots. \tag{1.4.6}$$

Of course the computations should not be carried out in this form, employing the inverse of $J(\mathbf{s}^{(\nu)})$, but rather by solving the linear systems in

$$J(\mathbf{s}^{(\nu)})\,\Delta\mathbf{s}^{(\nu)} = -\boldsymbol{\phi}(s^{(\nu)}), \qquad \mathbf{s}^{(\nu+1)} = \mathbf{s}^{(\nu)} + \Delta\mathbf{s}^{(\nu)}, \qquad \nu = 0, 1, \ldots. \tag{1.4.7}$$

The convergence of Newton's method is frequently quite rapid, even better in fact than the geometric type of convergence illustrated in Theorems 1.4.1 and 1.4.2. We do not present the best possible results here but shall be content to show that in a sense the previous type of error analysis applies with *arbitrarily small* values of λ. Thus we state the following.

THEOREM 1.4.3. *Let the function* $\boldsymbol{\phi}(\mathbf{s})$ *have a zero at* $\mathbf{s} = \mathbf{s}^*$, *continuous first derivatives in some neighborhood* $N_\rho(\mathbf{s}^*)$ *of* $\mathbf{s}^*$ *and nonsingular Jacobian at* $\mathbf{s}^*$, *that is,* $\det J(\mathbf{s}^*) \neq 0$. *Then for each* λ *in* $0 < \lambda < 1$ *there exists a positive number* ρ_λ *such that for any* $\mathbf{s}^{(0)} \in N_{\rho_\lambda}(\mathbf{s}^*)$ *the Newton iterates* (1.4.6) *converge to* $\mathbf{s}^*$ *with*

$$|\mathbf{s}^* - \mathbf{s}^{(\nu)}| \leq \lambda^\nu|\mathbf{s}^* - \mathbf{s}^{(0)}|.$$

Proof. As the terms in the function $\det J(\mathbf{s})$ are products of combinations of the derivatives $\partial\phi_i(\mathbf{s})/\partial s_j$, this function must be continuous on $N_\rho(\mathbf{s}^*)$. But since $\det J(\mathbf{s}^*) \neq 0$, it follows by the continuity that $\det J(\mathbf{s}) \neq 0$ on $N_{\rho'}(\mathbf{s}^*)$ for some positive $\rho' \leq \rho$. Now consider the matrix

$$B(\mathbf{s}, \mathbf{h}) \equiv I - J^{-1}(\mathbf{s})J(\mathbf{s} + \mathbf{h}),$$

which for each $\mathbf{s} \in N_{\rho'}(\mathbf{s}^*)$ is a continuous function of $\mathbf{h}$ for $\mathbf{s} + \mathbf{h} \in N_\rho(\mathbf{s}^*)$. If the elements of the inverse of the Jacobian are denoted by $J^{-1}(\mathbf{s}) \equiv (a_{ij}(\mathbf{s}))$ then we may write the *maximum absolute row sum* of $B(\mathbf{s}, \mathbf{h})$ as

$$\delta(\mathbf{s}, \mathbf{h}) \equiv \max_{1 \leq i \leq n} \left\{\left|1 - \sum_{k=1}^{n} a_{ik}(\mathbf{s}) \frac{\partial \phi_k(\mathbf{s} + \mathbf{h})}{\partial s_i}\right| + \sum_{j \neq i}\left|\sum_{k=1}^{n} a_{ik}(\mathbf{s}) \frac{\partial \phi_k(\mathbf{s} + \mathbf{h})}{\partial s_j}\right|\right\}.$$

It is clear that $\delta(\mathbf{s}, \mathbf{0}) = 0$ for each $\mathbf{s} \in N_{\rho'}(\mathbf{s}^*)$ since $B(\mathbf{s}, \mathbf{0}) = 0$ and $\delta(\mathbf{s}, \mathbf{h})$ is a continuous function of $\mathbf{h}$ whenever $B(\mathbf{s}, \mathbf{h})$ is. Thus, for any positive $\lambda < 1$, we can be assured that $\delta(\mathbf{s}, \mathbf{h}) \le \lambda < 1$, provided $\mathbf{s}$ is sufficiently close to $\mathbf{s}^*$ and $\mathbf{h}$ is sufficiently small.

The details that we actually use in the proof are slightly more complicated than the above, but the idea is the same. In place of $\delta(\mathbf{s}, \mathbf{h})$ we must introduce a function of the three vectors $\mathbf{s}$, $\mathbf{h}$ and $\boldsymbol{\theta} = (\theta_1, \ldots, \theta_n)$ with $0 < \theta_i < 1$:

$$\delta(\mathbf{s}, \mathbf{h}, \boldsymbol{\theta}) \equiv \max_{1 \le i \le n} \left\{ \left| 1 - \sum_{k=1}^{n} a_{ik}(\mathbf{s}) \frac{\partial \phi_k(\mathbf{s} + \theta_k \mathbf{h})}{\partial s_i} \right| + \sum_{j \ne 1} \left| \sum_{k=1}^{n} a_{ik}(\mathbf{s}) \frac{\partial \phi_k(\mathbf{s} + \theta_k \mathbf{h})}{\partial s_j} \right| \right\}.$$

Again $\delta(\mathbf{s}, \mathbf{0}, \boldsymbol{\theta}) = 0$ for $\mathbf{s} \in N_{\rho'}(\mathbf{s}^*)$ and $\delta(\mathbf{s}, \mathbf{h}, \boldsymbol{\theta})$ is continuous in $\mathbf{h}$ and $\boldsymbol{\theta}$ whenever $B(\mathbf{s}, \mathbf{h})$ is continuous in $\mathbf{h}$. Thus for any positive $\lambda < 1$ we can find a $\rho_\lambda > 0$ such that

$$\delta(\mathbf{s}, \mathbf{h}, \boldsymbol{\theta}) \le \lambda < 1, \tag{1.4.8a}$$

provided that

$$\mathbf{s} \in N_{\rho_\lambda}(\mathbf{s}^*), \qquad |\mathbf{h}| \le \rho_\lambda, \qquad \text{and} \quad 0 < \theta_i < 1, \qquad i = 1, 2, \ldots, n. \tag{1.4.8b}$$

(In general we may expect that $\rho_\lambda \le \rho'/2$.)

Now we shall show that the iterates in Equations (1.4.6) satisfy the basic inequality

$$|\mathbf{s}^* - \mathbf{s}^{(\nu+1)}| \le \lambda |\mathbf{s}^* - \mathbf{s}^{(\nu)}|, \qquad \nu = 0, 1, \ldots, \tag{1.4.9}$$

provided that $\mathbf{s}^{(0)} \in N_{\rho_\lambda}(\mathbf{s}^*)$. For an induction, assume it true for $\mathbf{s}^{(1)}, \ldots, \mathbf{s}^{(\nu)}$ and then from Equations (1.4.6) and $\boldsymbol{\phi}(\mathbf{s}^*) = 0$,

$$\begin{aligned} \mathbf{s}^* - \mathbf{s}^{(\nu+1)} &= \mathbf{s}^* - \mathbf{s}^{(\nu)} + J^{-1}(\mathbf{s}^{(\nu)}) \boldsymbol{\phi}(\mathbf{s}^{(\nu)}) \\ &= [\mathbf{s}^* - \mathbf{s}^{(\nu)}] - J^{-1}(\mathbf{s}^{(\nu)})[\boldsymbol{\phi}(\mathbf{s}^*) - \boldsymbol{\phi}(\mathbf{s}^{(\nu)})]. \end{aligned}$$

But by Taylor's theorem we have, with $\mathbf{h}^{(\nu)} \equiv \mathbf{s}^* - \mathbf{s}^{(\nu)}$,

$$\phi_i(\mathbf{s}^*) = \phi_i(\mathbf{s}^{(\nu)} + \mathbf{h}^{(\nu)}) = \phi_i(\mathbf{s}^{(\nu)}) + \sum_{j=1}^{n} \frac{\partial \phi_i(\mathbf{s}^{(\nu)} + \theta_i \mathbf{h}^{(\nu)})}{\partial s_j} h_j^{(\nu)},$$
$$0 < \theta_i < 1, \qquad i = 1, 2, \ldots, n.$$

Using this in the previous equation yields, componentwise,

$$\begin{aligned} s_i^* - s_i^{(\nu+1)} = {} & \left[1 - \sum_{k=1}^{n} a_{ik}(\mathbf{s}^{(\nu)}) \frac{\partial \phi_k(\mathbf{s}^{(\nu)} + \theta_k \mathbf{h}^{(\nu)})}{\partial s_i} \right] (s_i^* - s_i^{(\nu)}) \\ & + \sum_{j \ne i} \left[\sum_{k=1}^{n} a_{ik}(\mathbf{s}^{(\nu)}) \frac{\partial \phi_k(\mathbf{s}^{(\nu)} + \theta_k \mathbf{h}^{(\nu)})}{\partial s_j} \right] (s_j^* - s_j^{(\nu)}), \qquad 1 \le i \le n. \end{aligned}$$

Taking absolute values, we easily obtain

$$|\mathbf{s}^* - \mathbf{s}^{(\nu+1)}| \leq \delta(\mathbf{s}^{(\nu)}, \mathbf{s}^* - \mathbf{s}^{(\nu)}, \boldsymbol{\theta})|\mathbf{s}^* - \mathbf{s}^{(\nu)}|.$$

But obviously this result applies for $\nu = 0$ and so the induction is concluded. The theorem now follows, as in the proof of Theorem 1.4.1, from the recursive application of the basic inequality (1.4.9). ∎

If $\boldsymbol{\phi}(\mathbf{s})$ has continuous second derivatives then it can be shown that Newton's method is frequently second-order (see Problem 1.4.4), that is, that

$$|\mathbf{s}^* - \mathbf{s}^{(\nu+1)}| \leq M|\mathbf{s}^* - \mathbf{s}^{(\nu)}|^2, \qquad \nu = 0, 1, \ldots.$$

In fact this can easily be seen from the above proof since $\delta(\mathbf{s}, \mathbf{h}, \boldsymbol{\theta}) = \delta(\mathbf{s}, \mathbf{0}, \boldsymbol{\theta}) + \mathcal{O}(\mathbf{h}) = \mathcal{O}(\mathbf{h})$, using the expansion to next order in h. Further results on Newton's method are contained in Problems 1.4.6 and 1.4.7.

In the case of *scalar equations* there are numerous procedures that may yield convergent second-order or even higher-order iterations. A particularly effective such scheme for solving $s = g(s)$ is *Aitken's* δ^2*-method*, defined as

$$s_{\nu+1} = s_\nu - \frac{\delta_\nu^2}{\delta'_{\nu+1} - \delta_\nu}, \qquad \nu = 0, 1, \ldots, \tag{1.4.10a}$$

where

$$\begin{aligned} s'_{\nu+1} &= g(s_\nu), & s''_{\nu+1} &= g(s'_{\nu+1}); \\ \delta_\nu &= s'_{\nu+1} - s_\nu, & \delta'_{\nu+1} &= s''_{\nu+1} - s'_\nu. \end{aligned} \tag{1.4.10b}$$

The convergence properties of this scheme are thoroughly discussed in Isaacson and Keller (1966), pp. 103–108.

Problems

1.4.1 If $\mathbf{g}(\mathbf{s})$ has continuous first derivatives for all $\mathbf{s} \in N_\rho(\mathbf{s}^{(0)})$ and

$$\mu \equiv \max_{1 \leq i \leq n,\, \mathbf{s} \in N_\rho(\mathbf{s}^{(0)})} \sum_{j=1}^{n} |\partial g_i(\mathbf{s})/\partial s_j| < 1, \qquad |\mathbf{s}^{(0)} - \mathbf{g}(\mathbf{s}^{(0)})| \leq (1 - \mu)\rho,$$

show that the *conclusions* of Theorem 1.4.1 hold for the sequence $\mathbf{s}^{(\nu+1)} = \mathbf{g}(\mathbf{s}^{(\nu)})$.

1.4.2 If $\mathbf{g}(\mathbf{s}^*) = \mathbf{s}^*$ and $|\mathbf{g}(\mathbf{s}^*) - \mathbf{g}(\mathbf{s})| \leq \lambda|\mathbf{s}^* - \mathbf{s}|$, $\lambda < 1$ for all $\mathbf{s} \in N_\rho(\mathbf{s}^*)$, then $\mathbf{s}^{(\nu+1)} = \mathbf{g}(\mathbf{s}^{(\nu)}) \to \mathbf{s}^*$ for any $\mathbf{s}^{(0)} \in N_\rho(\mathbf{s}^*)$. [Note that this is weaker than the Lipschitz continuity of $\mathbf{g}(\mathbf{s})$.]

1.4.3 Let the scalar function $\phi(s)$ have a continuous first derivative which satisfies $0 < \gamma \leq \phi'(s) \leq \Gamma$ for all s. Show that the iterations

$$s_{\nu+1} = s_\nu - m\phi(s_\nu), \qquad \nu = 0, 1, \ldots,$$

converge to the root of $\phi(s) = 0$ for any s_0 and fixed m in

$$0 < m < \frac{2}{\Gamma}.$$

Show that the "best" choice for the parameter is

$$m = \frac{2}{\Gamma + \gamma},$$

in which case

$$\lambda = \frac{\Gamma - \gamma}{\Gamma + \gamma}$$

is a bound on the geometric convergence factor.

1.4.4 Show that Newton's method with initial guess $\mathbf{s}^{(0)}$ is actually second-order, provided that $\boldsymbol{\phi}(\mathbf{s})$ has continuous second derivatives which satisfy

$$\sum_{i,j,k} \left| \frac{\partial^2 \phi_k(\mathbf{s})}{\partial s_i\, \partial s_j} \right| \le \frac{\delta_0^2}{M} < 1,$$

where $\boldsymbol{\phi}(\mathbf{s}^*) = 0$, $\|J^{-1}(\mathbf{s})\| \le M$ for all $\mathbf{s} \in N_\rho(\mathbf{s}^*)$ and

$$\delta_0 = |\mathbf{s}^* - \mathbf{s}^{(0)}| < \rho.$$

1.4.5* Prove the following.

THEOREM 1.4.4. *Let $\boldsymbol{\phi}(\mathbf{s})$ have Lipschitz-continuous first derivatives on $N_\rho(\mathbf{s}^{(0)})$ and satisfy*

(a) $J(\mathbf{s}^{(0)}) \equiv \dfrac{\partial \boldsymbol{\phi}(\mathbf{s}^{(0)})}{\partial \mathbf{s}}$ *is nonsingular,* $\|J^{-1}(\mathbf{s}^{(0)})\| \le K_0$;

(b) $\|J(\mathbf{s}) - J(\mathbf{t})\| \le K_1|\mathbf{s} - \mathbf{t}|$ *for all* $\mathbf{s}$ *and* $\mathbf{t} \in N_\rho(\mathbf{s}^{(0)})$;

(c) $|J^{-1}(\mathbf{s}^{(0)})\boldsymbol{\phi}(\mathbf{s}^{(0)})| \le \eta$.

Further, let

(d) $K_0K_1\rho < 1, \qquad \eta \le (1 - K_0K_1\rho)\rho$.

Then $\boldsymbol{\phi}(\mathbf{s}) = 0$ has a unique root in $N_\rho(\mathbf{s}^{(0)})$. This root is the limit of the sequence $\{\mathbf{s}^{(\nu)}\}$ defined by

$$\mathbf{s}^{(0)} = \mathbf{s}^{(0)},$$

$$\mathbf{s}^{(\nu+1)} = \mathbf{s}^{(\nu)} - J^{-1}(\mathbf{s}^{(0)})\boldsymbol{\phi}(\mathbf{s}^{(\nu)}), \qquad \nu = 0, 1, 2, \ldots.$$

[HINT: Show that the mapping $\mathbf{s} = \mathbf{g}(\mathbf{s})$ with $\mathbf{g}(\mathbf{s}) \equiv \mathbf{s} - J^{-1}(\mathbf{s}^{(0)})\boldsymbol{\phi}(\mathbf{s})$ satisfies the hypothesis of Theorem 1.4.1.]

1.4.6 With $J(\mathbf{s}) \equiv \partial\boldsymbol{\phi}(\mathbf{s})/\partial\mathbf{s}$ continuous on a convex domain D, define $J(\mathbf{s}, \mathbf{t}) \equiv \int_0^1 J(\mathbf{s} + \theta[\mathbf{t} - \mathbf{s}])\, d\theta$. For all $\mathbf{s}$ and $\mathbf{t} \in D$ show that $J(\mathbf{s}, \mathbf{t}) = J(\mathbf{t}, \mathbf{s})$ and $\phi(\mathbf{s}) - \phi(\mathbf{t}) = J(\mathbf{s}, \mathbf{t})(\mathbf{s} - \mathbf{t})$.

If $\|J(\mathbf{s}) - J(\mathbf{t})\| \leq K|\mathbf{s} - \mathbf{t}|$ for all $\mathbf{s}$ and $\mathbf{t} \in D$, show that

$$\|J(\mathbf{s}) - J(\mathbf{s}, \mathbf{t})\| \leq \frac{K}{2}|\mathbf{s} - \mathbf{t}|.$$

1.4.7* Prove the following.

THEOREM 1.4.5. *Let the hypothesis of Theorem* 1.4.4 *be satisfied and in addition*

(d)′ $K_0K_1\rho < \frac{2}{3}$, $\quad \eta \leq (1 - \frac{3}{2}K_0K_1\rho)\rho$.

Then Newton's method,

$$\mathbf{s}^{(\nu+1)} = \mathbf{s}^{(\nu)} - J^{-1}(\mathbf{s}^{(\nu)})\boldsymbol{\phi}(\mathbf{s}^{(\nu)}), \qquad \nu = 0, 1, 2, \ldots,$$

with initial guess $\mathbf{s}^{(0)}$, *converges to the unique root of* $\boldsymbol{\phi}(\mathbf{s}) = 0$ *in* $N_\rho(\mathbf{s}^{(0)})$. [HINT: First show that $J(\mathbf{s})$ is nonsingular for all $\mathbf{s} \in N_\rho(\mathbf{s}^{(0)})$. Then show by induction, using Problem 1.4.6, that all $\mathbf{s}^{(\nu)} \in N_\rho(\mathbf{s}^{(0)})$. Use the unique root $\mathbf{s}^*$, established in Problem 1.4.5, and deduce, again using Problem 1.4.6, that

$$|\mathbf{s}^{(\nu+1)} - \mathbf{s}^*| \leq \frac{K_0K_1\rho}{2(1 - K_0K_1\rho)}|\mathbf{s}^{(\nu)} - \mathbf{s}^*|, \qquad \nu = 1, 2, \ldots.].$$

1.4.8* (Continuation of Problem 1.4.7.) Show that the errors in Newton's method under the conditions of Theorem 1.4.5 actually satisfy

$$|\mathbf{s}^{(\nu+1)} - \mathbf{s}^*| \leq \left(\frac{\lambda}{3}\right)^{2^\nu} \frac{\rho}{\lambda}, \qquad \nu = 1, 2, \ldots,$$

where $\lambda = K_0K_1\rho/2(1 - K_0K_1\rho)$.

SUPPLEMENTARY REFERENCES AND NOTES

Section 1.1 The theory of existence, uniqueness, and dependence of solutions on parameters for initial-value problems is quite well-developed. Accounts of this theory, in order of increasing sophistication, are to be found in Birkhoff and Rota (1962), Ince (1944), Coddington and Levinson (1955), and Hartman (1964).

Section 1.2 The theory of linear two-point boundary-value problems is treated in Coddington and Levinson (1955), pp. 284–312; Ince (1944), pp. 204–222, and Hartman (1964), pp. 407–412. The important special case of second-order equations receives a thorough treatment in all of these texts. Existence theory for general nonlinear problems is not fully developed. The case of periodic solutions is treated in Hartman (1964), pp. 413–435, and these methods can also be applied to more general boundary-value problems. Many other results are contained in journals; a few references related to the present study are R. Conti (1961), H. Keller (1966), M. Lees (1966), A. Lasota and Z. Opial (1966).

Section 1.3 The theory of Dahlquist (1956), (1959) is basic for a thorough understanding of numerical methods for initial-value problems. This theory is presented in great detail and extended in Henrici (1962), the standard reference in this field. Briefer accounts are contained in Isaacson and Keller (1966), Antosiewicz and Gautschi (1962), and their references.

Section 1.4 More general discussions of iterative methods for solving nonlinear systems are contained in the texts: Ostrowski (1960), Traub (1964), Isaacson and Keller (1966). A thorough study of the convergence of Newton's method is contained in Kantorovich and Akilov (1964) and the effect of roundoff or inaccurate evaluation of the functions and derivatives is given by Lancaster (1966). Roundoff effects in contracting maps are studied in detail by Urabe (1956). Additional iterative methods of interest are studied in Ortega and Rockoff (1966).

CHAPTER 2

INITIAL-VALUE METHODS (SHOOTING)

2.1 *Linear Second-Order Equations and Systems*

We consider first the single linear second-order equation

$$Ly \equiv -y'' + p(x)y' + q(x)y = r(x), \qquad a \leq x \leq b, \tag{2.1.1a}$$

subject to the general two-point boundary conditions

$$a_0 y(a) - a_1 y'(a) = \alpha, \quad b_0 y(b) + b_1 y'(b) = \beta, \quad |a_0| + |b_0| \neq 0. \tag{2.1.1b}$$

We assume the functions $p(x)$, $q(x)$, and $r(x)$ to be continuous on $[a, b]$ and require that *the homogeneous problem*

$$Lz = 0; \quad a_0 z(a) - a_1 z'(a) = 0, \quad b_0 z(b) + b_1 z'(b) = 0 \tag{2.1.2}$$

have only the trivial solution, $z(x) \equiv 0$. Then by the Alternative Theorem (see Problem 1.2.5) the boundary-value problem (2.1.1) has a solution which is unique. Of course we could have required (2.1.1) to have a unique solution, and then it would follow that (2.1.2) has only the trivial solution. We shall describe and analyze the initial-value or shooting method for computing accurate approximations to the solution of the linear boundary-value problem (2.1.1).

Two functions $y^{(1)}(x)$ and $y^{(2)}(x)$ are uniquely defined on $[a, b]$ as solutions of the respective initial-value problems

$$Ly^{(1)} = r(x); \quad y^{(1)}(a) = -\alpha c_1, \quad y^{(1)\prime}(a) = -\alpha c_0; \tag{2.1.3a}$$

and

$$Ly^{(2)} = 0; \quad y^{(2)}(a) = a_1, \quad y^{(2)\prime}(a) = a_0. \tag{2.1.3b}$$

Here c_0 and c_1 are any constants such that

$$a_1c_0 - a_0c_1 = 1. \tag{2.1.3c}$$

The fact that these problems have unique solutions on $[a, b]$ is an obvious consequence of Theorem 1.1.1. The function $y(x)$ defined by

$$y(x) = y(x; s) \equiv y^{(1)}(x) + sy^{(2)}(x), \qquad a \le x \le b, \tag{2.1.4a}$$

satisfies $a_0y(a) - a_1y'(a) = \alpha(a_1c_0 - a_0c_1) = \alpha$ and so will be a solution of problem (2.1.1) if s is chosen such that

$$\phi(s) \equiv b_0y(b; s) + b_1y'(b; s) - \beta = 0.$$

This equation is linear in s and has the single root

$$s = \frac{\beta - [b_0y^{(1)}(b) + b_1y^{(1)\prime}(b)]}{[b_0y^{(2)}(b) + b_1y^{(2)\prime}(b)]}, \tag{2.1.4b}$$

provided that $[b_0y^{(2)}(b) + b_1y^{(2)\prime}(b)] \neq 0$. However, if this quantity does vanish, then $z(x) \equiv y^{(2)}(x)$ would be a *nontrivial* solution of the homogeneous problem (2.1.2). Since this has been excluded, we see that the above construction of a solution of (2.1.1) is valid.

The initial-value or shooting method consists in simply carrying out the above procedure numerically; that is, we compute approximations to $y^{(1)}(x)$, $y^{(1)\prime}(x)$, $y^{(2)}(x)$, and $y^{(2)\prime}(x)$ and use them in Equations (2.1.4). To solve the initial-value problems (2.1.3) and (2.1.4) numerically, we first write them as equivalent first-order systems, say

$$\begin{pmatrix} y^{(1)} \\ v^{(1)} \end{pmatrix}' = \begin{pmatrix} v^{(1)} \\ pv^{(1)} + qy^{(1)} - r \end{pmatrix}, \quad y^{(1)}(a) = -\alpha c_1, \quad v^{(1)}(a) = -\alpha c_0; \tag{2.1.5a}$$

and

$$\begin{pmatrix} y^{(2)} \\ v^{(2)} \end{pmatrix}' = \begin{pmatrix} v^{(2)} \\ pv^{(2)} + qy^{(2)} \end{pmatrix}, \qquad y^{(2)}(a) = a_1, \qquad v^{(2)}(a) = a_0. \tag{2.1.5b}$$

Now any of the numerical methods discussed in Section 1.3 can be employed. Of course, the same method should be used to solve both systems (2.1.5a–b), and for convenience we will require that this be done on the net

$$x_j = a + jh, \qquad j = 0, 1, \ldots, J, \qquad h = \frac{b - a}{J}; \tag{2.1.6}$$

so that $x_J = b$.

Let the numerical solutions of (2.1.5) obtained on the net (2.1.6) be denoted, in obvious notation, by the net functions

$$\{Y_j^{(1)}\}, \qquad \{V_j^{(1)}\}, \qquad \{Y_j^{(2)}\}, \qquad \{V_j^{(2)}\}.$$

We assume that the scheme employed has an order of accuracy at least p and is stable. Then if the solutions of problems (2.1.5a–b) are sufficiently smooth, we are assured that for $0 \le j \le J$,

$$|y^{(\nu)}(x_j) - Y_j^{(\nu)}| = \mathcal{O}(h^p), \qquad |v^{(\nu)}(x_j) - V_j^{(\nu)}| = \mathcal{O}(h^p), \qquad \nu = 1, 2, \tag{2.1.7}$$

provided the initial data used in the calculations also satisfy these relations. At the point $x_0 = a$ the exact data can be used so that

$$Y_0^{(1)} = -\alpha c_1, \qquad v_0^{(1)} = -\alpha c_0; \qquad Y_0^{(2)} = a_1, \qquad V_0^{(2)} = a_0. \tag{2.1.8}$$

To approximate the solution, $y(x)$, of the boundary-value problem (2.1.1) we now form, by analogy with Equations (2.1.4),

$$Y_j = Y_j^{(1)} + s_h Y_j^{(2)}, \qquad j = 0, 1, \ldots, J, \tag{2.1.9a}$$

where

$$s_h = \frac{\beta - [b_0 Y_J^{(1)} + b_1 V_J^{(1)}]}{[b_0 Y_J^{(2)} + b_1 V_J^{(2)}]}. \tag{2.1.9b}$$

We may expect, in light of the estimates (2.1.7), that $|Y_j - y(x_j)| = \mathcal{O}(h^p)$. This is indeed true but, as we shall also show, there may be practical difficulties in actually computing an accurate approximation to $y(x)$ for a fixed small $h > 0$.

Let us define the errors in the numerical solutions by

$$e_j \equiv Y_j - y(x_j), \qquad e_j^{(\nu)} \equiv Y_j^{(\nu)} - y^{(\nu)}(x_j), \qquad \varepsilon_j^{(\nu)} \equiv V_j^{(\nu)} - v^{(\nu)}(x_j),$$
$$\nu = 1, 2; \qquad 0 \le j \le J.$$

Evaluate Equation (2.1.4a) at $x = x_j$ and subtract the result from Equation (2.1.9a) to get, with the above notation,

$$\begin{aligned} e_j &= e_j^{(1)} + s_h Y_j^{(2)} - s y^{(2)}(x_j) \\ &= [e_j^{(1)} + s e_j^{(2)}] + (s_h - s) Y_j^{(2)}, \quad 0 \le j \le J. \end{aligned} \tag{2.1.10a}$$

However, a small calculation, using Equations (2.1.4b) and (2.1.9b), reveals that

$$s_h - s = -\frac{[b_0 e_J^{(1)} + b_1 \varepsilon_J^{(1)}] + s[b_0 e_J^{(2)} + b_1 \varepsilon_J^{(2)}]}{(b_0 Y_J^{(2)} + b_1 V_J^{(2)})}. \tag{2.1.10b}$$

Recalling the estimates (2.1.7), we have

$$b_0 Y_J^{(2)} + b_1 V_J^{(2)} = (b_0 y^{(2)}(b) + b_1 y^{(2)\prime}(b)) + \mathcal{O}(h^p),$$

and so for sufficiently small h the denominator in Equation (2.1.10b) cannot vanish. From Equations (2.1.10) we now conclude that

$$|Y_j - y(x_j)| = \mathcal{O}(h^p), \qquad 0 \le j \le J.$$

It is no more difficult to demonstrate that

$$V_j \equiv V_j^{(1)} + s_h V_j^{(2)} = y'(x_j) + \mathcal{O}(h^p), \qquad 0 \le j \le J,$$

so we also have accurate approximations to the derivative of the solution, for sufficiently small h.

The above result can now be summarized as follows.

THEOREM 2.1.1. *Let the boundary-value problem* (2.1.1) *have a unique solution,* $y(x)$. *Let the numerical solutions of the initial-value problems* (2.1.5a–b), *using a stable method with order of accuracy* p *on the net* (2.1.6), *be denoted by* $\{Y_j^{(\nu)}\}$, $\{V_j^{(\nu)}\}$, $\nu = 1, 2$. *Then the net functions*

$$Y_j = Y_j^{(1)} + s_h Y_j^{(2)}, \qquad V_j = V_j^{(1)} + s_h V_j^{(2)}, \qquad 0 \le j \le J,$$

with s_h *as given by Equation* (2.1.9b) *and* h *sufficiently small, satisfy*

$$|y(x_j) - Y_j| = \mathcal{O}(h^p), \qquad |y'(x_j) - V_j| = \mathcal{O}(h^p), \qquad 0 \le j \le J,$$

provided the initial-value problems (2.1.5a–b) *have sufficiently smooth solutions.* ∎

A particularly important case of the general boundary-value problem (2.1.1) occurs when $a_0 = b_0 = 1$ and $a_1 = b_1 = 0$ in the boundary conditions (that is, y specified at both ends). The error expressions (2.1.10) yield in this special case

$$e_j = [e_j^{(1)} + se_j^{(2)}] - [e_J^{(1)} + se_J^{(2)}]\frac{Y_j^{(2)}}{Y_J^{(2)}}, \qquad 0 \le j \le J.$$

However, the initial condition imposed on $Y_j^{(2)}$ is now, from Equations (2.1.8) with $a_1 = 0$, just $Y_0^{(2)} = 0$. Thus we have $e_0 = e_J = 0$ and our approximate solution must be very accurate near both endpoints. Of course if the ratio $Y_j^{(2)}/Y_J^{(2)}$ becomes large in the interior, $a < x_j < b$, this accuracy may be greatly diminished. In particular, if the fixed number $y^{(2)}(b)$ is small, then $Y_J^{(2)} = y^{(2)}(b) + \mathcal{O}(h^p)$, which depends upon h, may be very close to zero for some particular mesh spacing.

Returning to the general problem, we see that a similar loss in accuracy or error magnification may occur whenever

$$Y_j^{(2)}/(b_0 Y_J^{(2)} + b_1 V_J^{(2)})$$

becomes very large. Of course since this ratio involves computed quantities, a practical assessment of this difficulty is possible.

Another very important source of difficulty can be loss of significant digits in forming the expressions in Equations (2.1.9a). Thus if $Y_j^{(1)}$ and $s_h Y_j^{(2)}$ are nearly equal and of opposite sign for some range of j values, we get cancellation of the leading digits. Unfortunately this situation can easily arise. For instance suppose that the solution of the initial-value problem (2.1.3a) grows in magnitude as $x \to b$ and that the boundary condition at $x = b$ has $b_1 = 0$ (that is, $y(b) = \beta$ is specified). Then if $|\beta| \ll |b_0 Y_J^{(1)}|$ we have, from Equation (2.1.9b), approximately

$$s_h \doteq -(Y_J^{(1)}/Y_J^{(2)}),$$

and so Equation (2.1.9a) becomes

$$Y_j \doteq Y_j^{(1)} - \left(\frac{Y_J^{(1)}}{Y_J^{(2)}}\right) Y_j^{(2)}.$$

Clearly the cancellation problem occurs here for x_j near b. Note that the solution $y^{(1)}(x)$ need not grow very fast, and in fact for $\beta = 0$, corresponding to the commonly-occurring boundary condition $y(b) = 0$, the difficulty is always potentially present. If either of the solutions, $y^{(1)}(x)$ or $y^{(2)}(x)$, changes rapidly near $x = b$, then these errors will not propagate very far from the endpoint.

The loss of significant digits can also occur with $b_1 \neq 0$ but is always easily observed. However, when employing an automatic digital computer *one must remember to test for this effect* as there may be no sign of it in a list of the leading digits of $\{Y_j\}$. If the loss of significance cannot be easily overcome by carrying additional figures, that is, by using double-precision arithmetic, then the parallel-shooting techniques of Section 2.4 can be employed.

There is no difficulty in extending the results of this section to the case in which $a_0 = 0$ but $a_1 \neq 0$ in the boundary conditions (2.1.1b). We pose this as Problem 2.1.1.

The initial-value method can easily be applied to problems which are much more general than problem (2.1.1). To illustrate such applications we shall consider a two-point boundary-value problem for a system of n coupled linear second-order differential equations. As we shall see, such problems can be solved in terms of the solution of $n + 1$ initial-value problems for the system. More general linear (and nonlinear) systems are treated in Section 2.3.

We write the system as

$$L\mathbf{y} \equiv -\mathbf{y}'' + P(x)\mathbf{y}' + Q(x)\mathbf{y} = \mathbf{r}(x), \qquad a \le x \le b, \qquad (2.1.11a)$$

where $\mathbf{y}(x)$ and $\mathbf{r}(x)$ are n-vectors and $P(x)$ and $Q(x)$ are nth-order matrices whose elements are continuous functions of x. The two-point boundary conditions are

$$\begin{aligned} A_0\mathbf{y}(a) - A_1\mathbf{y}'(a) &= \boldsymbol{\alpha}, \qquad \det A_0 \neq 0, \\ B_0\mathbf{y}(b) + B_1\mathbf{y}'(b) &= \boldsymbol{\beta}. \end{aligned} \qquad (2.1.11b)$$

Here A_0, A_1, B_0, and B_1 are nth-order constant matrices, $\boldsymbol{\alpha}$ and $\boldsymbol{\beta}$ are constant n-vectors and again we assume that the homogeneous problem, that is, $\boldsymbol{\alpha} \equiv \boldsymbol{\beta} \equiv \mathbf{r}(x) \equiv 0$ in (2.1.11), has only the trivial solution, $\mathbf{y}(x) \equiv 0$. An alternative theorem holds for such systems and so we know that a unique solution exists.

We now introduce $n + 1$ vector-valued functions $\mathbf{y}^{(\nu)}(x)$, $\nu = 0, 1, \ldots, n$, as the solutions of the corresponding initial-value problems

$$L\mathbf{y}^{(0)} = \mathbf{r}(x), \qquad A_0\mathbf{y}^{(0)}(a) - A_1\mathbf{y}^{(0)\prime}(a) = \boldsymbol{\alpha}, \qquad \mathbf{y}^{(0)\prime}(a) = \mathbf{0}; \tag{2.1.12a}$$

$$L\mathbf{y}^{(\nu)} = \mathbf{0}, \qquad A_0\mathbf{y}^{(\nu)}(a) - A_1\mathbf{y}^{(\nu)\prime}(a) = \mathbf{0}, \qquad \mathbf{y}^{(\nu)\prime}(a) = \mathbf{I}^{(\nu)}, \qquad \nu = 1, 2, \ldots, n. \tag{2.1.12b$^\nu$}$$

The vector $\mathbf{I}^{(\nu)}$ is the νth-unit vector, that is, the νth column of the nth-order identity matrix I. By Theorem 1.1.1 it follows that each initial-value problem in (2.1.12) has a unique solution on $[a, b]$. The solutions of problems (2.1.12b$^\nu$) form an nth-order matrix which we denote by

$$Y(x) \equiv (\mathbf{y}^{(1)}(x), \mathbf{y}^{(2)}(x), \ldots, \mathbf{y}^{(n)}(x)).$$

Then with an arbitrary n-vector, $\mathbf{s} = (s_1, s_2, \ldots, s_n)^T$, define the vector

$$\begin{aligned} \mathbf{y}(x; \mathbf{s}) &\equiv \mathbf{y}^{(0)}(x) + Y(x)\mathbf{s} \\ &\equiv \mathbf{y}^{(0)}(x) + \sum_{\nu=1}^{n} s_\nu \mathbf{y}^{(\nu)}(x). \end{aligned} \tag{2.1.13}$$

Now if $\mathbf{s}$ is a root of

$$\boldsymbol{\phi}(\mathbf{s}) \equiv B_0\mathbf{y}(b; \mathbf{s}) + B_1\mathbf{y}'(b; \mathbf{s}) - \boldsymbol{\beta} = 0, \tag{2.1.14}$$

then the function $\mathbf{y}(x) = \mathbf{y}(x; \mathbf{s})$ is the solution of the boundary-value problem (2.1.11).

Proceeding formally, with Equation (2.1.13) in Equation (2.1.14), we see that $\boldsymbol{\phi}(\mathbf{s})$ is linear in $\mathbf{s}$ and the system (2.1.14) has the unique solution

$$\mathbf{s} = [B_0 Y(b) + B_1 Y'(b)]^{-1}(\boldsymbol{\beta} - [B_0\mathbf{y}^{(0)}(b) + B_1\mathbf{y}^{(0)\prime}(b)]), \tag{2.1.15}$$

provided the indicated inverse exists. However, if the matrix

$$[B_0 Y(b) + B_1 Y'(b)] \tag{2.1.16}$$

is singular, then its columns must be linearly dependent. That is, since the columns are $B_0\mathbf{y}^{(\nu)}(b) + B_1\mathbf{y}^{(\nu)\prime}(b)$, $\nu = 1, 2, \ldots, n$, there must exist scalars a_ν, $\nu = 1, 2, \ldots, n$, not all vanishing, such that

$$\sum_{\nu=1}^{n} a_\nu[B_0\mathbf{y}^{(\nu)}(b) + B_1\mathbf{y}^{(\nu)\prime}(b)] = 0.$$

But then the function

$$\mathbf{z}(x) \equiv \sum_{\nu=1}^{n} a_\nu \mathbf{y}^{(\nu)}(x)$$

satisfies $B_0\mathbf{z}(b) + B_1\mathbf{z}'(b) = 0$ and so is a solution of the homogeneous boundary-value problem corresponding to (2.1.11). It has been assumed that this problem has only the trivial solution, $\mathbf{z}(x) \equiv 0$, and since

$$\mathbf{z}'(a) = \sum_{\nu=1}^{n} a_\nu \mathbf{I}^{(\nu)}$$

it follows that $a_\nu = 0$, $\nu = 1, 2, \ldots, n$. From this contradiction we conclude that Equation (2.1.14) indeed has the unique solution shown in Equation (2.1.15).

Numerical solutions of (2.1.11) are obtained by solving each of the $n + 1$ initial-value problems in (2.1.12) and forming, at the net points, the quantities corresponding to (2.1.13) and (2.1.15). The linear system (2.1.14) which must be solved is usually of relatively small order and so is not, in principle, a difficult numerical problem.

To estimate the accuracy of the initial-value method applied to the problem (2.1.11) we proceed as in the previous case (where $n = 1$). Thus we need not write the obvious first-order systems which are solved in place of (2.1.12). Rather we denote by $\mathbf{Y}_j$, $\mathbf{Y}_j^{(\nu)}$, $\mathbf{V}_j^{(\nu)}$, and $\mathbf{s}^{(h)}$ the numerical approximations to $\mathbf{y}(x_j)$, $\mathbf{y}^{(\nu)}(x_j)$, $\mathbf{y}^{(\nu)\prime}(x_j)$, and $\mathbf{s}$, respectively. If a stable method with order of accuracy at least p is employed to solve the initial value problems then we may assume that

$$|\mathbf{e}_j^{(\nu)}| \equiv |\mathbf{Y}_j^{(\nu)} - \mathbf{y}^{(\nu)}(x_j)| = \mathcal{O}(h^p), \qquad |\boldsymbol{\epsilon}_j^{(\nu)}| \equiv |\mathbf{V}_j^{(\nu)} - \mathbf{y}^{(\nu)\prime}(x_j)| = \mathcal{O}(h^p),$$
$$\nu = 0, 1, \ldots, n; \qquad 0 \le j \le J.$$

Now subtracting Equation (2.1.13) with $x = x_j$ from its numerical counterpart we get

$$\mathbf{e}_j \equiv \mathbf{Y}_j - \mathbf{y}(x_j) = \mathbf{e}_j^{(0)} + \sum_{\nu=1}^{n} s_\nu \mathbf{e}_j^{(\nu)} + \sum_{\nu=1}^{n} (s_\nu^{(h)} - s_\nu)\mathbf{Y}_j^{(\nu)}, \qquad 0 \le j \le J. \tag{2.1.17}$$

Similarly from Equation (2.1.14) and its numerical equivalent we find that

$$H(\mathbf{s}^{(h)} - \mathbf{s}) = -(B_0\mathbf{e}_J^{(0)} + B_1\boldsymbol{\epsilon}_J^{(0)}) + [B_0E + B_1\varepsilon]\mathbf{s}, \tag{2.1.18a}$$

where the matrix H is

$$H \equiv [B_0 Y(b) + B_1 Y'(b)] + [B_0E + B_1\varepsilon]. \tag{2.1.18b}$$

Here E is the matrix with columns $\mathbf{e}_J^{(\nu)}$, $\nu = 1, 2, \ldots, n$, and ε has the columns $\boldsymbol{\epsilon}_J^{(\nu)}$, $\nu = 1, 2, \ldots, n$. Thus for sufficiently small h the matrix H is nonsingular, since the matrix (2.1.16) is nonsingular (see Problem 2.1.3), and we conclude that

$$|\mathbf{Y}_j - \mathbf{y}(x_j)| = \mathcal{O}(h^p), \qquad 0 \le j \le J,$$

just as in the case with $n = 1$. A similar estimate applies for the first derivatives.

Combining these observations we may easily construct a proof of the following.

THEOREM 2.1.2. *Let the boundary-value problem* (2.1.11) *have a unique solution,* $\mathbf{y}(x)$. *Let the numerical solutions of problems* (2.1.12a, b$^\nu$), *using a stable method with order of accuracy p on the net* (2.1.6), *be denoted by* $\{\mathbf{Y}_j^{(\nu)}\}$, $\{\mathbf{V}_j^{(\nu)}\}$, $\nu = 0, 1, \ldots, n$. *Define the net functions*

$$\mathbf{Y}_j(\mathbf{s}) = \mathbf{Y}_j^{(0)} + \sum_{\nu=1}^{n} s_\nu \mathbf{Y}_j^{(\nu)}, \qquad \mathbf{V}_j(\mathbf{s}) = \mathbf{V}_j^0 + \sum_{\nu=1}^{n} s_\nu \mathbf{V}_j^{(\nu)}, \qquad 0 \le j \le J.$$

For sufficiently small h a unique vector $\mathbf{s}^{(h)} \equiv (s_1^{(h)}, s_2^{(h)}, \ldots, s_n^{(h)})$ *is defined as the solution of the linear system*

$$\boldsymbol{\phi}(\mathbf{s}) \equiv B_0\mathbf{Y}_J(\mathbf{s}) + B_1\mathbf{V}_J(\mathbf{s}) - \boldsymbol{\beta} = 0. \tag{2.1.19}$$

With this root the net functions $\{\mathbf{Y}_j(\mathbf{s}^{(h)})\}$ *and* $\{\mathbf{V}_j(\mathbf{s}^{(h)})\}$ *satisfy*

$$|\mathbf{y}(x_j) - \mathbf{Y}_j(\mathbf{s}^{(h)})| = \mathcal{O}(h^p), \qquad |\mathbf{y}'(s_j) - \mathbf{V}_j(s^{(h)})| = \mathcal{O}(h^p), \qquad 0 \le j \le J,$$

provided that the initial-value problems (2.1.12) *have sufficiently smooth solutions.* ▌

Although this theorem is similar to Theorem 2.1.1 and the error estimates, valid as $h \to 0$, are gratifying, it should not be inferred that the practical difficulties are as simple as they were in the case $n = 1$. Combining Equations (2.1.17) and (2.1.18), we see that if H is ill-conditioned for finite values of h we may lose accuracy. The cancellation difficulty in forming (2.1.13) is now even more important. It can occur, for instance, if any fixed corresponding components of $s_{\nu_1}\mathbf{y}^{(\nu_1)}(x)$ and $s_{\nu_2}\mathbf{y}^{(\nu_2)}(x)$ for fixed $\nu_1 \ne \nu_2$ are nearly equal, of opposite sign, and in magnitude larger than the other corresponding components for $\nu \ne \nu_1, \nu_2$. But again some of the sources of error may be eliminated by using the parallel-shooting methods of Section 2.4.

Problems

2.1.1 Replace the condition $\det A_0 \ne 0$ in Equations (2.1.11b) by the condition rank $[A_0, A_1] = n$ and show how to solve the resulting boundary-value problem (2.1.11).

2.1.2 Prove the Alternative Theorem for the linear second-order system (2.1.11). [HINT: Use the reduction to an initial-value problem for one system which has a unique solution.]

2.1.3* Show that $A + B$ is nonsingular if A is nonsingular and the elements of B are sufficiently small [see Isaacson and Keller (1966), Theorem 1.5, Chapter 1].

2.1.4 Use the above result to show that Equation (2.1.19) has a unique solution $\mathbf{s}^{(h)}$ for sufficiently small h. [HINT: Relate the coefficient matrix to H in Equation (2.1.18b).]

2.1.5 Use the initial-value method to solve the boundary-value problem

$$y'' = -100y, \qquad y(0) = 1, \qquad y(2\pi + \varepsilon) = 1.$$

If the linearly-independent solutions

$$y^{(1)}(x) = \cos 10x, \qquad y^{(2)}(x) \equiv 10^{-1} \sin 10x$$

are employed, show that, for small $|\varepsilon|$,

$$s = 5\varepsilon + \mathcal{O}(\varepsilon^3).$$

Discuss the possible computational difficulties caused by small $|\varepsilon|$.

2.1.6 Use the initial-value method to solve the boundary-value problem:

$$y'' = 100y, \qquad y(0) = 1, \qquad y(3) = \varepsilon + \cosh 30.$$

If the linearly-independent solutions

$$y^{(1)}(x) = \cosh 10x, \qquad y^{(2)}(x) = 10^{-1} \sinh 10x$$

are employed, show that

$$s = 10\varepsilon/\sinh 30.$$

Explain the difficulty in carrying out the numerical solution for small $|\varepsilon|$.

2.2 Nonlinear Second-Order Equations

We now consider second-order equations, which may be nonlinear, of the form

$$y'' = f(x, y, y'), \qquad a \le x \le b, \tag{2.2.1a}$$

subject to the general two-point boundary conditions

$$\begin{aligned} a_0 y(a) - a_1 y'(a) &= \alpha, \qquad a_i \ge 0, \\ b_0 y(b) + b_1 y'(b) &= \beta, \qquad b_i \ge 0; \qquad a_0 + b_0 > 0. \end{aligned} \tag{2.2.1b}$$

The function $f(x, y, z)$ in Equation (2.2.1a) will be assumed to satisfy the hypothesis of Theorem 1.2.2. This assures us that the boundary-value problem (2.2.1) has a solution which is unique.

The related initial-value problem that we consider is

$$u'' = f(x, u, u'), \qquad a \le x \le b, \tag{2.2.2a}$$

$$u(a) = a_1 s - c_1\alpha, \qquad u'(a) = a_0 s - c_0\alpha, \tag{2.2.2b}$$

where c_0 and c_1 are any constants such that

$$a_1 c_0 - a_0 c_1 = 1. \tag{2.2.2c}$$

The solution of this problem, which we denote by $u = u(x; s)$, will be a solution of the boundary-value problem (2.2.1) if and only if s is a root of

$$\phi(s) \equiv b_0 u(b; s) + b_1 u'(b; s) - \beta = 0. \tag{2.2.3}$$

Under the previously-stated conditions on f, it is shown in the proof of Theorem 1.2.2 that *the function $\phi(s)$ has a positive derivative which is bounded away from zero for all s*, and so Equation (2.2.3) has a unique root for any value of β.

The initial-value or shooting methods we now study essentially consist of iterative schemes for approximating the root of Equation (2.2.3). In particular we shall employ schemes of the type indicated in Section 1.4. These all require evaluations of the function $\phi(s)$ for a sequence of values of s. This is done by means of numerical solutions of a sequence of initial-value problems of the form (2.2.2). [Of course if $f(x, y, z)$ is linear in y and z, then $\phi(s)$ is linear in s and the root of $\phi(s) = 0$ could be determined by only two evaluations of $\phi(s)$. This is what was done in Section 2.1. However, a linear equation can also be solved by iteration and so the present procedures also apply to linear boundary-value problems.] Obviously we should try to employ very rapidly converging iteration schemes as this would reduce the number of initial-value problems that must be solved. We shall discuss the use of several different schemes, one of which is Newton's method and is particularly well-suited for the current class of problems. First we must consider some effects of using numerical approximations to $u(b; s)$ and $u'(b; s)$ in attempting to evaluate $\phi(s)$.

To solve the initial-value problem (2.2.2) numerically, we write it as a first-order system, with $v \equiv u'$;

$$\begin{aligned} u' &= v, & u(a) &= a_1 s - c_1\alpha; \\ v' &= f(x, u, v), & v(a) &= a_0 s - c_0\alpha. \end{aligned} \tag{2.2.4}$$

If we write the solution of this system as

$$u = u(x; s), \qquad v = v(x; s),$$

the function $\phi(s)$ as defined in (2.2.3) becomes

$$\phi(s) \equiv b_0 u(b; s) + b_1 v(b; s) - \beta. \tag{2.2.5}$$

Again, any of the methods in Section 1.3 can be used to solve the initial-value problem (2.2.4) numerically. We use the net in (2.1.6) and denote the numerical solution, for each value of s, by

$$U_j(s), \qquad V_j(s), \qquad 0 \le j \le J. \tag{2.2.6a}$$

If the numerical method has order of accuracy p and is stable, and if the solutions of (2.2.4) are sufficiently smooth, then

$$|U_j(s) - u(x_j; s)| = \mathcal{O}(h^p), \qquad |V_j(s) - v(x_j; s)| = \mathcal{O}(h^p), \qquad 0 \le j \le J. \tag{2.2.6b}$$

In terms of the numerical solution we define the function

$$\Phi(s) \equiv b_0 U_J(s) + b_1 V_J(s) - \beta, \tag{2.2.7}$$

and from Equations (2.2.6) and (2.2.5) it follows that

$$|\Phi(s) - \phi(s)| = \mathcal{O}(h^p). \tag{2.2.8}$$

Thus in general we cannot hope to evaluate $\phi(s)$ exactly but rather we get an $\mathcal{O}(h^p)$ approximation to it.

It is clear, from the above error bounds, that the solution of the boundary-value problem (2.2.1) can be approximated to an accuracy at least $\mathcal{O}(h^p)$ if the root $s = s^*$ of Equation (2.2.3) is known. However we can *at best* hope to approximate s^* to an accuracy of $\mathcal{O}(h^p)$, by virtue of the estimate (2.2.8). But then the solution of the boundary-value problem will also be approximated to this same accuracy by means of the following.

LEMMA 2.2.1. *Let $u(x; s)$ and $v(x; s)$ denote the solution of the initial-value problem* (2.2.4) *with* $\mathbf{f} \equiv (v, f(x, u, v))^T$ *satisfying the hypothesis of Theorem* 1.1.1. *Let* $\{U_j(s)\}$, $\{V_j(s)\}$ *denote the numerical solution of* (2.2.4) *on the net in* (2.1.6) *using a stable, p*th*-order-accurate scheme. Then if the solutions of* (2.2.4) *are sufficiently smooth, we have for any values s and t*

$$\left.\begin{aligned} |U_j(t) - u(x_j; s)| &\le \mathcal{O}(h^p) + \mathcal{O}(|s - t|), \\ |V_j(t) - v(x_j; s)| &\le \mathcal{O}(h^p) + \mathcal{O}(|s - t|), \end{aligned}\right\} \quad 0 \le j \le J. \tag{2.2.9}$$

Proof. From the hypothesis it follows by Theorem 1.1.1 that u and v are Lipschitz-continuous in s. But then

$$\begin{aligned} |U_j(t) - u(x_j; s)| &\le |U_j(t) - u(x_j; t)| + |u(x_j; t) - u(x_j; s)| \\ &\le \mathcal{O}(h^p) + K'|s - t|, \end{aligned}$$

where we have used the estimate (2.2.6b), which is valid, and K' is the Lipschitz constant for u. A similar result holds for v. ▮

If in (2.2.9) we set $s = s^*$ and $|s^* - t| = \mathcal{O}(h^p)$ we see that both the solution of the boundary-value problem (2.2.1) and its first derivative can be computed to within $\mathcal{O}(h^p)$ if s^* can be approximated to this accuracy.

To approximate the root of Equation (2.2.3) we write the equation in the form

$$s = g(s) \equiv s - m\phi(s), \qquad m \neq 0, \tag{2.2.10a}$$

and consider the functional iteration scheme: s_0 = arbitrary,

$$s_{\nu+1} = g(s_\nu), \qquad \nu = 0, 1, 2, \ldots. \tag{2.2.10b}$$

The fact that the sequence $\{s_\nu\}$ converges to the root under quite general circumstances is the content of the following.

THEOREM 2.2.1. *Let $f(x, y, y')$ in Equation (2.2.1a) satisfy the hypothesis of Theorem 1.2.2, and in addition for some positive constant N,*

$$\frac{\partial f}{\partial y} \leq N.$$

Then the function $\phi(s)$ in Equation (2.2.3) has a continuous derivative, $\dot{\phi}(s)$, which satisfies

$$0 < \gamma \leq \dot{\phi}(s) \leq \Gamma, \tag{2.2.11a}$$

where

$$\gamma \equiv b_0\left[a_1 + a_0\left(\frac{1 - e^{-2\mu L}}{2\mu}\right)\right] + b_1 a_0 e^{-2\mu L};$$

$$\Gamma \equiv \frac{e^{\mu L}}{2\sigma}\{[(\sigma - \mu)a_1 + a_0][(\sigma + \mu)b_1 + b_0]e^{\sigma L} - [(\sigma + \mu)a_1 - a_0][(\sigma - \mu)b_1 - b_0]e^{-\sigma L}\}; \tag{2.2.11b}$$

$$\mu \equiv \frac{M}{2}, \qquad \sigma \equiv \sqrt{\mu^2 + N}, \qquad L \equiv b - a.$$

For any s_0 and fixed m in

$$0 < m < 2/\Gamma \tag{2.2.11c}$$

the iterates in Equations (2.2.10b) converge to the root s^* of Equation (2.2.3). In particular, with the choice $m = 2/(\Gamma + \gamma)$, the iterates satisfy

$$|s_\nu - s^*| \leq \left[\frac{1 - (\gamma/\Gamma)}{1 + (\gamma/\Gamma)}\right]^\nu \frac{|\phi(s_0)|}{\gamma}; \qquad \nu = 1, 2, \ldots. \tag{2.2.12}$$

Proof. We first note that Theorem 1.1.2 is applicable to the system (2.2.4). Thus the functions $\xi(x) \equiv \partial u(x; s)/\partial s$ and $\xi'(x) \equiv \partial u'(x; s)/\partial s$, where $u(x; s)$ is the solution of (2.2.2), are continuous functions of s for all $x \in [a, b]$. But then

$$\dot{\phi}(s) = b_0\xi(b) + b_1\xi'(b)$$

is also continuous in s.

As in the proof of Theorem 1.2.2 we have the variational problem for $\xi(x)$,

$$\xi'' = p(x)\xi' + q(x)\xi, \qquad \xi(a) = a_1, \qquad \xi'(a) = a_0, \qquad a \leq x \leq b,$$

where now

$$|p(x)| \leq M, \qquad 0 < q(x) \leq N.$$

It has already been shown [see the inequalities (1.2.5)] that since $a_0 \geq 0$ and $a_1 \geq 0$,

$$\xi'(x) > a_0 e^{-2\mu(x-a)} > 0, \qquad \xi(x) > a_1 + a_0\left(\frac{1 - e^{-2\mu(x-a)}}{2\mu}\right) > 0,$$

and so clearly, with the notation in (2.2.11), $\dot{\phi}(s) \geq \gamma$. To obtain an upper bound we observe that now

$$\xi'' \leq M\xi' + N\xi.$$

It easily follows (see Problem 2.2.1) that $\xi(x) \leq \eta(x)$ and $\xi'(x) \leq \eta'(x)$, where $\eta(x)$ is the solution of

$$\eta'' = M\eta' + N\eta, \qquad \eta(a) = \xi(a), \qquad \eta'(a) = \xi'(a).$$

But this linear problem with constant coefficients has the solution

$$\eta(x) = \frac{e^{\mu(x-a)}}{2\sigma}\{[(\sigma - \mu)a_1 + a_0]e^{\sigma(x-a)} + [(\sigma + \mu)a_1 - a_0]e^{-\sigma(x-a)}\},$$

and so $\dot{\phi}(s) \leq \Gamma$ follows, establishing (2.2.11a).

We now use the mean-value theorem to deduce that

$$g(s) - g(t) = (s - t) - m[\phi(s) - \phi(t)] = [1 - m\dot{\phi}(t + \theta(s - t))](s - t),$$
$$0 < \theta < 1.$$

Thus $g(s)$ in Equations (2.2.10) satisfies a Lipschitz condition with constant λ given by

$$\lambda = \max(|1 - m\gamma|, |1 - m\Gamma|).$$

For any m in $0 < m < 2/\Gamma$, it follows that $\lambda < 1$, and thus Theorem 1.4.1 is applicable (with $\rho = \infty$) to the scheme (2.2.10b). With $m = 2/(\Gamma + \gamma)$ we obtain

$$\lambda = \frac{\Gamma - \gamma}{\Gamma + \gamma},$$

and part (d) of Theorem 1.4.1 yields (2.2.12). ∎

One of the nice features of this result is that it determines a specific value for m, in terms of known data and the two bounds M and N, which insures convergence. However, if we consider for the moment the case with $a_1 = b_1 = 0$, $a_0 = b_0 = 1$ (that is, function specified at both endpoints), then Equation (2.2.11b) yields

$$\frac{\gamma}{\Gamma} = \left(1 + 4\frac{N}{M^2}\right)^{1/2} e^{-M(b-a)} \frac{\sinh\,[M(b - a)/2]}{\sinh\,[(1 + (4N/M^2))^{1/2}M(b - a)/2]}.$$

From this we see that the convergence factor, $(\Gamma - \gamma)/(\Gamma + \gamma)$, in (2.2.12) may be very close to unity if the interval $(b - a)$ is "long" or the constants M or N are "large." In such cases many iterations may be required to obtain an accurate approximation to the root. However, we shall consider more rapidly convergent schemes shortly.

Let us now consider the effects on the above procedure which result from the fact that $\phi(s)$ cannot be evaluated exactly. That is, we employ the iteration scheme in Equations (2.2.10), using the numerical solutions (2.2.6) of (2.2.2) to evaluate the function $\phi(s)$. Recalling Equation (2.2.7), the actual iteration scheme is then: s_0 = arbitrary,

$$s_{\nu+1} = s_\nu - m\Phi(s_\nu), \qquad \nu = 0, 1, 2, \ldots. \tag{2.2.13}$$

The basic result can be stated as follows.

THEOREM 2.2.2. *Let $f(x, y, y')$ in Equation* (2.2.1*a*) *satisfy the hypothesis of Theorem* 2.2.1. *Let numerical solutions* (2.2.6) *of the initial-value problem* (2.2.4) *be computed as in Lemma* 2.2.1. *Let the sequence* $\{s_\nu\}$ *be defined by Equations* (2.2.13), *with* $\Phi(s)$ *as defined in Equation* (2.2.7). *Then for any m in* $0 < m < 2/\Gamma$ *we have, with* $y(x)$, *the solution of the boundary-value problem* (2.2.1), *for some positive* $\lambda = \lambda(m) < 1$,

$$|U_j(s_\nu) - y(x_j)| \le \mathcal{O}(h^p) + \mathcal{O}(\lambda^\nu), \qquad |V_j(s_\nu) - y'(x_j)| \le \mathcal{O}(h^p) + \mathcal{O}(\lambda^\nu),$$
$$0 \le j \le J, \qquad \nu = 0, 1, \ldots.$$

Proof. We may apply Lemma 2.2.1 with $t = s_\nu$ and $s = s^*$, the root of Equation (2.2.3), in which case $u(x; s^*) = y(x)$. The proof is then reduced to estimating $|s^* - s_\nu|$. But from the estimate (2.2.8), which is valid by hypothesis, we can write Equations (2.2.13) as

$$s_{\nu+1} = s_\nu - m\phi(s_\nu) + \mathcal{O}(h^p), \qquad \nu = 0, 1, \ldots.$$

It has been shown in Theorem 2.2.1 that $g(s) \equiv s - m\phi(s)$ is Lipschitz-continuous with constant $\lambda < 1$ for m in $0 < m \le 2/\Gamma$. Now Theorem 1.4.2 can be applied, with $\delta \equiv \mathcal{O}(h^p)$, to deduce that

$$|s^* - s_\nu| \le \lambda^\nu|s^* - s_0| + \mathcal{O}(h^p). \quad \blacksquare$$

It should be noted that the convergence factor λ, in the above theorem, is independent of the net spacing h. Thus for each h one can find an integer $\nu = \nu(h, p, \lambda)$ such that

$$\lambda^\nu \leq h^p,$$

and so, in principle, the total error can be made $\mathcal{O}(h^p)$.

The virtue of rapidly converging iteration schemes for solving Equation (2.2.3) is quite clear and we recall, from Theorem 1.4.3, that Newton's method may have arbitrarily small convergence factors, λ. Thus we indicate the proper way in which to employ this method in the present problem. The iterations $\{s_\nu\}$ are now defined by: s_0 = arbitrary,

$$s_{\nu+1} = s_\nu - \frac{\phi(s_\nu)}{\dot{\phi}(s_\nu)}, \qquad \nu = 0, 1, \ldots. \tag{2.2.14}$$

We have previously indicated how $\phi(s)$ in Equation (2.2.5) is to be evaluated, or rather approximated, by $\Phi(s)$, in terms of a numerical solution of the initial-value problem (2.2.4). We shall evaluate $\dot{\phi}(s)$ by means of a numerical solution of the variational problem corresponding to (2.2.4). This variational problem, or one equivalent to it, has already been employed in Theorem 2.2.1 to study the properties of $\dot{\phi}(s)$. If we introduce

$$\xi(x) \equiv \frac{\partial u(x; s)}{\partial s}, \qquad \eta(x) \equiv \frac{\partial v(x; s)}{\partial s},$$

where u and $v = u'$ are the solution of (2.2.4), then formal differentiation in this initial-value problem yields

$$\begin{aligned} \xi' &= \eta, & \xi(a) &= a_1; \\ \eta' &= p(x; s)\eta + q(x; s)\xi, & \eta(a) &= a_0, \end{aligned} \tag{2.2.15a}$$

where

$$p(x; s) \equiv \frac{\partial f(x, u(x; s), v(x; s))}{\partial v}, \qquad q(x; s) \equiv \frac{\partial f(x, u(x; s), v(x; s))}{\partial u}. \tag{2.2.15b}$$

Thus we have, from Equation (2.2.5),

$$\dot{\phi}(s) \equiv b_0\xi(b; s) + b_1\eta(b; s). \tag{2.2.16}$$

The numerical solution of the initial-value problem (2.2.15) should be evaluated on the net (2.1.6), along with the numerical solution of the initial-value problem (2.2.4) and by the same method. Then in many cases the coefficients $p(x)$ and $q(x)$ can be evaluated with little additional computation over that required to evaluate $f(x, u, v)$. Thus one iteration in Newton's method (2.2.14) requires the solution of two initial-value problems, but the amount of computation may frequently be less than that required for two iterations in the scheme (2.2.10).

Finally, we point out the obvious but frequently forgotten fact that "shooting" can be applied in either direction. This applies to linear problems as well. If the solutions of the initial-value problem grow from $x = a$ to $x = b$, then it is likely that the shooting method will be most effective in reverse; that is, using $x = b$ as the initial point. We call this procedure *reverse shooting*.

Problems

2.2.1 Let $\xi(x)$ and $\eta(x)$ satisfy on $[a, b]$:

$$\xi'' \leq M\xi' + N\xi,$$
$$\eta'' = M\eta' + N\eta; \quad \xi(a) = \eta(a) \geq 0, \qquad \xi'(a) = \eta'(a) \geq 0.$$

If the constants M and N satisfy $M \geq 0$ and $N \geq 0$, show that $\xi(x) \leq \eta(x)$ and $\xi'(x) \leq \eta'(x)$ on $[a, b]$. [HINT: "Solve" the differential inequality satisfied by $\zeta(x) \equiv \eta(x) - \xi(x)$. First derive the result that $[\zeta' + B\zeta] \geq 0$, where $B = (-M \pm \sqrt{M^2 + 4N})/2$. Then it follows that $\zeta(x) \geq 0$ on $[a, b]$ and so $[\zeta'' + M\zeta'] \geq 0$ which implies $\zeta'(x) \geq 0$.]

2.2.2* If $f(x, u, v)$ is as in Theorem 2.2.1 and in addition $\partial f(x, u, v)/\partial u$ and $\partial f(x, u, v)/\partial v$ are Lipschitz-continuous functions of u and v, show that $\dot{\phi}(s)$ as defined in Equation (2.2.16) is a Lipschitz-continuous function of s.

2.2.3 Use Theorem 2.2.1, Problem 2.2.2, and Problem 1.4.7 to show that Newton's method in (2.2.14) converges if $|b - a|$ is sufficiently small and some later iterate in (2.2.10) is used as initial guess.

2.2.4 Show that the choice $m = 2/(\Gamma + \gamma)$ in Theorem 2.2.1 is in general the best possible in light of the inequalities (2.2.11a).

2.2.5 Compute the solution of

$$y'' = \tanh y + \cos y'; \qquad y(0) - y'(0) = 1, \qquad y(1) + y'(1) = 0,$$

by shooting. Try using several different uniformly-spaced nets (that is, different values of h), contracting maps, and Newton's method.

2.2.6 Verify the applicability of the theoretical results to the problem formulated in Problem 2.2.5.

2.3 *Linear and Nonlinear Systems*

The application of initial-value techniques to rather general nonlinear boundary-value problems is immediate. First of all, the particular equation or set of equations can be replaced by an equivalent system of say n first-order equations of the form

$$\mathbf{y}' = \mathbf{f}(x, \mathbf{y}), \qquad a < x < b. \tag{2.3.1a}$$

The components, $y_j(x)$, of the n-vector $\mathbf{y}$ are either the original dependent variables or some derivatives of them or linear combinations of these functions. Assuming the original two-point boundary conditions to be linear, we can now write them in the form

$$A\mathbf{y}(a) + B\mathbf{y}(b) = \boldsymbol{\alpha}. \tag{2.3.1b}$$

Here A and B are constant square matrices of order n and $\boldsymbol{\alpha}$ is a specified n-vector. In the indicated reduction to (2.3.1) we have assumed that the highest-order derivatives of each dependent variable which occurred in the original system of differential equations did not occur in the boundary conditions. The n boundary conditions in Equation (2.3.1b) will be independent if the matrix (A, B), which is $n \times 2n$, has rank n. It should be noted that periodic boundary conditions are included as a special case in (2.3.1b).

For purposes of our general discussion we associate with the boundary-value problem (2.3.1) the initial-value problem

$$\text{(a)}\quad \mathbf{u}' = \mathbf{f}(x, \mathbf{u}), \qquad \text{(b)}\quad \mathbf{u}(a) = \mathbf{s}. \tag{2.3.2}$$

Under appropriate smoothness conditions on $\mathbf{f}(x, \mathbf{u})$ we are assured, as in Theorems 1.1.1 and 1.1.2, that a unique solution of (2.3.2) exists on $a \leq x \leq b$, say

$$\mathbf{u} = \mathbf{u}(\mathbf{s}; x),$$

and it depends Lipschitz-continuously or even differentiably on the components of $\mathbf{s}$. We now seek $\mathbf{s}$ such that $\mathbf{u}(\mathbf{s}; x)$ is a solution of the boundary-value problem (2.3.1). This occurs if and only if $\mathbf{s}$ is a root of the system of n equations

$$\boldsymbol{\phi}(\mathbf{s}) \equiv A\mathbf{s} + B\mathbf{u}(\mathbf{s}; b) - \boldsymbol{\alpha} = 0 \tag{2.3.3}$$

(see Theorem 1.2.4). In particular cases the initial values in Equation (2.3.2b) may be chosen quite conveniently, for instance so that say p of the conditions in (2.3.1b) are automatically satisfied. Then only $q = n - p$ undetermined components s_j would occur in the chosen replacement of (2.3.2b). These components must then be determined in order to satisfy the q conditions in (2.3.3) that are not identically satisfied. This procedure can be applied whenever p independent conditions in Equation (2.3.1b) involve only p components of $\mathbf{y}(a)$ [see Problem 1.2.12].

Under the hypothesis of Theorem 1.2.6, which we assume to hold, we are assured that Equation (2.3.3) has a unique root, or equivalently that the boundary-value problem (2.3.1) has a unique solution. Furthermore the root of Equation (2.3.3) is also the fixed point of a contracting map which we write as

$$\mathbf{s} = \mathbf{g}(\mathbf{s}) \equiv \mathbf{s} - Q[A\mathbf{s} + B\mathbf{u}(\mathbf{s}; b) - \boldsymbol{\alpha}]. \tag{2.3.4}$$

Here, since $(A + B)$ is assumed nonsingular, we take

$$Q \equiv (A + B)^{-1}.$$

(We recall that the most general boundary conditions (2.3.1b) do not necessarily satisfy this condition and the computations in such cases may converge with a different choice for Q.) Our formal procedure is now clear. To show that the solution of the boundary-value problem (2.3.1) can be computed to any desired accuracy, we approximate the root of Equation (2.3.4) by iterations and use a sufficiently accurate such approximation to "solve" the initial-value problem (2.3.2) numerically. Of course, in carrying out the implied iterations, a sequence of numerical solutions of (2.3.2) is required. We indicate below more details in this convergence proof which is quite similar to that in Section 2.2, and then we discuss the practical aspects of actually computing an accurate approximation. Newton's method is again advocated and we show that it is applicable in the present case.

Let one of the numerical methods of Section 1.3 be used to obtain a numerical solution of the initial-value problem (2.3.2) on the net (2.1.6). We denote the numerical solution at each net point x_j, for each value of $\mathbf{s}$, by the n-vectors

$$\mathbf{U}_j(\mathbf{s}), \qquad j = 0, 1, \ldots, J.$$

If the numerical method is stable and has order of accuracy p, and the solutions of the initial-value problem (2.3.2) are, as we assume, sufficiently smooth, then

$$|\mathbf{U}_j(\mathbf{s}) - \mathbf{u}(\mathbf{s}; x_j)| \leq \mathcal{O}(h^p), \qquad j = 0, 1, \ldots, J. \tag{2.3.5}$$

Now, just as in Lemma 2.2.1, it follows that for any two sets of initial data $\mathbf{s}$ and $\mathbf{t}$

$$|\mathbf{U}_j(\mathbf{t}) - \mathbf{u}(\mathbf{s}; x_j)| \leq \mathcal{O}(h^p) + \mathcal{O}(|\mathbf{s} - \mathbf{t}|), \qquad j = 0, 1, \ldots, J. \tag{2.3.6}$$

If $\mathbf{s}$ is a root of Equation (2.3.3) all we need do, by the estimate (2.3.6), is to find a $\mathbf{t}$ such that $|\mathbf{s} - \mathbf{t}| = \mathcal{O}(h^p)$, and then $\mathbf{U}_j(\mathbf{t})$ is accurate of order p.

From the proof of Theorem 1.2.6 we know that $\mathbf{g}(\mathbf{s})$ in (2.3.4) satisfies

$$\left\| \frac{\partial \mathbf{g}(\mathbf{s})}{\partial \mathbf{s}} \right\|_\infty \leq \lambda < 1. \tag{2.3.7}$$

Thus, as previously stated, $\mathbf{g}(\mathbf{s})$ is contracting [by Problem 1.2.11]. However, in terms of the numerical solution, we approximate $\mathbf{g}(\mathbf{s})$ by

$$\mathbf{G}(\mathbf{s}) \equiv \mathbf{s} - Q[A\mathbf{s} + B\mathbf{U}_J(\mathbf{s}) - \boldsymbol{\alpha}], \tag{2.3.8a}$$

and the iterations employed to obtain the root of Equation (2.3.4) are $\mathbf{s}^{(0)} =$ arbitrary,

$$\mathbf{s}^{(\nu+1)} = \mathbf{G}(\mathbf{s}^{(\nu)}), \qquad \nu = 0, 1, 2, \ldots. \tag{2.3.8b}$$

However, by (2.3.5), (2.3.4), and (2.3.8a), we have

$$|\mathbf{G}(\mathbf{s}) - \mathbf{g}(\mathbf{s})| = |QB(\mathbf{U}_J(s) - \mathbf{u}(\mathbf{s}; b))| \leq \mathcal{O}(h^p).$$

Thus the iterates (2.3.8b) actually satisfy

$$\mathbf{s}^{(\nu+1)} = \mathbf{g}(\mathbf{s}^\nu) + \mathcal{O}(h^p).$$

But since $\mathbf{g}(\mathbf{s})$ is contracting, it follows that Theorem 1.4.2 is applicable. Then if $\mathbf{s}^*$ is the root of Equation (2.3.4), the iterates in (2.3.8) must satisfy

$$|\mathbf{s}^* - \mathbf{s}^{(\nu)}| \leq \lambda^\nu|\mathbf{s}^* - \mathbf{s}^{(0)}| + \mathcal{O}(h^p).$$

Using this result with $\mathbf{t} = \mathbf{s}^{(\nu)}$ and $\mathbf{s} = \mathbf{s}^*$ in (2.3.6), we obtain what may be summarized as follows.

THEOREM 2.3.1. *Let* $\mathbf{f}(x, \mathbf{y})$, *A and B satisfy the hypothesis of Theorem* 1.2.6. *Further, let* $\mathbf{f}(x, \mathbf{y})$ *be so smooth that the numerical solutions* $\mathbf{U}_j(\mathbf{s})$ *of the initial-value problem* (2.3.2) *computed by a stable* pth-*order method on the net* (2.1.6) *satisfy* (2.3.5). *Then the sequence* $\mathbf{s}^{(\nu)}$ *defined in* (2.3.8) *is such that*

$$|\mathbf{U}_j(\mathbf{s}^{(\nu)}) - \mathbf{y}(x_j)| \leq \mathcal{O}(h^p) + \lambda^\nu|\mathbf{s}^* - \mathbf{s}^{(0)}|, \quad j = 0, 1, \ldots, J, \quad \nu = 0, 1, \ldots;$$

where $\mathbf{y}(x) \equiv \mathbf{u}(\mathbf{s}^*; x)$ *is the unique solution of the boundary-value problem* (2.3.1) *and* $\lambda < 1$ *satisfies the inequality* (1.2.18e). ∎

Of course the iteration scheme (2.3.8) is frequently not of practical value (when, for example, $1 - \lambda$ is too small), and we would generally prefer a higher-order method for greater efficiency. For this purpose we recall Newton's method for approximating a root of Equation (2.3.3) as follows: $\mathbf{s}^{(0)}$ = arbitrary,

$$\mathbf{s}^{(\nu+1)} = \mathbf{s}^{(\nu)} + \Delta\mathbf{s}^{(\nu)}, \qquad \nu = 0, 1, \ldots, \tag{2.3.9a}$$

where $\Delta\mathbf{s}^{(\nu)}$ is the solution of the nth-order linear algebraic system

$$\frac{\partial\boldsymbol{\phi}(\mathbf{s}^{(\nu)})}{\partial\mathbf{s}}\Delta\mathbf{s}^{(\nu)} = -\boldsymbol{\phi}(\mathbf{s}^{(\nu)}), \qquad \nu = 0, 1, \ldots. \tag{2.3.9b}$$

These iterates are well defined when the coefficient matrix, or Jacobian matrix, $\partial\boldsymbol{\phi}/\partial\mathbf{s}$ is nonsingular for each $\mathbf{s}^{(\nu)}$. If this can be established, then the question of convergence is relevant. However, in the present case, Theorem 1.4.3 becomes applicable when the Jacobian is nonsingular. Thus to justify Newton's method applied to Equation (2.3.3) we need only show that $\partial\boldsymbol{\phi}(\mathbf{s})/\partial\mathbf{s}$ is nonsingular.

Differentiating in Equation (2.3.3), we obtain the representation

$$\frac{\partial\boldsymbol{\phi}(\mathbf{s})}{\partial\mathbf{s}} = A + BW(\mathbf{s}; b). \tag{2.3.10}$$

Here the nth-order matrix $W(\mathbf{s}; x) \equiv \partial\mathbf{u}(\mathbf{s}; x)/\partial\mathbf{s}$ is the solution of the variational system, obtained by differentiating in (2.3.2),

$$\text{(a)}\quad W' = \frac{\partial\mathbf{f}(x, \mathbf{u}(\mathbf{s}; x))}{\partial\mathbf{u}} W, \qquad \text{(b)}\quad W(0) = I. \tag{2.3.11}$$

We may write Equation (2.3.10), recalling the assumption that $(A + B)$ is nonsingular, as

$$\frac{\partial\boldsymbol{\phi}(\mathbf{s})}{\partial\mathbf{s}} = (A + B)[I + (A + B)^{-1}B(W(\mathbf{s}; b) - I)].$$

But it is shown in the proof of Theorem 1.2.6 that

$$\|(A + B)^{-1}B(W(\mathbf{s}; b) - I)\|_\infty < 1,$$

and so we can conclude that the Jacobian matrix is nonsingular (see Isaacson and Keller (1966), p. 16) for all $\mathbf{s}$. Of course, the matrix (2.3.10) may very well be nonsingular even though $(A + B)$ is singular. Thus, Newton's method is frequently applicable when our present proof of the fact is not.

The calculational procedure for employing Newton's method consists in solving $n + 1$ initial-value problems for nth-order systems at each iteration. The single nonlinear system (2.3.2) is integrated numerically with $\mathbf{s} = \mathbf{s}^{(\nu)}$, say, and at the same time and on the same net the n linear systems in (2.3.11) are integrated using the currently-computed values of $\mathbf{U}_j(\mathbf{s}^{(\nu)})$ in place of $\mathbf{u}(\mathbf{s}^{(\nu)}; x_j)$ to evaluate the elements in $\partial\mathbf{f}(x, \mathbf{u})/\partial\mathbf{u}$. Then $\mathbf{s}^{(\nu+1)}$ is computed from (2.3.9) where, however, $\boldsymbol{\phi}(\mathbf{s})$ in Equation (2.3.3) and $\partial\boldsymbol{\phi}(\mathbf{s})/\partial\mathbf{s}$ in Equation (2.3.10) are approximated using the appropriate numerical solutions for $\mathbf{u}$ and W at $x_J = b$.

There may be great practical merit in applying Newton's method to linear systems of differential equations. Sometimes the iteration schemes are less sensitive to cancellation errors that can disturb and even invalidate direct applications of shooting, as pointed out in Section 2.1. More important, however, is the fact that the matrix $W(x)$, the solution of (2.3.11), is independent of $\mathbf{s}$ for linear problems. That is, if $\mathbf{f}(x, \mathbf{u})$ has the form

$$\mathbf{f}(x, \mathbf{u}) = K(x)\mathbf{u} + \mathbf{f}_0(x), \tag{2.3.12}$$

where $K(x)$ is an nth-order matrix, then clearly

$$\frac{\partial\mathbf{f}(x, \mathbf{u})}{\partial\mathbf{u}} = K(x)$$

and the variational problem (2.3.11) reduces to

$$\text{(a)}\quad W' = K(x)W, \qquad \text{(b)}\quad W(0) = I. \tag{2.3.13}$$

Recalling Equation (2.3.10), we then see that $\partial\boldsymbol{\phi}(\mathbf{s})/\partial\mathbf{s}$ is independent of $\mathbf{s}$ and hence the system of Equations (2.3.3) for the determination of $\mathbf{s}$ is *a linear system.* (This was shown before more directly, in Section 1.2.1.) Since $W(x) \equiv \partial\mathbf{u}(\mathbf{s}; x)/\partial\mathbf{s}$, and we have just shown that $\boldsymbol{\phi}(\mathbf{s})$ is linear, so that $\boldsymbol{\phi}(\mathbf{s}) = [\partial\boldsymbol{\phi}(\mathbf{s})/\partial\mathbf{s}]\mathbf{s} + \boldsymbol{\phi}(\mathbf{0})$, it follows that Equation (2.3.3) can be written as

$$[A + BW(b)]\mathbf{s} = \boldsymbol{\alpha} - B\mathbf{u}(\mathbf{0}; b). \tag{2.3.14}$$

Thus by solving the n systems (2.3.13) the coefficient matrix is determined, and by solving the single system (2.3.2) with $\mathbf{s} = \mathbf{0}$ and $\mathbf{f}$ as given by Equation (2.3.12), the inhomogeneous data are determined.

Now to solve a *sequence* of linear boundary-value problems, given by (2.3.1) and (2.3.12), in which the boundary conditions change, we need only use the new data A, B, and $\boldsymbol{\alpha}$ in Equation (2.3.14), solve the resulting linear algebraic system for $\mathbf{s}$ and then solve the initial-value problem (2.3.2) using this value of $\mathbf{s}$. If in addition the inhomogeneous term in the differential equation is to be changed [this is the term $\mathbf{f}_0(x)$ in Equation (2.3.12)], we need one additional integration of (2.3.2) with $\mathbf{s} = \mathbf{0}$, as above, to compute $\mathbf{u}(\mathbf{0}; b)$. In many linear problems of interest it is only the inhomogeneous data that are to be altered and in such cases the above-indicated procedures are very efficient. It may frequently be worthwhile, in such cases, to compute $W(x)$ with extra accuracy (that is, multiple precision).

Of course in most nonlinear problems of practical interest the hypothesis of Theorem 1.2.6 will not be satisfied. We have already seen, for instance, that the condition, $(A + B)$ nonsingular, is rather special. Nevertheless, the initial-value method we have indicated is very frequently applicable though the proofs are not. The practical problem is to determine some matrix Q, now $\neq (A + B)^{-1}$, so that (2.3.4) is contracting; see for example Problems 2.3.4 and 2.3.5. Much more general results are indicated in Section 5.5 and, for instance, in Conti (1958) or Losota and Opial (1966). Of course if the initial unknown vector $\mathbf{s}$ can be reduced to a $q = n - p$-dimensional vector, as previously discussed, this should be done. Then the system of Equations (2.3.3) reduces to q equations, the corresponding Jacobian matrix in Equation (2.3.10) is of order q and the matrix W is $n \times q$. Now only $q + 1$ systems of order n need be integrated for each iteration [see Problem 1.2.12]. Needless to say this reduction should be done for linear problems when applicable.

It may be a formidable task to apply any iteration scheme successfully when one does not have some reasonable estimate of the location of a zero of $\boldsymbol{\phi}(\mathbf{s})$. Unfortunately there are no universal procedures for obtaining such estimates. But patience in searching for an acceptable first estimate is frequently rewarded. It is not unusual in practice to devote ninety percent or more of the computing effort to locating a neighborhood of a root. In many problems of scientific interest, as opposed to textbook problems, there are parameters in the equations or boundary conditions, and solutions are required over an entire domain of these parameters. But then when a solution has been obtained for one set of parameter values we can frequently make small changes in the values of the parameters and obtain neighboring solutions with ease.

This device of altering a parameter can be so effective that several techniques for solving nonlinear problems are based on *introducing* such a parameter. For instance we may introduce a single parameter t in such a way that for $t = 1$ the problem is the one whose solution is desired and for $t = 0$ it is some problem that is easily solved. Now if the solution depends continuously on t we may be able to proceed in small steps Δt, starting from $t = 0$ as indicated above, to compute finally the solution for $t = 1$. The study of these methods is briefly considered in Section 5.5.1. Let it suffice to say here that two very powerful techniques for proving existence theorems—the continuity and Poincaré continuation methods—are based on this idea [see Bers, John, and Schechter (1964), pp. 238–240, 285, Lasota and Opial (1966) and Section 5.5].

Problems

2.3.1 Show that the necessary and sufficient condition for the n linear boundary conditions in (2.3.1b) to be linearly independent is that rank $(A, B) = n$ [here A and B are nth-order matrices while (A, B) is a matrix with n rows and $2n$ columns].

2.3.2 Show that a necessary and sufficient condition for rank $(A, B) = n$ is that there be two nth-order matrices, say P_1 and P_2, such that the nth-order matrix $AP_1 + BP_2$ is nonsingular.

2.3.3 Use the results stated in Problems 2.3.1 and 2.3.2 to deduce that a *necessary* condition for the applicability of Newton's method in solving (2.3.1), that is, in finding a root of Equation (2.3.3), is that the boundary conditions (2.3.1b) be linearly independent.

2.3.4* If $\partial\mathbf{f}(x, \mathbf{u})/\partial\mathbf{u}$ is Lipschitz-continuous in $\mathbf{u}$, the matrix

$$[A + BW(\mathbf{s}^{(0)}; b)]$$

in Equation (2.3.10) is nonsingular and $|\boldsymbol{\phi}(\mathbf{s}^{(0)})|$ is sufficiently small, then Equation (2.3.3) has a root if $|b - a|$ is sufficiently small. Use the result in Problem 1.4.5 of Chapter 1 to indicate a proof of the above and estimate the magnitudes of $|\boldsymbol{\phi}(\mathbf{s}^{(0)})|$ and $|b - a|$ in terms of the magnitudes of $|\mathbf{f}|$, $\|\partial\mathbf{f}/\partial\mathbf{u}\|$ and the Lipschitz constant.

2.3.5 Formulate the boundary-value problem

$$y'' + y = 0, \qquad y(0) = 0, \qquad y'(L) = \cos L$$

in terms of a first-order system. Show that the conditions of Theorems 1.2.5 and 1.2.6 are satisfied if $L < \log 2$. For $L > \log 2$, show that a contraction is obtained by using, in place of $Q = (A + B)^{-1} = I$, the matrix

$$Q = \begin{pmatrix} 1 & 0 \\ 0 & q \end{pmatrix},$$

with any q satisfying

$$0 < q \cos L < 2.$$

What happens when $\cos L = 0$? Try actual computations in the above with varying values of L.

2.4 Variants of Shooting; Parallel Shooting

In the previous sections we have discussed some of the difficulties which arise in the shooting method. Of those discussed the best known, perhaps because it is the most common, is that caused by the growth of solutions of the initial-value problems which must be solved. We recall that this causes a loss in accuracy in attempts to solve the corresponding algebraic or transcendental system. [Technically we may say that the problem of finding a root of $\boldsymbol{\phi}(\mathbf{s}) = 0$ is "ill-conditioned" or "not properly posed" in the sense of Section 3, Chapter 1, in Isaacson and Keller (1966).] As has been pointed out, greater accuracy in the computations may sometimes overcome this difficulty. This is not always practicable or desirable and so we shall describe variations of the shooting method that can frequently eliminate this difficulty. The finite-difference or integral-equation methods of Chapters 3 and 4 do not suffer from this particular growth problem, in principle.

There is another more striking phenomenon that can hamper the initial-value method when it is applied to nonlinear problems. We have already pointed out that many, if not most, boundary-value problems of interest do not satisfy the nice conditions imposed in results like our Theorems 2.2.2 and 2.3.1 which insure the applicability of shooting. In particular the functions $f(x, y, y')$ or $\mathbf{f}(x, \mathbf{y})$ may be unbounded in the x, $\mathbf{y}$-domain of interest. Then solutions of the corresponding initial-value problems may become singular in $[a, b]$ for some values of the initial data. If this occurs we cannot even carry out the integration from a to b in order to improve the initial data. Furthermore, the location of the singularity is frequently a very sensitive function of the initial data and thus great accuracy in the initial estimate of $s^{(0)}$ or $\mathbf{s}^{(0)}$ may be required in order to start an iterative scheme; see Problem 2.4.1. When this difficulty with singularities occurs it is quite apparent in the computations. The parallel shooting schemes presented below can frequently circumvent the trouble, or else a finite difference procedure can be employed.

As we shall see, parallel shooting is actually a combination of difference methods and initial-value methods.

The variant known as multiple or parallel shooting is designed to reduce the growth of the solutions of the initial-value problems which must be solved. This is done by dividing the interval $[a, b]$ into a number of subintervals, integrating appropriate initial-value problems over each subinterval, and then *simultaneously* adjusting all of the "initial" data in order to satisfy the boundary conditions and appropriate continuity conditions at the subdivision points. The variety of procedures afforded by this technique is limitless, and we consider only two rather obvious forms of it here. Of course the motivation in selecting subintervals and directions of integration is, as indicated above, to reduce the magnification of errors caused by growing solutions.

To illustrate we shall consider the boundary-value problem, on an interval $a \leq x \leq b$, for a system of n first-order equations:

$$\text{(a)} \quad \mathbf{y}' = \mathbf{f}(x, \mathbf{y}), \qquad \text{(b)} \quad A\mathbf{y}(a) + B\mathbf{y}(b) = \boldsymbol{\alpha}. \tag{2.4.1}$$

We divide this interval into J subintervals with the points x_j, say

$$a \equiv x_0 < x_1 < x_2 < \cdots < x_{J-1} < x_J \equiv b. \tag{2.4.2a}$$

We refer to the interval $x_{j-1} \leq x \leq x_j$ as the jth subinterval and denote its length by

$$\Delta_j \equiv x_j - x_{j-1}. \tag{2.4.2b}$$

Now on each subinterval we introduce the new independent variable t, the new dependent variable $\mathbf{y}_j(t)$, and the new n-vector functions $\mathbf{f}_j(t, \mathbf{z})$ by

$$\begin{gathered} t \equiv \frac{x - x_{j-1}}{\Delta_j}, \qquad \mathbf{y}_j(t) \equiv \mathbf{y}(x) = \mathbf{y}(x_{j-1} + t\Delta_j), \\ \mathbf{f}_j(t, \mathbf{z}) \equiv \Delta_j \mathbf{f}(x_{j-1} + t\Delta_j, \mathbf{z}), \\ \text{for} \quad x_{j-1} < x < x_j, \qquad j = 1, 2, \ldots, J. \end{gathered} \tag{2.4.3}$$

With these changes of variables the system (2.4.1a) becomes, on the jth subinterval,

$$\frac{d\mathbf{y}_j}{dt} = \mathbf{f}_j(t, \mathbf{y}_j(t)), \qquad 0 < t < 1; \qquad j = 1, 2, \ldots, J. \tag{2.4.4a}$$

The boundary conditions (2.4.1b) are now

$$A\mathbf{y}_1(0) + B\mathbf{y}_J(1) = \boldsymbol{\alpha}. \tag{2.4.4b}$$

Continuity of the solution of (2.4.1) at each interior point x_j is expressed by the conditions

$$\mathbf{y}_{j+1}(0) - \mathbf{y}_j(1) = 0, \qquad j = 1, 2, \ldots, J - 1. \tag{2.4.4c}$$

The J systems of n first-order equations (2.4.4a) can be written in the vector form

$$\frac{d}{dt}\mathbf{Y} = \mathbf{F}(t, \mathbf{Y}), \qquad 0 < t < 1, \tag{2.4.5a}$$

and the J sets of conditions (2.4.4b, c) can be written in the matrix-vector form

$$P\mathbf{Y}(0) + Q\mathbf{Y}(1) = \boldsymbol{\gamma}. \tag{2.4.5b}$$

Here we have introduced the nJ-dimensional vectors

$$\mathbf{Y}(t) \equiv \begin{pmatrix} \mathbf{y}_1(t) \\ \mathbf{y}_2(t) \\ \vdots \\ \mathbf{y}_J(t) \end{pmatrix}, \qquad \mathbf{F}(t, \mathbf{Y}) \equiv \begin{pmatrix} \mathbf{f}_1(t, \mathbf{y}_1) \\ \mathbf{f}_2(t, \mathbf{y}_2) \\ \vdots \\ \mathbf{f}_J(t, \mathbf{y}_J) \end{pmatrix}, \qquad \boldsymbol{\gamma} \equiv \begin{pmatrix} \boldsymbol{\alpha} \\ \mathbf{0} \\ \vdots \\ \mathbf{0} \end{pmatrix}; \tag{2.4.6}$$

and the square matrices of order nJ

$$P \equiv \begin{pmatrix} A & 0 & 0 & \cdots & 0 \\ 0 & I & 0 & \cdots & 0 \\ 0 & 0 & I & \cdots & 0 \\ \vdots & & & & \vdots \\ 0 & 0 & 0 & \cdots & I \end{pmatrix},$$

$$Q \equiv \begin{pmatrix} 0 & 0 & 0 & \cdots & B \\ -I & 0 & 0 & \cdots & 0 \\ 0 & -I & 0 & \cdots & 0 \\ \vdots & & & & \vdots \\ 0 & 0 & 0 & -I & 0 \end{pmatrix}. \tag{2.4.7}$$

In (2.4.7) the identity matrix I is of order n and the blocks of zeros are also of order n.

Under modest smoothness conditions on $\mathbf{f}(x, \mathbf{y})$, the boundary-value problem (2.4.1) for n first-order equations on $a \leq x \leq b$ can be shown to be equivalent to the boundary-value problem (2.4.5) for nJ first-order equations on $0 \leq t \leq 1$. Clearly one parallel shooting procedure for solving (2.4.1) consists in applying ordinary shooting, as in Section 2.3, to (2.4.5). For this purpose we must solve the initial-value problem

$$\frac{d\mathbf{U}}{dt} = \mathbf{F}(t, \mathbf{U}), \qquad \mathbf{U}(0) = \mathbf{s}, \tag{2.4.8a}$$

for a first-order system of nJ equations. We try to pick the nJ-dimensional initial vector, $\mathbf{s}$, such that

$$\boldsymbol{\Phi}(\mathbf{s}) \equiv P\mathbf{s} + Q\mathbf{U}(\mathbf{s}; 1) - \boldsymbol{\gamma} = \mathbf{0}. \tag{2.4.9}$$

Of course Newton's method is advocated for solving this system. To employ it we compute, along with the solutions of the initial-value problem (2.4.8a), the nJ-order matrix solution, $W(\mathbf{s}, t) \equiv \partial\mathbf{U}/\partial\mathbf{s}$, of the variational system

$$\frac{dW}{dt} = \frac{\partial\mathbf{F}(t, \mathbf{U}(\mathbf{s}; t))}{\partial\mathbf{Y}} W, \qquad W(0) = I. \tag{2.4.10a}$$

The iterates are then given by $\mathbf{s}^{(\nu+1)} = \mathbf{s}^{(\nu)} + \Delta\mathbf{s}^{(\nu)}$, where

$$[P + QW(\mathbf{s}^{(\nu)}; 1)]\,\Delta\mathbf{s}^{(\nu)} = -\boldsymbol{\Phi}(\mathbf{s}^{(\nu)}). \tag{2.4.11}$$

The computations involved in carrying out the above procedure are not nearly as complicated as they might seem. This is due to the special form of $\mathbf{F}(t, \mathbf{Y})$ defined in (2.4.6). In fact, if we introduce the n-vectors $\mathbf{s}_j$ such that $\mathbf{s} = (\mathbf{s}_1^T, \mathbf{s}_2^T, \ldots, \mathbf{s}_J^T)^T$, then the system (2.4.8a) is in fact

$$\frac{d\mathbf{u}_j(t)}{dt} = \mathbf{f}_j(t, \mathbf{u}_j), \qquad \mathbf{u}_j(0) = \mathbf{s}_j, \qquad j = 1, 2, \ldots, J. \tag{2.4.8b}$$

This is J first-order systems of n equations and each system is independent of the others. Thus these J initial-value problems can be solved independently of each other. In fact the solution of the set of initial-value problems (2.4.8) is ideally suited for computation on *parallel computers* (which are at present being designed). The same is of course true of the variational system (2.4.10) for the matrix W of order nJ. If we define the J matrices, W_j, of order n by

$$\frac{dW_j}{dt} = \frac{\partial\mathbf{f}_j(t, \mathbf{u}_j(\mathbf{s}_j; t))}{\partial\mathbf{u}_j} W_j, \qquad W_j(0) = I, \qquad j = 1, 2, \ldots, J, \tag{2.4.10b}$$

then $W(\mathbf{s}, t) = \operatorname{diag}\{W_j(\mathbf{s}_j; t)\}$. This follows since $W_j = \partial\mathbf{u}_j/\partial\mathbf{s}_j$ and $\partial\mathbf{u}_i/\partial\mathbf{s}_j \equiv 0$ if $i \neq j$. The only coupling of the systems (2.4.8b) and (2.4.10b) for different values of j occurs in the algebraic problem of solving the linear system (2.4.11). We examine below some questions relating to the nonsingularity of the coefficient matrix in Equation (2.4.11). In particular the explicit form of the inverse of this matrix can be deduced under special circumstances (see Problem 2.4.3).

We first observe that the nJ-order matrix $(P + Q)$ is nonsingular if the nth-order matrix $(A + B)$ is nonsingular. In fact, with the definitions

$$R \equiv (A + B)^{-1}, \qquad S \equiv -RB, \qquad T \equiv RA = I + S, \tag{2.4.12a}$$

it is easily verified that

$$(P + Q)^{-1} = \begin{pmatrix} R & S & S & \cdots & S \\ R & T & S & \cdots & S \\ R & T & T & \cdots & \vdots \\ \vdots & & & & S \\ R & T & T & \cdots & T \end{pmatrix}. \tag{2.4.12b}$$

A derivation of this result is indicated in Problem 2.4.2. Now the coefficient matrix in Equation (2.4.11) can be written as

$$(P + Q)[I + (P + Q)^{-1}Q(W(\mathbf{s}^{(\nu)}; 1) - I)],$$

and hence it is nonsingular if

$$\|(P + Q)^{-1}Q\|_\infty \, \|W(\mathbf{s}; 1) - I\|_\infty < 1. \tag{2.4.13}$$

Using (2.4.12) and (2.4.7) it follows that

$$\|(P + Q)^{-1}Q\|_\infty \leq M \equiv (J + 1)m, \tag{2.4.14a}$$

where

$$m \equiv \max\{\|(A + B)^{-1}B\|_\infty, \|(A + B)^{-1}A\|_\infty\}.$$

Furthermore, just as in the proof of Theorem 1.2.6, we find from (2.4.10b) and the form of W that

$$\|W(\mathbf{s}; 1) - I\|_\infty \leq \max_{1 \leq j \leq J} \left[\exp\left(\Delta_j \int_0^1 k(x_{j-1} + t\Delta_j)\, dt\right) - 1\right], \tag{2.4.14b}$$

where $k(x)$ is, as usual, the bound

$$k(x) \geq \left\|\frac{\partial \mathbf{f}(x, \mathbf{y})}{\partial \mathbf{y}}\right\|_\infty, \qquad a \leq x \leq b.$$

We see that taking all the Δ_j "small" reduces the bound in (2.4.14b). But since $\sum_1^J \Delta_j = b - a$, this implies "large" values for J which increases the bound in (2.4.14a). If we assume that $k(x)$ is bounded on $[a, b]$ and take subintervals of equal length, $\Delta_j = (b - a)/J$, $j = 1, 2, \ldots, J$, then the inequality (2.4.13) is satisfied if

$$b - a < \frac{J \ln [1 + (1/M)]}{\max_{a \leq x \leq b} k(x)}.$$

On the other hand, Theorem 1.2.6 is valid if we replace the hypothesis (1.2.18e) by the stronger condition

$$b - a < \frac{\ln [1 + (1/m)]}{\max_{a \leq x \leq b} k(x)}.$$

For large J and m the above inequalities are almost equivalent, since then

$$\ln\left(1 + \frac{1}{m}\right) \approx \frac{1}{m}, \qquad J \ln\left(1 + \frac{1}{(J + 1)m}\right) \approx \frac{J}{(J + 1)m}.$$

We should also observe that when (2.4.13) is satisfied, the proof of Theorem 1.2.6 applies to the boundary-value problem (2.4.5) to show that it has a unique solution.

What we have shown above is, essentially, that the parallel shooting procedure does not make matters *worse* than they were for the original initial-value problem. We have not been able to show, for instance, that the length of the interval $[a, b]$ over which our proof of existence and uniqueness holds, can be increased. But the purpose of the parallel shooting method is to overcome certain practical difficulties that occur in ordinary shooting. While actual computations verify its usefulness in special cases, we may also give some theoretical justification as follows. In solving the initial-value problems (2.3.2) and (2.4.8) numerically, errors are of course introduced. If the same stable scheme of order p, say, is used in each case with the same net spacing h, and the functions $\mathbf{f}(x, \mathbf{u})$ and $\mathbf{F}(t, \mathbf{U})$ are sufficiently smooth, then the numerical errors at points of the intervals $[a, b]$ and $[0, 1]$ can be bounded by

$$|\mathbf{u}(x_i) - \mathbf{u}_i| \le h^p M_1 \exp [K_1 |x_0 - x_i|],$$
$$|\mathbf{U}(t_i) - \mathbf{U}_i| \le h^p M_2 \exp [K_2 |t_0 - t_i|].$$

Some estimates of this form are derived in Section 1.3. The constants M_1 and M_2 depend in a simple algebraic manner on bounds for the functions $\mathbf{f}$ and $\mathbf{F}$ and possibly some of their derivatives. However, the constants K_1 and K_2 are essentially the Lipschitz constants for $\mathbf{f}$ and $\mathbf{F}$, respectively, or polynomials in these constants. Then if we take $\Delta_j = |b - a|/J$, $j = 1, 2, \ldots, J$, in the subdivision (2.4.2) it is clear from (2.4.3) and (2.4.6) that

$$K_2 \doteq K_1 \frac{|b - a|}{J}.$$

Then since solutions of (2.3.2) at $x = b$ correspond to those of (2.4.8) at $t = 1$, we see that the *bound on the error in the numerical solution is reduced exponentially by parallel shooting* [that is, the bound is proportional to $(\exp [K_1|b - a|])^{1/J}$ rather than to $(\exp [K_1|b - a|])$]. The constant M_2 is at worst larger than M_1 by a factor which is a small power of J. To show the detailed dependence of the quantities M_1, M_2, K_1, and K_2 on bounds for $\mathbf{f}$, $\mathbf{F}$, and their derivatives is rather complicated. Problems 1.3.3–6 contain some results which should clarify matters. More details can be found in Isaacson and Keller, Chapter 8, Sections 2 and 3. Very explicit error estimates are contained in Henrici (1962) and (1963).

A schematic diagram of the parallel shooting scheme presented above is shown in Figure 2.4.1. By analogy it is easy to indicate other parallel shooting schemes which may in many cases be more efficient. For instance, if we agree to take $J = 2N$, so that there is always an even number of subintervals, we can shoot from each odd-labeled subdivision point toward each of its two adjacent even-labeled points. A glance at Figure 2.4.2 for the special case $J = 4$ should suffice. Here there are only two n-vectors $\mathbf{s}_1$ and $\mathbf{s}_2$ to be determined [by applying continuity at x_2, that is $\mathbf{u}_2(1) = \mathbf{u}_3(1)$, and the appropriate boundary conditions on $\mathbf{u}_1(1)$ and $\mathbf{u}_4(1)$]. The continuity conditions at x_1 and x_3 are automatically satisfied. We do not bother with a detailed formulation of this case as the diagram should suffice. In this manner the systems replacing (2.4.9) and (2.4.11) are only half the order they would be if all shooting were in the same direction for the same number of subintervals.

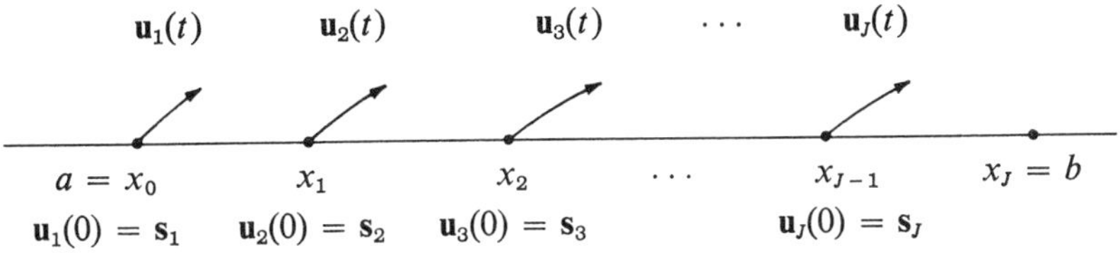

Figure 2.4.1

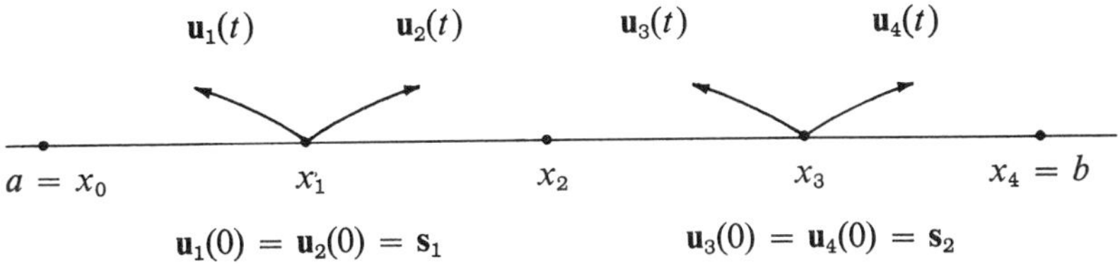

Figure 2.4.2

A rather common special case of the above modification occurs for $J = 2$ and is generally described in terms of shooting from each endpoint and matching at some interior point. In fact all of our schemes are just as well considered to be multiple matching generalizations of this well-known case.

If it is known that all solutions of a given system of differential equations decay in one direction then all integrations should be done in that direction. However if solutions grow more or less equally in both directions then there is no preferred direction and the more efficient process of shooting in both directions should be used.

A rough practical guide to the choice of subintervals to be used in parallel shooting can be based on the old-fashioned technique (when calculating with fixed point arithmetic) of rescaling the solution as it grows. That is, integrate the initial-value problem

$$\mathbf{y}' = \mathbf{f}(x, \mathbf{y}), \qquad \mathbf{y}(a) = \boldsymbol{\xi}$$

using some "typical" initial value $\boldsymbol{\xi}$ on some sufficiently fine net $\{x_\nu\}$ with $x_0 = a$. After each step of the integration test to see if the growth condition

$$\|\mathbf{y}(x_\nu)\| > R\|\mathbf{y}(x_0)\|,$$

where R is some selected reasonable growth factor, say $R = 10^3$ (the choice of R depends upon the machine word length, as well as other factors). When this test is passed, take x_ν to be one of the points in (2.4.2a), replace $\|\mathbf{y}(x_0)\|$ in the test by the new value, $\|\mathbf{y}(x_\nu)\|$, and continue the integration and testing. Some experimenting with choices of $\boldsymbol{\xi}$ and R may be required in very difficult cases.

2.4.1 *Power Series*

Power series are occasionally applied or suggested as a means for solving boundary-value problems. Their application is in fact a variant of the initial-value method. For example, if $\mathbf{f}(x, \mathbf{y})$ in Equation (2.4.1a) is an analytic function of the $n + 1$ variables $(x, \mathbf{y})$, then we may seek power-series solutions in the form

$$\mathbf{y}(x) = \sum_{k=0}^{\infty} \mathbf{a}_k(x - a)^k. \tag{2.4.15}$$

Using this in the expanded form of Equation (2.4.1a) and equating coefficients of like powers of $(x - a)$, we obtain (recursive) formulae for the coefficients $\mathbf{a}_k$, $k = 1, 2, \ldots$, in terms of the initial point a, and the leading coefficients $\mathbf{a}_0$ [i.e., $\mathbf{a}_k = \mathbf{a}_k(\mathbf{a}_0)$].

Now $\mathbf{a}_0$ is to be determined so that the boundary conditions (2.4.1b) are satisfied. This yields the system

$$A\mathbf{a}_0 + B\sum_{k=0}^{\infty} \mathbf{a}_k(b - a)^k - \boldsymbol{\alpha} = 0. \tag{2.4.16}$$

Since the $\mathbf{a}_k$ for $k \geq 1$ are functions of $\mathbf{a}_0$ the system (2.4.16) can be considered a transcendental system of equations in $\mathbf{a}_0$. [Of course if there are p independent conditions on $\mathbf{y}(a)$ in (2.4.1b), or equivalently on $\mathbf{a}_0$ in (2.4.1b), then the system is easily reduced to one containing only $q = n - p$ equations and unknowns.]

However, if power series are employed to solve the initial-value problem (2.3.2) then the resulting transcendental system $\boldsymbol{\phi}(\mathbf{s}) = 0$, with $\boldsymbol{\phi}(\mathbf{s})$ defined in (2.3.3), is identical to the system (2.4.16) if we set $\mathbf{s} \equiv \mathbf{a}_0$. In this sense the power-series procedure is but a special case of the initial-value method.

To justify the power-series method in any case requires a study of the convergence of the appropriate series (2.4.15) over the interval $a \leq x \leq b$. Thus, at best, the method has rather limited applicability. In actual calculations we do not use (2.4.15) but some truncated version, say,

$$\mathbf{u}_N(x) = \sum_{k=0}^{N} \mathbf{a}_k(x - a)^k. \tag{2.4.17}$$

The error committed in this way is formally analogous to the truncation error introduced when solving the initial-value problem (2.4.1) by some numerical scheme. While the calculations using (2.4.17) are always well-defined, the formal result as $N \to \infty$ may be nonsense if the radius of convergence is too small, that is, less than $b - a$. (As indicated in Section 2.3 the schemes based on direct numerical integration of the initial-value problem have no such limitation imposed by radius-of-convergence difficulties.)

However, there are several ways in which the power-series method may be salvaged. One way is by direct analytic continuation from a to b by expansions about a sequence of intermediate points, x_j, $j = 1, 2, \ldots, J$. This procedure is essentially the same as using the one-step integration scheme known as the Taylor expansion method to solve the initial-value problem (2.3.2). When the motivation is power series one would employ many terms in the expansions of form (2.4.17) and few "expansion points" x_j, that is, small J. Conversely, many closely-spaced "net points," x_j, and few terms in the Taylor series would be used in schemes motivated by numerical integration. The difference in point of view depends, of course, on which limiting process is intended, $\max_j |x_j - x_{j+1}| \to 0$, or $N \to \infty$; only in the latter case is analyticity required.

Some other ways in which power series may be used are suggested by the parallel shooting variants discussed above. For instance, the solution can be expanded about each endpoint and the two series matched at some common interior point. Then two unknown leading-coefficient vectors, say $\mathbf{a}_0$ and $\mathbf{b}_0$, must be determined by the matching conditions and the boundary conditions. Obviously the solution could be expanded about N interior points, matched at $N - 1$ interior points separating these expansion points, and the boundary conditions imposed to determine N unknown leading-coefficient vectors. For instance, the odd-labeled points in Figure 2.4.2 may be taken as the origins for expansions, and at the even-labeled interior points the appropriate series would be matched. The convergence problems and questions of accuracy in such power-series methods are untreated in the literature.

Problems

2.4.1 The problem $y'' = m^2 \sinh q^2 y$, $y(0) = y(1) = 0$ has the unique solution $y(x) \equiv 0$. Show that the initial-value problem

$$u'' = m^2 \sinh q^2 u, \qquad u(0) = 0, \qquad u'(0) = s,$$

has a singularity located at approximately

$$x_\infty = \frac{1}{qm} \log \frac{8m}{qs} + \cdots.$$

Thus in order that $x_\infty > 1$, that is, lie outside [0, 1], the initial parameter s must satisfy

$$s \leq \frac{8m}{q} e^{-mq}.$$

For $m = q = 4$ the initial parameter s must be accurate to within e^{-16}. [This problem is from B. A. Troesch (1960)]. [HINT: By integrating $y'y'' = [\frac{1}{2}(y')^2]'$, the solution can be obtained in the implicit form: $x = \int_0^u [\cdots]\, du$. Setting $u = \infty$, we obtain an expression for x_∞ which can be estimated by expanding the integrand appropriately.]

2.4.2 For the matrix $(P + Q)$ determined by Equation (2.4.7) verify the block factorization:

$$(P + Q) = \begin{pmatrix} (A+B) & B & B & \cdots & B \\ 0 & I & 0 & \cdots & 0 \\ & 0 & I & \cdots & 0 \\ \vdots & & & & \vdots \\ & & & & 0 \\ 0 & & \cdots & & I \end{pmatrix} \begin{pmatrix} I & & & \\ -I & I & & 0 \\ & \ddots & \ddots & \\ 0 & & -I & I \end{pmatrix}$$

Use the above to deduce the result in (2.4.12).

2.4.3 Using the fact that $W = \operatorname{diag}\{W_j\}$ and the assumption that the nth-order matrix $(A + BW_J \cdots W_1)$ is nonsingular, derive an explicit representation for the inverse of the nJ-order matrix $(P + QW)$. [HINT: Factor $(P + QW)$ into block-triangular form with first row $(A + BW_J \cdots W_1)$, $BW_J \cdots W_2, \ldots, BW_J$ in a manner similar to that of Problem 2.4.2. Find the inverse as the product of triangular matrices in analogy with (2.4.12).]

2.4.4 Formulate the parallel shooting scheme implied by Figure 2.4.2 (generalized to $J = 2N$) for the boundary-value problem (2.4.1). Examine the linear system obtained by applying Newton's method to solve for the nN-dimensional vector $\mathbf{s}$.

2.4.5 Consider the boundary-value problem of order $n = 2m$

$$\mathbf{y}' = \mathbf{f}(x, \mathbf{y}, \mathbf{z}), \qquad A\mathbf{y}(a) = \boldsymbol{\alpha}, \qquad \mathbf{z}' = \mathbf{g}(x, \mathbf{y}, \mathbf{z}), \qquad B\mathbf{z}(b) = \boldsymbol{\beta},$$

where $\mathbf{y}$, $\mathbf{z}$, $\mathbf{f}$, $\mathbf{g}$, $\boldsymbol{\alpha}$, and $\boldsymbol{\beta}$ are m-dimensional and the mth-order matrices A and B are nonsingular. Formulate the parallel shooting schemes of both types, Figures 2.4.1 and 2.4.2, for the same $J = 2N$ intervals for this problem. Compare the application of Newton's method in each case and formally determine the relevant inverse matrices. Discuss the solution when $\mathbf{f}$ and $\mathbf{g}$ are linear in $\mathbf{y}$ and $\mathbf{z}$.

SUPPLEMENTARY REFERENCES AND NOTES

Section 2.1 A technique for reducing the effect of cancellation errors in some linear boundary-value problems is given in Godunov (1961). This procedure has been made more practical by S. Conti (1966), whose scheme can be formulated as a special form of parallel shooting.

Section 2.2 Shooting methods for nonlinear second-order problems are described by Collatz (1960) and Fox (1957).

Section 2.3 The Newton method applied to shooting with somewhat specialized boundary conditions is described in Goodman and Lance (1956). Rather than solve a small-order linear system to obtain the corrections to the initial data they integrate (numerically) the adjoint system backwards.

Since the variational systems are linear, predictor-corrector schemes are better replaced by the corrector alone, properly solved [see Riley, Bennett, and McCormick (1967)].

Section 2.4 A limiting form of parallel shooting yields a finite-difference scheme (that is, when the modified Euler one-step method is used over each interval). Multiple shooting in various forms has been considered by, among others: Fox (1960), Morrison, Riley, and Zancanaro (1962), and Kalnins (1964). For linear problems, Conti's (1966) procedure is closely related. However, one of the most important features of parallel shooting—the exponential reduction in error growth—does not seem to have been recognized previously.

For an example of the use of power series in solving nonlinear boundary-value problems see Reiss, Greenberg, and Keller (1957) or Weinitishke (1960), who employs analytic continuation.

CHAPTER 3

FINITE-DIFFERENCE METHODS

3.1 Linear Second-Order Equations

We again consider the single linear second-order equation

$$Ly(x) \equiv -y'' + p(x)y' + q(x)y = r(x), \qquad a < x < b, \tag{3.1.1a}$$

which is now subject to the special two-point boundary conditions

$$y(a) = \alpha, \qquad y(b) = \beta. \tag{3.1.1b}$$

More general boundary conditions will be discussed later. A unique solution, $y(x)$, of (3.1.1) exists by the Corollary to Theorem 1.2.2 if $p(x)$, $q(x)$, $r(x)$ are continuous on $[a, b]$ and $q(x)$ is positive there. But since these functions are continuous on a closed bounded interval, there must exist positive constants P^*, Q^*, and Q_* such that

$$|p(x)| \leq P^*, \qquad 0 < Q_* \leq q(x) \leq Q^*, \qquad a \leq x \leq b. \tag{3.1.2}$$

(It should be observed that the conditions here imposed on $p(x)$, $q(x)$, and $r(x)$ are much more restrictive than those required in Section 2.1 for the application of initial-value methods.) We shall now study finite-difference methods for computing approximations to the solution of the boundary-value problem (3.1.1).

On the interval $[a, b]$ we place a uniform net, say

$$x_j = a + jh, \qquad j = 0, 1, \ldots, J + 1, \qquad h = \frac{b - a}{J + 1}. \tag{3.1.3}$$

To approximate $y(x)$ on this net we define a net function $\{u_j\}$ as the solution of a system of finite-difference equations which are in some sense an approximation to the problem (3.1.1). An obvious such difference formulation is

$$L_h u_j \equiv -\left(\frac{u_{j+1} - 2u_j + u_{j-1}}{h^2}\right) + p(x_j)\left(\frac{u_{j+1} - u_{j-1}}{2h}\right) + q(x_j)u_j \tag{3.1.4a}$$

$$= r(x_j), \qquad 1 \le j \le J;$$

$$u_0 = \alpha, \qquad u_{J+1} = \beta. \tag{3.1.4b}$$

At each interior net point, x_j, the derivatives in Equation (3.1.1a) have been replaced by corresponding centered difference quotients to give Equations (3.1.4a). We proceed to show that the linear system (3.1.4) can be solved by a simple algorithm, if h is sufficiently small, and then we estimate the accuracy of this numerical solution.

The result of multiplying the Equations (3.1.4a) by $h^2/2$ can be written as

$$\frac{h^2}{2} L_h u_j \equiv a_j u_{j-1} + b_j u_j + c_j u_{j+1} = \frac{h^2}{2} r(x_j), \qquad 1 \le j \le J, \tag{3.1.5a}$$

where

$$\begin{aligned} a_j &\equiv -\frac{1}{2}\left[1 + \frac{h}{2} p(x_j)\right], \\ b_j &\equiv \left[1 + \frac{h^2}{2} q(x_j)\right], \qquad 1 \le j \le J, \\ c_j &\equiv -\frac{1}{2}\left[1 - \frac{h}{2} p(x_j)\right], \end{aligned} \tag{3.1.5b}$$

Now the system (3.1.4) is, in vector notation,

$$A\mathbf{u} = \mathbf{r}, \tag{3.1.6a}$$

where we have introduced the J-dimensional vectors

$$\mathbf{u} \equiv \begin{pmatrix} u_1 \\ u_2 \\ \vdots \\ u_J \end{pmatrix}, \qquad \mathbf{r} \equiv \begin{pmatrix} r_1 \\ r_2 \\ \vdots \\ r_J \end{pmatrix} \equiv \frac{h^2}{2} \begin{pmatrix} r(x_1) \\ r(x_2) \\ \vdots \\ r(x_J) \end{pmatrix} - \begin{pmatrix} a_1\alpha \\ 0 \\ \vdots \\ 0 \\ c_J\beta \end{pmatrix}, \tag{3.1.6b}$$

and the J-order matrix

$$A = \begin{pmatrix} b_1 & c_1 & 0 & \cdot & \cdot & \cdot & 0 \\ a_2 & b_2 & c_2 & & & & \\ & \cdot & \cdot & \cdot & & & \cdot \\ \cdot & & \cdot & \cdot & \cdot & & \cdot \\ \cdot & & & \cdot & \cdot & \cdot & \cdot \\ \cdot & & & & a_{J-1} & b_{J-1} & c_{J-1} \\ 0 & \cdot & \cdot & \cdot & 0 & a_J & b_J \end{pmatrix}. \tag{3.1.6c}$$

A matrix of the form (3.1.6c), with nonzero elements only on the diagonal and the two adjacent codiagonals, is called a *tridiagonal* matrix. This special form frequently permits a very efficient application of the Gaussian elimination procedure, accounting for the occurrence of the known zero elements.

Let us derive this procedure quite formally, for solving any system of the form (3.1.6), and then justify its application when the coefficients (3.1.5b) are used. We assume that A is nonsingular and can be factored into the product

$$A = LU,$$

where

$$L \equiv \begin{pmatrix} \beta_1 & 0 & \cdot & \cdot & 0 \\ a_2 & \beta_2 & & & \cdot \\ 0 & \cdot & \cdot & & \cdot \\ \cdot & & \cdot & \cdot & 0 \\ 0 & \cdot & 0 & a_J & \beta_J \end{pmatrix}, \qquad U \equiv \begin{pmatrix} 1 & \gamma_1 & 0 & \cdot & 0 \\ 0 & 1 & \gamma_2 & & \cdot \\ \cdot & & \cdot & \cdot & 0 \\ \cdot & & & \cdot & \gamma_{J-1} \\ 0 & \cdot & \cdot & 0 & 1 \end{pmatrix}.$$

It follows that the β_j and γ_j must satisfy

$$\begin{aligned} &\beta_1 = b_1, \qquad \gamma_1 = c_1/\beta_1; \\ &\beta_j = b_j - a_j\gamma_{j-1}, \qquad j = 2, 3, \ldots, J; \\ &\gamma_j = c_j/\beta_j, \qquad j = 2, 3, \ldots, J-1. \end{aligned} \tag{3.1.7a}$$

The system (3.1.6a) can now be replaced by an equivalent pair of systems, say

$$L\mathbf{z} = \mathbf{r}, \qquad U\mathbf{u} = \mathbf{z}.$$

But since L and U are triangular, the solutions of these systems are easily obtained as

$$z_1 = r_1/\beta_1, \qquad z_j = (r_j - a_j z_{j-1})/\beta_j, \qquad j = 2, 3, \ldots, J; \tag{3.1.7b}$$

and

$$u_J = z_J, \qquad u_j = z_j - \gamma_j u_{j+1}, \qquad j = J-1, J-2, \ldots, 1. \tag{3.1.7c}$$

Thus, in summary, our formal procedure for solving Equation (3.1.6a) with coefficient matrix of the form (3.1.6c) is to compute the quantities β_j and γ_j in (3.1.7a), then evaluate the "intermediate solution," z_j, in (3.1.7b); and finally, u_j is given in (3.1.7c).

The only way in which the direct factorization procedure (3.1.7) can fail is for some β_j to vanish. This difficulty can frequently be avoided by means of the following.

THEOREM 3.1.1. *Let the elements of A in* (3.1.6c) *satisfy*

$$|b_j| > |a_j| + |c_j|, \qquad j = 1, 2, \ldots, J; \qquad (a_1 \equiv c_J \equiv 0).$$

Then A is nonsingular and the quantities β_j and γ_j in (3.1.7a) *are bounded by*

$$\text{(a)}\quad |\gamma_j| < 1, \qquad \text{(b)}\quad |b_j| - |a_j| \le |\beta_j| \le |b_j| + |a_j|.$$

Proof. If $\beta_j \neq 0$ for $1 \le j \le J$ the factorization $A = LU$ is valid. Then $\det A = (\det L)(\det U) = \beta_1\beta_2 \cdots \beta_J \neq 0$, so that A is nonsingular.

From the hypothesis, $|\gamma_1| = |c_1/b_1| < 1$. For an inductive proof of (a) assume $|\gamma_i| < 1$ for $i \le j - 1$. But we know that

$$\gamma_j = c_j/(b_j - a_j\gamma_{j-1})$$

and thus

$$|\gamma_j| \le \frac{|c_j|}{|b_j| - |a_j|\,|\gamma_{j-1}|} \le \frac{|c_j|}{|b_j| - |a_j|} < 1,$$

so part (a) follows. Now we use $|\gamma_j| < 1$ in $\beta_j = b_j - a_j\gamma_{j-1}$ and take absolute values to conclude part (b). ∎

Returning to our finite-difference problem, we have the following obvious corollary.

COROLLARY. *Let $p(x)$ and $q(x)$ satisfy* (3.1.2) *and the net spacing h satisfy*

$$h \le \frac{2}{P^*}.$$

Then the finite-difference system (3.1.4) *or* (3.1.6) *with coefficients given in* (3.1.5) *has a unique solution, $\{u_j\}$, given by* (3.1.7).

Proof. From (3.1.2) and (3.1.5b) we have

$$|b_j| \ge 1 + \frac{h^2}{2} Q_*.$$

But if $h \le 2/P^*$, then

$$|a_j| = \frac{1}{2}\left[1 + \frac{h}{2}p(x_j)\right], \qquad |c_j| = \frac{1}{2}\left[1 - \frac{h}{2}p(x_j)\right],$$

and so

$$|a_j| + |c_j| = 1.$$

Now the hypothesis of Theorem 3.1.1 is satisfied and the corollary follows. ∎

To estimate the error in the numerical solution of boundary-value problems by finite-difference methods, we can employ a stability theory exactly analogous to that used in Section 1.3 for numerical solutions of initial-value problems. We first define the *local truncation errors*, $\tau_j[v]$, in L_h, as an approximation to L, for any smooth function $v(x)$, by

$$\tau_j[v] \equiv L_h v(x_j) - Lv(x_j), \qquad 1 \le j \le J. \tag{3.1.8}$$

If $v(x)$ has continuous fourth derivatives on $[a, b]$, then for L as defined in (3.1.1a) and L_h as defined in (3.1.4a),

$$\begin{aligned}\tau_j[v] = -&\left[\frac{v(x_j + h) - 2v(x_j) + v(x_j - h)}{h^2} - v''(x_j)\right] \\ &+ p(x_j)\left[\frac{v(x_j + h) - v(x_j - h)}{2h} - v'(x_j)\right] \qquad (3.1.9) \\ = -&\frac{h^2}{12}\,[v''''(\xi_j) - 2p(x_j)v'''(\eta_j)], \qquad 1 \le j \le J.\end{aligned}$$

Here ξ_j and η_j are values in $[x_{j-1}, x_{j+1}]$ which come from the application of Taylor's theorem. Thus from the first lines of (3.1.9) we find that L_h *is consistent with* L; that is, $\tau_j[v] \to 0$ as $h \to 0$, for all functions $v(x)$ having a continuous second derivative on $[a, b]$. Further, L_h *has second-order accuracy* (in approximating L) for functions, $v(x)$, with continuous fourth derivatives on $[a, b]$. If $p(x) \equiv 0$ it is not difficult to devise fourth-order-accurate difference approximations to L (see Problem 3.1.1).

The linear difference operator L_h is said to be *stable* if for sufficiently small h there is a positive constant M, independent of h, such that

$$|v_j| \le M\left\{\max\,(|v_0|, |v_{J+1}|) + \max_{1 \le i \le J} |L_h v_i|\right\}, \qquad 0 \le j \le J + 1, \tag{3.1.10}$$

for all net functions $\{v_j\}$. We note that stability is solely a property of the finite-difference approximation L_h. Before exploiting the consequences of stability we shall establish this property for the operator in (3.1.4a). We have the following.

THEOREM 3.1.2. *Let L_h be defined by* (3.1.4a) *or* (3.1.5) *where $p(x)$ and $q(x)$ satisfy* (3.1.2). *Then for all net spacing h such that*

$$h \le \frac{2}{P^*},$$

and all net functions $\{v_j\}$, the inequality (3.1.10) *holds with the constant*

$$M = \max\,(1, 1/Q_*);$$

that is, L_h is stable.

Proof. From the definition in (3.1.5a) we have the identity

$$b_j v_j \equiv -a_j v_{j-1} - c_j v_{j+1} + \frac{h^2}{2} L_h v_j, \qquad 1 \leq j \leq J.$$

Since $h \leq 2/P^*$, it follows from (3.1.5b) and (3.1.2) that

$$b_j \geq 1 + \frac{h^2}{2} Q_* > 1, \qquad |a_j| + |c_j| = 1.$$

Taking absolute values in the above identity, we now get

$$\left(1 + \frac{h^2}{2} Q_*\right)|v_j| \leq \max_{0 \leq i \leq J+1} |v_i| + \frac{h^2}{2} \max_{1 \leq i \leq J} |L_h v_i|, \qquad 1 \leq j \leq J.$$

If $\max_{0 \leq i \leq J+1} |v_i|$ occurs for i in $1 \leq i \leq J$, then the inequality above implies, if we choose the appropriate index j, that

$$\max_{0 \leq i \leq J+1} |v_i| \leq \frac{1}{Q_*} \max_{1 \leq i \leq J} |L_h v_i|.$$

In this case clearly for any j in $1 \leq j \leq J$, with $M \equiv \max(1, 1/Q_*)$,

$$|v_j| \leq \frac{1}{Q_*} \max_{1 \leq i \leq J} |L_h v_i| \leq M \max_{1 \leq i \leq J} |L_h v_i|$$
$$\leq M\left\{\max(|v_0|, |v_{J+1}|) + \max_{1 \leq i \leq J} |L_h v_i|\right\}.$$

But if $\max_{0 \leq i \leq J+1} |v_i|$ occurs for $i = 0$ or $i = J + 1$ the result (3.1.10) follows trivially since $M \geq 1$. ∎

One consequence of stability is the fact that the difference equations (3.1.4) have a unique solution. This follows since the homogeneous system can have only the trivial solution, $u_j \equiv 0$. We have previously given a constructive proof of this result in the Corollary to Theorem 3.1.1, under the same hypothesis. A more important application of stability is in obtaining the following error estimate.

THEOREM 3.1.3. *Let $p(x)$ and $q(x)$ satisfy* (3.1.2) *and h satisfy $h \leq 2/P^*$. Then the numerical solution $\{u_j\}$ of* (3.1.4) *and solution $y(x)$ of* (3.1.1) *satisfy, with $M \equiv \max(1, 1/Q_*)$,*

$$|u_j - y(x_j)| \leq M \max_{1 \leq i \leq J} |\tau_i[y]|, \qquad 0 \leq j \leq J + 1.$$

If $y(x)$ has four continuous derivatives on $[a, b]$, then

$$|u_j - y(x_j)| \leq M \frac{h^2}{12}(M_4 + 2P^* M_3), \qquad 0 \leq j \leq J + 1,$$

where $M_\nu \equiv \max_{a \leq x \leq b} |d^\nu y(x)/dx^\nu|$; $\nu = 3, 4$.

Proof. From (3.1.4a) and (3.1.1), evaluated at $x = x_j$, we have by the linearity of L_h and the definition (3.1.8)

$$L_h[u_j - y(x_j)] = r(x_j) - L_h y(x_j) = Ly(x_j) - L_h y(x_j) = -\tau_j[y].$$

But from Equations (3.1.1b) and (3.1.4b),

$$u_0 - y(x_0) = u_{J+1} - y(x_{J+1}) = 0.$$

Now we apply (3.1.10) with $v_j \equiv u_j - y(x_j)$ and recall (3.1.9) to obtain the theorem. ∎

In Problems 3.1.2 and 3.1.3 we show how the above results can be extended to apply to Equation (3.1.1a), subject to more general boundary conditions.

3.1.1 *Difference Corrections and $h \to 0$ Extrapolation*

The difference scheme (3.1.4) has been shown to yield an approximation to the solution of the boundary-value problem (3.1.1) to within an error that is $\mathcal{O}(h^2)$. For somewhat more special equations a slightly improved difference scheme yields $\mathcal{O}(h^4)$ errors (see Problem 3.3.1). We shall briefly examine two ways in which, with additional calculations, the scheme (3.1.4) can be made to yield $\mathcal{O}(h^4)$ accuracy. These error-reduction procedures are Richardson's *deferred approach to the limit* or, as we prefer to call it, *extrapolation to zero mesh-width*, and Fox's (1957) method of *difference corrections.*

The theoretical basis for both methods is the same, namely that there exists some function $e(x)$, independent of the net spacing h, such that the error has the form

$$[y(x_j) - u_j] = h^2 e(x_j) + \mathcal{O}(h^4), \qquad 0 \le j \le J + 1. \tag{3.1.11}$$

Suppose we can compute $\{E_j\}$, an $\mathcal{O}(h^2)$ approximation to $\{e(x_j)\}$; then clearly

$$\bar{u}_j \equiv u_j + h^2 E_j$$

is an $\mathcal{O}(h^4)$ approximation to $y(x_j)$ on the net. This is essentially the difference-correction method and there may be various ways in which the E_j can be determined.

For the $h \to 0$ extrapolation we solve (3.1.4) twice, with the net spacings h and $h/2$. Let the respective solutions of these difference problems be denoted by $\{u_j(h)\}$ and $\{u_j(h/2)\}$. For any point x common to both nets, say $x = jh = 2j(h/2)$, we have from (3.1.11)

$$\tfrac{1}{3}\{4[y(x) - u_{2j}(h/2)] - [y(x) - u_j(h)]\} = \mathcal{O}(h^4).$$

Thus an $\mathcal{O}(h^4)$ approximation to $y(x)$ on the net with spacing h is given by

$$\bar{u}_j \equiv \tfrac{4}{3}u_{2j}(h/2) - \tfrac{1}{3}u_j(h), \qquad 0 \le j \le J + 1.$$

A derivation of (3.1.11) is contained in the proof of the following.

THEOREM 3.1.4. *Let $p(x)$ and $q(x)$ satisfy* (3.1.2) *and h satisfy $h < 2/P^*$. In addition let $p(x)$, $q(x)$, and $r(x)$ be so smooth*† *that $y(x)$, the solution of the boundary-value problem* (3.1.1), *has a continuous sixth derivative on $[a, b]$. Then* (3.1.11) *holds with $\{u_j\}$, the solution of* (3.1.4), *and $e(x)$, the solution of the boundary-value problem*

$$Le(x) = \theta(x), \qquad a < x < b, \tag{3.1.12a}$$

$$e(a) = e(b) = 0; \tag{3.1.12b}$$

where

$$\theta(x) \equiv -\tfrac{1}{12}[y^{(iv)}(x) - 2p(x)y^{(iii)}(x)]. \tag{3.1.13}$$

Proof. We first note, by the Corollary to Theorem 1.2.2, that the boundary-value problem (3.1.12) has a unique solution. Further, since $\theta(x)$ has a continuous second derivative this solution, $e(x)$, must have a continuous fourth derivative on $[a, b]$.

Now define the net function $\{e_j\}$ by

$$e_j \equiv h^{-2}[y(x_j) - u_j], \qquad 0 \le j \le J + 1.$$

Using (3.1.1a), (3.1.4a), and (3.1.8) we obtain

$$\begin{aligned} L_h e_j &= h^{-2}[L_h y(x_j) - L_h u_j] \\ &= h^{-2}[L_h y(x_j) - Ly(x_j)] + h^{-2}[Ly(x_j) - L_h u_j] \\ &= h^{-2}\tau_j[y], \qquad 1 \le j \le J. \end{aligned}$$

But by using the continuity of $y^{(vi)}(x)$ the evaluation (3.1.9) of the local truncation error can be done more precisely to yield

$$\tau_j[y] = h^2\theta(x_j) + \mathcal{O}(h^4).$$

Thus we have

$$L_h e_j = \theta(x_j) + \mathcal{O}(h^2), \qquad 1 \le j \le J. \tag{3.1.14a}$$

In addition, Equations (3.1.1b) and (3.1.4b) yield

$$e_0 = e_{J+1} = 0. \tag{3.1.14b}$$

† It is sufficient that $p(x)$, $q(x)$, and $r(x)$ have continuous fourth derivatives on $[a, b]$.

From Equation (3.1.12a) evaluated at $x = x_j$ and Equations (3.1.14a) we get now

$$\begin{aligned} L_h[e(x_j) - e_j] &= L_h e(x_j) - Le(x_j) - \mathcal{O}(h^2) \\ &= \tau_j[e] + \mathcal{O}(h^2) \\ &= \mathcal{O}(h^2), \qquad 1 \le j \le J, \end{aligned}$$

where we have used the smoothness of $e(x)$ in estimating $\tau[e]$. But Theorem 3.1.2 is applicable and so from the stability of L_h, that is, from (3.1.10) with v_j replaced by $e(x_j) - e_j$, we deduce that

$$|e(x_j) - e_j| \le \mathcal{O}(h^2).$$

Then multiplying by h^2 yields (3.1.11). ∎

An important by-product of this proof is the determination of various difference problems of the form (3.1.14), whose solutions can be used in the difference-correction procedure. Clearly as in the above proof any net function $\{E_j\}$ defined by

$$L_h E_j = \Theta_j, \qquad 1 \le j \le J, \tag{3.1.15a}$$

$$E_0 = E_{J+1} = 0, \tag{3.1.15b}$$

satisfies, as shown in Problem 3.1.7,

$$|e(x_j) - E_j| \le M|\theta(x_j) - \Theta_j| + \mathcal{O}(h^2).$$

Thus we need only determine Θ_j such that

$$\theta(x_j) - \Theta_j = \mathcal{O}(h^2) \tag{3.1.16}$$

and then the difference-corrected solution

$$\bar{u}_j \equiv u_j + h^2 E_j$$

is an $\mathcal{O}(h^4)$ approximation to $y(x_j)$.

One way in which Θ_j to satisfy (3.1.16) can be obtained is to evaluate $\theta(x)$ in terms of $y'(x)$ and $y(x)$. That is, using the fact that

$$y''(x) = p(x)y' + q(x)y - r(x),$$

and differentiating twice, we obtain a relation of the form

$$\theta(x) = A(x)y'(x) + B(x)y(x) + C(x), \tag{3.1.17a}$$

where $A(x)$, $B(x)$, and $C(x)$ are expressions given in terms of $p(x)$, $q(x)$, $r(x)$ and their first two derivatives [see Problem 3.1.4]. Now we claim that

$$\Theta_j \equiv A(x_j)\frac{(u_{j+1} - u_{j-1})}{2h} + B(x_j)u_j + C(x_j), \qquad j = 1, 2, \ldots, J \tag{3.1.17b}$$

satisfies (3.1.16). Clearly from (3.1.11), since $e(x)$ is sufficiently smooth,

$$\frac{u_{j+1} - u_{j-1}}{2h} = y'(x_j) + h^2 e'(x_j) + \mathcal{O}(h^2),$$

and the result easily follows.

A more direct way to define Θ_j so as to satisfy (3.1.16) is to form some obvious finite-difference analog of $\theta(x)$ defined in (3.1.13), using the numerical solution u_j in place of the actual solution $y(x_j)$. This is in fact the procedure employed by Fox (1957) in originating the difference-correction method. From Equations (3.1.15a) we note that the Θ_j are only required at interior net points. But since third- and fourth-order derivatives of $y(x)$ occur in $\theta(x)$ it is clear that unsymmetric differences, or some other device, must be used near the endpoints of the interval. One trick devised by Fox for this purpose is simply to compute the numerical solution u_j at points exterior to $[a, b]$ and then to use centered differences.

The above procedures can frequently be employed to obtain approximate solutions of higher-order accuracy than the $\mathcal{O}(h^4)$ indicated here. Of course even more differentiability is required of the solution and the error expansion (3.1.11) is replaced by, say,

$$[y(x_j) - u_j] = h^2 e(h, x_j) + \mathcal{O}(h^{2m}),$$

where $m \geq 2$. If $e(h, x_j)$ can be expressed as a polynomial in h^2 of degree at most $m - 1$, then the problem is reduced to obtaining sufficiently accurate approximations to the "coefficient" functions. We pose the details as an exercise. This extension of the difference correction or $h \to 0$ extrapolation seems to be the only theoretically justified procedure for obtaining arbitrary (high) order accuracy by finite differences (at present).

Problems

3.1.1 Find coefficients A_j, B_j, C_j so that

$$L_h u_j \equiv A_j u_{j-1} + B_j u_j + C_j u_{j+1}$$

approximates

$$Ly(x) \equiv -y''(x) + q(x)y(x),$$

to fourth order for all solutions $v(x)$ of $Lv = 0$ which have six continuous derivatives.

[HINT: expand $L_h v(x_j)$ about x_j by Taylor's theorem to sixth order in h. But since $Lv = 0$ we have

$$\tau[v] = L_h v - Lv \equiv L_h v - Lv - (Lv)' - (Lv)''.]$$

3.1.2 In place of (3.1.1b) let us impose the boundary conditions

$$\begin{array}{llll} a_0 y(a) - a_1 y'(a) = \alpha, & a_0 \geq 0, & a_1 \geq 0, & \\ b_0 y(b) + b_1 y'(b) = \beta, & b_0 \geq 0, & b_1 \geq 0, & \end{array} \quad a_0 + b_0 \neq 0.$$

Now let the difference equations be

$$L_h u_j = r(x_j), \qquad 0 \leq j \leq J + 1,$$

$$a_0 u_0 - a_1 \frac{u_1 - u_{-1}}{2h} = \alpha, \qquad b_0 u_{J+1} + b_1 \frac{u_{J+2} - u_J}{2h} = \beta,$$

where L_h is as defined in (3.1.4a). Show that for $h < 2/P^*$ these equations have a unique solution and derive explicit formulas for computing this solution based on Theorem 3.1.1 (that is, extend the Corollary to Theorem 3.1.1).

3.1.3 Use the results of the above problem to estimate the error in the numerical scheme described there.

3.1.4 Determine the functions $A(x)$, $B(x)$, and $C(x)$ to be used in (3.1.17a) by differentiating the equation $y'' = p(x)y' + q(x)y - r(x)$ twice and eliminating y'' and y'''. Note that $\theta(x)$ is defined in (3.1.13).

3.1.5 To solve the boundary-value problem

$$Ly \equiv y'' - q(x)y = r(x), \qquad y(a) = y(b) = 0,$$

use a difference scheme of the form: $u_0 = u_{J+1} = 0$;

$$\begin{aligned} L_h u &\equiv \frac{[u_{j+1} - 2u_j + u_{j-1}]}{h^2} - [\alpha_1 q_{j+1} u_{j+1} + \alpha_0 q_j u_j + \alpha_{-1} q_{j-1} u_{j-1}] \\ &= [\alpha_1 r_{j+1} + \alpha_0 r_j + \alpha_{-1} r_{j-1}], \qquad j = 1, 2, \ldots, J; \end{aligned}$$

where $h = (b - a)/(J + 1)$ and $q_{j+1} = q(x_{j+1})$ etc.

(a) Determine α_1, α_0, α_{-1} such that the local truncation error is $\mathcal{O}(h^4)$, for the solution $y(x)$, assuming y^{iv}, q^{iv}, and r^{iv} continuous. Note that $y^{\text{iv}} - [qy]'' = r''$.

(b) If $q(x) \geq Q_* > 0$, show that, for sufficiently small h,

$$|u_j - y(x_j)| \leq \frac{h^4}{720} \frac{2M_6 + 5N_4 + 5R_4}{Q_*}, \qquad 0 \leq j \leq J + 1,$$

where $M_6 \geq |y^{\text{vi}}|$, $N_4 \geq |(qy)^{\text{iv}}|$, $R_4 \geq |r^{\text{iv}}|$ on $[a, b]$.

3.1.6 Consider the boundary-value problem

$$(a(x)y')' - p(x)y' - q(x)y = r(x); \qquad y(a) = y(b) = 0$$

and the corresponding difference problem

$$\left\{a\left(x_j + \frac{h}{2}\right)\left[\frac{u_{j+1} - u_j}{h^2}\right] - a\left(x_j - \frac{h}{2}\right)\left[\frac{u_j - u_{j-1}}{h^2}\right]\right\}$$

$$- p(x_j)\left[\frac{u_{j+1} - u_{j-1}}{2h}\right] - q(x_j)u_j = r(x_j),$$

$$j = 1, 2, \ldots, J, \qquad u_0 = u_{J+1} = 0.$$

(a) If y^{iv} and a''' are continuous, show that the truncation error in this scheme is $\mathcal{O}(h^2)$.

(b) If $q(x) \geq Q_*$ and $A^* \geq a(x) \geq A_* > 0$, show that

$$|u_j - y(x_j)| \leq \frac{A^*}{A_* Q_*} \|\tau\|,$$

provided that $A_* - (h/2)p(x_j) \geq 0$ for $j = 1, 2, \ldots, J$. [HINT: Proceed as in the proof of Theorem 3.1.1 to demonstrate stability of the difference scheme but now divide by $|a_j| + |c_j| = a_j + c_j \geq 2A_*$ before bounding the coefficients.]

3.1.7 From (3.1.12), (3.1.13), and (3.1.15) deduce that

$$L_h(e(x_j) - E_j) = [\theta(x_j) - \Theta_j] + \mathcal{O}(h^2), \qquad 1 \leq j \leq J;$$

$$e(a) - E_0 = 0, \qquad e(b) - E_{J+1} = 0.$$

Then apply Theorem 3.1.2 to get

$$|e(x_j) - E_j| \leq M|\theta(x_j) - \Theta_j| + \mathcal{O}(h^2).$$

3.2 Nonlinear Second-Order Equations

We now apply finite-difference methods to boundary-value problems of the form

$$\mathcal{L}y(x) \equiv -y'' + f(x, y, y') = 0, \qquad a < x < b, \tag{3.2.1a}$$

$$y(a) = \alpha, \qquad y(b) = \beta. \tag{3.2.1b}$$

In order that Theorem 1.2.2 be valid we assume that $f(x, y, z)$ has continuous derivatives which satisfy

$$\left|\frac{\partial f}{\partial z}\right| \leq P^*, \qquad 0 < Q_* \leq \frac{\partial f}{\partial y} \leq Q^*, \tag{3.2.2}$$

for some positive constants P^*, Q^*, and Q_*.

The uniform net (3.1.3) will be used and on this net an obvious difference approximation to the boundary-value problem (3.2.1) is given by

$$\mathscr{L}_h u_j \equiv -\left(\frac{u_{j+1} - 2u_j + u_{j-1}}{h^2}\right) + f\left(x_j, u_j, \frac{u_{j+1} - u_{j-1}}{2h}\right) = 0, \qquad 1 \le j \le J, \tag{3.2.3a}$$

$$u_0 = \alpha, \qquad u_{J+1} = \beta. \tag{3.2.3b}$$

Again centered-difference quotients have replaced derivatives at interior net points. The resulting difference equations (3.2.3) are in general nonlinear and we shall employ iterative methods to solve them. In fact one of these techniques also yields a proof of the existence and uniqueness of a solution of the difference problem. But first we consider the question of convergence and error estimates for the numerical solution.

The local truncation error, $\tau_j[v]$, is defined and evaluated essentially as in (3.1.9) to yield

$$\begin{aligned}\tau_j[v] &\equiv \mathscr{L}_h v(x_j) - \mathscr{L} v(x_j)\\ &= -\frac{h^2}{12}\left[v''''(\xi_j) - 2\,\frac{\partial f(x_j, v(x_j), v'(\eta_j))}{\partial z}\, v'''(\zeta_j)\right], \qquad 1 \le j \le J.\end{aligned} \tag{3.2.4}$$

Here $v(x)$ is any function with a continuous fourth derivative on $[a, b]$ and ξ_j, η_j, and ζ_j are points in $[x_{j-1}, x_{j+1}]$. The definition and proof of stability for $\mathscr{L}_h$ are contained in the following.

THEOREM 3.2.1. *Let $\mathscr{L}_h$ be defined by* (3.2.3a) *where $f(x, y, z)$ has continuous derivatives satisfying* (3.2.2). *Then for all net spacing h such that*

$$h \le \frac{2}{P^*},$$

$\mathscr{L}_h$ is stable in the sense that for all net functions $\{v_j\}$ and $\{w_j\}$

$$|v_j - w_j| \le M\left\{\max\left(|v_0 - w_0|, |v_{J+1} - w_{J+1}|\right) + \max_{1 \le i \le J} |\mathscr{L}_h v_i - \mathscr{L}_h w_i|\right\}, \qquad 0 \le j \le J + 1,$$

where $M = \max(1, 1/Q_*)$.

Proof. By the mean-value theorem we easily obtain

$$\frac{h^2}{2}[\mathscr{L}_h v_i - \mathscr{L}_h w_i] = a_i[v_{i-1} - w_{i-1}] + b_i[v_i - w_i] + c_i[v_{i+1} - w_{i+1}], \tag{3.2.5a}$$

where

$$a_i \equiv -\frac{1}{2}\left\{1 + \frac{h^2}{2}\,\partial f\left(x_i, w_i + \theta_i[v_i - w_i], \frac{w_{i+1} - w_{i-1}}{2h}\right.\right.$$
$$\left.\left. + \theta_i\frac{[v_{i+1} - w_{i+1}] - [v_{i-1} - w_{i-1}]}{2h}\right)\Big/\partial z\right\},$$

$$b_i \equiv \left\{1 + \frac{h^2}{2}\,\partial f\left(x_i, w_i + \theta_i[v_i - w_i], \frac{w_{i+1} - w_{i-1}}{2h}\right.\right.$$
$$\left.\left. + \theta_i\frac{[v_{i+1} - w_{i+1}] - [v_{i-1} - w_{i-1}]}{2h}\right)\Big/\partial y\right\}, \qquad (3.2.5b)$$

$$c_i \equiv -1 - a_i, \qquad 0 < \theta_i < 1, \qquad 1 \le j \le J.$$

But from (3.2.2) and $h \le 2/P^*$ we get

$$|a_i| + |c_i| = 1, \qquad b_i \ge 1 + h^2Q_*/2$$

and the proof now follows exactly as that of Theorem 3.1.2. ∎

The error estimate can now be stated as the following.

COROLLARY. *Let $f(x, y, z)$ have continuous derivatives which satisfy* (3.2.2). *Then the numerical solution $\{u_j\}$ of the difference equations* (3.2.3) *and the solution $y(x)$ of the boundary-value problem* (3.2.1) *satisfy, with $M \equiv \max(1, 1/Q_*)$,*

$$|u_j - y(x_j)| \le M \max_{1\le i\le J} |\tau_i[y]|, \qquad 0 \le j \le J + 1.$$

If $y(x)$ has a continuous fourth derivative on $[a, b]$, then

$$|u_j - y(x_j)| \le M\frac{h^2}{12}(M_4 + 2P^*M_3), \qquad 0 \le j \le J + 1.$$

Proof. The proof is essentially the same as that of Theorem 3.1.3 with the evaluation (3.2.4) for $\tau_j[y]$. ∎

Whenever $f(x, y, z)$ is not linear in y and z the difference Equations (3.2.3) form a nonlinear system. To solve such a system or indeed to demonstrate existence of a solution we could apply the general procedures indicated in Section 1.4. We shall in fact employ a functional iteration scheme for which the Contracting Mapping theorem is valid.

To devise our iteration scheme we multiply (3.2.3a) by $h^2/2$, add ωu_j to each side and, assuming $\omega \ne -1$, the result can be written as

$$u_j = (1 + \omega)^{-1}\left[\tfrac{1}{2}(u_{j+1} + u_{j-1}) + \omega u_j - \frac{h^2}{2}f\left(x_j, u_j, \frac{u_{j+1} - u_{j-1}}{2h}\right)\right],$$
$$1 \le j \le J, \quad (3.2.6a)$$

$$u_0 = \alpha, \qquad u_{J+1} = \beta. \qquad (3.2.6b)$$

Clearly this system is equivalent to that in (3.2.3) for any finite $\omega \neq -1$. It is in the canonical form

$$\mathbf{u} = \mathbf{g}(\mathbf{u}), \qquad \mathbf{u} \equiv (u_0, u_1, \ldots, u_{J+1})^T, \tag{3.2.6c}$$

where the components $g_j(\mathbf{u})$ are defined by the right-hand sides of Equations (3.2.6a, b). The functional iteration implied by (3.2.6) is

$$u_j^{(0)} = \text{arbitrary}, \qquad 1 \leq j \leq J; \tag{3.2.7a}$$

$$u_j^{(\nu+1)} = (1+\omega)^{-1}\left[\frac{1}{2}(u_{j+1}^{(\nu)} + u_{j-1}^{(\nu)}) + \omega u_j^{(\nu)} - \frac{h^2}{2} f\left(x_j, u_j^{(\nu)}, \frac{u_{j+1}^{(\nu)} - u_{j-1}^{(\nu)}}{2h}\right)\right],$$

$$1 \leq j \leq J, \qquad \nu = 0, 1, \ldots; \tag{3.2.7b}$$

$$u_0^{(\nu)} = \alpha, \qquad u_{J+1}^{(\nu)} = \beta, \qquad \nu = 0, 1, \ldots. \tag{3.2.7c}$$

This is an explicit iteration scheme [in contrast to implicit schemes where linear systems must be solved; see Newton's method (3.2.10) below]. The fact that the iterates $\{u_j^{(\nu)}\}$ converge to a solution of (3.2.3) for an appropriate choice of ω is but part of the content of the following.

THEOREM 3.2.2. *Let $f(x, y, z)$ have continuous derivatives which satisfy* (3.2.2). *Then the system of difference equations* (3.2.3) *have a unique solution for each h such that*

$$h \leq \frac{2}{P^*}.$$

This solution is the limit of the iterates $\{u_j^{(\nu)}\}$ as $\nu \to \infty$ defined by (3.2.7) *with any finite ω satisfying*

$$\omega \geq \frac{h^2}{2} Q^*.$$

The convergence factor for this scheme is at most

$$\lambda(\omega) \equiv 1 - \frac{(h^2/2)Q_*}{1+\omega}.$$

Proof. Let $\mathbf{g}(\mathbf{u})$ be as defined in (3.2.6); then by the definition of $\mathscr{L}_h$ we can write

$$g_j(\mathbf{u}) \equiv \begin{cases} \alpha, & j = 0, \\ u_j - (1+\omega)^{-1}\dfrac{h^2}{2}\mathscr{L}_h u_j, & 1 \leq j \leq J, \\ \beta, & j = J+1. \end{cases}$$

Now Taylor's theorem as used in (3.2.5) implies, for any pair of net functions $\{v_j\}$ and $\{w_j\}$, that

$$g_j(\mathbf{v}) - g_j(\mathbf{w}) = \begin{cases} 0, & j = 0; \\ (1+\omega)^{-1}[-a_j(v_{j-1} - w_{j-1}) + (1 + \omega - b_j) \\ \qquad \times (v_j - w_j) - c_j(v_{j+1} - w_{j+1})], & 1 \le j \le J; \\ 0, & j = J + 1. \end{cases}$$

The coefficients here are defined in (3.2.5b). Thus with $h \le 2/P^*$ and $\omega \ge h^2Q^*/2$ we have $-a_j \ge 0$, $(1 + \omega - b_j) \ge 0$ and $-c_j \ge 0$ so that upon taking absolute values,

$$|g_j(\mathbf{v}) - g_j(\mathbf{w})| \le \frac{1 + \omega - a_j - b_j - c_j}{1 + \omega} \max_i |v_i - w_i| \le \lambda(\omega) \max_i |v_i - w_i|,$$
$$0 \le j \le J + 1.$$

That is, in the maximum norm, $\mathbf{g}(\mathbf{u})$ is Lipschitz-continuous,

$$|\mathbf{g}(\mathbf{v}) - \mathbf{g}(\mathbf{w})| \le \lambda(\omega)|\mathbf{v} - \mathbf{w}|,$$

with constant $\lambda(\omega) < 1$ for $\omega \ge h^2Q^*/2$. The theorem now follows from Theorem 1.4.1 (with $\rho = \infty$) and the fact that (3.2.6) and (3.2.3) have identical solutions for finite positive ω. ▮

It is important to observe that the scheme (3.2.7) converges for *any initial estimate* $u_j^{(0)}$, $1 \le j \le J$, provided $\omega \ge h^2Q^*/2$. As far as our analysis shows the best value for ω is the limiting value

$$\omega = \omega^* \equiv \frac{h^2}{2} Q^*.$$

We easily see that $\lambda(\omega) > \lambda(\omega^*)$ if $\omega > \omega^*$. Of course, even with this choice the convergence may be quite slow since

$$\lambda(\omega) = 1 - \mathcal{O}(h^2)$$

can be very close to unity. Thus we may wish to consider alternative iteration schemes for solving (3.2.3) and of these Newton's method can be shown to be applicable.

To define the Newton iterations we first write Equations (3.2.3a) in the form

$$\boldsymbol{\phi}(\mathbf{u}) = 0, \tag{3.2.8a}$$

where

$$\mathbf{u} \equiv \begin{pmatrix} u_1 \\ u_2 \\ \vdots \\ u_J \end{pmatrix}, \qquad \boldsymbol{\phi}(\mathbf{u}) \equiv \begin{pmatrix} \phi_1(\mathbf{u}) \\ \phi_2(\mathbf{u}) \\ \vdots \\ \phi_J(\mathbf{u}) \end{pmatrix}, \qquad \phi_j(\mathbf{u}) \equiv \frac{h^2}{2} \mathscr{L}_h u_j,$$

$$1 \le j \le J. \quad (3.2.8b)$$

Now the Jacobian of $\boldsymbol{\phi}(\mathbf{u})$ is easily found to be the Jth-order tridiagonal matrix

$$A(\mathbf{u}) \equiv \frac{\partial \boldsymbol{\phi}(\mathbf{u})}{\partial \mathbf{u}} = \begin{pmatrix} B_1(\mathbf{u}) & C_1(\mathbf{u}) & 0 & \cdots & & 0 \\ A_2(\mathbf{u}) & B_2(\mathbf{u}) & C_2(\mathbf{u}) & \cdots & & \\ \vdots & & & & & \vdots \\ & & A_{J-1}(\mathbf{u}) & B_{J-1}(\mathbf{u}) & C_{J-1}(\mathbf{u}) & \\ 0 & \cdots & 0 & A_J(\mathbf{u}) & B_J(\mathbf{u}) & \end{pmatrix}, \quad (3.2.9a)$$

where

$$A_j(\mathbf{u}) \equiv -\frac{1}{2}\left[1 + \frac{h}{2}\frac{\partial f}{\partial z}\left(x_j, u_j, \frac{u_{j+1} - u_{j-1}}{2h}\right)\right], \qquad 2 \le j \le J;$$

$$B_j(\mathbf{u}) \equiv \left[1 + \frac{h^2}{2}\frac{\partial f}{\partial y}\left(x_j, u_j, \frac{u_{j+1} - u_{j-1}}{2h}\right)\right], \qquad 1 \le j \le J; \qquad (3.2.9b)$$

$$C_j(\mathbf{u}) \equiv -\frac{1}{2}\left[1 - \frac{h}{2}\frac{\partial f}{\partial z}\left(x_j, u_j, \frac{u_{j+1} - u_{j-1}}{2h}\right)\right], \qquad 1 \le j \le J - 1.$$

In computing $\phi_1(\mathbf{u})$, $\phi_J(\mathbf{u})$, $A_J(\mathbf{u})$, $B_1(\mathbf{u})$, $B_J(\mathbf{u})$, and $C_1(\mathbf{u})$ we use $u_0 \equiv \alpha$ and $u_{J+1} \equiv \beta$ so that (3.2.3b) is always satisfied. Now, with any initial estimate $\mathbf{u}^{(0)}$ of the quantities u_j, $1 \le j \le J$, we define

$$\mathbf{u}^{(\nu+1)} = \mathbf{u}^{(\nu)} + \Delta\mathbf{u}^{(\nu)}, \qquad \nu = 0, 1, 2, \ldots, \qquad (3.2.10a)$$

where $\Delta\mathbf{u}^{(\nu)}$ is the solution of

$$A(\mathbf{u}^{(\nu)})\,\Delta\mathbf{u}^{(\nu)} = -\boldsymbol{\phi}(\mathbf{u}^{(\nu)}), \qquad \nu = 0, 1, 2, \ldots. \qquad (3.2.10b)$$

If the function $f(x, y, z)$ satisfies (3.2.2) and $h \le 2/P^*$, it follows as in the proof of the Corollary to Theorem 3.1.1 that the linear system (3.2.10b) has a unique solution, $\Delta\mathbf{u}^{(\nu)}$, for any $\mathbf{u}^{(\nu)}$. Further, this solution is easily computed by means of the algorithm described in (3.1.7) for solving (3.1.6). Thus Newton's method can be applied in attempting to solve (3.2.3) or its equivalent (3.2.8). In order for this method to converge it is sufficient (see Theorem 1.4.3) that the initial iterate $\mathbf{u}^{(0)}$ be "close" to the solution. But by Theorem

3.2.2 we can obtain close estimates to the solution by applying the iteration scheme (3.2.7). A reasonable procedure is thus to determine the initial Newton iterate by first computing several iterates of the explicit scheme.

The treatment of more general linear boundary conditions than those in (3.2.1b) is easily included in appropriate modifications of the difference scheme (3.2.3); see Problems 3.2.1, 3.2.2. The difference-correction and extrapolation-to-zero-mesh-width procedures can frequently be applied to the nonlinear problems treated here. The basic result required is an extension of Theorem 3.1.4 contained in Problem 3.2.3.

Difference schemes and corresponding iterations are frequently suggested by analytical attempts to solve or approximate the solution of nonlinear boundary-value problems. When there is some evidence that the analytical procedures are effective there may be advantages in such approaches. However, the replacement of derivatives by difference quotients finally yields a difference scheme which must be examined independently.

To illustrate, we consider a common analytical procedure, which is to linearize about some approximate solution and then solve the linearized problem to obtain a correction. Of course this procedure can be continued to yield an iteration scheme and we formulate it that way. Thus let $y^{(\nu)}(x)$ be some approximation to the solution of (3.2.1) with say $y^{(\nu)}(a) = \alpha$ and $y^{(\nu)}(b) = \beta$. Then if $e^{(\nu)}(x) = y(x) - y^{(\nu)}(x)$ is the error, we have

$$-[y^{(\nu)} + e^{(\nu)}]'' + f(x, [y^{(\nu)} + e^{(\nu)}], [y^{(\nu)} + e^{(\nu)}]') = 0.$$

Assuming $e^{(\nu)}(x)$ to be small and $f(x, y, z)$ sufficiently smooth, we have

$$-[y^{(\nu)} + e^{(\nu)}]'' + \frac{\partial f}{\partial z}(x, y^{(\nu)}, y^{(\nu)'})e^{(\nu)'} + \frac{\partial f}{\partial y}(x, y^{(\nu)}, y^{(\nu)'})e^{(\nu)} + f(x, y^{(\nu)}, y^{(\nu)'}) = \mathcal{O}((e^{(\nu)})^2 + (e^{(\nu)'})^2).$$

Neglecting the second-order terms on the right-hand side above, we get the linearized problem for $e^{(\nu)}(x)$

$$-e^{(\nu)''} + \frac{\partial f}{\partial z}(x, y^{(\nu)}, y^{(\nu)'})e^{(\nu)'} + \frac{\partial f}{\partial y}(x, y^{(\nu)}, y^{(\nu)'})e^{(\nu)} = y^{(\nu)''} - f(x, y^{(\nu)}, y^{(\nu)'}), \tag{3.2.11a}$$

$$e^{(\nu)}(a) = e^{(\nu)}(b) = 0. \tag{3.2.11b}$$

Of course this procedure is just a form of Newton's method (applied to a nonlinear operator equation in a function space rather than in a finite dimensional space). A study of the convergence of $y^{(\nu)}(x) + e^{(\nu)}(x)$ as $\nu \to \infty$ is beyond our present subject [see, for example, Kantorovich and Akilov (1964), pp. 735–739]. However, let us assume that it converges in the anticipated second-order fashion [that is, $e^{(\nu+1)} = \mathcal{O}(e^{(\nu)2})$].

To carry out the above scheme numerically, we need only solve the sequence of linear problems (3.2.11). If we employ the difference technique of Section 3.1, with $y^{(\nu)\prime}(x_j)$ replaced by centered difference quotients in the coefficients f, $\partial f/\partial y$, and $\partial f/\partial z$, then *we obtain exactly the system* (3.2.10) with $\Delta u_j^{(\nu)}$ approximating $e^{(\nu)}(x_j)$ and $u_j^{(\nu)}$ approximating $y^{(\nu)}(x_j)$. Thus our accuracy can be no better than that of the numerical solution $\{u_j\}$ determined by the difference scheme (3.2.3). Of course we could use initial-value methods as in section 2.1 for solving (3.2.11). In this case greater accuracy in approximating $e^{(\nu)}(x_j)$ is easily obtained.

Another analytical scheme to approximate the solution of (3.2.1), based on solving linear boundary-value problems, follows from writing (3.2.1a) as

$$y'' - \omega y = f(x, y, y') - \omega y.$$

Then with arbitrary $y^{(0)}(x)$ satisfying $y^{(0)}(a) = \alpha$, $y^{(0)}(b) = \beta$ we define $\{y^{(\nu)}(x)\}$ by

$$y^{(\nu+1)''} - \omega y^{(\nu+1)} = f(x, y^{(\nu)}, y^{(\nu)\prime}) - \omega y^{(\nu)}, \tag{3.2.12a}$$

$$y^{(\nu+1)}(a) = \alpha, \qquad y^{(\nu+1)}(b) = \beta. \tag{3.2.12b}$$

For a range of values of ω it can be shown (see Section 4.1) that $y^{(\nu)}(x) \to y(x)$ if $f(x, y, z)$ satisfies appropriate conditions. If the difference method of Section 3.1 is used to approximate the solution of (3.2.12), we obtain an implicit iteration scheme for solving the difference equations (3.2.3). The resulting procedure is related to but not quite the same as that in (3.2.7).

Problems

3.2.1 Replace (3.2.1b) by

$$a_0 y(a) - a_1 y'(a) = \alpha, \qquad b_0 y(b) + b_1 y'(b) = \beta,$$
$$a_0 > 0, \quad a_1 \geq 0, \quad b_0 \geq 0, \quad b_1 \geq 0.$$

Apply (3.2.3a) for $j = 0$ and $j = J + 1$ and approximate the above by

$$a_0 u_0 - a_1 \frac{u_1 - u_{-1}}{2h} = \alpha, \qquad b_0 u_{J+1} + b_1 \frac{u_{J+2} - u_J}{2h} = \beta.$$

Prove convergence of $\{u_j\} \to \{y(x_j)\}$ if $y(x)$ has four continuous derivatives and (3.2.1a) holds on $a \leq x \leq b$.

3.2.2 Modify (3.2.7) to apply to the difference scheme of Problem 3.2.1 and prove the analog of Theorem 3.2.2.

3.2.3 State and prove an extension of Theorem 3.1.4 which applies to the solution $y(x)$ of (3.2.1) and numerical solution $\{u_j\}$ of (3.2.3).

3.2.4 Consider, in place of (3.2.3), the difference equations

$$u_0 = \alpha, \qquad u_{J+1} = \beta,$$

$$\frac{u_{j+1} - 2u_j + u_{j-1}}{h^2} = f\left(x_j, \frac{u_{j+1} + u_{j-1}}{2}, \frac{u_{j+1} - u_{j-1}}{2h}\right),$$

$$j = 1, 2, \ldots, J.$$

(a) Show that $|u_j - y(x_j)| = \mathcal{O}(h^2)$, where $y(x)$ is the four times continuously differentiable solution of (3.2.1), (3.2.2) holds, $h \leq 2/P^*$, and $y''''(x)$, $\partial f/\partial y$, and $\partial f/\partial y'$ are continuous.

(b) Under the above assumptions, prove convergence of the iterations: $u_0^{(\nu+1)} = u_{J+1}^{(\nu+1)} = 0$,

$$u_j^{(\nu+1)} = \frac{1}{2}\left[u_{j+1}^{(\nu)} + u_{j-1}^{(\nu)}\right] - \frac{h^2}{2} f\left(x_j, \frac{u_{j+1}^{(\nu)} + u_{j-1}^{(\nu)}}{2}, \frac{u_{j+1}^{(\nu)} - u_{j-1}^{(\nu)}}{2h}\right),$$

$$j = 1, 2, \ldots, J.$$

Does this converge for all net spacing?

3.2.5 Show in detail that the centered difference approximations to the system of linear differential equations (3.2.11) yield a linear algebraic system identical to that in (3.2.10).

3.3 *Linear and Nonlinear Systems*

The difference methods discussed in Sections 3.1 and 3.2 are quite special as they were devised for second-order (scalar) equations. It would not be difficult to write down obvious generalizations that might be applicable to systems of coupled second-order equations. But as even this is rather limited, we shall consider here, as in Section 2.3, the general systems of n first-order equations subject to linear two-point boundary conditions

$$\text{(a)} \quad L\mathbf{y} \equiv \mathbf{y}' - \mathbf{f}(x, \mathbf{y}) = 0; \qquad \text{(b)} \quad A\mathbf{y}(a) + B\mathbf{y}(b) = \boldsymbol{\alpha}. \tag{3.3.1}$$

As previously indicated, a very general class of problems can be reduced to this form, including, of course, various systems of linear differential equations and periodic boundary conditions.

In the present discussion we take the net points on $[a, b]$ as

$$x_j = a + jh, \qquad j = 0, 1, \ldots, J; \qquad h = \frac{b-a}{J}. \tag{3.3.2}$$

The use of nonuniform spacing causes no difficulties and we employ the uniform net (3.3.2) merely for notational convenience. We denote by the

n-dimensional vectors $\mathbf{u}_j$ approximations to the corresponding values of the solution $\mathbf{y}(x_j)$ of (3.3.1) at the points of our net. One obvious system of difference equations for the determination of these approximations is

$$L_h\mathbf{u}_j \equiv \frac{\mathbf{u}_j - \mathbf{u}_{j-1}}{h} - \mathbf{f}\left(x_{j-1/2}, \frac{\mathbf{u}_j + \mathbf{u}_{j-1}}{2}\right) = 0, \qquad j = 1, 2, \ldots, J; \tag{3.3.3a}$$

and the boundary conditions become

$$A\mathbf{u}_0 + B\mathbf{u}_J - \boldsymbol{\alpha} = 0. \tag{3.3.3b}$$

The scheme in (3.3.3a) is known as the centered-difference method when used for the equation (3.3.1a) subject to initial conditions. The nonlinear term in (3.3.3a) might have been chosen as

$$\tfrac{1}{2}[\mathbf{f}(x_j, \mathbf{u}_j) + \mathbf{f}(x_{j-1}, \mathbf{u}_{j-1})],$$

and the resulting scheme is called the modified Euler method. We find our choice more convenient, if not necessarily more accurate. Some discussion of the above choice is to be found in Problem 3.3.3.

The Equations (3.3.3), of $J + 1$ sets of n equations each, are the difference equations whose solution is to approximate that of (3.3.1) on the net. We shall examine the problem of solving these difference equations and shall also estimate their accuracy in approximating $\mathbf{y}(x_j)$. First, we rewrite them in a more uniform notation. Let the $n(J + 1)$-dimensional vector $\mathbf{U}$ be defined by

$$\mathbf{U} \equiv \begin{pmatrix} \mathbf{u}_0 \\ \mathbf{u}_1 \\ \vdots \\ \mathbf{u}_J \end{pmatrix}.$$

Then Equations (3.3.3) can be written as the system of $n(J + 1)$ equations

$$\boldsymbol{\Phi}(\mathbf{U}) \equiv \begin{pmatrix} A\mathbf{u}_0 + B\mathbf{u}_J - \boldsymbol{\alpha} \\ hL_h\mathbf{u}_1 \\ \vdots \\ hL_h\mathbf{u}_J \end{pmatrix} = \mathbf{0}. \tag{3.3.4}$$

We now see one basic difference, at least in point of view, between the initial-value methods and the finite-difference methods. In initial-value methods some unknowns, the initial values, are somehow determined first; then, in a definite order the other unknowns are determined recursively so as to be accurate approximations to solutions of the differential equations, and only when the last variables are computed are the boundary conditions employed. In finite-difference schemes no particular variables are preferred and the differential equations and boundary conditions are presumably treated simultaneously. [In some iterative attempts at solving the system (3.3.4) one might proceed recursively guessing at $\mathbf{u}_0$, say, then solving the equations in (3.3.3a), exactly or approximately, in the order $j = 1, 2, \ldots, J$ and finally checking (3.3.3b) to change the value of $\mathbf{u}_0$. But this procedure is then in fact an initial-value method; it is in general a rather poor one, since there are much better schemes than the centered-difference method for integrating the system (3.3.1a) over the net (3.3.2).] From this point of view the parallel shooting schemes of Section 2.4 are combinations of initial-value and finite-difference methods. The integration of the initial-value problem over each interval of length Δ_j corresponds to the use of (3.3.3a) over the corresponding interval. Indeed we shall see great similarities in some of our present analysis to that in Section 2.4.

It is instructive to consider systems of linear differential equations first, and so we take

$$\mathbf{f}(x, \mathbf{y}) \equiv K(x)\mathbf{y} + \mathbf{f}_0(x), \tag{3.3.5}$$

where $K(x)$ is an nth-order matrix with continuous elements on $[a, b]$. Using this form we find that the system of difference equations (3.3.4) reduces to the linear form

$$\mathbf{\Phi}(\mathbf{U}) \equiv \mathscr{L}_h\mathbf{U} - \boldsymbol{\gamma} = 0. \tag{3.3.6a}$$

Here the $n(J + 1)$th-order matrix $\mathscr{L}_h$ and vector $\boldsymbol{\gamma}$ are

$$\mathscr{L}_h \equiv \begin{pmatrix} A & 0 & 0 & \cdots & 0 & B \\ -L_1 & R_1 & 0 & & \cdots & 0 \\ 0 & -L_2 & R_2 & 0 & \cdots & 0 \\ \vdots & & & & & \vdots \\ 0 & 0 & 0 & \cdots & -L_J & R_J \end{pmatrix},$$

$$\boldsymbol{\gamma} \equiv \begin{pmatrix} \boldsymbol{\alpha} \\ h\mathbf{f}_0(x_{1/2}) \\ h\mathbf{f}_0(x_{3/2}) \\ \vdots \\ h\mathbf{f}_0(x_{J-1/2}) \end{pmatrix}, \tag{3.3.6b}$$

where we have used the notation for nth-order matrices

$$R_j \equiv I - \frac{h}{2} K(x_{j-1/2}), \qquad L_j \equiv I + \frac{h}{2} K(x_{j-1/2}), \qquad j = 1, 2, \ldots, J. \tag{3.3.6c}$$

Under somewhat general conditions we can show that the matrix $\mathscr{L}_h$ is nonsingular. We have in fact the following.

THEOREM 3.3.1. *Let* $\mathbf{f}(x, \mathbf{y})$ *be given by Equation* (3.3.5), *where* $K(x)$ *is continuous on* $[a, b]$ *and satisfies*

$$\max_{a \le x \le b} \|K(x)\|_\infty \le K. \tag{3.3.7a}$$

Let the matrices A and B satisfy

$$(A + B) \text{ nonsingular}. \tag{3.3.7b}$$

Finally let the interval $[a, b]$ *be so small that*

$$|b - a| < \frac{\ln [1 + (1/m)]}{K^*}; \quad \begin{cases} K^* \equiv K/(1 - h_0 K/_2) \\ m \equiv \|(A + B)^{-1} B\|_\infty . \end{cases} \tag{3.3.7c}$$

Then the boundary-value problem (3.3.1) *and, for all net spacing* $h < h_0 < 2/K$, *the finite-difference equations* (3.3.3) *have unique solutions for all* $\propto$ *and bounded integrable* $\mathbf{f}_0(x)$.

Proof. The fact that the linear boundary-value problem (3.3.1), with (3.3.5), has a unique solution follows from Theorem 1.2.5. It only remains to show that the matrix $\mathscr{L}_h$ in (3.3.6) is nonsingular for small h.

Since $\|K(x)\|_\infty \le K$ by (3.3.7a), we are assured that for $h < 2/K$ the matrices L_j and R_j in (3.3.6c) are nonsingular (see Isaacson and Keller (1966), p. 16 or pp. 135–136). We define nth-order matrices P_j and the $n(J + 1)$th-order diagonal matrix D as

$$P_j \equiv R_j^{-1} L_j, \qquad j = 1, 2, \ldots, J; \qquad D \equiv \begin{pmatrix} I & 0 & \cdots & 0 \\ 0 & R_1 & & \vdots \\ \vdots & & & 0 \\ 0 & \cdots & 0 & R_J \end{pmatrix} \tag{3.3.8a}$$

and then form

$$D^{-1}\mathscr{L}_h = \begin{pmatrix} A & 0 & 0 & \cdots & 0 & B \\ -P_1 & I & 0 & & \cdots & 0 \\ & & & & & \vdots \\ \vdots & & & & & \\ & & & & & 0 \\ 0 & & \cdots & & -P_J & I \end{pmatrix}.$$

Elementary block transformations can be employed to triangularize the above matrix. Specifically, with the definitions

$$T_1 \equiv \begin{pmatrix} I & 0 & \cdots & 0 \\ P_1 & I & & \\ \vdots & & & \vdots \\ 0 & & \cdots & I \end{pmatrix}, \qquad T_2 \equiv \begin{pmatrix} I & 0 & 0 & \cdots & 0 \\ 0 & I & 0 & \cdots & \\ 0 & P_2 & I & \cdots & \vdots \\ \vdots & & & & \\ 0 & & \cdots & & I \end{pmatrix}, \ldots,$$

$$T_J \equiv \begin{pmatrix} I & & \cdots & & 0 \\ & & & & \vdots \\ \vdots & & & & \\ & & & I & 0 \\ 0 & & \cdots & P_J & I \end{pmatrix}, \tag{3.3.8b}$$

we obtain the triangular matrix

$$D^{-1}\mathscr{L}_h T_J \cdots T_1 = \begin{pmatrix} (A + B\pi_1) & B\pi_2 & \cdots & B\pi_J & B \\ 0 & I & & \cdots & 0 \\ \vdots & & & & \vdots \\ 0 & & \cdots & 0 & I \end{pmatrix}. \tag{3.3.8c}$$

Here we have used the abbreviations

$$\pi_J \equiv P_J; \qquad \pi_j = \pi_{j+1}P_j, \qquad j = 1, 2, \ldots, J - 1. \tag{3.3.8d}$$

Since D and the T_j, $j = 1, 2, \ldots, J$ are nonsingular, it follows from (3.3.8c) that $\mathscr{L}_h$ is nonsingular if and only if the nth-order matrix $(A + B\pi_1)$ is nonsingular. More simply $D^{-1}\mathscr{L}_h$ can be factored as in Problem 2.4.2.

Since $(A + B)$ is nonsingular by (3.3.7b), we write

$$(A + B\pi_1) = (A + B)[I + (A + B)^{-1}B(\pi_1 - I)].$$

With $h < 2/K$ it can be shown, as in Problem 3.3.1, that

$$\|\pi_1 - I\|_\infty \le e^{K^*|b-a|} - 1.$$

Then, using (3.3.7c), we see that

$$\|(A + B)^{-1}B(\pi_1 - I)\|_\infty \le m(e^{K^*|b-a|} - 1) < 1,$$

and so $(A + B\pi_1)$ is nonsingular. ∎

The above derivation of (3.3.8) is essentially the same as that used to deduce (2.4.12). The estimate of $\|\pi_1 - I\|_\infty$ is related to that in (1.2.17), but is slightly more involved. By writing down the inverse of $\mathscr{L}_h$ we obtain an explicit representation for the solution of the finite-difference problem (3.3.3) for the linear case where $\mathbf{f}$ is given by (3.3.5). From (3.3.8b) we easily find this inverse in the form:

$$\mathscr{L}_h^{-1} = T_J \cdots T_1 \begin{pmatrix} (A + B\pi_1)^{-1} & \cdots & 0 \\ 0 & I \quad \cdots & 0 \\ \vdots & & \vdots \\ 0 & \cdots & I \end{pmatrix} \begin{pmatrix} I & -B\pi_2 & \cdots & -B\pi_J & -B \\ 0 & I & & \cdots & 0 \\ \vdots & & & & \vdots \\ 0 & & \cdots & 0 & I \end{pmatrix} D^{-1}. \tag{3.3.9}$$

Again we point out that the condition $(A + B)$ nonsingular can be eliminated. For h sufficiently small, it can be shown that $(A + B\pi_1)$ is nonsingular if $(A + B\Omega_a^b\{K\})$ is nonsingular (see Problem 3.3.7). But this latter condition is, recalling the proof of Theorem 1.2.5 and remarks following, just the necessary and sufficient condition for the existence of a unique solution to (3.3.1) and (3.3.5).

Next we turn to the nonlinear case and show, under conditions quite similar to those of Theorem 1.2.6, that a unique solution of the difference equations (3.3.3) exists. This is done by means of an iteration scheme which could also be employed in practice to compute the difference approximation. These results are the content of the following.

THEOREM 3.3.2. *Let* $\mathbf{f}(x, \mathbf{u})$ *satisfy on* R: $a \le x \le b$, $|\mathbf{u}| < \infty$ *the conditions*

(a) $\mathbf{f}(x, \mathbf{u})$ *continuous*;

(b) $\dfrac{\partial f_i(x, \mathbf{u})}{\partial u_j}$ *continuous*, $i, j = 1, 2, \ldots, n$; (3.3.10)

(c) $\left\|\dfrac{\partial \mathbf{f}(x, \mathbf{u})}{\partial \mathbf{u}}\right\|_\infty \le K.$

Furthermore, let the matrices A and B, the scalar K and the length of the interval, $|b - a|$, *satisfy*

(d) $(A + B)$ *nonsingular*;

(e) $K|b - a| \le \ln\left(1 + \dfrac{\lambda}{m}\right)$; (3.3.10)

for some λ *in* $0 < \lambda < 1$ *where*

$$m \equiv \max\{\|(A + B)^{-1}B\|_\infty, \|(A + B)^{-1}A\|_\infty\}.$$

Then the boundary-value problem (3.3.1) *and, for all* $h = |b - a|/J$, *the finite-difference problems* (3.3.3) *have solutions which are unique.*

In particular the solution of the difference equations (3.3.3) *is the limit of the sequence* $\{\mathbf{U}^{(\nu)}\}$, *with* $\mathbf{U}^{(0)}$ *arbitrary, defined by*

$$\begin{aligned}
&\text{(a)}\quad \mathbf{u}_j^{(\nu+1)} - \mathbf{u}_{j-1}^{(\nu+1)} = h\mathbf{f}\left(x_{j-1/2}, \frac{\mathbf{u}_j^{(\nu)} + \mathbf{u}_{j-1}^{(\nu)}}{2}\right), \qquad j = 1, 2, \ldots, J,\\
&\text{(b)}\quad A\mathbf{u}_0^{(\nu+1)} + B\mathbf{u}_J^{(\nu+1)} = \boldsymbol{\alpha}.
\end{aligned} \tag{3.3.11}$$

Proof. Obviously the hypothesis of Theorem 1.2.6 in Chapter 1 is implied by (3.3.10), and so the existence and uniqueness of the solution to the problem (3.3.1) are established.

We shall write the difference equations (3.3.3) in a form from which it can be shown that the iteration scheme in (3.3.11) comes from a contracting map. First we observe that by setting $h = 0$ in (3.3.6) the matrices R_j and L_j reduce to the nth-order identity, and $\mathscr{L}_h$ becomes simply

$$\mathscr{L}_0 \equiv (P + Q), \tag{3.3.12}$$

where P and Q are the $n(J + 1)$th-order matrices defined by (2.4.7). It follows from (3.3.10d) that $\mathscr{L}_0$ is nonsingular, and in fact $\mathscr{L}_0^{-1}$ is given by (2.4.12). Thus the difference equations (3.3.3) or their equivalent form (3.3.4) have the same solutions as the system

$$\mathbf{U} = \mathbf{U} - \mathscr{L}_0^{-1}\boldsymbol{\Phi}(\mathbf{U}) \equiv \mathbf{G}(\mathbf{U}). \tag{3.3.13}$$

We now leave to Problem 3.3.2 the simple demonstration that, with the same $\mathbf{U}^{(0)}$, the iterates $\mathbf{U}^{(\nu+1)} = \mathbf{G}(\mathbf{U}^{(\nu)})$ are identical with those in (3.3.11).

It only remains to show that $\mathbf{G}(\mathbf{U})$ is contracting. A direct computation gives

$$\frac{\partial \mathbf{G}}{\partial \mathbf{U}} = \frac{h}{2}\mathscr{L}_0^{-1}\begin{pmatrix} 0 & 0 & \cdots & & 0\\ K_1 & K_1 & & \cdots & \\ 0 & K_2 & K_2 & \cdots & \\ \vdots & & & & \vdots\\ 0 & & \cdots & K_J & K_J \end{pmatrix}, \tag{3.3.14}$$

where the nth-order matrices K_j are defined by

$$K_j \equiv \frac{\partial \mathbf{f}[x_{j-1/2}, (\mathbf{u}_j + \mathbf{u}_{j-1})/2]}{\partial \mathbf{u}}, \qquad j = 1, 2, \ldots, J.$$

Using (2.4.12) we obtain, on multiplying out in the above and taking the norm,

$$\left\|\frac{\partial \mathbf{G}}{\partial \mathbf{U}}\right\|_\infty \leq \frac{h}{2}\cdot 2KJm.$$

Here we have used (3.3.10c) to bound $\|K_j\|_\infty \le K$. Since $Jh = |b - a|$, it now follows from (3.3.10e) that

$$\left\|\frac{\partial \mathbf{G}}{\partial \mathbf{U}}\right\|_\infty \le m \ln\left(1 + \frac{\lambda}{m}\right) < \lambda < 1. \quad \blacksquare \tag{3.3.15}$$

In many cases the iteration scheme (3.3.11) will converge too slowly to be of practical value, and so we consider the application of Newton's method for solving (3.3.3) or equivalently (3.3.4). With some initial guess, $\mathbf{U}^{(0)}$, of the solution we now compute the sequence $\{\mathbf{U}^{(\nu)}\}$ by

$$\mathbf{U}^{(\nu+1)} = \mathbf{U}^{(\nu)} + \Delta\mathbf{U}^{(\nu)}, \qquad \nu = 0, 1, 2, \ldots,$$

where $\Delta\mathbf{U}^{(\nu)}$ is the solution of the linear algebraic system

$$\frac{\partial \mathbf{\Phi}(\mathbf{U}^{(\nu)})}{\partial \mathbf{U}} \Delta\mathbf{U}^{(\nu)} = -\mathbf{\Phi}(U^{(\nu)}).$$

As usual in our study, we need only verify the nonsingularity of the coefficient matrix in this system to insure the convergence of Newton's method (since Theorem 1.4.3 is applicable). However, using the definition of $\mathbf{G}(\mathbf{U})$ in (3.3.13), it follows that

$$\frac{\partial \mathbf{\Phi}(\mathbf{U})}{\partial \mathbf{U}} = \mathscr{L}_0\left[I - \frac{\partial \mathbf{G}(\mathbf{U})}{\partial \mathbf{U}}\right].$$

From (3.3.15) we deduce that the matrix $[I - \partial\mathbf{G}/\partial\mathbf{U}]$ is nonsingular and hence, under the hypothesis of Theorem 3.3.2, Newton's method is applicable.

We finally turn to the question of the accuracy of the finite-difference solution. Conditions which insure us that the error is $\mathcal{O}(h^2)$ are contained in

THEOREM 3.3.3. *Let the hypothesis of Theorem* 3.3.2 *hold and in addition let* $\mathbf{f}(x, \mathbf{u})$ *have two continuous derivatives with respect to* x *and* $\mathbf{u}$ *on* $a \le x \le b$, $|\mathbf{u}| < \infty$. *Then the solution* $\mathbf{y}(x)$ *of the boundary-value problem* (3.3.1) *and the approximate solution* $\mathbf{u}_j$ *defined by the difference problem* (3.3.3) *satisfy*

$$|\mathbf{y}(x_j) - \mathbf{u}_j| = \mathcal{O}(h^2), \qquad j = 0, 1, \ldots, J.$$

Proof. Under the conditions on $\mathbf{f}$ we are assured that $\mathbf{y}(x)$ has three continuous derivatives and thus

$$\frac{\mathbf{y}(x_j) - \mathbf{y}(x_{j-1})}{h} - \mathbf{y}'(x_{j-1/2}) = \mathcal{O}(h^2),$$

$$\frac{\mathbf{y}(x_j) + \mathbf{y}(x_{j-1})}{2} - \mathbf{y}(x_{j-1/2}) = \mathcal{O}(h^2).$$

From (3.3.1a) at $x = x_{j-1/2}$ and (3.3.3a) we obtain, with $\mathbf{e}_j \equiv \mathbf{y}(x_j) - \mathbf{u}_j$,

$$\frac{\mathbf{e}_j - \mathbf{e}_{j-1}}{h} - K_j \frac{\mathbf{e}_j + \mathbf{e}_{j-1}}{2} = \mathcal{O}(h^2), \qquad j = 1, 2, \ldots, J.$$

Here each row of the matrix K_j is a corresponding row of $\partial\mathbf{f}/\partial\mathbf{u}$ evaluated at some point $(x, \mathbf{u})$ with $x = x_{j-1/2}$,

$$\mathbf{u} = \frac{\theta}{2}(\mathbf{u}_j + \mathbf{u}_{j-1}) + \frac{(1-\theta)}{2}(\mathbf{y}(x_j) + \mathbf{y}(x_{j-1})), \qquad 0 < \theta < 1.$$

From (3.3.1b) and (3.3.3b) we obtain

$$A\mathbf{e}_0 + B\mathbf{e}_J = 0.$$

Combining the above results into matrix form using $\mathcal{L}_0$ of (3.3.12) and $\mathbf{E} \equiv (\mathbf{e}_0^T, \mathbf{e}_1^T, \ldots, \mathbf{e}_J^T)^T$, we get

$$\mathcal{L}_0\mathbf{E} = \frac{h}{2}\begin{pmatrix} 0 & 0 & & \cdots & 0 \\ K_1 & K_1 & & \cdots & \\ 0 & K_2 & K_2 & \cdots & \vdots \\ \vdots & & & & \\ 0 & \cdots & & K_J & K_J \end{pmatrix}\mathbf{E} + \mathcal{O}(h^3).$$

This can be written, in an obvious notation, as

$$\mathbf{E} = \frac{h}{2}\mathcal{L}_0^{-1}K\mathbf{E} + \mathcal{L}_0^{-1}\mathcal{O}(h^3).$$

The coefficient matrix $h/2\mathcal{L}_0^{-1}K$, which enters here, is quite similar to $\partial\mathbf{G}/\partial\mathbf{U}$ in (3.3.14). The elements of the K_j are not evaluated at the same points in the two cases but by (3.3.10) we find, just as in (3.3.15), that

$$\left\|\frac{h}{2}\mathcal{L}_0^{-1}K\right\|_\infty < \lambda.$$

Taking norms in the error equation, we find that

$$\|\mathbf{E}\|_\infty \le \lambda\|\mathbf{E}\|_\infty + \mathcal{O}(h^2)$$

and, since $\lambda < 1$,

$$\|\mathbf{E}\|_\infty \le \mathcal{O}(h^2).$$

We have set $\|\mathcal{L}_0^{-1}\mathcal{O}(h^3)\|_\infty = \mathcal{O}(h^2)$ since $\|\mathcal{L}_0^{-1}\|_\infty \le (J+2)m$ and $\mathcal{O}(h^3)$ represents a vector. ▌

3.3.1 *Difference Corrections and* $h \to 0$ *Extrapolation for Systems*

We can frequently improve the $\mathcal{O}(h^2)$ accuracy of the finite difference solution, $\mathbf{u}_j$, defined in (3.3.3) by the application of difference corrections or $h \to 0$ extrapolation. These procedures are justified in the present case if we can show that there is some vector-valued function, $\mathbf{e}(x)$, such that the exact solution, $\mathbf{y}(x)$, of (3.3.1) and the numerical solution, $\mathbf{u}_j$, are related by

$$\mathbf{y}(x_j) = \mathbf{u}_j + h^2\mathbf{e}(x_j) + \mathcal{O}(h^4), \qquad j = 0, 1, \ldots, J. \tag{3.3.16}$$

A derivation of this result is quite similar to that of (3.1.11) (see Theorem 3.1.4) and is based on the following.

THEOREM 3.3.4. *Let* $\mathbf{f}(x, \mathbf{y})$, A *and* B *satisfy the hypothesis of Theorem* 3.3.3 *and in addition* $\mathbf{f}$ *be so smooth that the solution* $\mathbf{y}(x)$ *of* (3.3.1) *has a continuous fifth derivative*† *on* $[a, b]$. *Then* (3.3.16) *holds with* $\{\mathbf{u}_j\}$ *the solution of* (3.3.3) *and* $\mathbf{e}(x)$ *the solution of the linear boundary-value problem:*

$$\begin{aligned} &\text{(a)} \quad \mathbf{e}' = K(x)\mathbf{e} + \boldsymbol{\theta}(x), \qquad a < x < b, \\ &\text{(b)} \quad A\mathbf{e}(a) + B\mathbf{e}(b) = 0; \end{aligned} \tag{3.3.17}$$

where

$$K(x) \equiv \frac{\partial \mathbf{f}(x, \mathbf{y}(x))}{\partial \mathbf{y}}, \qquad \boldsymbol{\theta}(x) \equiv \frac{1}{24}\mathbf{y}^{('''')}(x) - \frac{1}{4}K(x)\mathbf{y}^{(''')}(x). \tag{3.3.18}$$

Proof. Let $\mathbf{v}(x)$ be the unique solution of the initial-value problem

$$\mathbf{v}' = K(x)\mathbf{v} + \boldsymbol{\theta}(x), \qquad \mathbf{v}(a) = \mathbf{0}.$$

Then define the n-vector $\boldsymbol{\alpha}$ by

$$\boldsymbol{\alpha} = -B\mathbf{v}(b).$$

Now the problem (3.3.17) is reduced to solving the boundary-value problem

$$\mathbf{w}' = K(x)\mathbf{w}, \qquad A\mathbf{w}(a) + B\mathbf{w}(b) = \boldsymbol{\alpha}$$

since then

$$\mathbf{e}(x) = \mathbf{w}(x) + \mathbf{v}(x).$$

But Theorem 1.2.5 implies that the boundary-value problem for $\mathbf{w}$ has a unique solution. Thus we conclude that the problem (3.3.17) also has a unique solution.

Since $\mathbf{y}(x)$ has a continuous fifth derivative we see that $K(x)$ must have at least a continuous third derivative and so $\boldsymbol{\theta}(x)$ has a continuous second

† We need only require a Lipschitz-continuous fourth derivative and the theorem is valid. However, we are not seeking the weakest conditions.

derivative. But then $\mathbf{e}(x)$, a solution of (3.3.17a), must have a continuous third derivative.

With the definitions

$$\mathbf{e}_j \equiv h^{-2}[\mathbf{y}(x_j) - \mathbf{u}_j], \qquad j = 0, 1, \ldots, J \tag{3.3.19}$$

we obtain from (3.3.3a) and (3.3.1a) evaluated at $x = x_{j-1/2}$

$$\mathbf{e}_j = \mathbf{e}_{j-1} + h^{-1}\left[\mathbf{f}\left(x_{j-1/2}, \mathbf{y}(x_{j-1/2})\right) - \mathbf{f}\left(x_{j-1/2}, \frac{\mathbf{u}_j + \mathbf{u}_{j-1}}{2}\right)\right] + \frac{h}{24}\mathbf{y}^{(\prime\prime\prime)}(x_{j-1/2}) + \mathcal{O}(h^3).$$

By Theorem 3.3.3 we have $[\mathbf{y}(x_j) - \mathbf{u}_j] = \mathcal{O}(h^2)$, and a Taylor expansion yields

$$\frac{\mathbf{y}(x_j) + \mathbf{y}(x_{j-1})}{2} - \mathbf{y}(x_{j-1/2}) = \frac{h^2}{4}\mathbf{y}^{(\prime\prime)}(x_{j-1/2}) + \mathcal{O}(h^4).$$

Then several Taylor expansions yield

$$\left[\mathbf{f}\left(x_{j-1/2}, \mathbf{y}(x_{j-1/2})\right) - \mathbf{f}\left(x_{j-1/2}, \frac{\mathbf{u}_j + \mathbf{u}_{j-1}}{2}\right)\right]$$
$$= \frac{\partial \mathbf{f}(x_{j-1/2}, \mathbf{y}(x_{j-1/2}))}{\partial \mathbf{y}}\left[h^2\left(\frac{\mathbf{e}_j + \mathbf{e}_{j-1}}{2}\right) - \frac{h^2}{4}\mathbf{y}^{(\prime\prime)}(x_{j-1/2})\right] + \mathcal{O}(h^4),$$

and, using this above, we find that

$$\mathbf{e}_j = \mathbf{e}_{j-1} + \frac{h}{2}K(x_{j-1/2})(\mathbf{e}_j + \mathbf{e}_{j-1}) + h\boldsymbol{\theta}(x_{j-1/2}) + \mathcal{O}(h^3),$$
$$j = 1, 2, \ldots, J. \tag{3.3.20a}$$

From (3.3.1b) and (3.3.3b) we get

$$A\mathbf{e}_0 + B\mathbf{e}_J = 0. \tag{3.3.20b}$$

We have already shown that $\mathbf{e}(x)$ has three continuous derivatives. So it now follows, by applying the analysis in the proof of Theorem 3.3.3 to the boundary-value problem (3.3.17) and the difference problem (3.3.20), that

$$\mathbf{e}(x_j) - \mathbf{e}_j = \mathcal{O}(h^2), \qquad j = 0, 1, \ldots, J.$$

Recalling the definition (3.3.19), we finally deduce (3.3.16). ▮

The applications of the results contained in (3.3.16) and (3.3.17) are quite similar to the corresponding results in Section 3.1.1. For the $h \to 0$

extrapolation we find, in an obvious notation, that

$$\bar{\mathbf{u}}_j = \frac{4}{3}\mathbf{u}_{2j}\left(\frac{h}{2}\right) - \frac{1}{3}\mathbf{u}_j(h)$$

satisfies

$$|\mathbf{y}(x_j) - \bar{\mathbf{u}}_j| = \mathcal{O}(h^4), \qquad j = 0, 1, \ldots, J.$$

Thus with two computations of the form (3.3.3), using $h = [b - a]/J$ and $h = [b - a]/2J$, respectively, we obtain an $\mathcal{O}(h^4)$ approximation.

The difference correction is again based on computing an $\mathcal{O}(h^2)$ accurate approximation to $\mathbf{e}(x)$ as defined in (3.3.17). However, in addition to the fact that $\boldsymbol{\theta}(x)$ is not known exactly, we also see that the coefficient $K(x)$ cannot be precisely evaluated (since $y(x)$ is in general unknown). Of course, if we employ the difference solution $\mathbf{u}_j$, we get

$$\frac{\partial \mathbf{f}(x_j, \mathbf{u}_j)}{\partial \mathbf{y}} = K(x_j) + \mathcal{O}(h^2).$$

It can be shown that with this approximation and some $\mathcal{O}(h^2)$ approximation, $\boldsymbol{\Theta}_j$, to $\boldsymbol{\theta}(x_j)$, a difference problem whose solutions are $\mathcal{O}(h^2)$ approximations to $\mathbf{e}(x_j)$ can be formulated (see Problem 3.3.4). Of course as is implied by the name of this procedure we may approximate $\boldsymbol{\theta}(x_j)$ by using appropriate difference quotients of the numerical solution to replace the derivatives in (3.3.18). When $\mathbf{f}(x, \mathbf{y})$ is linear in $\mathbf{y}$ the difficulty with $K(x)$ does not occur, and $\boldsymbol{\theta}(x)$ can be written in terms of $\mathbf{y}(x)$ and higher derivatives of given functions. [This is analogous to the treatment of the second-order scalar case, where $\theta(x)$ in (3.1.13) was reduced to (3.1.17a).]

Problems

3.3.1* Let $\pi_1 = (R_J^{-1}L_J)\cdots(R_1^{-1}L_1)$, where R_j and L_j are as defined in (3.3.6c) and $h < h_0 < 2/K$. Show that $\|\pi_1 - I\| \le e^{K^*|b-a|} - 1$. [HINT: It can be shown (see Isaacson and Keller [1966], p. 15) that

$$R_j^{-1} = I + \sum_{\nu=1}^{\infty} ((h/2)K(x_{j-1/2}))^{\nu}.$$

Using this J times, we get $\pi_1 = I + E$, where E contains first and higher powers of h but no terms independent of h. Then estimate

$$\|\pi_1 - I\|_\infty = \|E\|_\infty = (1 + \|E\|_\infty) - 1,$$

using the facts that $\|K(x)\|_\infty < K$ and

$$\|R_j^{-1}\| \le \frac{1}{1 - (h/2)K}, \qquad \|R_j^{-1} - I\|_\infty \le \frac{1}{1 - (h/2)K} - 1,$$

$$\left(\frac{1 + (h/2)K}{1 - (h/2)K}\right)^J \le e^{K^*|b-a|} \quad ; \mathrm{K}^* \equiv \frac{k}{1 - hK/2}$$

since $h = |b - a|/J$. See also Problems 3.3.7 and 1.2.8.]

3.3.2 Using the matrix $\mathscr{L}_0$ of (3.3.12) and the vector function $\mathbf{\Phi}$ defined in (3.3.4), define $\mathbf{F}(\mathbf{U}) \equiv \mathscr{L}_0\mathbf{U} - \mathbf{\Phi}(\mathbf{U})$. Then from (3.3.13) we have that $\mathbf{G}(\mathbf{U}) = \mathscr{L}_0^{-1}\mathbf{F}(\mathbf{U})$. Verify that the iterates in (3.3.11) are simply

$$\mathscr{L}_0\mathbf{U}^{(\nu+1)} = \mathbf{F}(\mathbf{U}^{(\nu)}).$$

3.3.3 Examine the validity of Theorems 3.3.2 and 3.3.3 if, in (3.3.3a), the term $\mathbf{f}(x_{j-1/2}, (\mathbf{u}_j + \mathbf{u}_{j-1})/2)$ is replaced by $\frac{1}{2}[\mathbf{f}(x_j, \mathbf{u}_j) + \mathbf{f}(x_{j-1}, \mathbf{u}_{j-1})]$.

3.3.4 Let a difference problem for approximating the solution $\mathbf{e}(x)$ of the boundary-value problem (3.3.17) be of the form

$$\frac{\mathbf{E}_j - \mathbf{E}_{j-1}}{h} = [K(x_{j-1/2}) + \mathcal{O}(h^2)]\frac{\mathbf{E}_j + \mathbf{E}_{j-1}}{2} + [\boldsymbol{\theta}(x_j) + \mathcal{O}(h^2)],$$
$$j = 1, 2, \ldots, J;$$
$$A\mathbf{E}_0 + B\mathbf{E}_J = 0.$$

Show that under the hypothesis of Theorem 3.3.1, for h sufficiently small,

$$|\mathbf{e}(x_j) - \mathbf{E}_j| = \mathcal{O}(h^2), \qquad j = 0, 1, \ldots, J.$$

3.3.5 Newton's method (in function space) for solving the boundary-value problem (3.3.1) consists in determining a sequence of functions $\{\mathbf{y}^{(\nu)}(x)\}$, where $\mathbf{y}^{(\nu+1)}(x) = \mathbf{y}^{(\nu)}(x) + \mathbf{e}^{(\nu)}(x)$ and $\mathbf{e}^{(\nu)}(x)$ is the solution of the linear boundary-value problem

$$\mathbf{e}^{(\nu)\prime}(x) - \mathbf{f}_{\mathbf{y}}(x, \mathbf{y}^{(\nu)}(x))\mathbf{e}^{(\nu)}(x) = \mathbf{f}(x, \mathbf{y}^{(\nu)}(x)) - \mathbf{y}^{(\nu)\prime}(x),$$
$$A\mathbf{e}^{(\nu)}(a) + B\mathbf{e}^{(\nu)}(b) = \boldsymbol{\alpha} - A\mathbf{y}^{(\nu)}(a) - B\mathbf{y}^{(\nu)}(b).$$

Show that the centered-difference approximation to this system on the net (3.3.2) yields the same algebraic problem as that obtained by employing the usual Newton procedure to solve the algebraic system (3.3.3) [see formulation after Equation (3.3.15)].

3.3.6 In the parallel shooting formulation (2.4.5) use the apparently very crude approximation to (2.4.5a)

$$\mathbf{Y}(1) - \mathbf{Y}(0) = \mathbf{F}\left(\frac{1}{2}, \frac{\mathbf{Y}(1) + \mathbf{Y}(0)}{2}\right).$$

Show that with (2.4.5b) the resulting difference approximation to the problem (2.3.1) [or equivalently (2.4.1)] is identical to the centered-difference scheme (3.3.3) provided $\Delta_j = h$ for $j = 1, 2, \ldots, J$. Thus, in a sense, difference methods can be obtained as special cases of parallel shooting techniques.

3.3.7 Show that π_1, defined in (3.3.8d), satisfies

$$\pi_1 = \mathop{\Omega}_a^b \{K\} + \mathcal{O}(h^2)$$

if $K(x)$ is twice continuously differentiable on $[a, b]$. [HINT: Define $S_0 = I$,

$$S_j - S_{j-1} = \frac{h}{2} K(x_{j-1/2})(S_j + S_{j-1}), \qquad j = 1, 2, \ldots, J,$$

and note that $S_J = \pi_1$. Show that $\{S_j\}$ converges to the solution of

$$Y' = K(x)Y, \; Y(a) = I.]$$

SUPPLEMENTARY REFERENCES AND NOTES

Section 3.1 The derivation of difference approximations to linear differential operators is discussed by Fox (1957), Kantorovich and Krylov (1958), and Babuška, Práger, and Vitásek (1966). Detailed studies of the properties of the coefficient matrices for various difference schemes are given by Varga (1962). These result in more precise error estimates. Variational methods leading to difference equations, or other Ritz approximations, are discussed by Farrington, Gregory, and Taub (1957), Kantorovich and Krylov (1958), and Varga (1965).

Our convergence proof uses neither the maximum principle nor any fancy properties of matrices. With them we could get stronger results, such as allowing $q(x) = 0$.

It is not clear (or perhaps even true) that an arbitrarily high-order difference approximation to the linear differential operator results in correspondingly high-order-accurate approximations to the solution. Thus *repeated* difference corrections and $h \to 0$ extrapolations are at present the only theoretically justified means for obtaining arbitrarily high-order accuracy using difference methods, see for example Pereyra (1966).

Section 3.2 A very thorough study of difference methods applied to $y'' = f(x, y)$ with $f_y \geq 0$ is contained in Henrici (1962), and Lees (1966) allows $f_y < 0$. Difference corrections are also studied in these works, and interesting generalizations are discussed by Pereyra (1966). See also Brown (1962), who employs nonuniform nets. For $f_y \geq 0$ and $f_{yy} < 0$ (that is, concave nonlinearities), Kalaba (1959) shows the applicability of Newton's method (in the continuous form), and Wendroff (1966) justifies it for the difference equations as well. Such applications of Newton's method are sometimes called "quasilinearization" as in Bellman and Kalaba (1965), but the new term seems superfluous. A general study of the equivalence of differencing and then applying Newton's method with the application of Newton's method and then differencing is given by Ortega and Rheinboldt (1966).

Section 3.3 Difference methods for first-order systems are rarely treated in the literature. However, for many of the special boundary conditions occurring in practical problems (for example, separated end conditions) the difference equations are in block-tridiagonal form and hence are easily solved. We do not (in general) advocate using three-point difference equations for first-order systems (that is, $\mathbf{u}_{j+1} - \mathbf{u}_{j-1} = 2h\mathbf{f}(x_j, \mathbf{u}_j)$), since stability problems may arise. Also, special treatment is required at the endpoints to obtain as many equations as unknowns. The application of Galerkin's method to first-order systems has been studied by Urabe (1966), and is in a sense a generalization of difference methods using piecewise linear approximating functions (see Appendix 1). Sylvester and Meyer (1965) employ Newton's method, presumably without being aware of the fact, to the difference scheme (3.3.3); see Problem 3.3.5 to clarify their approach.

CHAPTER 4

INTEGRAL-EQUATION METHODS

4.1 *Green's Functions; Equivalent Integral Equations*

Many of the numerical methods in Section 1.3 were suggested by first replacing the initial-value problem by an integral equation and then applying quadrature formulae to the integral equation. The same can be done for boundary-value problems, but the derivation of equivalent integral equations is somewhat more complicated. It is based on the determination of the Green's function for a *linear* boundary-value problem.

The Green's function for a linear boundary-value problem is roughly analogous to the inverse of the coefficient matrix in a linear system of equations. In brief, let the boundary-value problem be, for example,

$$Ly(x) \equiv (p(x)y'(x))' - q(x)y(x) = r(x), \qquad a \leq x \leq b, \tag{4.1.1a}$$

$$y(a) = 0, \qquad y(b) = 0; \tag{4.1.1b}$$

where $p(x) > 0$ and $q(x) \geq 0$. Then the Green's function, $g(x, \xi)$, is a function such that the solution of (4.1.1) is given by

$$y(x) = -\int_a^b g(x, \xi)r(\xi)\, d\xi. \tag{4.1.2}$$

This representation is to be valid for all inhomogeneous terms $r(x)$ satisfying appropriate smoothness conditions. Thus the Green's function, g, is determined by the differential operator L in (4.1.1a) and the boundary conditions (4.1.1b).

It is not difficult to show that $g(x, \xi)$ is uniquely defined by the following conditions:

(a) $g(x, \xi)$ is continuous in x for fixed ξ and $g(a, \xi) = g(b, \xi) = 0$. (4.1.3)

(b) The first and second x-derivatives of $g(x, \xi)$ are continuous for $x \neq \xi$ and at $x = \xi$ the jump condition,

$$\left.\frac{dg(x, \xi)}{dx}\right|_{x=\xi_-}^{x=\xi_+} = -\frac{1}{p(\xi)}, \quad \text{is satisfied.}$$

(c) $Lg(x, \xi) = 0$ for all $x \neq \xi$.

We now have the basic theorem.

THEOREM 4.1.1. *The function* $y(x)$ *is a solution of* (4.1.1) *if and only if it is given by Equation* (4.1.2), *where* $g(x, \xi)$ *is the Green's function as defined in* (4.1.3).

Proof. The proof of this theorem follows by forming Ly from Equation (4.1.2) and using (4.1.3b) when differentiating to get (4.1.1). Then assuming $y(x)$ to satisfy (4.1.1), we multiply this equation by $g(\xi, x)$ and integrate with respect to x, using partial integration to get (4.1.2) with x and ξ interchanged. The details can be found in Courant-Hilbert, Vol. I (1953), pp. 535 et seq., and are left to the reader as an exercise. ∎

If the boundary conditions (4.1.1b) are replaced by inhomogeneous conditions, a simple modification of Equation (4.1.2) yields the solution (see Problem 4.1.1). More general boundary conditions are easily treated as in Problem 4.1.2.

To construct the Green's function we first determine two nontrivial solutions $y_1(x)$ and $y_2(x)$ of the homogeneous equation $Ly = 0$ such that $y_1(a) = 0$ and $y_2(b) = 0$. Then we have

$$g(x, \xi) = A\begin{cases} y_1(x)y_2(\xi), & x < \xi; \\ y_2(x)y_1(\xi), & x > \xi. \end{cases} \tag{4.1.4a}$$

The quantity A is determined from condition (4.1.3b) as

$$A = -\{p(\xi)[y_2'(\xi)y_1(\xi) - y_1'(\xi)y_2(\xi)]\}^{-1}. \tag{4.1.4b}$$

However, since

$$\frac{d}{d\xi}A^{-1} = p[y_2''y_1 - y_1''y_2] + p'[y_2'y_1 - y_1'y_2] = y_1Ly_2 - y_2Ly_1 = 0,$$

we see that A is indeed a constant.

We now consider several ways in which Green's functions can be used to reduce *nonlinear* boundary-value problems to integral equations. Let us first treat the problem

$$Ly = f(x, y, y'), \tag{4.1.5a}$$

$$y(a) = y(b) = 0, \tag{4.1.5b}$$

where L is as defined in (4.1.1a). Under the assumption that this problem has a solution $y(x)$, it follows from Theorem 4.1.1, by letting $r(x) \equiv f(x, y(x), y'(x))$, that the solution satisfies

$$y(x) = -\int_a^b g(x, \xi)f(\xi, y(\xi), y'(\xi))\, d\xi. \tag{4.1.6}$$

Conversely any solution $y(x)$ of Equation (4.1.6) satisfies (4.1.5) and hence the equivalence of the boundary-value problem (4.1.5) with the integro-differential equation (4.1.6) is demonstrated. If the boundary conditions (4.1.5b) are replaced by inhomogeneous conditions, then the corresponding integral equation (4.1.6) is modified by the addition of an inhomogeneous term (see Problem 4.1.1).

We have assumed above that the operator L was given as defined in (4.1.1a). However, by simply changing the definition of $f(x, y, y')$ we could obtain a different operator, a corresponding Green's function and hence an integral equation different from (4.1.6) but still equivalent to (4.1.5). As an illustration of this procedure we consider the problem:

$$L_0 y \equiv y''(x) = f(x, y), \qquad y(0) = y(1) = 0. \tag{4.1.7}$$

The Green's function for L_0, subject to the indicated boundary conditions, is easily found to be

$$g_0(x, \xi) = \begin{cases} x(1-\xi), & x < \xi; \\ (1-x)\xi, & x > \xi. \end{cases}$$

Thus the problem (4.1.7) is equivalent to the integral equation

$$y(x) = -\int_0^1 g_0(x, \xi) f(\xi, y(\xi))\, d\xi. \tag{4.1.8}$$

However, by subtracting $k^2 y$ from each side of the differential equation in (4.1.7) we have the equivalent boundary-value problem

$$L_k y \equiv y''(x) - k^2 y(x) = f(x, y(x)) - k^2 y(x), \qquad y(0) = y(1) = 0. \tag{4.1.9}$$

The Green's function for L_k and the indicated boundary conditions is now

$$g_k(x, \xi) = \frac{1}{k \sinh k} \begin{cases} \sinh kx \sinh k(1-\xi), & x < \xi; \\ \sinh k(1-x) \sinh k\xi, & x > \xi. \end{cases} \tag{4.1.10}$$

Thus from (4.1.9) we obtain the integral equation

$$y(x) = \int_0^1 g_k(x, \xi)[k^2 y(\xi) - f(\xi, y(\xi))]\, d\xi, \tag{4.1.11}$$

which must be equivalent to the boundary-value problem (4.1.7). The usefulness of this procedure in the present case can be demonstrated by actually using (4.1.11) to solve (4.1.7). We have the following.

THEOREM 4.1.2. *Let f_y be continuous and satisfy $0 \le \partial f(x, y)/\partial y \le N$ for $x \in [0, 1]$ and all y. Then a unique solution of* (4.1.7) *exists and for any k*

such that $k^2 \geq N$ it is given by the limit of the convergent sequence of functions

$$y^{(0)}(x) \equiv 0, \tag{4.1.12a}$$

$$y^{(\nu+1)}(x) = \int_0^1 g_k(x, \xi)[k^2 y^{(\nu)}(\xi) - f(\xi, y^{(\nu)}(\xi))]\, d\xi, \qquad \nu = 0, 1, \ldots . \tag{4.1.12b}$$

Proof. The functions $y^{(\nu)}(x)$ are uniformly continuous on [0, 1] by the continuity of the Green's function. Thus we need only show that they form a Cauchy sequence to conclude that they have a uniformly-continuous limit function $y(x)$. To do this we form

$$\begin{aligned} e^{(\nu+1)}(x) &\equiv y^{(\nu+1)}(x) - y^{(\nu)}(x) \\ &= \int_0^1 g_k(x, \xi)\left[k^2 - \frac{\partial f}{\partial y}(\xi, y^{(\nu)}(\xi) - \theta(\xi)e^{(\nu)}(\xi))\right] e^{(\nu)}(\xi)\, d\xi, \end{aligned}$$

where Taylor's theorem has been used and $0 \leq \theta(\xi) \leq 1$. We note that $g_k(x, \xi) \geq 0$, and since $k^2 \geq N$ the bracket in the integrand satisfies $0 \leq [k^2 - (\partial f/\partial y)] \leq k^2$. Then, calling

$$\|e^{(\nu)}\| \equiv \max_{0 \leq x \leq 1} |e^{(\nu)}(x)|,$$

we obtain

$$\begin{aligned} |e^{(\nu+1)}(x)| &\leq \int_0^1 g_k(x, \xi) k^2\, d\xi \cdot \|e^{(\nu)}\| \\ &= \left(1 - \frac{\cosh k(\frac{1}{2} - x)}{\cosh (k/2)}\right) \cdot \|e^{(\nu)}\| \\ &\leq \left(1 - \frac{1}{\cosh (k/2)}\right) \cdot \|e^{(\nu)}\| \equiv \mu_k \|e^{(\nu)}\|. \end{aligned}$$

Since this holds for all $x \in [0, 1]$, it follows that

$$\|e^{(\nu+1)}\| \leq \mu_k \|e^{(\nu)}\|; \qquad \nu = 1, 2, \ldots,$$

where $\mu_k \equiv (1 - 1/\cosh (k/2)) < 1$. We deduce from this that $\|e^{(\nu+1)}\| \leq \mu_k^\nu \|e^{(1)}\|$ and hence that $\{y^{(\nu)}(x)\}$ is a Cauchy sequence.

We may then take the limit in (4.1.12b) as $\nu \to \infty$ and find that the limit function $y(x)$ satisfies the integral equation (4.1.11). Since this equation is equivalent to (4.1.7), we have demonstrated the existence of a solution to the boundary-value problem. If two solutions $u(x)$ and $y(x)$ exist they both satisfy (4.1.11), and by the above argument

$$\|u - y\| \leq \mu_k \|u - y\|.$$

Thus $\|u - y\| = 0$ and uniqueness follows. ∎

The sequence of functions defined in (4.1.12) are called the *Neumann iterates* for the integral equation (4.1.11) with initial iterate zero. The corresponding iterates for any continuous nonzero initial estimate $y^{(0)}(x)$ would also converge to a solution. It should be observed that the iterates $y^{(\nu)}(x)$ can also be defined as the solutions of the sequence of linear boundary-value problems:

$$y^{(\nu+1)''}(x) - k^2 y^{(\nu+1)}(x) = f(x, y^{(\nu)}(x)) - k^2 y^{(\nu)}(x), \qquad y^{(\nu+1)}(0) = 0,$$
$$y^{(\nu+1)}(1) = 0.$$

Of course this is but a special case of the scheme suggested in (3.2.12). Thus, Theorem 4.1.2 yields a proof of convergence for this scheme in the case indicated.

We must recall that the existence of a unique solution of (4.1.7), provided $\partial f/\partial y > 0$, is implied by Theorem 1.2.2. However, the above proof of this fact is *constructive* and suggests rather obvious numerical methods for approximating the solution. Theorem 4.1.2 is in fact a contracting mapping theorem for an integral operator on a function space, quite analogous to Theorem 1.4.1. For a more general discussion from the viewpoint of functional analysis see Collatz (1960), pp. 34–42.

It is important to consider systems of integral equations by means of which quite general boundary-value problems can be treated. For example (4.1.6) is a special case of the integro-differential equation

$$y(x) = \int_a^b F[x, \xi; y(\xi), y'(\xi)]\, d\xi. \tag{4.1.13a}$$

Under appropriate smoothness conditions on the functions $y(x)$ and $F_x(x, \xi; y, y')$, the derivative $y'(x)$ must satisfy

$$y'(x) = \int_a^b F_x[x, \xi; y(\xi), y'(\xi)]\, d\xi. \tag{4.1.13b}$$

Thus we see that boundary-value problems of the form (4.1.5) can be replaced by a system of integral equations.

As another example, we consider a system of boundary-value problems of the general form

$$\left.\begin{aligned} &\text{(a)}\quad L_\nu y_\nu(x) \equiv (p_\nu(x) y_\nu'(x))' - q_\nu(x) y_\nu(x) = f_\nu(x; y_1, \ldots, y_n), \\ &\qquad\qquad\qquad\qquad\qquad\qquad\qquad\qquad a \le x \le b, \\ &\text{(b)}\quad y_\nu(a) = 0, \qquad y_\nu(b) = 0, \end{aligned}\right\} \quad 1 \le \nu \le n. \tag{4.1.14}$$

where $p_\nu(x) > 0$ and $q_\nu(x) \ge 0$ for all ν.

If $g_\nu(x, \xi)$ is the Green's function for the differential operator L_ν, subject to the homogeneous boundary conditions (4.1.14b), then the solution of (4.1.14) must satisfy

$$y_\nu(x) = -\int_a^b g_\nu(x, \xi) f_\nu(\xi; y_1(\xi), \ldots, y_n(\xi))\, d\xi, \qquad 1 \le \nu \le n. \tag{4.1.15}$$

Conversely we see that any twice continuously differentiable solution of (4.1.15) is a solution of (4.1.14). (It is assumed here that all functions $p_\nu'(x)$, $q_\nu(x)$, and $f_\nu(x; y_1, \ldots, y_n)$ are continuous.) If the boundary conditions are not homogeneous then, as before, inhomogeneous terms must be included. If more general boundary conditions than those in (4.1.14b) are to be treated (but with the y_ν still uncoupled), we need only use the appropriate Green's functions. [If the boundary conditions are coupled the situation is considerably more complicated. However, if we associate the νth pair of conditions containing $y_\nu(a)$ and $y_\nu(b)$ with the νth equation in (4.1.14a) and consider $y_\mu(a)$ and $y_\mu(b)$ for $\mu \neq \nu$ as inhomogeneous terms, an equivalent system of integral equations is easily derived. This system then involves values of the dependent variables at the endpoints, and a theoretical analysis is rather complicated. But there are frequently no practical difficulties in the numerical solution of such systems by the methods in Section 4.2.]

The systems (4.1.13) and (4.1.15) are special cases of a rather general system of n coupled integral equations which can be written as

$$\mathbf{y}(x) = \int_a^b \mathbf{F}(x, \xi; \mathbf{y}(\xi))\, d\xi. \tag{4.1.16}$$

A constructive existence and uniqueness proof for this system is easily obtained if the integrand satisfies appropriate conditions. We have in fact the following.

THEOREM 4.1.3. *Let the continuous functions* $F_i(x, \xi; \mathbf{y})$ *have continuous derivatives,* $\partial F_i(x, \xi; \mathbf{y})/\partial y_j$ *on* $a \le x,\ \xi \le b$, $|\mathbf{y}| < \infty$, *for* $i, j = 1, 2, \ldots, n$. *Further for any vector-valued function* $\mathbf{z}(\xi)$ *continuous on* $a \le \xi \le b$ *let*

$$\int_a^b \sum_{j=1}^n \left| \frac{\partial F_i}{\partial y_j}(x, \xi; \mathbf{z}(\xi)) \right| d\xi \le \mu < 1, \qquad i = 1, 2, \ldots, n, \qquad a \le x \le b. \tag{4.1.17}$$

Then the integral equation (4.1.16) *has a unique continuous solution* $\mathbf{y}(x)$, *which is the limit of the convergent sequence of functions*

$$\text{(a)}\quad \mathbf{y}^{(0)}(x) = 0$$
$$\text{(b)}\quad \mathbf{y}^{(\nu+1)}(x) = \int_a^b \mathbf{F}(x, \xi; \mathbf{y}^{(\nu)}(\xi))\, d\xi, \qquad \nu = 0, 1, \ldots. \tag{4.1.18}$$

Proof. The proof is quite similar to that of Theorem 4.1.2, using the Contracting Mapping principle. The details are left to Problem 4.1.4. ∎

Problems

4.1.1 Let $y_0(x)$ be the solution of the boundary-value problem

$$Ly_0 = 0; \qquad y_0(a) = \alpha, \qquad y_0(b) = \beta.$$

Then show that the solution of the problem

$$Ly = r(x), \qquad y(a) = \alpha, \qquad y(b) = \beta$$

is

$$y(x) = y_0(x) - \int_a^b g(x, \xi)r(\xi)\, d\xi.$$

Here L is as defined in (4.1.1a) and $g(x, \xi)$ is the Green's function in (4.1.2). The function $y_0(x)$ can be determined as a linear combination of $y_1(x)$ and $y_2(x)$ as used in (4.1.4).

4.1.2 Define the Green's function for the differential operator L of (4.1.1a), subject to the boundary conditions

$$a_0 y(a) + a_1 y'(a) = 0, \qquad b_0 y(b) - b_1 y'(b) = 0.$$

4.1.3 Prove that the Green's function $g(x, \xi)$, as defined in (4.1.3) for the operator L in (4.1.1a), is symmetric; that is,

$$g(x, \xi) = g(\xi, x).$$

[This is always true of *self-adjoint* differential operators; that is, if for all u and v vanishing at $x = a$ and $x = b$, $\int_a^b uLv\, dx = \int_a^b vLu\, dx$.]

4.1.4 Carry out the details in the proof of Theorem 4.1.3. [HINT: If $\mathbf{e}^{(\nu)}(\xi) \equiv \mathbf{u}^{(\nu+1)}(\xi) - \mathbf{u}^{(\nu)}(\xi)$ is continuous, then

$$F_i(x, \xi; \mathbf{u}^{(\nu)}(\xi) + \mathbf{e}^{(\nu)}(\xi)) = F_i(x, \xi; \mathbf{u}^{(\nu)}(\xi)) + \sum_j \frac{\partial F_i}{\partial y_j}(x, \xi; \mathbf{u}^{(\nu)}(\xi) + \theta_i(\xi)\mathbf{e}^{(\nu)}(\xi))e_j^{(\nu)}(\xi),$$

where $\theta_i(\xi)$ in $0 \le \theta_i \le 1$ is a continuous function of ξ on $[a, b]$.]

4.1.5 Show that the Green's function for $Lu \equiv -u'' - 4u$, subject to $u(0) = u'(1) = 0$, is

$$g(x, t) = \frac{1}{2\cos 2}\begin{cases} \sin 2x \cos 2(1 - \xi), & x < \xi, \\ \cos 2(1 - x)\sin 2\xi, & x > \xi. \end{cases}$$

4.2 *Numerical Solution of Integral Equations*

We now consider numerical methods for approximating the solution of integral equations. If these methods are effective, then they also furnish approximations to the solution of corresponding boundary-value problems. We shall first consider the special integral equation (4.1.11), obtained from the boundary-value problem (4.1.7) or equivalently from (4.1.9), and then discuss the obvious extensions to more general systems of integral equations of the forms (4.1.15) and (4.1.16).

The net points x_j, $j = 1, 2, \ldots, J$, are *not* required to be equally spaced, but we take them such that

$$0 \leq x_1 < x_2 \cdots < x_J \leq 1.$$

The net function $\{u_j\}$ which is to approximate $y(x_j)$ is defined as the solution of the system of equations

$$u_i = \sum_{j=1}^{J} \alpha_j g_k(x_i, x_j)[k^2 u_j - f(x_j, u_j)], \qquad i = 1, 2, \ldots, J. \tag{4.2.1}$$

The quantities α_j are the coefficients for a quadrature formula over [0, 1] with nodes at the points x_j. Thus for each value $x = x_i$ we use the same approximation to the corresponding integral in (4.1.11). This is just a convenience and is more efficient for storage purposes. More "accurate" schemes based, for example, on weighted Gaussian-type quadrature formulae, do not have this feature and will be discussed later.

We require that the quadrature scheme used in (4.2.1) satisfy

$$\begin{aligned} &\text{(a)} \quad \alpha_j \geq 0, \qquad j = 1, 2, \ldots, J. \\ &\text{(b)} \quad \lim_{J \to \infty} E_J\{F(x)\} = 0, \end{aligned} \tag{4.2.2}$$

for all $F(x)$ which are continuous on [0, 1]. Here we have defined the quadrature error in the formula using the coefficients α_j and nodes x_j as

$$E_J\{F(\xi)\} \equiv \int_0^1 F(\xi)\, d\xi - \sum_{j=1}^{J} \alpha_j F(x_j). \tag{4.2.3}$$

Thus we want nonnegative coefficients and convergence for all *continuous* integrands. (This last condition (4.2.2b) is imposed because $g_k(x, \xi)$, and hence the integrand in Equation (4.1.11), is continuous on $0 \leq \xi \leq 1$ but has a discontinuous first derivative at $\xi = x$.) We say that the quadrature scheme has *degree of precision* N if $E_J\{\xi^n\} = 0$ for $n = 0, 1, \ldots, N$ and $E_J\{\xi^{N+1}\} \neq 0$. Then all polynomials of degree N or less are integrated exactly by the scheme.

Let us consider first the question of convergence of the approximate solution, u_j, to the exact solution of the integral equation (4.1.11). Later we treat the problem of solving this nonlinear system by iteration. The local truncation errors, τ_i, of the approximation (4.2.1) are defined by

$$\tau_i = E_J\{g_k(x_i, \xi)[k^2 y(\xi) - f(\xi, y(\xi))]\}, \qquad i = 1, 2, \ldots, J. \tag{4.2.4}$$

Now we can state the basic error estimate as a theorem.

THEOREM 4.2.1. *Let the quadrature formula with nodes* $\{x_j\}$ *and coefficients* $\{\alpha_j\}$ *satisfy* (4.2.2) *and let* f *and* $\partial f/\partial y$ *be continuous and satisfy* $0 \leq \partial f/\partial y \leq M$ *for all* $x \in [0, 1]$ *and all* y. *Then for any* k *such that* $k^2 \geq M$ *there exists an integer* J_k *and positive quantities* ε_{Jk} *such that for all* $J \geq J_k$

$$|u_i - y(x_i)| \leq \left(\varepsilon_{Jk} + \cosh\frac{k}{2}\right)\|\tau\|, \qquad i = 1, 2, \ldots, J. \tag{4.2.5a}$$

Here $\{u_i\}$ *is the solution of* (4.2.2), $y(x)$ *is the solution of* (4.1.11) *and*

$$\lim_{J\to\infty} \varepsilon_{Jk} = 0, \qquad \lim_{J\to\infty} \|\tau\| = 0. \tag{4.2.5b}$$

Proof. We use the definitions

$$e_j \equiv u_j - y(x_j), \qquad \|e\| \equiv \max_{1\leq j\leq J} |e_j|, \qquad \|\tau\| \equiv \max_{1\leq j\leq J} |\tau_j|.$$

Then, using Taylor's theorem and the continuity of $\partial f/\partial y$, we deduce from (4.2.1) and (4.2.4) that

$$e_i = \sum_{j=1}^{J} \alpha_j g_k(x_i, x_j)[k^2 - \tilde{f}_{y,j}]e_j - \tau_i, \qquad i = 1, 2, \ldots, J.$$

Here $\tilde{f}_{y,j}$ is an appropriate intermediate value of $\partial f/\partial y$. Taking absolute values above and recalling that $\alpha_j \geq 0$, $0 \leq k^2 - f_y \leq k^2$, and $g(x, \xi) \geq 0$, we get

$$|e_i| \leq \left[\sum_{j=1}^{J} \alpha_j k^2 g_k(x_i, x_j)\right] \cdot \|e\| + \|\tau\|, \qquad i = 1, 2, \ldots, J. \tag{4.2.6}$$

From the definition (4.2.3) it follows that

$$\left[\sum_{j=1}^{J} \alpha_j k^2 g_k(x_i, x_j)\right] = \int_0^1 g_k(x_i, \xi)k^2\, d\xi - E_J\{k^2 g_k(x_i, \xi)\},$$

$$\leq \left[1 - \frac{1}{\cosh(k/2)}\right] + k^2|E_J\{g_k(x_i, \xi)\}|, \qquad i = 1, 2, \ldots, J.$$

However, the functions $g_k(x_i, \xi)$ are continuous functions of ξ on [0, 1] for all $i = 1, 2, \ldots, J$. Thus by the convergence property (4.2.2b) we can make $|E_J\{g_k(x_i, \xi)\}|$ arbitrarily small by taking J sufficiently large. Let J_k be an integer such that for all $J \geq J_k$:

$$|E_J\{g_k(x_i, \xi)\}| < \frac{1}{k^2 \cosh(k/2)}, \qquad i = 1, 2, \ldots, J.$$

Now define the quantities ε_{Jk} by

$$\varepsilon_{Jk} \equiv \left(\frac{1}{\cosh(k/2)} - k^2\|E_J\{g_k(x, \xi)\}\|\right)^{-1} - \cosh\frac{k}{2},$$

where

$$\|E_J\{g_k(x, \xi)\}\| = \max_{1\leq i\leq J} |E_J\{g_k(x_i, \xi)\}|.$$

Clearly $\varepsilon_{Jk} \geq 0$ for $J \geq J_k$ and $\lim_{J\to\infty} \varepsilon_{Jk} = 0$. Combining the above, we have, for $J \geq J_k$,

$$\left[\sum_{j=1}^{J} \alpha_j k^2 g_k(x_i, x_j)\right] \leq 1 - \frac{1}{\varepsilon_{Jk} + \cosh(k/2)}; \qquad i = 1, 2, \ldots, J.$$

This result in (4.2.6) implies that

$$\|e\| \leq \left(\varepsilon_{Jk} + \cosh\frac{k}{2}\right)\|\tau\|,$$

and the theorem clearly follows. ∎

It should be noted that the error bound in (4.2.5) estimates the order of convergence of the approximate solution to the exact solution by the order of convergence of the quadrature formula applied to the integrand in Equation (4.1.11). To get more detailed estimates of the error we must specify a particular quadrature scheme and then examine the corresponding quadrature error for integrands of the indicated form.

A particularly simple example is furnished by employing the *trapezoidal rule* with equally-spaced nodes. Thus we take $x_j = jh$, $h = 1/J$, $\alpha_1 = \alpha_J = \frac{1}{2}$, and $\alpha_j = 1$, $j = 2, 3, \ldots, J - 1$. For brevity let us introduce the notation

$$G(x, \xi) \equiv g_k(x, \xi)[k^2 y(\xi) - f(\xi, y(\xi))],$$

where $y(x)$ is the solution of (4.1.11). Since Theorem 4.2.1 is valid for the trapezoidal rule we need only estimate the errors in applying this rule to approximate

$$\int_0^1 G(x_i, \xi)\, d\xi.$$

These errors are

$$E_J\{G(x_i, \xi)\} = \int_0^1 G(x_i, \xi)\, d\xi - \sum_{j=1}^{J} \alpha_j G(x_i, x_j), \qquad i = 1, 2, \ldots, J.$$

But by introducing $\alpha_{ij} = (1 - \frac{1}{2}\delta_{ij})\alpha_j$ we can write them as

$$E_J\{G(x_i, \xi)\} = \left\{\int_0^{x_i} G(x_i, \xi)\, d\xi - \sum_{j=1}^{i} \alpha_{ij} G(x_i, x_j)\right\}$$
$$+ \left\{\int_{x_i}^1 G(x_i, \xi)\, d\xi - \sum_{j=i}^{J} \alpha_{ij} G(x_i, x_j)\right\}, \qquad 1 \leq i \leq J.$$

Thus for $0 < x_i < 1$ the error is the sum of two quadrature errors employing the trapezoidal rule over $[0, x_i]$ and $[x_i, 1]$, respectively. More important, however, is the fact that the discontinuity of the first derivative of the integrand only occurs at the endpoints of these intervals. The same is true for

$x_i = 0$ and $x_i = 1$. Assuming sufficient differentiability of $f(x, y(x))$, we can apply the usual error estimates for the trapezoidal rule [see Isaacson and Keller (1966), p. 318] to each of the bracketed terms above to get

$$|E_J\{G(x_i, \xi)\}| \leq h^2 M,$$

where

$$M = \frac{1}{12} \max_{0 \leq x \leq 1} \left[\max_{\substack{\xi \in [0, x], \\ \xi \in [x, 1]}} \left| \frac{d^2 G(x, \xi)}{d\xi^2} \right| \right].$$

Using this result in Theorem 4.2.1 we find that *the trapezoidal rule yields accuracy which is at least* $\mathcal{O}(1/J^2)$.

It can be shown that the conditions in (4.2.2) are satisfied by any quadrature formula with positive coefficients whose degree of precision becomes unbounded as the number of nodes becomes infinite. Thus Theorem 4.2.1 is valid for ordinary Gaussian quadrature formulae. It is also valid for various composite quadrature formulae such as Simpson's rule. However, since the integrand in (4.1.11) does not, in general, have continuous derivatives, the usual estimates of the error in such higher-order accurate schemes can not be employed. Indeed, in some exceptional cases, the use of the trapezoidal rule in (4.2.1) may yield more accurate approximations to the exact solution of (4.1.11) than would be obtained by using Simpson's rule or Gaussian quadrature (with the same number of nodes). But these higher-order schemes are frequently quite accurate even when the standard error estimates are not valid [see Davis and Rabinowitz (1967), p. 122].

To circumvent this difficulty, caused by the presence of the Green's function in the integrand, we can employ special weighted quadrature formulae. That is, for each x_i we approximate an integral of the form

$$\int_0^1 g_k(x_i, \xi)\phi(\xi)\, d\xi, \qquad i = 1, 2, \ldots, J \tag{4.2.7a}$$

by a sum of the form

$$\sum_{j=1}^{J} \beta_{ij}\phi(x_j), \qquad i = 1, 2, \ldots, J. \tag{4.2.7b}$$

Again we require the equivalent of (4.2.2); nonnegative coefficients $\beta_{ij} \geq 0$ and convergence for all continuous functions $\phi(x)$,

$$\lim_{J\to\infty} E_{i,J}\{\phi(\xi)\} \equiv \lim_{J\to\infty} \left[\int_0^1 g_k(x_i, \xi)\phi(\xi)\, d\xi - \sum_{j=1}^{J} \beta_{ij}\phi(x_j)\right] = 0,$$
$$i = 1, 2, \ldots, J. \tag{4.2.8}$$

In (4.2.7b) we have actually introduced J different weighted quadrature formulae and so must now retain J^2 coefficients, β_{ij}, $i, j = 1, 2, \ldots, J$. The

ith scheme in (4.2.7b) is said to have degree of precision N if $E_{i,J}\{\xi^n\} = 0$ for $n = 1, 2, \ldots, N$ and $E_{i,J}\{\xi^{N+1}\} \neq 0$. If the degree of precision is at least zero, then

$$\sum_{j=1}^{J} \beta_{ij} = \int_0^1 g_k(x_i, \xi)\, d\xi = k^{-2}\left(1 - \frac{\cosh k(\frac{1}{2} - x_i)}{\cosh k/2}\right). \tag{4.2.9}$$

We do not study here the derivation of such schemes. Let it suffice to say that with J nodes, weighted quadrature formulae (4.2.7b) with degrees of precision $N_i \leq J$ can be determined [see Isaacson and Keller (1966), pp. 331–334].

In place of (4.2.1), we now consider the approximating system

$$u_i = \sum_{j=1}^{J} \beta_{ij}[k^2 u_j - f(x_j, u_j)], \qquad i = 1, 2, \ldots, J. \tag{4.2.10}$$

It is straightforward to prove a result similar to Theorem 4.2.1. In fact, if we require that (4.2.9) hold, then we can drop the small quantities corresponding to ε_{Jk} in (4.2.5). We leave the statement and proof of this result to Problem 4.2.1.

The application of either of the above procedures to systems of integral equations of the form (4.1.15) is quite clear. When using the weighted quadrature formulae, however, we must determine nJ^2 coefficients, that is, J^2 of them, for each Green's function $g_\nu(x, \xi)$, $\nu = 1, 2, \ldots, n$. For the more general systems of the form (4.1.16), the weighted formulae may not be appropriate. Thus we approximate their solutions by the set of n-vectors $\{\mathbf{u}_j\}$ which satisfy

$$\mathbf{u}_i = \sum_{j=1}^{J} \alpha_j \mathbf{F}(x_i, x_j; \mathbf{u}_j), \qquad i = 1, 2, \ldots, J. \tag{4.2.11}$$

A convergence proof for this case is again analogous to that of Theorem 4.2.1. We have the following.

THEOREM 4.2.2. *Let the quadrature formula with nodes* $\{x_j\}$ *and coefficients* $\{\alpha_j\}$ *satisfy* (4.2.2). *Let* $\mathbf{F}(x, \xi; \mathbf{y})$ *satisfy the hypothesis of Theorem* 4.1.3. *Then there exists an integer* J_0 *and positive quantities* $\varepsilon_J < 1 - \mu$ *such that for each* $J \geq J_0$:

$$|\mathbf{u}_i - \mathbf{y}(x_i)| \leq \frac{1}{1 - \mu - \varepsilon_J} \|\tau\|, \qquad i = 1, 2, \ldots, J.$$

Here $\{\mathbf{u}_i\}$ *is the solution of* (4.2.11), $\mathbf{y}(x)$ *is the solution of* (4.1.16), $\lim_{J\to\infty} \varepsilon_J = 0$ *and*

$$\|\tau\| = \max_{\substack{1 \leq i \leq J,\\ 1 \leq \nu \leq n}} E_J\{F_\nu(x_i, \xi; \mathbf{y}(\xi))\}.$$

Proof. The proof follows closely that of Theorem 4.2.1, using the expansion in Problem 4.1.4; we leave the details to Problem 4.2.2. ▌

We now turn to the problems of solving the nonlinear systems (4.2.1), (4.2.10), or (4.2.11). Of course part of our task is to show that these systems have solutions which are unique. Under appropriate conditions the obvious functional iteration schemes can be used to demonstrate these results. The details should be clear by now, so we are content to sketch the procedure as applied to the system (4.2.1).

The hypothesis of Theorem 4.2.1 is assumed to hold, and we take $k^2 \geq M$ and $J \geq J_k$. A sequence of net functions $\{u_j^{(\nu)}\}$, $\nu = 0, 1, 2, \ldots$ is defined as follows:

$$\text{(a)}\quad u_i^{(0)} = \text{arbitrary}, \qquad i = 1, 2, \ldots, J;$$

$$\text{(b)}\quad u_i^{(\nu+1)} = \sum_{j=1}^{J} \alpha_j g_k(x_i, x_j)[k^2 u_j^{(\nu)} - f(x_j, u_j^{(\nu)})], \qquad i = 1, 2, \ldots, J. \tag{4.2.12}$$

We first prove that $\{u_j^{(\nu)}\}$ is a Cauchy sequence (in ν, for each j) by the procedures used in Theorems 4.1.2 and 4.2.1. That is, with the definitions $e_i^{(\nu+1)} \equiv u_i^{(\nu+1)} - u_i^{(\nu)}$, we get from (4.2.12b) for $\nu + 1$ and ν

$$|e_i^{(\nu+1)}| = \left|\sum_{j=1}^{J} \alpha_j g_k(x_i, x_j)[k^2 - f_{y,j}]e_j^{(\nu)}\right| \leq \left[\sum_{j=1}^{J} \alpha_j g_k(x_i, x_j)k^2\right] \|e^{(\nu)}\|$$

$$\leq \left(1 - \frac{1}{\varepsilon_{Jk} + \cosh(k/2)}\right)\|e^{(\nu)}\| \leq \mu_k \|e^{(\nu)}\|, \qquad i = 1, 2, \ldots, J.$$

Thus we get

$$\|e^{(\nu+1)}\| \leq \mu_k \|e^{(\nu)}\|,$$

and since $\mu_k < 1$ the $\{u_j^{(\nu)}\}$ forms a Cauchy sequence whose limit, say $\{u_j\}$, satisfies (4.2.1). Existence of a solution is thus demonstrated and uniqueness easily follows since for any other solution $\{v_j\}$ we obtain

$$\|u - v\| \leq \mu_k \|u - v\|,$$

and hence $\{u_j\} \equiv \{v_j\}$. With essentially no more difficulty the systems (4.2.10) and (4.2.11) can be treated analogously.

It should be stressed that the indicated functional iterations converge for *any* initial guess at the solution. But the rate of convergence may be quite slow; in the above case μ_k can be very near unity. In seeking more-rapidly-convergent schemes Newton's method is again a likely choice. To derive it for the system (4.2.1) we replace u_j by $u_j^{(\nu)} + \Delta u_j^{(\nu)}$, expand in powers of Δu

and drop all second- and higher-order terms to get the linear system

$$\Delta u_i^{(\nu)} = \sum_{j=1}^{J} \alpha_j g_k(x_i, x_j)[k^2 - f_y(x_j, u_j^{(\nu)})]\, \Delta u_j^{(\nu)} + r_i^{(\nu)}, \qquad i = 1, 2, \ldots \tag{4.2.13a}$$

with

$$r_i^{(\nu)} \equiv \sum_{j=1}^{J} \alpha_j g_k(x_i, x_j)[k^2 u_j^{(\nu)} - f(x_j, u_j^{(\nu)})] - u_i^{(\nu)}. \tag{4.2.13b}$$

The scheme is then

$$u_i^{(0)} = \text{arbitrary}, \qquad u_i^{(\nu+1)} = u_j^{(\nu)} + \Delta u_i^{(\nu)}, \qquad \nu = 0, 1, \ldots,$$

where the $\Delta u_i^{(\nu)}$ are as defined in (4.2.13). To show that this linear system is nonsingular, so that Newton's method is well defined, we need only observe that the coefficient matrix is diagonally dominant (provided $k^2 \geq M$ and $J \geq J_k$, as in Theorem 4.2.1). A "sufficiently close" initial estimate to insure convergence (by Theorem 1.4.3) can be obtained by using the scheme (4.2.12) for several iterations.

Finally we point out that Newton's method as defined above is also suggested by first linearizing the integral equation (4.1.11) about some approximate solution, say $y^{(\nu)}(x)$, to obtain a correction $\Delta y^{(\nu)}(x)$ as the solution of

$$\Delta y^{(\nu)}(x) = \int_0^1 g_k(x, \xi)[k^2 - f_y(\xi, y^{(\nu)}(\xi))]\, \Delta y^{(\nu)}(\xi)\, d\xi + r^{(\nu)}(x), \tag{4.2.14a}$$

where

$$r^{(\nu)}(x) = \int_0^1 g_k(x, \xi)[k^2 y^{(\nu)}(\xi) - f(\xi, y^{(\nu)}(\xi))]\, d\xi - y^{(\nu)}(x). \tag{4.2.14b}$$

Then if the linear integral equation (4.2.14) is numerically approximated, in the obvious way, we obtain (4.2.13). Thus even if the continuous iterates $y^{(\nu+1)}(x) = y^{(\nu)}(x) + \Delta y^{(\nu)}(x)$, $\nu = 0, 1, \ldots$ converge quite rapidly to the solution $y(x)$ of (4.1.11), the accuracy of the corresponding numerical approximation is determined as in Theorem 4.2.1. These considerations are exactly analogous to those in (3.2.11) et seq. The reader is urged to consider Newton's method for both the discrete problem (4.2.11) and the corresponding system of integral equations (4.1.16) [see Problems 4.2.3 and 4.2.4].

Problems

4.2.1 Prove the following THEOREM 4.2.1′: *Let the quadrature formula* (4.2.7b) *have nonnegative coefficients and degree of precision at least zero. Let* $f(x, y)$ *satisfy the hypothesis of Theorem* 4.2.1. *Then for any* $k^2 > M$ *the solutions* $\{u_j\}$ *of* (4.2.10) *and* $y(x)$ *of* (4.1.11) *satisfy:*

$$|u_i - y(x_i)| \leq \cosh\frac{k}{2} \max_{1 \leq j \leq J} |E_{j,J}\{[k^2 y(\xi) - f(\xi, y(\xi))]\}|, \qquad i = 1, 2, \ldots, J.$$

4.2.2 Carry out the details in the proof of Theorem 4.2.2.

4.2.3 Formulate the application of Newton's method for the solution of the nonlinear system (4.2.11). Under the conditions of Theorems 4.1.3 and 4.2.2, show that for sufficiently large J the relevant coefficient matrix is always nonsingular.

4.2.4 Derive the linear integral equations resulting from the application of Newton's method to the nonlinear system of integral equations (4.1.16). Let the continuous iterates defined in this way be approximated by employing a quadrature formula to evaluate the integrals on some net. Can the algebraic problem of Problem 4.2.3 be obtained in this way?

4.2.5 (a) Replace the sequence of linear differential equations in (3.2.12) by a corresponding sequence of centered difference equations. Write these equations in the vector form

$$A\mathbf{v}^{(\nu+1)} = \mathbf{F}(\mathbf{v}^{(\nu)})$$

and determine the matrix A and vector function $\mathbf{F}(\mathbf{v})$.

(b) Write the iteration scheme (4.2.12) for solving the numerical approximation (4.2.1) to the integral equation (4.1.11) in the vector form

$$\mathbf{u}^{(\nu+1)} = B\mathbf{G}(\mathbf{u}^{(\nu)}),$$

and determine the matrix B and vector function $\mathbf{G}(\mathbf{u})$.

(c) Discuss the relation between the two algebraic systems above if the original boundary-value problems to be solved were identical and $\omega = k^2$. Is $B = A^{-1}$ when the trapezoidal rule is used in (4.1.11)? Compare operational counts in evaluating $\mathbf{v}^{(\nu+1)}$ and $\mathbf{u}^{(\nu+1)}$ when A is tridiagonal and B has essentially J^2 nonzero elements (that is, $h = 1/J$).

4.2.6 With a little more care in bounding the integral in the proof of Theorem 4.1.2, we may allow f_y to be "slightly" negative.

THEOREM 4.2.3. *Let f_y be continuous and satisfy, for some $k > 0$ and θ in $0 < \theta \leqslant 1$,*

$$k^2 \frac{1-\theta}{1-\cosh k/2} \leqslant \frac{\partial f(x, y)}{\partial y} \leqslant k^2 \tag{4.2.15}$$

for all $x \in [0, 1]$ and all y. Then a unique solution of (4.1.7) *exists and is given as the limit of the sequence of iterates $\{y^{(\nu)}(x)\}$ defined in* (4.1.12). [HINT: Use the bound $(k^2 - f_y) \leqslant (k^2 - \min f_y)$. The convergence factor is now $\mu = 1 - \theta/\cosh(h/2)$.] Much better results allowing $f_y < 0$ are given by M. Lees (1966).

4.2.7 The requirement that f satisfy appropriate conditions for all y is very severe. It is not difficult to eliminate such conditions. For example, one such weakening of the hypothesis can be stated as follows.

THEOREM 4.2.4. *Let $M > 0$ and $k > 0$ be such that*

$$|f(x, y)| \leqslant \frac{k^2}{\cosh k/2 - 1} M \tag{4.2.16}$$

for all $x \in [0, 1]$ and y in $|y| \leqslant M$. In addition let f_y satisfy (4.2.15) *with some $\theta \in (0, 1]$ for all $x \in [0, 1]$ and y in $|y| \leqslant M$. Then* (4.1.7) *has a unique solution satisfying*

$$|y(x)| \leqslant M,$$

and it is the limit of the iterates $\{y^{(\nu)}(x)\}$ in (4.1.12). [HINT: By induction show that $|y^{(\nu)}(x)| \leqslant M$, and then the proof of Theorem 4.2.3 is applicable.]

SUPPLEMENTARY REFERENCES AND NOTES

Section 4.1 Conversion of a boundary-value problem to an integral equation is one of the standard devices used in proving existence theorems [see Hartman (1964) and Courant-Hilbert (1953)]. There are procedures other than using Green's functions for replacing differential equation problems by equivalent integral equations. In particular, for first-order systems, an integration over the interval in question yields a system of integral equations subject to the original boundary conditions (which can frequently be incorporated into the integral equations).

Section 4.2 The theory of linear integral equations [as in Courant-Hilbert (1953)] leads to approximation methods not based on numerical quadrature. Some of these procedures are numerical, however, in the sense that large-scale computations are required for their implementation [see Kantorovich and Krylov (1958)].

CHAPTER 5

EIGENVALUE PROBLEMS

5.1 Introduction; Sturm-Liouville Problems

It is easily shown that a linear boundary-value problem may have nonunique solutions. This occurs in fact if and only if the corresponding homogeneous boundary-value problem has a nontrivial solution (see Theorem 1.2.3). If the coefficients of the equation and/or of the boundary conditions depend upon a parameter, it is frequently of interest to determine the value or values of the parameter for which such nontrivial solutions exist. These special parameter values are called eigenvalues and the corresponding nontrivial solutions are called eigenfunctions. A particularly simple and standard example is furnished by the homogeneous problem

$$y'' + \lambda y = 0; \qquad y(a) = y(b) = 0.$$

For each of the parameter values (eigenvalues)

$$\lambda = \lambda_n \equiv [n\pi/(b-a)]^2, \qquad n = 1, 2, \ldots$$

there exists a corresponding nontrivial solution (eigenfunction)

$$y(x) = y_n(x) \equiv c_n \sin \lambda_n^{1/2}(x-a), \qquad n = 1, 2, \ldots.$$

We note that the nth eigenfunction is nonunique to within an arbitrary constant factor c_n.

A fairly general class of eigen-problems, which includes many of the cases that occur in applied mathematics, is the Sturm-Liouville problems:

$$\begin{aligned} &\text{(a)} \quad Ly + \lambda r(x)y \equiv (p(x)y')' - q(x)y + \lambda r(x)y = 0, \\ &\text{(b)} \quad a_0 y(a) - a_1 p(a)y'(a) = 0, \qquad b_0 y(b) + b_1 p(b)y'(b) = 0. \end{aligned} \tag{5.1.1}$$

Here $p(x) > 0$, $r(x) > 0$, and $q(x) \geq 0$ while $p'(x)$, $q(x)$, and $r(x)$ are continuous on $[a, b]$. The constants a_ν and b_ν are nonnegative and at least one of each pair does not vanish. It is known that for such problems there exists an infinite sequence of nonnegative eigenvalues (see Problem 1.1)

$$0 \leq \lambda_1 < \lambda_2 < \lambda_3 \cdots. \tag{5.1.2a}$$

In addition there exist corresponding eigenfunctions, $y_n(x)$, which are twice continuously differentiable and satisfy the orthogonality relations:

$$\int_a^b y_n(x)y_m(x)r(x)\,dx = \delta_{nm}, \qquad n, m = 1, 2, \ldots. \tag{5.1.2b}$$

The normalization condition, (5.1.2b) with $n = m$, serves to make the eigenfunctions unique to within sign. It is also known that the nth eigenfunction has $n - 1$ distinct zeros in $a < x < b$. A proof of these results based on initial-value problems is contained in Coddington and Levinson (1955), pp. 189–190, 211–213.

We may relate the solution of (5.1.1) to an initial-value problem (just as in Section 2.2 for boundary-value problems). For any fixed λ we consider

$$\begin{aligned} &\text{(a)} \quad Lu + \lambda r(x)u = 0; \\ &\text{(b)} \quad a_0u(a) - a_1p(a)u'(a) = 0, \qquad c_0u(a) - c_1p(a)u'(a) = 1. \end{aligned} \tag{5.1.3}$$

Here c_0 and c_1 are any constants such that $(a_1c_0 - a_0c_1) \neq 0$. Then the two initial conditions in Equations (5.1.3b) are linearly independent and a unique nontrivial solution of the initial-value problem (5.1.3) exists. We denote this solution by $u(\lambda; x)$ and consider the (in general) transcendental equation

$$\phi(\lambda) \equiv b_0u(\lambda; b) + b_1p(b)u'(\lambda; b) = 0. \tag{5.1.4}$$

Clearly each eigenvalue λ_n in (5.1.2a) must satisfy this equation. Also every root, λ^*, of this equation is an eigenvalue of (5.1.1) and the corresponding solution $u(\lambda^*; x)$ of (5.1.3) is a corresponding eigenfunction of (5.1.1). We shall show, in Theorem 5.2.1, that the roots of (5.1.4) are simple. Note that the present analysis differs from the corresponding discussion of Section 1.2 only in that now a parameter in the equation must be adjusted and the adjoined initial condition remains fixed while previously the added initial condition was varied to satisfy the second boundary condition.

Of course the present considerations apply to eigenvalue problems more general than those in (5.1.1), for example, to problems in which the eigenvalue parameter λ enters into all of the coefficients of the equation and in the boundary conditions perhaps in a nonlinear way. Extensions to homogeneous systems of, say, m second-order equations with m parameters are also clearly suggested. We briefly discuss nonlinear eigenvalue problems in Section 5.5, that is, nonlinear boundary-value problems in which a parameter must be determined so that a solution exists (trivial or not).

To approximate the eigenvalues and eigenfunctions for problems of the form (5.1.1), and various generalizations of these problems, we may apply numerical methods which are exactly analogous to those used in Chapters 2, 3, and 4. But the proofs of convergence and estimates of the errors are now not always as easy to obtain as they were for the boundary-value problems. As with the boundary-value problems, the initial-value methods are more generally applicable and can easily yield higher-order-accurate approximations. The application of integral equation methods, in Section 5.4, automatically leads us to consider eigenvalue problems for integral equations which may not be associated with any particular Sturm-Liouville (or other) differential equation.

There are very important approximation methods for linear eigenvalue problems that are based on *variational principles*. These include some relevant numerical methods. But we do not discuss them here because the development of the required background should be done more carefully than can be done in our allotted space [see Collatz (1960), pp. 202–222].

Problems

5.1.1 (a) Show that the eigenvalues of (5.1.1) must be *nonnegative*. [HINT: Multiplying Equation (5.1.1a) by $y(x)$ and integrating, we get (since $r(x) > 0$ and $y(x) \not\equiv 0$)

$$\lambda = -\int_a^b y(x)Ly(x)\,dx \Big/ \int_a^b r(x)y^2(x)\,dx.$$

Use partial integration and the boundary conditions (5.1.1b) to show that the numerator is nonnegative since $p(x) > 0$, $q(x) \geq 0$, $a_0a_1 \geq 0$, and $b_0b_1 \geq 0$.]

(b) Show that the eigenvalues are *positive* if $a_0^2 + b_0^2 \neq 0$.

5.1.2 If $\lambda_n \neq \lambda_m$ are eigenvalues with corresponding eigenfunctions $y_n(x)$ and $y_m(x)$ of the problem (5.1.1), show that (5.1.2b) holds (for $n \neq m$). [HINT: Use partial integration in

$$\int_a^b (y_nLy_m - y_mLy_n)\,dx + (\lambda_m - \lambda_n)\int_a^b y_ny_mr(x)\,dx = 0.]$$

5.1.3 Use initial-value problems as in (5.1.3) to find the eigenvalues and eigenfunctions for the problem

$$y'' + \lambda y = 0, \qquad y'(a) = 0, \qquad y(b) = 0.$$

For example, require $y'(a) = 0$, $y(a) = 1$ and determine λ such that $y(b) = 0$.

5.2 Initial-Value Methods for Eigenvalue Problems

We consider the Sturm-Liouville problem (5.1.1), which has been reduced to finding the roots of Equation (5.1.4) and corresponding solutions of the initial-value problem (5.1.3). To approximate the solution of the initial-value problem (5.1.3) we first replace it by an equivalent first-order system, such as

$$u'(x) = \frac{v(x)}{p(x)}, \qquad u(a) = \frac{a_1}{a_1c_0 - a_0c_1};$$
$$v'(x) = [q(x) - \lambda r(x)]u(x), \qquad v(a) = \frac{a_0}{a_1c_0 - a_0c_1}. \tag{5.2.1}$$

The unique solution of this problem, for any λ, is denoted by

$$u(\lambda; x), \qquad v(\lambda; x);$$

and the equation (5.1.4) can be written as

$$\phi(\lambda) \equiv b_0u(\lambda; b) + b_1v(\lambda; b) = 0. \tag{5.2.2}$$

Now on some net, say a uniform net, for convenience, $x_j = jh + a$, $h = (b - a)/J$, we determine an approximate solution of (5.2.1) by one of the numerical procedures of Section 1.3. We denote this numerical solution for any fixed λ by

$$U_j(\lambda), \qquad V_j(\lambda); \qquad j = 0, 1, 2, \ldots.$$

If, as we assume, the numerical scheme is stable and has order of accuracy r and the solution of (5.2.1) is sufficiently smooth, then:

$$|U_j(\lambda) - u(\lambda; x_j)| \le Mh^r, \qquad |V_j(\lambda) - v(\lambda; x_j)| \le Mh^r. \tag{5.2.3}$$

Here the constant M is a bound on certain higher-order derivatives of u or v and thus depends upon the parameter λ; thus we write $M = M(\lambda)$. To approximate $\phi(\lambda)$ we use the numerical solution and define

$$\Phi(\lambda) \equiv b_0U_J(\lambda) + b_1V_J(\lambda). \tag{5.2.4}$$

From (5.2.2) and (5.2.3) it follows that, with $B \equiv b_0 + b_1 > 0$,

$$|\Phi(\lambda) - \phi(\lambda)| \le BM(\lambda)h^r. \tag{5.2.5}$$

It is clear that, just as in Section 2.2, we can expect at best to approximate any eigenfunction to an accuracy of $\mathcal{O}(h^r)$. In fact we note that since the right-hand side of the system (5.2.1) is linear in λ, it follows by the remarks after (1.1.3) (or Theorem 1.1.3), that $u(\lambda; x)$ and $v(\lambda; x)$ are Lipschitz-continuous in λ. Then clearly for any λ and μ

$$\begin{aligned} |U_j(\lambda) - u(\mu; x_j)| &\le |U_j(\lambda) - u(\lambda; x_j)| + |u(\lambda; x_j) - u(\mu; x_j)| \\ &\le M(\lambda)h^r + K|\lambda - \mu|, \end{aligned} \tag{5.2.6a}$$

where K is the appropriate Lipschitz constant. A similar result applies for $v(\mu; x)$; that is,

$$|V_j(\lambda) - v(\mu; x_j)| \le M(\lambda)h^r + K|\lambda - \mu|. \tag{5.2.6b}$$

(This result is the analog of Lemma 2.2.1.) Thus, if $\mu = \lambda_n$ is an eigenvalue of (5.1.1) and $|\lambda - \lambda_n| = \mathcal{O}(h^r)$, then we actually obtain $\mathcal{O}(h^r)$ approximations to the eigenfunction $y_n(x) = u(\lambda_n; x)$, and its first derivative, $y_n'(x) = v(\lambda_n; x)/p(x)$.

To devise an iteration scheme for computing a root of (5.2.2) we first examine the derivative $d\phi(\lambda)/d\lambda \equiv \dot{\phi}(\lambda)$ at any root $\lambda = \lambda_n$. We have

THEOREM 5.2.1. *The roots* $\lambda = \lambda_n$ *of Equation* (5.2.2) *are simple; that is,* $\dot{\phi}(\lambda_n) \ne 0$, $n = 1, 2, \ldots$.

Proof. It follows from (5.2.1) and Theorem 1.1.3 that $\xi(\lambda; x) \equiv \partial u(\lambda; x)/\partial\lambda$ and $\eta(\lambda; x) \equiv \partial v(\lambda; x)/\partial\lambda$ exist, are continuous, and form the solution of the variational problem

$$\begin{aligned} \xi' &= p^{-1}(x)\eta, \qquad \xi(a) = 0; \\ \eta' &= [q(x) - \lambda r(x)]\xi - r(x)u(\lambda; x), \qquad \eta(a) = 0. \end{aligned} \tag{5.2.7}$$

Setting $\lambda = \lambda_n$, an eigenvalue of (5.1.1) and hence a root of Equation (5.2.2), we find that ξ satisfies

$$L\xi + \lambda_n r(x)\xi = -r(x)y_n(x), \qquad a_0\xi(a) - a_1 p(a)\xi'(a) = 0.$$

Now suppose that $\dot{\phi}(\lambda_n) = b_0\xi(b) + b_1 p(b)\xi'(b) = 0$. Then ξ is the solution of a linear *boundary*-value problem whose corresponding homogeneous problem has the nontrivial solution $y_n(x)$. Thus by the complete Alternative Theorem the inhomogeneous term, $-r(x)y_n(x)$, must be orthogonal to $y_n(x)$ (since L is self-adjoint) [see Courant and Hilbert (2) pp. 355–356]. This orthogonality condition now states that

$$\int_a^b r(x)y_n^2(x)\,dx = 0.$$

However this is impossible since $y_n(x) \not\equiv 0$ and $r(x) > 0$ [or see for instance (5.1.2b)]. Thus it follows that $\dot{\phi}(\lambda_n) \ne 0$. ∎

We now proceed to show that, at least in principle, every root of Equation (5.2.2) can be determined by iteration. For this purpose we may employ a different parameter, m_n, for each root and we write Equation (5.2.2), for each $n = 1, 2, \ldots$, as

$$\lambda = g_n(\lambda) \equiv \lambda - m_n\phi(\lambda). \tag{5.2.8a}$$

Then we consider the obvious functional iteration schemes, $\lambda_n^{(0)}$ = arbitrary and

$$\lambda_n^{(\nu+1)} = g_n(\lambda_n^{(\nu)}), \qquad \nu = 0, 1, 2, \ldots. \tag{5.2.8b}$$

For appropriate choices of m_n and $\lambda_n^{(0)}$ the sequence $\{\lambda_n^{(\nu)}\}$ will converge to the nth eigenvalue λ_n. This is essentially the content of (compare with Theorem 2.2.1)

THEOREM 5.2.2. *For each $n = 1, 2, \ldots$ there are constants $\rho_n > 0$, γ_n and Γ_n with $\gamma_n\Gamma_n > 0$ such that for any $\lambda_n^{(0)}$ in*

$$|\lambda_n - \lambda_n^{(0)}| \leq (1 - K_n)\rho_n$$

and any m_n in

$$\begin{aligned} 0 < m_n < \frac{2}{\Gamma_n} \quad & \textit{if} \quad \Gamma_n > 0, \\ \frac{2}{\Gamma} < m_n < 0 \quad & \textit{if} \quad \Gamma_n < 0, \end{aligned} \tag{5.2.9a}$$

the iterates (5.2.8b) *converge to the eigenvalue λ_n. With the choice $m_n = 2/(\Gamma_n + \gamma_n)$ the iterates satisfy*

$$|\lambda_n^{(\nu)} - \lambda_n| \leq \left[\frac{1 - (\gamma_n/\Gamma_n)}{1 + (\gamma_n/\Gamma_n)}\right]^\nu \frac{|\phi(\lambda_n^{(0)})|}{|\gamma_n|}, \qquad \nu = 1, 2, \ldots.$$

The magnitudes of γ_n and Γ_n are given by

$$\begin{aligned} |\gamma_n| &= \min_{|\lambda_n - \lambda| \leq \rho_n} |b_0\xi(\lambda; b) + b_1 p(b)\eta(\lambda; b)|, \\ |\Gamma_n| &= \max_{|\lambda_n - \lambda| \leq \rho_n} |b_0\xi(\lambda; b) + b_1 p(b)\eta(\lambda; b)|, \end{aligned} \tag{5.2.9b}$$

where ξ, η is the solution of (5.2.7) *and sign $\gamma_n \equiv$ sign $\Gamma_n \equiv$ sign $\dot{\phi}(\lambda_n)$. The constant K_n is $K_n \equiv \min(|1 - m_n\gamma_n|, |1 - m_n\Gamma_n|)$.*

Proof. By Theorem 5.2.1 we know that $\dot{\phi}(\lambda_n) \neq 0$ for each $n = 1, 2, \ldots$. But since the solution ξ, η of (5.2.7) depends continuously on λ it follows that

$$\dot{\phi}(\lambda) = b_0\xi(\lambda; b) + b_1 p(b)\eta(\lambda; b) \tag{5.2.10}$$

is also a continuous function of λ. Then there is a neighborhood of each root in which $\dot{\phi}(\lambda)$ is bounded away from zero, let us say

$$\dot{\phi}(\lambda) \neq 0 \quad \text{for} \quad \lambda \text{ in } |\lambda - \lambda_n| \leq \rho_n; \qquad n = 1, 2, \ldots.$$

Now we define γ_n and Γ_n as follows [that is, as in Equation (5.2.9a)]:

$$\text{sign}\,\gamma_n \equiv \text{sign}\,\Gamma_n \equiv \text{sign}\,\dot{\phi}(\lambda_n),$$
$$|\gamma_n| \equiv \min_{|\lambda-\lambda_n|\le\rho_n} |\dot{\phi}(\lambda)|, \qquad |\Gamma_n| \equiv \max_{|\lambda-\lambda_n|\le\rho_n} |\dot{\phi}(\lambda)|.$$

Of course it is clear from (5.1.2a) and the continuity of $\phi(\lambda)$ that $(-1)^n\dot{\phi}(\lambda_n)$ has a fixed sign; thus the pairs (γ_n, Γ_n) will alternate in sign.

From the mean-value theorem it follows that for any λ and μ with g_n as defined in (5.2.8a)

$$\begin{aligned} g_n(\lambda) - g_n(\mu) &= (\lambda - \mu) - m_n[\phi(\lambda) - \phi(\mu)], \\ &= [1 - m_n\dot{\phi}(\mu + \theta(\lambda - \mu))](\lambda - \mu), \qquad 0 < \theta < 1. \end{aligned}$$

Now let us restrict λ and μ to within ρ_n of λ_n and take m_n as specified in (5.2.9a). Then clearly $m_n\dot{\phi}\,(\mu + \theta(\lambda - \mu)) > 0$ and in this ρ_n neighborhood of λ_n the function $g_n(\lambda)$ satisfies a Lipschitz condition with constant

$$K_n = \max\,(|1 - m_n\gamma_n|, \qquad |1 - m_n\Gamma_n|) < 1,$$

since

$$0 < |\gamma_n| \le |\Gamma_n| \qquad \text{and} \qquad 0 < |m_n| < 2/|\Gamma_n|.$$

(With the choice $m_n = 2/(\Gamma_n + \gamma_n)$ we get $K_n = |\Gamma_n - \gamma_n|/|\Gamma_n + \gamma_n|$.)

Now the hypothesis of Theorem 1.4.1 applies, and the proof is concluded by applying this theorem. ∎

Of course we cannot evaluate $\phi(\lambda)$ exactly but rather the approximation $\Phi(\lambda)$ in (5.2.4). Thus the sequence of numerical approximations to the eigenvalue λ_n is actually defined in terms of $\lambda_n^{(0)}$ by

$$\lambda_n^{(\nu+1)} = \lambda_n^{(\nu)} - m_n\Phi(\lambda_n^{(\nu)}), \qquad \nu = 0, 1, \ldots. \tag{5.2.11}$$

But if $\lambda_n^{(0)}$ is sufficiently close to λ_n we can obtain, by this procedure, approximations to the nth eigenvalue, its corresponding eigenvector and the derivative of this eigenvector which are accurate to $\mathcal{O}(h^r)$. These results are from the following.

THEOREM 5.2.3. *Let* $|\lambda_n^{(0)} - \lambda_n| \le \rho_{n,0} \equiv \rho_n - M_nh^r$, *where* ρ_n *and* m_n *are as defined in Theorem* 5.2.2 *and*

$$M_n \equiv |m_n|B \max_{|\lambda-\lambda_n|\le\rho_n} M(\lambda).$$

Let $\lambda_n^{(\nu)}$ *be defined by* (5.2.11), *where* $U_j(\lambda_n^{(\nu)})$ *and* $V_j(\lambda_n^{(\nu)})$ *are numerical solutions of* (5.2.1) *using a stable scheme with order of accuracy r on the net* $x_j = a + jh$.

Then for some eigenfunction $y_n(x)$ *of* (5.1.1) *belonging to the eigenvalue* λ_n:

$$\left.\begin{aligned}&\text{(a)}\quad |\lambda_n^{(\nu)} - \lambda_n| \le \mathcal{O}(h^r) + \mathcal{O}(K_n^\nu),\\ &\text{(b)}\quad |U_j(\lambda_n^{(\nu)}) - y_n(x_j)| \le \mathcal{O}(h^r) + \mathcal{O}(K_n^\nu),\\ &\text{(c)}\quad |V_j(\lambda_n^{(\nu)}) - p(x_j)y_n'(x_j)| \le \mathcal{O}(h^r) + \mathcal{O}(K_n^\nu),\end{aligned}\right\}\begin{aligned}&\nu = 1, 2, \dots,\\ &j = 1, 2, \dots, J,\end{aligned} \tag{5.2.12}$$

where $0 < K_n < 1$ *is as defined in Theorem* 5.2.2.

Proof. We first note that by (5.2.5) and (5.2.8a) the iteration scheme (5.2.11) is of the form

$$\lambda_n^{(\nu+1)} = g_n(\lambda_n^{(\nu)}) + \delta_n^{(\nu)},$$

where

$$|\delta_n^{(\nu)}| \le M_n h^r.$$

But from the hypothesis and Theorem 5.2.2 it follows that Theorem 1.4.2 is applicable. Then from conclusion (b) of that theorem we get (5.2.12a).

Since the eigenfunction of (5.1.1) corresponding to an eigenvalue λ_n is nonunique to within an arbitrary scalar factor, we can take $y_n(x)$ such that

$$y_n(a) = u(\lambda_n; a), \qquad y_n'(a) = p^{-1}(a)v(\lambda_n; a),$$

where $u(\lambda_n; x)$, $v(\lambda_n; x)$ is the solution of (5.2.1) with $\lambda = \lambda_n$. Then clearly $y_n(x) \equiv u(\lambda_n; x)$, $y_n'(x) = p^{-1}(x)v(\lambda_n; x)$ and using (5.2.12a) in (5.2.6) with $\lambda = \lambda_n^{(\nu)}$ and $\mu = \lambda_n$ we obtain (5.2.12b–c). ∎

This result is of course quite similar to Theorem 2.2.2. Also, since K_n is independent of h, we can define an integer $\nu^* = \nu(h, r, K_n)$ such that

$$K_n^{\nu^*} \le h^r.$$

Then the error bounds in (5.2.12) are all $\mathcal{O}(h^r)$ for $\nu \ge \nu^*$.

While the above results furnish a theoretical justification for the initial-value method applied to Sturm-Liouville eigenvalue problems, they do not give practical estimates for the important parameters m_n or initial guesses $\lambda_n^{(0)}$. In practice these could be obtained by computing $\phi(\lambda)$ [really $\Phi(\lambda)$] to locate, roughly, sign changes and estimates of the slope $\dot{\phi}(\lambda)$ near these sign changes. If (as is most likely in such computations) a sequence of eigenvalues is desired, we can employ the value $m_{n+1} = -m_n$ as a reasonable (first) estimate for m_{n+1}.

Of course higher-order iteration schemes would be preferable, particularly when many eigenvalues are to be computed, and Newton's method is a reasonable choice. Thus in place of (5.2.8) we would use, given $\lambda_n^{(0)}$,

$$\lambda_n^{(\nu+1)} = \lambda_n^{(\nu)} - \frac{\phi(\lambda_n^{(\nu)})}{\dot{\phi}(\lambda_n^{(\nu)})}, \qquad \nu = 0, 1, 2, \dots. \tag{5.2.13}$$

The derivative $\dot{\phi}(\lambda)$ is given by (5.2.10), in terms of the solution of the variational problem (5.2.7). Thus while solving the initial-value problem (5.2.1) numerically to compute $\Phi(\lambda_n^{(\nu)})$, the approximation to $\phi(\lambda_n^{(\nu)})$, we also solve (5.2.7) numerically to compute an approximation to $\dot{\phi}(\lambda_n^{(\nu)})$. It is quite clear in this case that very little extra computation is required to solve the variational problem, as all the coefficients must be evaluated previously in order to solve (5.2.1). Thus Newton's method is quite efficient for such problems.

Problems

5.2.1 Calculate the first seven eigenvalues of the problem

$$y'' + \lambda y = 0, \qquad y'(0) = y(2) = 0,$$

using shooting with: (a) functional iteration and (b) Newton's method. [See following problem for part (a).]

5.2.2 Show in detail how Theorem 5.2.2 applies in the above when the initial conditions used are

$$u'(0) = 0, \qquad u(0) = 1.$$

Verify in particular that the iterations (5.2.8) converge for the nth eigenvalue, provided that

$$0 < |m_n| < \frac{\sqrt{(2n-1)\pi}}{2}, \qquad (-1)^n m_n > 0, \qquad n = 1, 2, \ldots;$$

and the initial guess $\lambda_n^{(0)}$ is such that

$$|\sqrt{\lambda_n} - \sqrt{\lambda_n^{(0)}}| < \pi/4.$$

[Try using the value $m_n = (-1)^n\sqrt{(2n-1)\pi}/2$ in the calculations of part (a) in Problem 5.2.1.]

5.2.3 The nth positive zero, k_n, of the Bessel function $J_0(x)$ is the square root of the nth eigenvalue of

$$\frac{1}{x}(xy')' + k^2 y = 0, \qquad y'(0) = y(1) = 0.$$

Compute the first four such roots by the shooting method and compare them to the first four zeros of the solution of the initial-value problem

$$\frac{1}{x}(xJ_0')' + J_0 = 0; \qquad J(0) = 1, \qquad J_0'(0) = 0.$$

5.2.4 Show that Newton's method, as formulated in (5.2.13), for finding the eigenvalues of the problem

$$(*) \qquad y'' + \lambda^2 y = 0, \qquad y(0) = y(1) = 0$$

can be equivalent to using Newton's method for computing the roots of

$$\sin \lambda = 0$$

or of

$$\frac{\sin \lambda}{\lambda} = 0.$$

[HINT: Consider the initial-value problem

$$u' = \lambda v, \qquad u(0) = 0; \qquad v' = -\lambda u, \qquad v(0) = 1;$$

and the more obvious system equivalent to (*) suggested by (5.2.1).]

5.3 *Finite-Difference Methods for Eigenvalue Problems*

To study difference methods for eigen-problems we shall consider a slightly-less-general problem than that treated in the previous section. More general problems will then be covered in the exercises at the end of the section. Some of the present analysis in fact easily extends to partial differential equations. We consider first the problem

$$\begin{aligned} &\text{(a)}\quad Ly + \lambda y \equiv (p(x)y')' - q(x)y + \lambda y = 0;\\ &\text{(b)}\quad y(a) = 0, \qquad y(b) = 0; \end{aligned} \tag{5.3.1}$$

where $p(x) > 0$, $p'(x)$ and $q(x)$ are continuous, and $q(x) \geq 0$. This is a special case of (5.1.1) and so the eigenvalues and eigenvectors have the properties stated in and after (5.1.2).

Now we employ the uniform net

$$x_j = a + jh, \qquad j = 0, 1, \ldots, J + 1; \qquad h \equiv \frac{(b - a)}{J + 1},$$

and replace (5.3.1) by some finite-difference equations of the form

$$\begin{aligned} &\text{(a)}\quad L_h u_j + \Lambda u_j = 0, \qquad j = 1, 2, \ldots, J;\\ &\text{(b)}\quad u_0 = 0, \qquad u_{J+1} = 0. \end{aligned} \tag{5.3.2}$$

Here L_h is a "difference approximation" of the linear differential operator L, which for the present we take to be

$$L_h u_j \equiv \frac{1}{h}\left\{p\left(x_j + \frac{h}{2}\right)\left[\frac{u_{j+1} - u_j}{h}\right] - p\left(x_j - \frac{h}{2}\right)\left[\frac{u_j - u_{j-1}}{h}\right]\right\} - q(x_j)u_j. \tag{5.3.3a}$$

This can be written as

$$h^2 L_h u_j \equiv a_j u_{j-1} + b_j u_j + c_j u_{j+1}; \tag{5.3.3b}$$

$$a_j \equiv p\left(x_j - \frac{h}{2}\right), \qquad c_j \equiv p\left(x_j + \frac{h}{2}\right), \qquad b_j = -[a_j + c_j + h^2 q(x_j)]. \tag{5.3.3c}$$

The net function $\{u_j\}$ is intended to be an approximation, on the net, to some eigenfunction of (5.3.1), and the scalar Λ in (5.3.2) is intended to be an approximation to some corresponding eigenvalue λ.

Let us formulate the system (5.3.2) in vector form. We introduce the J-dimensional column vector $\mathbf{u}$ and Jth order square matrix A by

$$\mathbf{u} \equiv \begin{pmatrix} u_1 \\ u_2 \\ \vdots \\ u_J \end{pmatrix}; \tag{5.3.4a}$$

$$\begin{aligned} A \equiv (a_{i,j}), \qquad a_{j,j-1} &= -a_j, \qquad a_{j,j} = -b_j, \\ a_{j,j+1} = -c_j, \qquad a_{i,j} &= 0 \quad \text{if} \quad |i - j| > 1. \end{aligned} \tag{5.3.4b}$$

Then multiplying Equation (5.3.2a) by $(-h^2)$ and using (5.3.3) and (5.3.4) we obtain

$$A\mathbf{u} - h^2 \Lambda \mathbf{u} = 0. \tag{5.3.5}$$

Thus we find that our difference problem (5.3.2) is just an eigenvalue-eigenvector problem for the matrix A. But from Equations (5.3.3c) we note that $c_j = a_{j+1}$ for $j = 1, 2, \ldots, J - 1$, and so from (5.3.4b) we have $a_{j,j+1} = a_{j,j-1}$. Thus we conclude that A is a *symmetric* tridiagonal matrix. It is also rather easy to deduce that A is *positive definite* but since we do not use this fact directly it is posed as Problem 5.3.5.

There are very effective procedures for calculating the eigenvalues and eigenvectors of symmetric tridiagonal matrices. In particular we advocate using the Sturm sequence property of the principal minors of the matrix

$$(A - \mu I)$$

to locate the eigenvalues and then using the method of bisection to determine them accurately [see Wilkinson, pp. 299–302]. If many eigenvalues are to be determined a more rapidly converging iteration scheme may be employed. After accurate approximations to the required eigenvalues are obtained, the corresponding eigenvectors can be computed quite effectively by "inverse iteration"; that is, with $\mathbf{u}_0$ arbitrary and μ the approximation to an eigenvalue $h^2\Lambda$, we define a sequence of vectors $\{\mathbf{u}_\nu\}$ by

$$(A - \mu I)\mathbf{u}_{\nu+1} = \mathbf{u}_\nu, \qquad \nu = 0, 1, \ldots .$$

Of course the matrix $A - \mu I$ is near-singular and so one must take some care in solving these systems, that is, using Gauss elimination with maximal pivots and *not* the simple factorization (3.1.7). But in practice, only two or three iterations are required to determine the eigenvectors accurately [see Wilkinson, pp. 321–323].

Let us consider now the possible accuracy of the determination of any particular eigenvalue of (5.3.1). We first note that there are only J eigenvalues of the system (5.3.5). Thus, there is a problem of identifying what exact eigenvalue, λ, is being approximated by what approximate eigenvalue, Λ. Clearly, for any fixed J there are denumerably many exact eigenvalues which cannot be approximated well by any of the Λ. These points will be clarified in our error estimates.

First however we must define the error in "approximating" L by L_h. For any sufficiently smooth function, $\phi(x)$, say in the present case with four continuous derivatives, we define at the points of the net the local truncation errors

$$\tau_j\{\phi\} \equiv L_h\phi(x_j) - L\phi(x_j), \qquad j = 1, 2, \ldots, J. \tag{5.3.6a}$$

It follows that if $p'''(x)$ is continuous, then with L_h as defined in (5.3.3a)

$$\tau_j\{\phi\} = \mathcal{O}(h^2). \tag{5.3.6b}$$

The verification of this fact is contained in Problem 5.3.1.

Now let λ be some fixed eigenvalue of (5.3.1) with corresponding eigenfunction $y(x)$. From the definition (5.3.6a) and (5.3.1) we then have at the points of the net

$$\begin{aligned} &\text{(a)} \quad L_h y(x_j) + \lambda y(x_j) = \tau_j\{y\}, \qquad j = 1, 2, \ldots, J; \\ &\text{(b)} \quad y(x_0) = 0, \qquad y(x_{J+1}) = 0. \end{aligned} \tag{5.3.7}$$

Using the matrix A defined by (5.3.4b) and (5.3.3c), the above system can be written in vector form as

$$A\mathbf{y} - h^2\lambda\mathbf{y} = -h^2\boldsymbol{\tau}\{y\}. \tag{5.3.8}$$

Here we have introduced the J-dimensional vectors

$$\mathbf{y} \equiv \begin{pmatrix} y(x_1) \\ y(x_2) \\ \vdots \\ y(x_J) \end{pmatrix}, \qquad \boldsymbol{\tau}\{y\} \equiv \begin{pmatrix} \tau_1\{y\} \\ \tau_2\{y\} \\ \vdots \\ \tau_J\{y\} \end{pmatrix}.$$

Finally, let us recall the norms for any J-dimensional vector $\mathbf{v}$ with components v_j:

$$\|\mathbf{v}\| \equiv (\mathbf{v}, \mathbf{v})^{1/2} \equiv \left(\sum_{j=1}^{J} v_j^2\right)^{1/2}, \qquad \|\mathbf{v}\|_\infty \equiv \max_{1 \le j \le J} |v_j|. \tag{5.3.9}$$

For fixed λ, either the matrix $(A - h^2\lambda I)$ is nonsingular or else $h^2\lambda$ is an eigenvalue of A. In the former case we find from (5.3.8) that

$$\mathbf{y} = -h^2(A - h^2\lambda I)^{-1}\boldsymbol{\tau}\{y\},$$

and taking norms implies

$$\|\mathbf{y}\| \le h^2\|(A - h^2\lambda I)^{-1}\| \cdot \|\boldsymbol{\tau}\{y\}\|. \tag{5.3.10}$$

Since A is symmetric, so is $(A - h^2\lambda I)^{-1}$, and the natural matrix norm above is the spectral norm. That is, in terms of the eigenvalues $h^2\Lambda_j$ of A,

$$\|(A - h^2\lambda I)^{-1}\| = \max_{1 \le j \le J}\left(\frac{1}{h^2|\Lambda_j - \lambda|}\right).$$

Using this result we obtain, from (5.3.10),

$$\min_{1 \le j \le J} |\Lambda_j - \lambda| \le \frac{\|\boldsymbol{\tau}\{y\}\|}{\|\mathbf{y}\|}.$$

Here we have assumed that the net points x_j are not all at nodes of the eigenfunction $y(x)$. This merely requires that h be sufficiently small, since the nth eigenfunction has at most $n - 1$ internal nodes. If $h^2\lambda$ is an eigenvalue of A the relation above holds trivially. Thus we have deduced in general the following.

THEOREM 5.3.1. *For each fixed eigenvalue λ of* (5.3.1) *with corresponding eigenfunction $y(x)$, there exists an eigenvalue, say $h^2\Lambda$, of A such that for h sufficiently small*

$$|\Lambda - \lambda| \le \frac{\|\boldsymbol{\tau}\{y\}\|}{\|\mathbf{y}\|}. \quad \blacksquare \tag{5.3.11}$$

From this result we can obtain estimates of the error in the maximum norm $\|\boldsymbol{\tau}\{y\}\|_\infty$. For this purpose, we use the normalized eigenfunction, that is, $\int_a^b y^2(x)\,dx = 1$. Then since $y(x_0) = y(x_{J+1}) = 0$, we have, by the normalization,

$$h\|\mathbf{y}\|^2 = \left\{\sum_{i=1}^{J} hy^2(x_j) + \frac{y^2(x_0) + y^2(x_{J+1})}{2}h - \int_{x_0}^{x_{J+1}} y^2(x)\,dx\right\} + 1$$

$$= 1 + \frac{b-a}{12}y''(\xi)h^2, \qquad \xi \in [a, b].$$

Here we have employed the error in the trapezoidal rule [see Isaacson and

Keller (1966), p. 339]. We also see that, since $(J + 1)h = b - a$, we have

$$h\|\boldsymbol{\tau}\|^2 = h \sum_{i=1}^{J} \tau_j^2\{y\} \le (b - a)\|\boldsymbol{\tau}\{y\}\|_\infty^2 .$$

Using the above results, we deduce from (5.3.11) that

$$|\Lambda - \lambda| \le \sqrt{b - a}\, \|\boldsymbol{\tau}(y)\|_\infty \Big/ \left(1 + \frac{b - a}{12} y''(\xi)h^2\right)^{1/2} . \qquad (5.3.12)$$

Now (5.3.6b) implies that

$$|\Lambda - \lambda| \le \mathcal{O}(h^2). \qquad (5.3.13)$$

Thus we find that as $h \to 0$ *any fixed eigenvalue*, λ, *of the problem* (5.3.1) *is approximated by some eigenvalue of the difference problem* (5.3.2) *with an error that is* $\mathcal{O}(h^2)$. Of course for any fixed h the bound (5.3.12) is still applicable. Then if λ is chosen as one of the very large eigenvalues, $|\Lambda - \lambda|$ must also be large for all Λ. But this is consistent with (5.3.12), since $\|\boldsymbol{\tau}\{y\}\|_\infty$ then becomes large. We recall that the higher eigenfunctions oscillate more rapidly in $[a, b]$ and thus in general will have derivatives of increasing magnitude. In the present case the fourth-order derivatives of $y(x)$ enter into the determination of the $\tau_j\{y\}$.

The procedures above can be extended to derive possibly higher-order-accurate approximations than the $\mathcal{O}(h^2)$ one in (5.3.13). In fact the only property of the matrix A that was used in deriving (5.3.11) and (5.3.12) is its symmetry. Thus we can state the more general result:

THEOREM 5.3.2. *Let the difference operator* L_h *in* (5.3.2) *be such that in the equivalent vector system* (5.3.5) *the coefficient matrix* A *is symmetric. Then for each eigenvalue* λ *of* (5.3.1) *and corresponding normalized eigenfunction* $y(x)$ *there exists an eigenvalue* $h^2\Lambda$ *of* A *such that* (5.3.11) *and* (5.3.12) *hold.* ∎

The above results can be generalized and extended in a number of important directions. Consider first the slightly more general eigenvalue problem

$$Ly + \lambda r(x)y = 0, \qquad y(a) = y(b) = 0. \qquad (5.3.14)$$

Here L is as defined in (5.3.1a) and $r(x) > 0$ is continuous on $[a, b]$. On the net $\{x_j\}$ we consider the more general difference approximation $u_0 = u_{J+1} = 0$;

$$L_h u_j + \Lambda M_h u_j = 0, \qquad j = 1, 2, \ldots, J. \qquad (5.3.15)$$

Here $L_h u_j$ could be as defined in (5.3.3a) and $M_h u_j$ is some other difference expression which is an approximation to $r(x_j)y(x_j)$ (see Problem 5.3.2). If

M_h is represented by a tridiagonal positive definite matrix then the Sturm-sequence and inverse-iteration methods previously mentioned can be used to find the eigenvalues and eigenvectors of the system (5.3.15) or (5.3.17a) [see Wilkinson, pp. 340–341]. The local truncation error for this approximation is now defined, for an eigenvector $y(x)$ belonging to the eigenvalue λ, as

$$\tau_j\{\lambda, y\} \equiv [L_h y(x_j) - Ly(x_j)] + \lambda[M_h y(x_j) - r(x_j)y(x_j)], \qquad 1 \le j \le J. \tag{5.3.16}$$

Now using (5.3.14) at the net points x_j we have

$$L_h y(x_j) + \lambda M_h y(x_j) = \tau_j\{\lambda, y\}, \qquad 1 \le j \le J.$$

This system and (5.3.15) can be written in vector form as

$$\begin{aligned} &\text{(a)} \quad A\mathbf{u} - h^2\Lambda B\mathbf{u} = 0; \\ &\text{(b)} \quad A\mathbf{y} - h^2\lambda B\mathbf{y} = h^2\boldsymbol{\tau}\{\lambda, y\}. \end{aligned} \tag{5.3.17}$$

Here $\mathbf{u}$, $\mathbf{y}$, and $\boldsymbol{\tau}$ are as previously defined and A and B are the matrices corresponding to the difference operators $-h^2L_h$ and M_h respectively. We now have the following.

THEOREM 5.3.3. *Let L_h and M_h in* (5.3.15) *be such that A is symmetric and B is symmetric-positive-definite. Then for each eigenvalue λ of* (5.3.14) *and corresponding normalized eigenfunction $y(x)$, there exists an eigenvalue $h^2\Lambda$ of $B^{-1}A$ such that*

$$|\Lambda - \lambda| \le \|B^{-1}\| \frac{\|\boldsymbol{\tau}\{\lambda, y\}\|}{\|\mathbf{y}\|}.$$

Proof. By the properties of A and B it is known that there exists a basis for the J-dimensional space, $\{\mathbf{e}_j\}$, such that

$$A\mathbf{e}_j = h^2\Lambda_j B\mathbf{e}_j, \qquad (B\mathbf{e}_j, \mathbf{e}_k) = \delta_{jk}; \qquad j, k = 1, 2, \ldots, J.$$

Now let $\mathbf{y}$ be expanded in these basis vectors, say

$$\mathbf{y} = \sum_{j=1}^{J} \xi_j \mathbf{e}_j.$$

Then using this representation in (5.3.17b), we form

$$\frac{h^4(B^{-1}\boldsymbol{\tau}, \boldsymbol{\tau})}{(B\mathbf{y}, \mathbf{y})} = \sum_{j=1}^{J} h^4(\Lambda_j - \lambda)^2 \frac{\xi_j^2}{\sum_{k=1}^{J} \xi_k^2}.$$

On the right we have a sum with nonnegative coefficients of sum unity and, hence, for some $\Lambda_j = \Lambda$ we have

$$|\Lambda - \lambda| \le \frac{(B^{-1}\boldsymbol{\tau}, \boldsymbol{\tau})^{1/2}}{(B\mathbf{y}, \mathbf{y})^{1/2}}.$$

But for any matrix M and vector $\mathbf{x}$ we have (in any norm)

$$\|M\| \geq \frac{|(M\mathbf{x}, \mathbf{x})|}{(\mathbf{x}, \mathbf{x})}.$$

Thus $(B^{-1}\boldsymbol{\tau}, \boldsymbol{\tau})^{1/2} \leq \|B^{-1}\|^{1/2} \cdot \|\boldsymbol{\tau}\|$ and the spectral norm of B^{-1} is implied here. Further, since B is symmetric-positive-definite, $\|B^{-1}\| = 1/\beta_1$, where β_1 is the least eigenvalue of B. It is also well known that, for any vector $\mathbf{x} \neq 0$,

$$\beta_1 \leq \frac{(B\mathbf{x}, \mathbf{x})}{(\mathbf{x}, \mathbf{x})},$$

and the theorem now follows. ∎

This more general result allows greater flexibility in devising difference approximations which have higher-order accuracy (see Problem 5.3.4).

Problems

5.3.1 Use Taylor's theorem to find an expression for

$$L\phi(x) - L_h\phi(x),$$

where L is as defined in (5.3.1a), L_h is as defined in (5.3.3a), and p, q, ϕ are as smooth as desired.

5.3.2 With L and L_h as above, find at least two difference operators M_h such that

$$[L\phi + \lambda r(x)\phi] - [L_h\phi + \lambda M_h\phi] = \mathcal{O}(h^2).$$

5.3.3 Find explicit expressions for the eigenvalues and eigenvectors of the difference scheme

$$u_{j-1} - 2u_j + u_{j+1} = h^2\Lambda u_j, \qquad 1 \leq j \leq N;$$
$$u_0 = u_{N+1} = 0, \qquad h = \pi/(N+1).$$

Compare them with the eigenvalues and eigenfunctions of

$$y'' + \lambda y = 0, \qquad y(0) = y(\pi) = 0.$$

[HINT: Seek a solution of the difference equations in the form $u_j = \alpha^j$, and obtain two values $\alpha_\pm$ for which such solutions exist. Take a linear combination $u_j = A\alpha_+^j + B\alpha_-^j$ and determine the coefficients and Λ to satisfy the boundary conditions, $u_0 = u_{N+1} = 0$.]

5.3.4 (a) Devise a difference scheme for $y'' + \lambda y = 0$, $y(0) = y(1) = 0$ which yields $\mathcal{O}(h^4)$ approximations to the eigenvalues. Does this procedure necessarily yield such accuracy when applied to $y'' + \lambda r(x)y = 0$?

(b) Find the eigenvalues of the difference scheme and verify directly, with a comparison to $\lambda_n = n^2\pi^2$, that they are actually $\mathcal{O}(h^4)$. What about the eigenvectors?

5.3.5 For the matrix A as defined by (5.3.4b) and (5.3.3c), form the inner product $(A\mathbf{u}, \mathbf{u}) = \sum_i \sum_j a_{ij}u_iu_j$ and show that $(A\mathbf{u}, \mathbf{u}) > 0$ if $(\mathbf{u}, \mathbf{u}) > 0$; that is, show that A is positive-definite. [HINT: By proper arrangement of the sum show that

$$(A\mathbf{u}, \mathbf{u}) = \sum_{j=1}^{J} h^2(q_ju_j^2 + b_j(u_j - u_{j-1})^2) + c_Ju_J^2.]$$

5.3.6 Compute the first four eigenvalues of the example in Problem 5.2.3 by using finite differences. Vary the number of net points used and compare the results with those obtained by shooting with the same net spacing.

5.4 *Eigenvalue Problems for Integral Equations*

In Section 4.1 we showed how a boundary-value problem may be replaced by equivalent integral equations with the aid of various Green's functions. This procedure can also be used to convert eigenvalue problems for differential equations into eigenvalue problems for integral equations. Thus let $g(x, \xi)$ be the Green's function for the differential operator L in (5.1.1a) subject to the boundary conditions (5.1.1b). Then, treating the term $\lambda r(x)y(x)$ in (5.1.1a) as an inhomogeneous term, we see that any solution of (5.1.1) must satisfy the integral equation

$$y(x) = \lambda \int_a^b g(x, \xi)r(\xi)y(\xi)\, d\xi. \tag{5.4.1}$$

Conversely, if $y(x)$ is a nontrivial solution of (5.4.1) for some value of λ, we easily find by differentiation that this function must be an eigenfunction of (5.1.1) corresponding to the eigenvalue λ.

It is not difficult to show that the Green's function is symmetric, that is,

$$g(x, \xi) = g(\xi, x)$$

if the operator L and boundary conditions are self-adjoint. This is in fact the case for the problem at hand [see Courant and Hilbert (1953), p. 354]. Since $r(x) > 0$ on $[a, b]$ we can introduce a new variable and new kernel defined by

$$z(x) \equiv \sqrt{r(x)}\, y(x), \qquad k(x, \xi) \equiv \sqrt{r(x)}\, g(x, \xi)\sqrt{r(\xi)},$$

and, on multiplying Equation (5.4.1) by $\sqrt{r(x)}$, we obtain the *symmetric integral equation*

$$z(x) = \lambda \int_a^b k(x, \xi)z(\xi)\, d\xi. \tag{5.4.2}$$

Of course, the term "symmetric" applies to the kernel which satisfies, by virtue of the symmetry of the Green's function,

$$k(x, \xi) = k(\xi, x). \tag{5.4.3}$$

We shall now examine some numerical methods for solving general symmetric integral-equation problems of the form (5.4.2), not only those that arise from differential-equation problems. To insure that we include the latter problems we simply assume that the symmetric kernel, $k(x, y)$, is continuous in $a \le x,\, y \le b$. Under these conditions it is known that all the eigenvalues of (5.4.2) are real and denumerable in number, and that the corresponding eigenfunctions are continuous and orthogonal [see Courant and Hilbert (1953), pp. 122–132, and Problems 5.4.1 and 5.4.2]. It is obvious from Equation (5.4.2) that $\lambda = 0$ cannot be an eigenvalue, since the eigenfunctions are nontrivial by definition.

The numerical methods to be employed are essentially those of Section 4.2. Thus we introduce the arbitrarily-spaced net points ξ_j, $j = 1, 2, \ldots, J$ such that

$$a \le \xi_1 < \xi_2 < \cdots < \xi_J \le b.$$

These points are the nodes for a quadrature formula of the form

$$Q_J\{f(\xi)\} \equiv \sum_{j=1}^{J} \alpha_j f(\xi_j), \tag{5.4.4a}$$

which is to approximate an integral of the form

$$Q\{f(\xi)\} \equiv \int_a^b f(\xi)\, d\xi. \tag{5.4.4b}$$

Again we require that the quadrature scheme satisfy

$$\begin{aligned}
&\text{(a)}\quad \alpha_j > 0, \qquad j = 1, 2, \ldots, J \\
&\text{(b)}\quad \lim_{J\to\infty} E_J\{f(\xi)\} \equiv \lim_{J\to\infty} [Q\{f(\xi)\} - Q_J\{(\xi)\}] = 0,
\end{aligned} \tag{5.4.5}$$

for all functions $f(\xi)$ continuous on $[a, b]$.

To approximate the eigenvalues λ, and corresponding eigenfunctions $z(x)$, of (5.4.2), we seek nontrivial net functions $\{u_j\}$ and parameters, Λ, which satisfy the algebraic system

$$u_i = \Lambda \sum_{j=1}^{J} \alpha_j k(x_i, \xi_j) u_j, \qquad i = 1, 2, \ldots, J. \tag{5.4.6}$$

This is clearly a matrix eigenvalue problem, but in its present form the appropriate matrix is not necessarily symmetric. However, the system is easily symmetrized, in exact analogy with the treatment of (5.4.1), since the quadrature coefficients α_j are required to be positive. Thus we introduce the quantities

$$v_j \equiv \sqrt{\alpha_j}\, u_j, \qquad b_{ij} \equiv \sqrt{\alpha_i}\, k(x_i, x_j)\sqrt{\alpha_j}, \qquad i, j = 1, 2, \ldots, J \tag{5.4.7}$$

and (5.4.6) becomes, on multiplication by $\sqrt{\alpha_i}$,

$$v_i = \Lambda \sum_{j=1}^{J} b_{ij} v_j, \qquad i = 1, 2, \ldots, J. \tag{5.4.8a}$$

In matrix notation this can be written as

$$\mathbf{v} = \Lambda B \mathbf{v}, \tag{5.4.8b}$$

where the vector $\mathbf{v}$ and J-order matrix B are defined by

$$\mathbf{v} \equiv \begin{pmatrix} v_1 \\ v_2 \\ \vdots \\ v_J \end{pmatrix}, \qquad B \equiv (b_{ij}). \tag{5.4.9}$$

It is quite clear from the symmetry of the kernel, shown in Equation (5.4.3), that $b_{ij} = b_{ji}$ and so B is symmetric.

The computational problem is thus reduced to finding the eigenvalues, Λ^{-1}, and eigenvectors, $\mathbf{v}$, of the symmetric matrix B. Of course this matrix, in contrast to A in (5.3.4b), does not have any particularly simple structure (other than symmetry) to facilitate these computations. But there are procedures, by now standard, for reducing symmetric matrices to the tridiagonal form by means of a sequence of orthogonal transformations. In particular, the scheme devised by Householder is quite efficient and relatively accurate [see Wilkinson (1965), pp. 290–293]. The procedure of Section 5.3 can then be used to compute the eigenvalues and eigenvectors of the equivalent tridiagonal matrix. By transforming these eigenvectors with the inverses of the above orthogonal matrices, the eigenvectors of B are finally obtained (*ibid.*, p. 333).

We turn now to the problem of estimating the accuracy of the approximate eigenvalues, Λ. Our results are quite similar to the estimates obtained in Section 5.3. The basic result can be stated as follows.

THEOREM 5.4.1. *Let λ be any fixed eigenvalue of the integral equation* (5.4.2) *and $z(x)$ be a corresponding normalized eigenfunction, that is, $\int_a^b z^2(x)\,dx = 1$. Then for J sufficiently large there exists an eigenvalue Λ^{-1} of the matrix B in* (5.4.9) *such that*

$$|\Lambda^{-1} - \lambda^{-1}|^2 \leq \frac{\sum_{i=1}^{J} \alpha_i E_J^2\{k(x_i, \xi)z(\xi)\}}{1 + E_J\{z^2(\xi)\}}. \tag{5.4.10}$$

Proof. From the definition of the quadrature error, $E_J\{\cdot\}$, in (5.4.5b), we can write Equation (5.4.2) at each point $x = x_i$ as

$$z(x_i) = \lambda \sum_{j=1}^{J} \alpha_j k(x_i, \xi_j)z(\xi_j) + \lambda E_J\{k(x_i, \xi)z(\xi)\}, \qquad i = 1, 2, \ldots, J.$$

Multiplying through by $\lambda^{-1}\sqrt{\alpha_i}$ and introducing the quantities

$$w_i \equiv \sqrt{\alpha_i}\, z(x_i), \qquad \tau_i \equiv \sqrt{\alpha_i}\, E_J\{k(x_i, \xi)z(\xi)\},$$

we obtain in vector form the system

$$(B - \lambda^{-1}I)\mathbf{w} = -\boldsymbol{\tau}.$$

Now by the argument that led to Theorem 5.3.1 we conclude that for some eigenvalue Λ^{-1} of the symmetric matrix B

$$|\Lambda^{-1} - \lambda^{-1}| \leq \frac{\|\boldsymbol{\tau}\|}{\|\mathbf{w}\|}.$$

Finally, the normalization of $z(x)$ implies

$$\|\mathbf{w}\|^2 = \sum_{i=1}^{J} \alpha_i z^2(x_i) = 1 + E_J\{z^2(\xi)\},$$

and the theorem follows. ∎

We thus find that any fixed eigenvalue of the symmetric integral equation (5.4.2) can be approximated to an accuracy proportional to the error in the quadrature formula employed. Since the quadrature scheme is to satisfy (5.4.5b), then (5.4.10) implies, as $J \to \infty$,

$$|\Lambda^{-1} - \lambda^{-1}| \leq [\sqrt{b - a} + 0(1)] \max_{1 \leq i \leq J} |E_J\{k(x_i, \xi)z(\xi)\}|. \tag{5.4.11}$$

But we must recall that the integrands, $k(x_i, \xi)z(\xi)$, may not have the smoothness properties required for the validity of the usual error expressions for standard quadrature formulae. For instance, if, as might seem reasonable, we use a J-point Gaussian quadrature scheme, then the familiar error estimate is valid only if the integrand has $2J$ continuous derivatives. Of course, as we said before, such higher-order-accurate schemes can still furnish accurate approximations.

Problems

5.4.1 The kernel $k(x, \xi)$ in Equation (5.4.2) is said to be *positive-definite* if

$$\int_a^b \int_a^b k(x, \xi)\phi(x)\phi(\xi)\, dx\, d\xi > 0$$

for all continuous nonvanishing functions on $[a, b]$. Prove that all eigenvalues of Equation (5.4.2) are *positive* if the kernel is positive-definite.

5.4.2 Prove that with $a_0^2 + b_0^2 \neq 0$ in Equation (5.1.1b) the kernel in Equation (5.4.1) is positive-definite when symmetrized, that is, that $k(x, \xi) \equiv g(x, \xi)\sqrt{r(x)r(\xi)}$ is positive-definite. [HINT: Since $r(x) > 0$ it suffices to show that, for all continuous $\psi(x)$ on $[a, b]$,

$$\int_a^b \int_a^b g(x, \xi)\psi(x)\psi(\xi)\, dx\, d\xi > 0.$$

But $\int_a^b g(x, \xi)\psi(\xi)\, d\xi \equiv y(x)$ must be the solution of $Ly(x) = -\psi(x)$ subject to the boundary conditions (5.1.1b). Now proceed as in Problem 5.1.1.]

5.5 *Generalized Eigenvalue Problems*

We discuss here a class of problems which embraces all of those previously studied. These are boundary-value problems, nonlinear in general, which contain two types of parameters (or rather of parameter sets). The first set, which we denote by an m-vector $\boldsymbol{\lambda} \equiv (\lambda_1, \lambda_2, \ldots, \lambda_m)^T$, correspond to an eigenvalue in that they are to be determined so that the boundary-value problem has a solution (trivial or not in the nonlinear case). The second set of parameters, denoted by a q-vector, $\boldsymbol{\sigma} \equiv (\sigma_1, \sigma_2, \ldots, \sigma_q)^T$, represent given data on which the solution and eigenvalue depend. (This dependence is usually sought for some specified set of $\boldsymbol{\sigma}$ values.) The boundary-value problem is now formulated as

$$\mathbf{y}' = \mathbf{f}(x, \mathbf{y}; \boldsymbol{\lambda}, \boldsymbol{\sigma}), \qquad a < x < b, \tag{5.5.1a}$$

$$E\mathbf{y}(a) + F\mathbf{y}(b) = \boldsymbol{\gamma}. \tag{5.5.1b}$$

Here $\mathbf{y}$ and $\mathbf{f}$ are n-vectors, E and F are matrices with $(m + n)$ rows and n columns and $\boldsymbol{\gamma}$ is an $(n + m)$-vector.

We have made a slight concession to simplicity in the rather special linear boundary conditions (5.5.1b). With little extra effort we could allow E, F, and $\boldsymbol{\gamma}$ to depend upon $\boldsymbol{\lambda}$ and $\boldsymbol{\sigma}$. In fact, it would be almost as easy to impose very general (nonlinear) boundary conditions of the form

$$\mathbf{g}(\mathbf{y}(a), \mathbf{y}(b), \boldsymbol{\lambda}, \boldsymbol{\sigma}) = 0,$$

where $\mathbf{g}$ is an $(n + m)$-vector function of the indicated $(2n + m + q)$ arguments. This generalization is posed as Problem 5.5.1. Standard eigenvalue problems of the form (5.1.1) can be put into the form (5.5.1) by adjoining some normalization condition which makes the eigenfunction unique.

It should be observed that we include the possibility $m = 0$, in which case there is no eigenvalue to be determined. Also the parameter $\boldsymbol{\sigma}$ may have been *artificially introduced* into the problem to facilitate its numerical solution by means of a continuity procedure. We shall just proceed formally in this section as detailed proofs would not really clarify matters. But all of our discussion can be made quite rigorous.

An obvious approach to the study of the boundary-value problem (5.5.1) stems from a consideration of the related initial-value problem

$$\mathbf{u}' = \mathbf{f}(x, \mathbf{u}; \boldsymbol{\lambda}, \boldsymbol{\sigma}), \tag{5.5.2a}$$

$$\mathbf{u}(a) = \mathbf{s}. \tag{5.5.2b}$$

Assuming modest smoothness conditions on $\mathbf{f}$ with respect to x and $\mathbf{u}$ for all $\boldsymbol{\lambda}$ and $\boldsymbol{\sigma}$ in appropriate sets, we can be assured that this initial-value problem has a unique solution, say

$$\mathbf{u} \equiv \mathbf{u}(\mathbf{s}, \boldsymbol{\lambda}, \boldsymbol{\sigma}; x). \tag{5.5.3}$$

If this solution exists on $a \leq x \leq b$ we define the $(n + m)$-vector function

$$\boldsymbol{\phi}(\mathbf{s}, \boldsymbol{\lambda}, \boldsymbol{\sigma}) \equiv E\mathbf{u}(\mathbf{s}, \boldsymbol{\lambda}, \boldsymbol{\sigma}; a) + F\mathbf{u}(\mathbf{s}, \boldsymbol{\lambda}, \boldsymbol{\sigma}; b) - \boldsymbol{\gamma}, \tag{5.5.4a}$$

and then consider the system of equations

$$\boldsymbol{\phi}(\mathbf{s}, \boldsymbol{\lambda}, \boldsymbol{\sigma}) = 0. \tag{5.5.4b}$$

For a fixed value of $\boldsymbol{\sigma}$ this represents $(n + m)$ equations in as many unknowns $(\mathbf{s}, \boldsymbol{\lambda})$. Let $(\mathbf{s}_0, \boldsymbol{\lambda}_0)$ be a root corresponding to the fixed parameter value $\boldsymbol{\sigma} = \boldsymbol{\sigma}_0$. Then it is clear that

$$\mathbf{y}_0(x) \equiv \mathbf{u}(\mathbf{s}_0, \boldsymbol{\lambda}_0, \boldsymbol{\sigma}_0; x)$$

is a solution of the boundary-value problem (5.5.1) corresponding to the eigenvalue $\boldsymbol{\lambda} = \boldsymbol{\lambda}_0$ for the parameter value $\boldsymbol{\sigma} = \boldsymbol{\sigma}_0$.

Conversely, if the problem (5.5.1) with $\boldsymbol{\sigma} = \boldsymbol{\sigma}_0$ has an eigenvalue and corresponding solution, then the equation (5.5.4) has some root which we may call $(\mathbf{s}_0, \boldsymbol{\lambda}_0)$ when $\boldsymbol{\sigma} = \boldsymbol{\sigma}_0$. If the function $\mathbf{f}(x, \mathbf{u}; \boldsymbol{\lambda}, \boldsymbol{\sigma})$ is, for example, continuous in x and continuously differentiable with respect to $\mathbf{u}$, $\boldsymbol{\lambda}$, and $\boldsymbol{\sigma}$ in some open $n + m + q + 1$-dimensional domain containing $\mathbf{y}_0(x)$, $\boldsymbol{\lambda}_0$, $\boldsymbol{\sigma}_0$, $a \leq x \leq b$, then it is easy to prove that the initial-value problem (5.5.2) will have a unique solution which exists on $a \leq x \leq b$ for all initial data $\mathbf{s}$ sufficiently close to $\mathbf{s}_0 \equiv \mathbf{y}_0(a)$, provided that $(\boldsymbol{\lambda}, \boldsymbol{\sigma})$ is also close to $(\boldsymbol{\lambda}_0, \boldsymbol{\sigma}_0)$. This is the basic type of result which justifies the use of the shooting method in attempts to solve problems of the form (5.5.1) (or, more accurately, in attempts to compute accurate approximations to the solutions). Obviously, if we can somehow guess at approximations $(\mathbf{s}^{(0)}, \boldsymbol{\lambda}^{(0)})$ close to $(\mathbf{s}_0, \boldsymbol{\lambda}_0)$, then an iterative procedure applied to the systems (5.5.4) with $\boldsymbol{\sigma} = \boldsymbol{\sigma}_0$ may converge to the desired root.

Let us formulate the application of Newton's method for computing a root of

$$\boldsymbol{\phi}(\mathbf{s}, \boldsymbol{\lambda}, \boldsymbol{\sigma}_0) = 0.$$

Given the initial guess $(\mathbf{s}^{(0)}, \boldsymbol{\lambda}^{(0)})$ we define the sequence $\{\mathbf{s}^{(\nu)}, \boldsymbol{\lambda}^{(\nu)}\}$ by

$$\mathbf{s}^{(\nu+1)} = \mathbf{s}^{(\nu)} + \Delta\mathbf{s}^{(\nu)}, \qquad \boldsymbol{\lambda}^{(\nu+1)} = \boldsymbol{\lambda}^{(\nu)} + \Delta\boldsymbol{\lambda}^{(\nu)}, \qquad \nu = 0, 1, 2, \ldots, \tag{5.5.5a}$$

where $\Delta\mathbf{s}^{(\nu)}$, $\Delta\boldsymbol{\lambda}^{(\nu)}$ are determined by the linear system

$$\frac{\partial\boldsymbol{\phi}(\cdot)}{\partial\mathbf{s}}\Delta\mathbf{s}^{(\nu)} + \frac{\partial\boldsymbol{\phi}(\cdot)}{\partial\boldsymbol{\lambda}}\Delta\boldsymbol{\lambda}^{(\nu)} = -\boldsymbol{\phi}(\mathbf{s}^{(\nu)}, \boldsymbol{\lambda}^{(\nu)}, \boldsymbol{\sigma}_0), \qquad \nu = 0, 1, 2, \ldots. \tag{5.5.5b}$$

The argument in $\boldsymbol{\phi}(\cdot)$ is always that of the right-hand side. Each system in (5.5.5b) is of order $(n + m)$ and $\partial\boldsymbol{\phi}/\partial\mathbf{s}$ is $(n + m) \times n$ while $\partial\boldsymbol{\phi}/\partial\boldsymbol{\lambda}$ is $(n + m) \times m$. These coefficient matrices can be obtained in terms of the solutions of appropriate variational problems. That is, we define the $n \times n$ matrix

$$\frac{\partial\mathbf{u}(\mathbf{s}, \boldsymbol{\lambda}, \boldsymbol{\sigma}; x)}{\partial\mathbf{s}} \equiv W(\mathbf{s}, \boldsymbol{\lambda}, \boldsymbol{\sigma}; x) \tag{5.5.6a}$$

and the $n \times m$ matrix

$$\frac{\partial\mathbf{u}(\mathbf{s}, \boldsymbol{\lambda}, \boldsymbol{\sigma}; x)}{\partial\boldsymbol{\lambda}} \equiv V(\mathbf{s}, \boldsymbol{\lambda}, \boldsymbol{\sigma}; x). \tag{5.5.6b}$$

Formally differentiating in the initial-value problem (5.5.2), we obtain the variational problems

$$W' = A(x; \mathbf{s}, \boldsymbol{\lambda}, \boldsymbol{\sigma})W, \qquad W(a) = I, \tag{5.5.7a}$$

$$V' = A(x; \mathbf{s}, \boldsymbol{\lambda}, \boldsymbol{\sigma})V + G(x; \mathbf{s}, \boldsymbol{\lambda}, \boldsymbol{\sigma}), \qquad V(a) = 0, \tag{5.5.7b}$$

where the $n \times n$ matrix A and $n \times m$ matrix G are

$$\begin{aligned} A(x; \mathbf{s}, \boldsymbol{\lambda}, \boldsymbol{\sigma}) &\equiv \frac{\partial\mathbf{f}}{\partial\mathbf{u}}(x, \mathbf{u}(\mathbf{s}, \boldsymbol{\lambda}, \boldsymbol{\sigma}; x); \boldsymbol{\lambda}, \boldsymbol{\sigma}), \\ G(x; \mathbf{s}, \boldsymbol{\lambda}, \boldsymbol{\sigma}) &\equiv \frac{\partial\mathbf{f}}{\partial\boldsymbol{\lambda}}(x, \mathbf{u}(\mathbf{s}, \boldsymbol{\lambda}, \boldsymbol{\sigma}; x); \boldsymbol{\lambda}, \boldsymbol{\sigma}). \end{aligned} \tag{5.5.8}$$

Now it follows from differentiating in (5.5.4a) and using (5.5.6) and (5.5.7) that

$$\frac{\partial\boldsymbol{\phi}(\mathbf{s}, \boldsymbol{\lambda}, \boldsymbol{\sigma})}{\partial\mathbf{s}} = E + FW(\mathbf{s}, \boldsymbol{\lambda}, \boldsymbol{\sigma}; b), \qquad \frac{\partial\boldsymbol{\phi}(\mathbf{s}, \boldsymbol{\lambda}, \boldsymbol{\sigma})}{\partial\boldsymbol{\lambda}} = FV(\mathbf{s}, \boldsymbol{\lambda}, \boldsymbol{\sigma}; b). \tag{5.5.9a}$$

Thus introducing the $(n + m)$-order matrix Q by

$$Q(\mathbf{s}, \boldsymbol{\lambda}, \boldsymbol{\sigma}) \equiv [E + FW(\mathbf{s}, \boldsymbol{\lambda}, \boldsymbol{\sigma}; b), FV(\mathbf{s}, \boldsymbol{\lambda}, \boldsymbol{\sigma}; b)], \tag{5.5.9b}$$

the system (5.5.5*b*) can be written as

$$Q(\mathbf{s}^{(\nu)}, \boldsymbol{\lambda}^{(\nu)}, \boldsymbol{\sigma}_0)\begin{pmatrix}\Delta\mathbf{s}^{(\nu)}\\ \Delta\boldsymbol{\lambda}^{(\nu)}\end{pmatrix} = -\boldsymbol{\phi}(\mathbf{s}^{(\nu)}, \boldsymbol{\lambda}^{(\nu)}, \boldsymbol{\sigma}_0), \qquad \nu = 0, 1, 2, \ldots . \tag{5.5.10}$$

In brief, to apply Newton's method we integrate the nonlinear initial-value problem (5.5.2) with $(\mathbf{s}, \boldsymbol{\lambda}, \boldsymbol{\sigma}) = (\mathbf{s}^{(\nu)}, \boldsymbol{\lambda}^{(\nu)}, \boldsymbol{\sigma}_0)$. This solution is used to evaluate the matrices A and G in (5.5.8) as well as $\boldsymbol{\phi}(\mathbf{s}^{(\nu)}, \boldsymbol{\lambda}^{(\nu)}, \boldsymbol{\sigma}_0)$ in (5.5.4a). Then the systems (5.5.7) are solved [these are $(n + m)$ linear initial-value problems for first-order systems of n equations] and used to evaluate $Q(\mathbf{s}^{(\nu)}, \boldsymbol{\lambda}^{(\nu)}, \boldsymbol{\sigma}_0)$ as in (5.5.9b). Finally the linear algebraic system (5.5.10), of order $(n + m)$, is solved to yield the corrections $\Delta\mathbf{s}^{(\nu)}$ and $\Delta\boldsymbol{\lambda}^{(\nu)}$ from which $\mathbf{s}^{(\nu+1)}$ and $\boldsymbol{\lambda}^{(\nu+1)}$ are computed by (5.5.5a). For efficiency in storage and computing time the systems in (5.5.2) and (5.5.7) should be integrated simultaneously as one large system. We point out that in many problems, especially those motivated by physical applications, the quantities W and V in (5.5.6) are of independent interest.

In order to validate the application of Newton's method we would have to show (at least) that the matrix $Q(\mathbf{s}, \boldsymbol{\lambda}, \boldsymbol{\sigma})$ is nonsingular at each point $(\mathbf{s}^{(\nu)}, \boldsymbol{\lambda}^{(\nu)}, \boldsymbol{\sigma}_0)$. This can be done under appropriate conditions, and in fact the set-up above can actually be used to prove existence and (local) uniqueness of solutions to the problem (5.5.1). Roughly, the idea is to determine an initial guess $(\mathbf{s}^{(0)}, \boldsymbol{\lambda}^{(0)})$ such that $Q(\mathbf{s}^{(0)}, \boldsymbol{\lambda}^{(0)}, \boldsymbol{\sigma}_0)$ is nonsingular, and

$$\|Q^{-1}(\mathbf{s}^{(0)}, \boldsymbol{\lambda}^{(0)}, \boldsymbol{\sigma}_0)\| \le K_0, \qquad |\boldsymbol{\phi}(\mathbf{s}^{(0)}, \boldsymbol{\lambda}^{(0)}, \boldsymbol{\sigma}_0)| \le \eta/K_0. \tag{5.5.11a}$$

Further, the function $\mathbf{f}$ is assumed so smooth that

$$\|Q(\mathbf{s}, \boldsymbol{\lambda}, \boldsymbol{\sigma}_0) - Q(\mathbf{s}', \boldsymbol{\lambda}', \boldsymbol{\sigma}_0)\| \le K_1\left|\begin{pmatrix}\mathbf{s}\\ \boldsymbol{\lambda}\end{pmatrix} - \begin{pmatrix}\mathbf{s}'\\ \boldsymbol{\lambda}'\end{pmatrix}\right|, \tag{5.5.11b}$$

for all $(\mathbf{s}, \boldsymbol{\lambda})^T$ and $(\mathbf{s}', \boldsymbol{\lambda}')^T$ within a distance ρ of $(\mathbf{s}^{(0)}, \boldsymbol{\lambda}^{(0)})^T$. Finally, if the above constants satisfy

$$K_0K_1\rho < 1, \qquad \eta \le (1 - K_0K_1\rho)\rho, \tag{5.5.11c}$$

then $\boldsymbol{\phi}(\mathbf{s}, \boldsymbol{\lambda}, \boldsymbol{\sigma}_0) = 0$ has a unique root within ρ of $(\mathbf{s}^{(0)}, \boldsymbol{\lambda}^{(0)})$. This final result follows by contracting maps and is, in a slightly different notation, Theorem 1.4.4 [in Problem 1.4.5]. In fact by strengthening (5.5.11c) to require that

$$K_0K_1\rho < \tfrac{2}{3}, \qquad \eta \le (1 - \tfrac{3}{2}K_0K_1\rho)\rho, \tag{5.5.11d}$$

it follows from Theorem 1.4.5 [in Problem 1.4.7] that the Newton iterates (5.5.5) converge quadratically to the root.

Of course if a stable, accurate-of-order-p numerical method is used to solve the initial-value problems (5.5.2) and (5.5.7) then we can show that for sufficiently fine net spacing h the numerical determination of a root of (5.5.4) is within $\mathcal{O}(h^p)$ of an exact root. Then a solution of the generalized eigenvalue problem (5.5.1) is determined, along with an eigenvalue, to within an error that is $\mathcal{O}(h^p)$. Given the above-mentioned existence proof, our results about the shooting method follow somewhat as in the proof of Theorem 2.3.1, but are in fact very much more general. [For example, $A + B$ in Section 2.3 need not be nonsingular; $Q(\mathbf{s}) \equiv A + BW(\mathbf{s}, b)$ should satisfy conditions similar to those in (5.5.11a–c).]

5.5.1 Poincaré Continuation; Continuity Methods

Suppose that the generalized eigenvalue problem (5.5.1) has a solution for the parameter value $\boldsymbol{\sigma} = \boldsymbol{\sigma}_0$. Then under reasonable smoothness assumptions we may expect that a solution exists for $\boldsymbol{\sigma}$ near $\boldsymbol{\sigma}_0$ and that in fact the two solutions should be close, in some sense. This is frequently, but not always, the case. When it is true, an obvious continuity procedure is suggested for computing solutions over an entire domain of $\boldsymbol{\sigma}$ values. (In fact, as has been previously suggested, the parameter $\boldsymbol{\sigma}$ may have been introduced into the problem in order to apply some sort of continuity technique; see Problems 5.5.4–5.5.6.) The theoretical justification for the basic assumption is, as we show, but a slight extension of the Poincaré continuation procedure which is well known in the study of periodic solutions of differential equations [see Coddington and Levinson (1955), pp. 348–350].

Let the solution of (5.5.1) with $\boldsymbol{\sigma} = \boldsymbol{\sigma}_0$ be denoted by $\mathbf{y} = \mathbf{y}_0(x)$, $\boldsymbol{\lambda} = \boldsymbol{\lambda}_0$. Then with the choice $\mathbf{s} = \mathbf{s}_0 \equiv \mathbf{y}_0(a)$ we have, using (5.5.4),

$$\boldsymbol{\phi}(\mathbf{s}_0, \boldsymbol{\lambda}_0, \boldsymbol{\sigma}_0) = 0.$$

Under the previously-mentioned differentiability conditions on $\mathbf{f}(x, \mathbf{u}; \boldsymbol{\lambda}, \boldsymbol{\sigma})$ we observe from (5.5.9) that the Jacobian of $\boldsymbol{\phi}$ with respect to the $(n + m)$ variables $(\mathbf{s}, \boldsymbol{\lambda})$ is

$$\det \frac{\partial \boldsymbol{\phi}(\mathbf{s}, \boldsymbol{\lambda}, \boldsymbol{\sigma})}{\partial(\mathbf{s}, \boldsymbol{\lambda})} = \det Q(\mathbf{s}, \boldsymbol{\lambda}, \boldsymbol{\sigma}).$$

Let us assume now that $Q(\mathbf{s}_0, \boldsymbol{\lambda}_0, \boldsymbol{\sigma}_0)$ is nonsingular. [Note that this is *not* the assumption made above, Equation (5.5.11a), but would follow from (5.5.11a–c) since then $Q(\mathbf{s}, \boldsymbol{\lambda}, \boldsymbol{\sigma}_0)$ is nonsingular in some sphere including $(\mathbf{s}_0, \boldsymbol{\lambda}_0)$ and $(\mathbf{s}^{(0)}, \boldsymbol{\lambda}^{(0)})$.] Then, since $\boldsymbol{\phi}(\mathbf{s}, \boldsymbol{\lambda}, \boldsymbol{\sigma})$ is continuous in $\boldsymbol{\sigma}$, we may apply the implicit-function theorem to conclude that there exist an $\varepsilon > 0$

and unique continuous functions

$$\mathbf{s}(\boldsymbol{\sigma}), \qquad \boldsymbol{\lambda}(\boldsymbol{\sigma}), \tag{5.5.12a}$$

such that

$$\mathbf{s}(\boldsymbol{\sigma}_0) = \mathbf{s}_0, \qquad \boldsymbol{\lambda}(\boldsymbol{\sigma}_0) = \boldsymbol{\lambda}_0, \tag{5.5.12b}$$

and

$$\boldsymbol{\phi}(\mathbf{s}(\boldsymbol{\sigma}), \boldsymbol{\lambda}(\boldsymbol{\sigma}), \boldsymbol{\sigma}) = 0, \tag{5.5.12c}$$

for all $\boldsymbol{\sigma}$ in the sphere $N_\varepsilon(\boldsymbol{\sigma}_0) \equiv \{\boldsymbol{\sigma} \mid |\boldsymbol{\sigma} - \boldsymbol{\sigma}_0| < \varepsilon\}$. Finally, recalling (5.5.2)–(5.5.4), we find that a family of solutions to the generalized eigenvalue problem (5.5.1) is given by

$$\mathbf{y}(\boldsymbol{\sigma}; x) = \mathbf{u}(\mathbf{s}(\boldsymbol{\sigma}), \boldsymbol{\lambda}(\boldsymbol{\sigma}), \boldsymbol{\sigma}; x) \tag{5.5.13a}$$

with corresponding eigenvalue

$$\boldsymbol{\lambda} = \boldsymbol{\lambda}(\boldsymbol{\sigma}) \tag{5.5.13b}$$

for all $\boldsymbol{\sigma} \in N_\varepsilon(\boldsymbol{\sigma}_0)$.

When this continuation theorem is valid it is clear that the solution for a given value of $\boldsymbol{\sigma}$ will be a close approximation to the solution for the parameter values $\boldsymbol{\sigma} \pm \Delta\boldsymbol{\sigma}$ if $|\Delta\boldsymbol{\sigma}|$ is sufficiently small. Thus in whatever iteration scheme is employed to solve (5.5.4b), a sequence of close initial estimates of roots for a sequence of $\boldsymbol{\sigma}$ values can be generated as soon as one solution is known for a given value of $\boldsymbol{\sigma}$. This clarifies one form of continuity method in which $\boldsymbol{\sigma}$ is introduced into the problem in such a way, for example, that for $\boldsymbol{\sigma} = 0$ the resulting problem is easily solved (perhaps with an explicit solution). Starting values are thus furnished and one seeks to solve on a sequence of $\boldsymbol{\sigma}$ values whose final value is such that (5.5.1) becomes the problem whose solution was originally required. The need for introducing such continuity parameters may only become apparent after other attempts have proven fruitless.

If solutions of (5.5.1) are to be sought for a sequence of $\boldsymbol{\sigma}$ values then it would seem likely that even "better" initial guesses could be furnished by some sort of extrapolation procedure. This is of course justified if it is known that the functions in (5.5.12a) have more smoothness properties than just continuity. But the implicit-function theorem endows the "root" $(\mathbf{s}(\boldsymbol{\sigma}), \boldsymbol{\lambda}(\boldsymbol{\sigma}))$ with the same continuity and differentiability properties in $\boldsymbol{\sigma}$ as those enjoyed by $\boldsymbol{\phi}(\mathbf{s}, \boldsymbol{\lambda}, \boldsymbol{\sigma})$ as a function of $\boldsymbol{\sigma}$, $\mathbf{s}$, and $\boldsymbol{\lambda}$. Thus, for instance, with the continuously-differentiable $\mathbf{f}$ we have assumed, it follows that $\mathbf{s}(\boldsymbol{\sigma})$ and $\boldsymbol{\lambda}(\boldsymbol{\sigma})$ are continuously differentiable on $N_\varepsilon(\boldsymbol{\sigma}_0)$. Now suppose that a solution is known for some $\boldsymbol{\sigma} \in N_\varepsilon(\boldsymbol{\sigma}_0)$ and $\Delta\boldsymbol{\sigma}$ is such that $\boldsymbol{\sigma} \pm \Delta\boldsymbol{\sigma} \in N_\varepsilon(\boldsymbol{\sigma}_0)$. Then reasonable approximations to the roots at these new parameter values are given by the first two terms in the Taylor expansions

$$\begin{aligned} \mathbf{s}(\boldsymbol{\sigma} \pm \Delta\boldsymbol{\sigma}) &\doteq \mathbf{s}(\boldsymbol{\sigma}) \pm \frac{\partial \mathbf{s}(\boldsymbol{\sigma})}{\partial \boldsymbol{\sigma}} \Delta\boldsymbol{\sigma} \equiv \mathbf{s}^{(0)}(\boldsymbol{\sigma} \pm \Delta\boldsymbol{\sigma}), \\ \boldsymbol{\lambda}(\boldsymbol{\sigma} \pm \Delta\boldsymbol{\sigma}) &\doteq \boldsymbol{\lambda}(\boldsymbol{\sigma}) \pm \frac{\partial \boldsymbol{\lambda}(\boldsymbol{\sigma})}{\partial \boldsymbol{\sigma}} \Delta\boldsymbol{\sigma} \equiv \boldsymbol{\lambda}^{(0)}(\boldsymbol{\sigma} \pm \Delta\boldsymbol{\sigma}). \end{aligned} \tag{5.5.14}$$

In the case of a scalar parameter (or if $\Delta\boldsymbol{\sigma}$ has only one nonzero component) a rather obvious approach to the estimates in (5.5.14) is obtained by approximating the required derivatives by difference quotients. The resulting formulas are exactly those obtained by using linear extrapolation from two solutions. Of course this procedure is not as accurate as that in (5.5.14) and, perhaps more important, requires a knowledge of two solutions for close values of $\boldsymbol{\sigma}$. However, as we shall show, it is quite efficient to get very accurate approximations to the matrices $\partial\mathbf{s}/\partial\boldsymbol{\sigma}$ and $\partial\boldsymbol{\lambda}/\partial\boldsymbol{\sigma}$ using only one solution. Then larger steps, $\Delta\boldsymbol{\sigma}$, can be employed or (equivalently) more accurate initial guesses are obtained for smaller $|\Delta\boldsymbol{\sigma}|$.

To obtain the derivatives occurring in (5.5.14) we observe that the family of solutions (5.5.13) employed in (5.5.4) yields the result (5.5.12c) which is an identity in $\boldsymbol{\sigma}$ on $N_\varepsilon(\boldsymbol{\sigma}_0)$. Differentiating this identity with respect to $\boldsymbol{\sigma}$ gives us the system

$$\frac{\partial \boldsymbol{\phi}(\cdot)}{\partial \mathbf{s}} \frac{\partial \mathbf{s}}{\partial \boldsymbol{\sigma}} + \frac{\partial \boldsymbol{\phi}(\cdot)}{\partial \boldsymbol{\lambda}} \frac{\partial \boldsymbol{\lambda}}{\partial \boldsymbol{\sigma}} = -\frac{\partial \boldsymbol{\phi}(\cdot)}{\partial \boldsymbol{\sigma}},$$

where $(\cdot) \equiv (\mathbf{s}(\boldsymbol{\sigma}), \boldsymbol{\lambda}(\boldsymbol{\sigma}), \boldsymbol{\sigma})$. Recalling (5.5.9), this system can be written as

$$Q(\mathbf{s}(\boldsymbol{\sigma}), \boldsymbol{\lambda}(\boldsymbol{\sigma}), \boldsymbol{\sigma}) \begin{pmatrix} \partial \mathbf{s}(\boldsymbol{\sigma})/\partial \boldsymbol{\sigma} \\ \partial \boldsymbol{\lambda}(\boldsymbol{\sigma})/\partial \boldsymbol{\sigma} \end{pmatrix} = -FZ(\boldsymbol{\sigma}; b). \tag{5.5.15}$$

Here we have used (5.5.4a) and (5.5.2) to write

$$\frac{\partial \boldsymbol{\phi}(\cdot)}{\partial \boldsymbol{\sigma}} = FZ(\boldsymbol{\sigma}; b), \tag{5.5.16a}$$

where $Z(\boldsymbol{\sigma}; x) = \partial \mathbf{u}(\mathbf{s}(\boldsymbol{\sigma}), \boldsymbol{\lambda}(\boldsymbol{\sigma}), \boldsymbol{\sigma}; x)/\partial \boldsymbol{\sigma}$ is the n-rowed-by-q-columned matrix solution of the variational system

$$Z' = A(x; \mathbf{s}(\boldsymbol{\sigma}), \boldsymbol{\lambda}(\boldsymbol{\sigma}), \boldsymbol{\sigma})Z + H(x; \boldsymbol{\sigma}), \qquad Z(a) = 0. \tag{5.5.16b}$$

The n-th order matrix A is defined in (5.5.8) and the $n \times q$ matrix H is

$$H(x; \boldsymbol{\sigma}); \equiv \frac{\partial \mathbf{f}}{\partial \boldsymbol{\sigma}} (x, \mathbf{u}(\mathbf{s}(\boldsymbol{\sigma}), \boldsymbol{\lambda}(\boldsymbol{\sigma}), \boldsymbol{\sigma}; x); \boldsymbol{\lambda}(\boldsymbol{\sigma}), \boldsymbol{\sigma}). \tag{5.5.16c}$$

We recall that the nonsingularity of $Q(\mathbf{s}(\boldsymbol{\sigma}), \boldsymbol{\lambda}(\boldsymbol{\sigma}), \boldsymbol{\sigma})$ for a given $\boldsymbol{\sigma}$ value is our sufficient condition for Poincaré continuation, in which case (5.5.15) has a unique solution.

In summary we advocate the continuity method using the initial estimates in (5.5.14). The required derivatives are obtained by solving the system (5.5.15). This is done, for a given value of $\boldsymbol{\sigma}$, only *after* a sufficiently-accurate solution as in (5.5.12) or (5.5.13) has been computed. To obtain the coefficient matrix, Q, and the inhomogeneous term, $-FZ$, we must solve the three variational initial-value systems (5.5.7a), (5.5.7b), and (5.5.16b) only once using the accurate solution to evaluate the matrices in (5.5.8) and (5.5.16c). It should be recalled that if Newton's method is being employed, then a procedure for evaluating the matrix Q is already required and the only additional complication is in solving for $Z(\boldsymbol{\sigma}; b)$. But again, as with W and V, the components of Z may be of independent physical interest.

Many problems of the form (5.5.1) are such that the above procedures can be greatly simplified. For instance, the boundary conditions (5.5.1b) may require that $y_1(a) = y_2(a) = \cdots = y_r(a) = 0$ for some $r < n$. Then in (5.5.2b) we take $s_1 = s_2 = \cdots = s_r = 0$ and the system (5.5.4) reduces to $m + n - r$ equations in the as many unknowns $\boldsymbol{\lambda}$, $s_{r+1}, \ldots, s_n$. All subsequent calculations are correspondingly reduced.

The parallel shooting techniques of Section 2.4 can be applied, in an obvious manner, to the generalized eigenvalue problems. The finite-difference methods, as in Section 3.3, can also be employed, but their theoretical justification is considerably more complicated.

Problems

5.5.1 Consider the generalized eigenvalue problem posed by (5.5.1a) and

$$\mathbf{g}(\mathbf{y}(a), \mathbf{y}(b), \boldsymbol{\lambda}, \boldsymbol{\sigma}) = 0. \qquad (*)$$

Determine the corresponding transcendental system that results from using the initial-value problem (5.5.2) to solve this boundary-value problem. Determine the linear variational problems whose solutions yield the coefficient matrix for Newton's method in this case. [We assume that $\mathbf{g}(\mathbf{u}, \mathbf{v}, \boldsymbol{\lambda}, \boldsymbol{\sigma})$ has continuous derivatives $\partial\mathbf{g}/\partial\mathbf{u}$ and $\partial\mathbf{g}/\partial\mathbf{v}$ for all arguments.]

5.5.2 Determine a sufficient condition for Poincaré continuation in a neighborhood, $|\boldsymbol{\sigma} - \boldsymbol{\sigma}_0| < \varepsilon$, of a parameter value $\boldsymbol{\sigma}_0$ for which (5.5.1a) and (*) have a solution. Find the system which replaces (5.5.15) for the determination of $\partial\mathbf{s}(\boldsymbol{\sigma}_0)/\partial\boldsymbol{\sigma}$ and $\partial\boldsymbol{\lambda}(\boldsymbol{\sigma}_0)/\partial\boldsymbol{\sigma}$. How is the inhomogeneous term computed?

5.5.3 Let $\boldsymbol{\lambda}$ and $\boldsymbol{\sigma}$ in Equation (5.5.1a) be scalars (that is one-dimensional), and let Equation (5.5.1b) reduce to

$$y_1(a) = \cdots = y_n(a) = 0, \qquad y_1(b) = 0.$$

Formulate the initial-value procedure, Newton's method and accurate continuity method for this generalized eigenvalue problem. [Note that there is no initial parameter **s** to be introduced, just as in the Sturm-Liouville case treated in Section 5.2.]

5.5.4 The boundary-value problem

$$\text{(a)} \quad y'' + \frac{e^y}{2} = 0, \qquad y'(0) = 0, \qquad y(1) = 0$$

has *two* positive solutions. Consider the related boundary-value problem

$$\text{(b)} \quad z'' + (1 - \sigma)\frac{\pi^2}{4} z + \sigma\frac{e^z}{2} = 0, \qquad z'(0) = z(1) = 0,$$

which for $\sigma = 0$ has a one-parameter family of solutions

$$z(0, x) = a \cos \frac{\pi x}{2}.$$

Does the Poincaré continuation result hold for problem (b) in the neighborhood of $\sigma = 0$? What about the simpler problem

$$\text{(c)} \quad z'' + \sigma\frac{e^z}{2} = 0, \qquad z'(0) = z(1) = 0.$$

Try using the continuity method on (b) and (c) to get from $\sigma = 0$ to $\sigma = 1$, and compute both positive solutions of (a) in this way.

5.5.5 Consider the boundary-value problem

$$\mathbf{y}' = \mathbf{f}(x, \mathbf{y}), \qquad A\mathbf{y}(a) + B\mathbf{y}(b) = \boldsymbol{\alpha}, \tag{5.5.17}$$

where $\mathbf{f}$ and $\partial\mathbf{f}/\partial\mathbf{y}$ are bounded and continuous for $x \in [a, b]$, $|\mathbf{y}| < \infty$, and $A + B$ is nonsingular. This problem can be imbedded in the one-parameter family of problems

$$\mathbf{z}' = \sigma\mathbf{f}(x, \mathbf{z}), \qquad A\mathbf{z}(a) + B\mathbf{z}(b) = \boldsymbol{\alpha}. \tag{5.5.18}$$

(a) Show by Poincaré continuation that problem (5.5.18) has a solution $\mathbf{z}(x, \sigma)$ for each σ in some interval $0 \le \sigma \le \varepsilon$.

(b) If $\|\partial\mathbf{f}/\partial\mathbf{y}\| \le K$, show that in the above interval we may take ε satisfying

$$\varepsilon < \frac{1}{K|b - a|} \log\left(1 + \frac{1}{\|(A + B)^{-1}B\|}\right).$$

[HINT: Use Theorem 1.2.6.] Note that (5.5.18) is almost equivalent to scaling the independent variable in (5.5.17) by σ. This is actually the case for autonomous systems. Then our continuity procedure consists in stretching the interval to the desired length.

5.5.6 Suppose that $(A + B)$ in Problem 5.5.5 is singular but that for some matrix P† with positive eigenvalues $(A + BP)$ is nonsingular. Show that, with the real matrix†

$$M \equiv \frac{1}{b - a} \log P,$$

the one-parameter family of problems

$$\mathbf{z}' = \sigma \mathbf{f}(x, \mathbf{z}) + (1 - \sigma) M\mathbf{z}, \qquad A\mathbf{z}(a) + B\mathbf{z}(b) = \boldsymbol{\alpha} \tag{5.5.19}$$

has a solution $\mathbf{z}(x, \sigma)$ for each σ in some interval $0 \leq \sigma \leq \varepsilon$. Note that part (a) of Problem 5.5.5 results if $P \equiv I$ in the above.

SUPPLEMENTARY REFERENCES AND NOTES

Section 5.1. Thorough treatments of Sturm-Liouville eigenvalue problems are given in Ince (1944) and Coddington and Levinson (1955). The latter also studies much more general eigenvalue problems. The elegant and powerful variational formulation of self-adjoint eigenvalue problems is described in Collatz (1960), Kantorovich and Krylov (1958), and Courant and Hilbert (1937). A theory of eigenvalue problems for linear first-order systems is contained in Atkinson (1964).

Section 5.3 Variational methods play a large role in the study of difference methods for Sturm-Liouville eigenvalue problems. Farrington, Gregory, and Taub (1957) use them to get higher-order-accurate difference equations which are automatically symmetric. Then by Theorem 5.3.2 the eigenvalues are also higher-order-accurate approximations. Weinberger (1962) obtains upper and lower bounds on the eigenvalues by constructing special difference equations, and Birkhoff et al. (1966) improve this technique to obtain more accurate bounds as well as estimates for the accuracy of the eigenvector approximation. The error in difference approximations to the eigenvectors is also studied by Gary (1965).

Section 5.4 One of the basic methods for studying Sturm-Liouville eigenvalue problems is based on replacing them by equivalent integral equations and then developing the Fredholm Theory [see Courant and Hilbert (1953)]. The error in "difference" approximations to the eigenvalues of integral equations is briefly discussed in H. Keller (1965). A thorough study of such methods is contained in an excellent paper of H. Wielandt (1956).

† The condition on P need only be that $\log P$ exist. For convenience in calculations, we have required the logarithm to be real. For a definition of $\log P$, see Coddington and Levinson (1955), pp. 65–66.

Section 5.5 It may be possible to continue a solution even though det $Q(\mathbf{s}_0, \boldsymbol{\lambda}_0, \boldsymbol{\sigma}_0) = 0$ (that is, when the implicit function theorem fails). Continuation of periodic solutions of autonomous differential equations is an example of this situation [see Coddington and Levinson (1955), pp. 352–353]. A rather complicated example containing most of the exceptional cases which is solved by shooting with Aitkens' δ^2-method (rather than Newton's method, which we usually advocate) is contained in Keller and Wolfe (1965). A brief account of several different numerical studies of this same problem is given in Keller and Reiss (1965).

It is frequently a very practical procedure to reduce eigenvalue problems to ordinary boundary value problems by introducing the m new variables λ (x) as solutions of $\lambda' = 0$. The enlarged system (5.5.1) of m + n first order equations can then be solved as in Chapters 2 or 3.

Continuation is currently an active area of research. One very powerful new idea is to continue in a parameter such as "arclength" which is intrinsic to the problem. This type of parameter can be introduced in many ways and it is a natural procedure to use in connection with Newton's method. See in particular §5.6 or Appendix C.7.

CHAPTER 6

PRACTICAL EXAMPLES AND COMPUTATIONAL EXERCISES

Introduction

We present here details of some problems that were solved by the methods of the previous chapters or by very similar procedures. All of the problems are of independent interest and were originally studied for purposes other than to serve as illustrations for this monograph. In fact our theorems on existence, uniqueness, and convergence do not generally apply here (usually because of the presence of some type of singularity). But as the computations show, the behavior of the numerical procedures is just as if the theory were valid. We could have presented more general theoretical results to cover some of these and many other cases. However, it is likely that many, if not most, problems to be met in practice will be of this type in that they do not fit some relevant theory exactly. One of the main tasks confronting applied mathematicians in general is to judge correctly which approximation techniques are valid in the absence of complete theoretical justification. This is perhaps most necessary in the application of numerical methods since they are so easily applied in this developing age of digital computers. Our examples are intended to indicate judgments of this type as well as to illustrate modifications in the methods of the text. Some computations are suggested in each section and it would be instructive to try to duplicate the results in the text.

6.1 Shooting; Lubrication Theory

This problem in the theory of lubrication concerns the flow of a viscous compressible fluid through a very narrow gap and was treated by J. D. Cole, H. B. Keller, and P. Saffman [1967]. The programming of the numerical work was done by W. H. Mitchell and all the calculations reported here were performed in about 50 seconds on an I.B.M. 7094 in the W. H. Booth Computing Center at the California Institute of Technology.

The pressure distribution in the lubricant (that is, viscous fluid) between a cylindrical bearing and a plane surface results in a vertical thrust on the bearing. This thrust and the details of the pressure distribution are of interest in the theory of lubrication. Under the usual assumptions of this theory, for a compressible viscous isothermal film, the pressure $p(\phi)$ is found to satisfy the first-order ordinary differential equation

$$\frac{d}{d\phi}\left(\frac{p(\phi)}{p_0}\right) = \beta\cos^2\phi - \alpha\left(\frac{p_0}{p(\phi)}\right)\cos^4\phi, \qquad -\frac{\pi}{2} < \phi < \frac{\pi}{2}. \tag{6.1.1}$$

Here ϕ is defined by

$$x = (2Rh)^{1/2}\tan\phi$$

where, as shown in Figure 6.1.1, R is the radius of the cylinder and h is the gap width. The parameters α and β are defined by

$$\alpha \equiv \frac{12\mu F}{p_0\rho_0}\left(\frac{2R}{h^5}\right)^{1/2}, \qquad \beta \equiv \frac{6\mu U}{p_0}\left(\frac{2R}{h^3}\right)^{1/2}, \tag{6.1.2}$$

where ω is the angular speed of the cylinder, $U \equiv R\omega$, μ is the coefficient of viscosity of the lubricant, p_0 and ρ_0 are the pressure and density of the lubricant far from the gap (that is, at $|x| = \infty$), and F is the mass flux of lubricant through the gap per unit length of the cylinder. We have assumed that

$$p\left(-\frac{\pi}{2}\right) = p\left(\frac{\pi}{2}\right) = p_0, \tag{6.1.3}$$

so that the lubricant is driven through the gap solely by the rotation of the cylinder and not by any external pressure gradients.

In general, β and p_0 are known, and it is required to find the pressure distribution, $p(\phi)$, and the mass flux, α, such that the boundary-value problem (6.1.1, 6.1.3) is satisfied. Then, writing the vertical thrust on the bearing as $T = \sqrt{2Rh}\, p_0\tau$, we have, to within higher-order terms in h/R,

$$\tau = \int_{-\pi/2}^{\pi/2}\left[\frac{p(\phi)}{p_0} - 1\right]\sec^2\phi\, d\phi. \tag{6.1.4}$$

The problem thus formulated may be termed a nonlinear eigenvalue problem. For $\beta \ll 1$ and $\beta \gg 1$, approximate solutions can be obtained by ordinary perturbation and singular perturbation techniques, respectively. In the intermediate range of parameter values, numerical solutions are required.

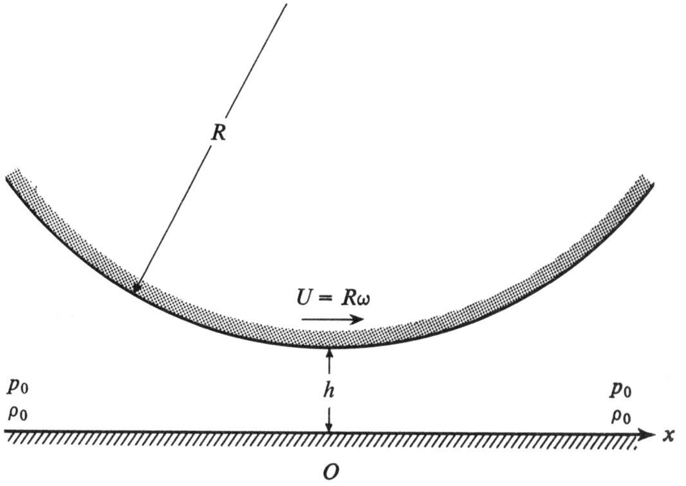

FIGURE 6.1.1 Sketch of the gap between a rotating cylinder and fixed plane

To facilitate comparisons with some of this analytical work it is convenient to introduce new variables, with $p_1 \equiv p(0)$, as follows:

$$\theta \equiv \phi - \frac{\pi}{2}, \qquad y(\theta) \equiv \frac{p(\phi)}{p_0},$$
$$\varepsilon \equiv \left(\frac{p_1}{p_0}\right)\beta^{-1}, \qquad \lambda \equiv \left(\frac{p_0}{p_1}\right)\alpha\beta^{-1}. \tag{6.1.5}$$

It is clear that the solution of (6.1.1) and (6.1.3) may be extended periodically, with period π, to $0 \le \phi \le \pi$, so that the problem becomes

$$\varepsilon\frac{dy}{d\theta} = \sin^2\theta - \lambda\frac{\sin^4\theta}{y}, \qquad -\frac{\pi}{2} < \theta < \frac{\pi}{2}, \tag{6.1.6}$$

and

$$y\left(\frac{\pi}{2}\right) = y\left(-\frac{\pi}{2}\right) = 1. \tag{6.1.7}$$

The boundary conditions (6.1.3) now serve to determine p_0/p_1, since

$$y(0) = \frac{p_0}{p_1}\,(=y(\pi)). \tag{6.1.8}$$

We shall solve the nonlinear eigenvalue problem (6.1.6–7) by the simple shooting technique using Newton's method. Thus we consider the initial-value problem

$$\text{(a)} \quad \varepsilon \frac{du}{d\theta} = \sin^2 \theta - \lambda \frac{\sin^4 \theta}{u},$$
$$\text{(b)} \quad u\left(\frac{\pi}{2}\right) = 1 \tag{6.1.9}$$

on the interval $-\pi/2 \leq \theta \leq \pi/2$. If $u = u(\lambda, \theta)$ is the solution of (6.1.9) for a fixed value of ε, and we can find $\lambda = \lambda(\varepsilon)$ such that

$$\phi(\lambda) \equiv u\left(\lambda, -\frac{\pi}{2}\right) - 1 = 0, \tag{6.1.10}$$

then clearly $y(\theta) \equiv u(\lambda, \theta)$ is a solution of (6.1.6–7) with the corresponding pair of parameter values ε, $\lambda(\varepsilon)$. Given some initial estimate λ_0 of the root of (6.1.10), we define the sequence $\{\lambda_\nu\}$ by Newton's method

$$\lambda_{\nu+1} = \lambda_\nu - \frac{\phi(\lambda_\nu)}{[d\phi(\lambda_\nu)/d\lambda]}, \qquad \nu = 0, 1, 2, \ldots. \tag{6.1.11}$$

If the solution $u(\lambda, \theta)$ of (6.1.9) is continuously differentiable with respect to λ, then, with

$$v(\lambda, \theta) \equiv \frac{\partial u(\lambda, \theta)}{\partial \lambda}, \tag{6.1.12}$$

we obtain the variational problem

$$\text{(a)} \quad \varepsilon \frac{dv}{d\theta} = \left(\frac{\lambda v}{u} - 1\right) \frac{\sin^4 \theta}{u},$$
$$\text{(b)} \quad v\left(\frac{\pi}{2}\right) = 0. \tag{6.1.13}$$

Now clearly, from (6.1.10) and (6.1.12),

$$\frac{d\phi(\lambda)}{d\lambda} = v\left(\lambda, -\frac{\pi}{2}\right),$$

and so the Newton iterates become simply

$$\lambda_{\nu+1} = \lambda_\nu + \frac{1 - u(\lambda_\nu, -\pi/2)}{v(\lambda_\nu, -\pi/2)}, \qquad \nu = 0, 1, 2, \ldots. \tag{6.1.14}$$

In the calculations we use the net

$$\theta_j = \pi/2 - j\Delta\theta$$

with spacings $\Delta\theta = \pi/102$ for $\varepsilon \geq 1/10$, and

$$\Delta\theta = \pi/502$$

for $\varepsilon \leq 1/10$. On these nets we solve the problems (6.1.9) and (6.1.13) as a first-order system of two equations using the Runge-Kutta scheme (1.3.14) to start (that is, to compute at θ_0, θ_1, θ_2, and θ_3) and the modified Adams method (1.3.16) to compute on $\theta_4 \geq \theta_j \geq -\pi/2$. Very little experimentation was required to find that the integration should be performed in the backward manner already indicated by (6.1.9b). (Theoretical considerations also suggested this as being more stable.)

For ε "large" the perturbation solution in powers of $1/\varepsilon$ implies that $\lambda \approx 4/3$. The calculations were started with $\varepsilon = 2$, using $\lambda = \lambda_0 = 1$ in the numerical integration of (6.1.9) and (6.1.13), and the sequence (6.1.14) was terminated when

$$|\lambda_\nu - \lambda_{\nu-1}| < 5 \times 10^{-6}.$$

This first run required $\nu = 4$ iterations, and all subsequent cases used only $\nu = 3$ or $\nu = 2$ iterations. When convergence was obtained for a given value of ε, the converged value, say $\lambda(\varepsilon)$ (or rather a rounded approximation to it), is used as the initial estimate, λ_0, for the next, slightly smaller, value of ε. Some of the sequences of Newton iterates obtained in this way are shown in Table 6.1.1, p. 158 (only half the ε values used in $1/10 \leq \varepsilon \leq 1$ are shown in this table).

To check the accuracy of the numerical work we compare some of the numerical results with corresponding results of the singular perturbation approximation. In particular this analytical procedure gives

$$\lambda(\varepsilon) = 1 + 2\varepsilon^2 - 16\varepsilon^4 + \cdots, \tag{6.1.15}$$

and with a numerical integration of some auxiliary differential equations (to determine the approximate coefficients) we get

$$y(0) \doteq 20.63\varepsilon^2 - 600\varepsilon^4 + \cdots. \tag{6.1.16}$$

In Table 6.1.2 (p. 158) the numerical solutions are compared with these approximations.

Of course the singular perturbation formulae become more accurate as $\varepsilon \to 0$. On the other hand, it is to be expected that the numerical calculations

TABLE 6.1.1

Some Newton iterates for determination by shooting of the parameter values $\lambda = \lambda(\varepsilon)$ in the nonlinear eigenvalue problem (6.1.6) *and* (6.1.7).

	ε	λ_0	λ_1	λ_2	λ_3	λ_4
$\Delta\theta = \pi/102$	2	1.000 0	1.291 5953	1.307 8512	1.307 8921	1.307 8921
	1	1.307 9	1.293 2455	1.293 2779	1.293 2779	
	$\frac{1}{2}$	1.258 1	1.215 9251	1.216 0673	1.216 0673	
	$\frac{1}{3}$	1.175 6	1.141 0717	1.141 1758	1.141 1758	
	$\frac{1}{4}$	1.114 0	1.092 9940	1.093 0389	1.093 0389	
	$\frac{1}{5}$	1.077 0	1.064 6403	1.064 6550	1.064 6550	
	$\frac{1}{6}$	1.055 0	1.047 2529	1.047 2579		
	$\frac{1}{7}$	1.041 0	1.035 9506	1.035 9524		
	$\frac{1}{8}$	1.031 7	1.028 2207	1.028 2215		
	$\frac{1}{9}$	1.025 2	1.022 7130	1.022 7133		
	$\frac{1}{10}$	1.020 5	1.018 6566	1.018 6567		
$\Delta\theta = \pi/502$	$\frac{1}{10}$	1.000 0	1.018 6417	1.018 6568	1.018 6568	
	$\frac{1}{20}$	1.018 7	1.004 9022	1.004 9048		
	$\frac{1}{30}$	1.004 9	1.002 2029	1.002 2029		
	$\frac{1}{40}$	1.002 2	1.001 2438	1.001 2438		
	$\frac{1}{50}$	1.001 2	1.000 7975	1.000 7975		
	$\frac{1}{60}$	1.000 8	1.000 5543	1.000 5543		
	$\frac{1}{70}$	1.000 6	1.000 4075	1.000 4075		
	$\frac{1}{80}$	1.000 4	1.000 3121	1.000 3121		
	$\frac{1}{90}$	1.000 3	1.000 2466	1.000 2466		
	$\frac{1}{100}$	1.000 2	1.000 1998	1.000 1998		

TABLE 6.1.2

Comparison of singular perturbation approximations (6.1.15) *and* (6.1.16) *with corresponding results from the numerical calculations.*

		$\varepsilon = \frac{1}{5}$	$\varepsilon = \frac{1}{10}$	$\varepsilon = \frac{1}{20}$	$\varepsilon = \frac{1}{100}$
λ	Singular Perturbation	1.0540 0000	1.0184 0000	1.0049 0000	1.0001 9984
	Numerical Calculations	1.0646 5511	1.0186 5686	1.0049 0484	1.0001 9988
$y(0)$	Singular Perturbation	−.1348 000	.1463 0000	.0478 2500	.0020 5700
	Numerical Calculation	.4129 7201	.1616 4435	.0481 1537	.0020 5708

become less accurate (for fixed net spacing) as $\varepsilon \to 0$. Thus we conclude from the comparison in Table 6.1.2 that our numerical results are quite accurate.

In Figure 6.1.2 we present some graphs of solutions of the nonlinear eigenvalue problem (6.1.6–7) for several values of ε.

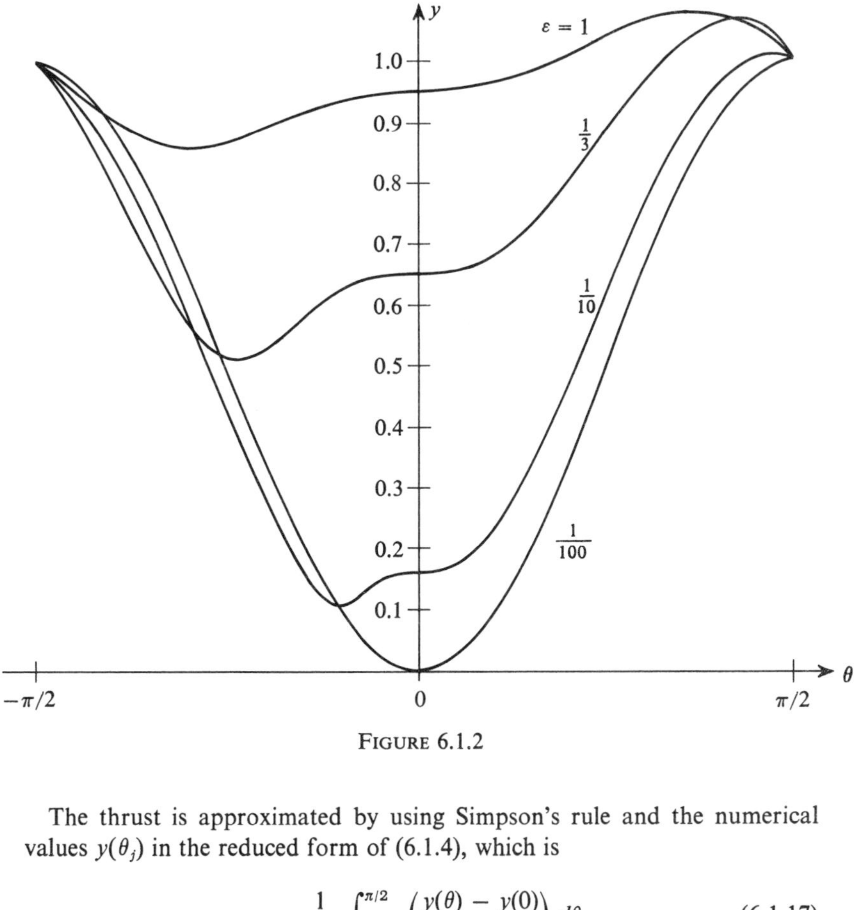

FIGURE 6.1.2

The thrust is approximated by using Simpson's rule and the numerical values $y(\theta_j)$ in the reduced form of (6.1.4), which is

$$\tau = \frac{1}{y(0)} \int_{-\pi/2}^{\pi/2} \left(\frac{y(\theta) - y(0)}{\sin^2 \theta} \right) d\theta. \tag{6.1.17}$$

The strange numbers of net points used were specially chosen so that this numerical integration could be done without the need for end-corrections. The singularity of the integrand at $\theta = 0$ is easily treated by simply eliminating the two panels over $-\Delta\theta \le \theta \le \Delta\theta$ in the quadrature formula. In this particular case the accuracy of Simpson's rule is maintained, since for the exact solution we find that

$$\frac{dy(0)}{d\theta} = \frac{d^2y(0)}{d\theta^2} = \frac{d^4y(0)}{dy^4} = \frac{d^5y(0)}{d\theta^5} = 0, \qquad \frac{d^3y(0)}{d\theta^3} = 2\varepsilon.$$

The computed value of $\tau y(0)$ from (6.1.17), using the above procedure with $\varepsilon = 10^{-2}$, was

$$\tau y(0) = 2.993.$$

From the singular perturbation procedure a crude approximation to the thrust is obtained as

$$\tau y(0) \approx \pi(1 - 10\varepsilon). \tag{6.1.18}$$

With $\varepsilon = 10^{-2}$, this gives

$$\tau y(0) \approx 2.827,$$

which is within 6% of the above numerical value. For smaller ε the formula (6.1.18) is expected to improve rapidly; it agrees best with the computed values of thrust for the smaller values of ε.

6.1.1 Computing Exercise; Forced Flow

If the lubricant is forced through the gap by an external pressure gradient, for example

$$p_0 = p\left(-\frac{\pi}{2}\right) \neq p\left(\frac{\pi}{2}\right) = p_2,$$

then the function $y(\theta)$ defined in (6.1.5) will be discontinuous at $\theta = 0$. The boundary conditions (6.1.7) still apply, but the differential equation (6.1.6) now holds only in the two intervals $-\pi/2 < \theta < 0$ and $0 < \theta < \pi/2$. The discontinuity condition at $\theta = 0$ becomes

$$y(0_-) = Ry(0^+), \qquad R \equiv \frac{p_2}{p_0}. \tag{6.1.19}$$

Solve the forced-flow case numerically by integrating (6.1.9) and (6.1.13) from $\theta = \pi/2$ to $\theta = 0_+$. Then apply the jump conditions

$$\begin{aligned} &\text{(a)}\quad u(0_-) = Ru(0_+), \\ &\text{(b)}\quad v(0_-) = Rv(0_+) \end{aligned} \tag{6.1.20}$$

and continue the integration from $\theta = 0_-$ to $\theta = -\pi/2$. Finally use (6.1.14) to find $\lambda = \lambda(\varepsilon, R)$ by Newton's method. [How is Equation (6.1.20b) derived?]

We note that for $R = 1$ the solution reduces to that in the case previously treated. Thus values of $\lambda(\varepsilon, 1)$ are given by the final entry in each row of Table 6.1.1. Using continuity in R, these furnish initial guesses at $\lambda(\varepsilon, 1 \pm \Delta R)$. This procedure can be continued.

A more efficient procedure than the obvious continuity method just suggested can be given (see Section 5.5.1). It is based on the fact that quite accurate approximations to $\lambda(\varepsilon, R \pm \Delta R)$ can be obtained from the linear approximation (that is, the first two terms in the Taylor expansion)

$$\lambda(\varepsilon, R \pm \Delta R) \doteq \lambda(\varepsilon, R) \pm \Delta R \frac{\partial \lambda}{\partial R}(\varepsilon, R). \tag{6.1.21}$$

The parameter $\lambda(\varepsilon, R)$ is of course determined such that the function $u = u(\varepsilon, R; \theta)$ satisfies the differential equation (6.1.9a) on $(0, \pi/2)$ and $(-\pi/2, 0)$ and the boundary conditions (6.1.9b), (6.1.20a), and (6.1.10). We write this latter condition now as

$$u\left(\lambda(\varepsilon, R), R, -\frac{\pi}{2}\right) - 1 = 0.$$

Differentiating with respect to R, we find that

$$\frac{\partial \lambda}{\partial R} = -\left(\frac{\partial u}{\partial R}\Big/\frac{\partial u}{\partial \lambda}\right)_{\theta = -\pi/2}$$

Thus introducing the variational problem for $w \equiv \partial u/\partial R$,

$$\text{(a)} \quad \varepsilon \frac{dw}{d\theta} = \left(\frac{\lambda w}{u} - 1\right)\frac{\sin^4 w}{u}, \qquad \theta \in (0, \pi/2), \qquad \theta \in (-\pi/2, 0),$$

$$\text{(b)} \quad w\left(\frac{\pi}{2}\right) = 0, \tag{6.1.22}$$

$$\text{(c)} \quad w(0_-) = Rw(0_+) + u(0_+),$$

we recall that $v = \partial u/\partial \lambda$ in (6.1.13). Hence, we have

$$\frac{\partial \lambda}{\partial R} = -\frac{w(-\pi/2)}{v(-\pi/2)}. \tag{6.1.23}$$

After λ, $u(\theta)$ and $v(\theta)$ have been determined for a given (ε, R) (that is, after the iterations have converged) *one integration* of the linear problem (6.1.22) will yield $\partial\lambda/\partial R$ by means of (6.1.23). Compare the step sizes ΔR that can be taken with this procedure with those using ordinary continuity methods.

Note that we could just as well consider the accurate continuity method with respect to the parameter ε and use

$$\lambda(\varepsilon \pm \Delta\varepsilon, R) \doteq \lambda(\varepsilon, R) \pm \Delta\varepsilon \frac{\partial\lambda(\varepsilon, R)}{\partial\varepsilon}. \tag{6.1.24}$$

Work out the detailed formulation for determining the derivative $\partial\lambda/\partial\varepsilon$. Try computing with these initial guesses.

Compare all the above with the general treatment of Section 5.5.1, using the notation $\boldsymbol{\sigma} = \binom{\varepsilon}{R}$ for the parameters.

6.2 *Finite Differences; Biophysics*

A class of problems concerning the diffusion of, say, oxygen into a cell in which an enzyme-catalyzed reaction occurs has been formulated and studied by means of singular perturbation theory by J. D. Murray. We present here a difference method for one particular such problem. The programming of the numerical work was done by J. Steadman, and all the calculations reported here were performed in about 20 seconds on a Control Data 6600 at the A.E.C. Computing and Applied Mathematics Center of the Courant Institute of Mathematical Sciences at New York University.

The diffusion-kinetics equation governing the steady concentration C of some substrate in an enzyme-catalyzed reaction has the general form

$$\nabla(D \nabla C) = g(C).$$

Here D is the molecular-diffusion coefficient of the substrate in the medium containing, say, uniformly-distributed bacteria and $g(C)$ is proportional to the reaction rate of the enzyme-substrate reaction. We consider the case with constant diffusion coefficient, D_0, in a spherical cell with the Michaelis–Menten-theory reaction rate. In dimensionless variables the diffusion-kinetics equation can now be written as

$$Ly \equiv (x^2y')' = x^2f(y), \qquad 0 < x < 1, \tag{6.2.1}$$

where

$$x \equiv \frac{r}{R}, \quad y(x) \equiv \frac{C(r)}{C_0}, \quad \varepsilon \equiv \frac{D_0C_0}{(nqR^2)}, \quad f(y) \equiv \varepsilon^{-1}\frac{y(x)}{y(x) + k}, \quad k \equiv \frac{k_m}{C_0}.$$

Here R is the radius of the cell, C_0 is the constant concentration of the substrate in $r > R$, k_m is the Michaelis constant, q is the maximum rate at which each bacterium can operate, and n is the number of bacteria. Typical ranges for both dimensionless parameters, ε and k, are, in some cases, 10^{-3} to 10^{-1}.

Assuming the cell membrane to have infinite permeability, it follows that

$$y(1) = 1. \tag{6.2.2a}$$

Further, from the assumed continuity and symmetry of $y(x)$ with respect to $x = 0$, we must have

$$y'(0) = 0. \tag{6.2.2b}$$

Equations (6.2.1) and (6.2.2) form a nonlinear second-order two-point boundary-value problem for which we seek *positive* solutions for various values of ε and k.

For small values of ε a singular perturbation approximation to the solution has been given by J. D. Murray. For larger parameter values, specifically when $6\varepsilon k > 1$, an iteration scheme with $y^{(0)}(x) \equiv 0$ and

$$Ly^{(\nu+1)} = x^2 f(y^{(\nu)}), \qquad \frac{d}{dx} y^{(\nu+1)}(0) = 0, \qquad y^{(\nu+1)}(1) = 1, \qquad \nu = 0, 1, \ldots, \tag{6.2.3a}$$

can be shown to converge to a positive solution $y(x)$ in the alternating manner

$$0 \equiv y^{(0)}(x) \le y^{(2)}(x) \le \cdots \le y(x) \le \cdots \le y^{(3)}(x) \le y^{(1)}(x) \equiv 1. \tag{6.2.3b}$$

Less-precise upper and lower bounds on the exact solution can be derived for the case of arbitrary (small) parameter values. The uniqueness of *positive* solutions is also easily demonstrated. We do not present any of these details here but shall show how accurate *finite-difference* approximations can be obtained and shall illustrate some of the typical difficulties that can arise.

With the net spacing $h = 1/(J + 1)$ and net points

$$x_j = jh,$$

we define the difference operator, L_h, which is to approximate L in (6.2.1), by

$$h^2 L_h v_j \equiv x_{j-1/2}^2 v_{j-1} - [x_{j-1/2}^2 + x_{j+1/2}^2] v_j + x_{j+1/2}^2 v_{j+1}. \tag{6.2.4}$$

Now let u_j be the numerical approximation to $y(x_j)$ and require that, in place of (6.2.1),

$$L_h u_j = x_j^2 f(u_j), \qquad j = 0, 1, 2, \ldots, J. \tag{6.2.5}$$

The boundary conditions (6.2.2a–b) are replaced by, respectively,

$$\begin{aligned} &\text{(a)} \quad u_{J+1} = 1, \\ &\text{(b)} \quad u_1 - u_{-1} = 0. \end{aligned} \tag{6.2.6}$$

Here and in (6.2.5), for $j = 0$, we have introduced the notation u_{-1} to represent an approximation to the extension of the solution $y(x)$ to $y(-h)$. Using (6.2.6) in (6.2.5), we can eliminate u_{-1} and u_{J+1} to obtain the $J + 1$ difference equations

$$\begin{aligned}
\phi_0(\mathbf{u}) &\equiv u_0 - u_1 = 0;\\
\phi_j(\mathbf{u}) &= [-x_{j-1/2}^2 u_{j-1} + (x_{j-1/2}^2 + x_{j+1/2}^2)u_j - x_{j+1/2}^2 u_{j+1}]\\
&\qquad + h^2 x_j^2 f(u_j) = 0, \qquad 1 \le j < J; \qquad (6.2.7)\\
\phi_J(\mathbf{u}) &\equiv [-x_{J-1/2}^2 u_{J-1} + (x_{J-1/2}^2 + x_{J+1/2}^2)u_J]\\
&\qquad - x_{J+1/2}^2 + h^2 x_J^2 f(u_J) = 0.
\end{aligned}$$

To solve these nonlinear difference equations Newton's method is employed. Thus, as in (3.2.9), the Jacobian matrix of $\boldsymbol{\phi}(\mathbf{u})$ is

$$A(\mathbf{u}) \equiv \frac{\partial \boldsymbol{\phi}(\mathbf{u})}{\partial \mathbf{u}} = \begin{pmatrix} B_0 & C_0 & & \cdots & 0 \\ A_1 & B_1 & C_1 & \cdots & \vdots \\ \vdots & & & & \\ & \cdots & A_{J-1} & B_{J-1} & C_{J-1} \\ 0 & \cdots & & A_J & B_J \end{pmatrix}, \qquad (6.2.8a)$$

where explicitly

$$\begin{aligned}
B_0(\mathbf{u}) &\equiv \frac{\partial \phi_0}{\partial u_0} = 1, \qquad C_0 \equiv \frac{\partial \phi_0}{\partial u_1} = -1,\\
A_j(\mathbf{u}) &\equiv \frac{\partial \phi_j}{\partial u_{j-1}} = -x_{j-1/2}^2, \qquad 1 \le j \le J,\\
B_j(\mathbf{u}) &\equiv \frac{\partial \phi_j}{\partial u_j} = (x_{j-1/2}^2 + x_{j+1/2}^2) + h^2 x_j^2 \frac{\varepsilon^{-1}k}{(u_j + k)^2}, \qquad 1 \le j \le J, \qquad (6.2.8b)\\
C_j(\mathbf{u}) &\equiv \frac{\partial \phi_j}{\partial u_{j+1}} = -x_{j+1/2}^2, \qquad 1 \le j \le J - 1.
\end{aligned}$$

Then the iterates $\mathbf{u}^{(\nu)} \equiv (u_0^{(\nu)}, u_1^{(\nu)}, \ldots, u_J^{(\nu)})^T$ are computed from

$$\mathbf{u}^{(\nu+1)} = \mathbf{u}^{(\nu)} + \Delta\mathbf{u}^{(\nu)}, \qquad \nu = 0, 1, 2, \ldots, \qquad (6.2.9a)$$

where each $\Delta\mathbf{u}^{(\nu)}$ is the solution of the corresponding linear system

$$A(\mathbf{u}^{(\nu)})\, \Delta\mathbf{u}^{(\nu)} = -\boldsymbol{\phi}(\mathbf{u}^{(\nu)}), \qquad \nu = 0, 1, 2, \ldots. \qquad (6.2.9b)$$

These linear systems are each of order $J + 1$ but have the simple tridiagonal form discussed in Section 3.1. Thus they are easily solved using the simple recursions derived there [see Equations (3.1.6–7)]. It is of interest to note that while only $B_j(\mathbf{u})$ actually varies with $\mathbf{u}$, the factorization of $A(\mathbf{u}^{(\nu)})$ must be done anew for each iteration. To start the iterations we use a simple approximation suggested by the analytical bounds alluded to above, namely

$$u_j^{(0)} = (1 - \varepsilon k)x_j^2 + \varepsilon k; \qquad j = 0, 1, \ldots, J + 1. \tag{6.2.10}$$

It is usually necessary, in order for Newton's method to converge to a desired solution, that the initial iterate, $u_j^{(0)}$, be "sufficiently close" to this solution. We were rather lucky that on the first attempt, using $\varepsilon = k = 0.1$ and (6.2.10) with $h = 10^{-2}$, the procedure (6.2.9) converged in only three iterations. The details are illustrated in Figure 6.2.1, where the initial guess ($\nu = 0$) and the first three iterates $u_j^{(\nu)}$ are plotted.

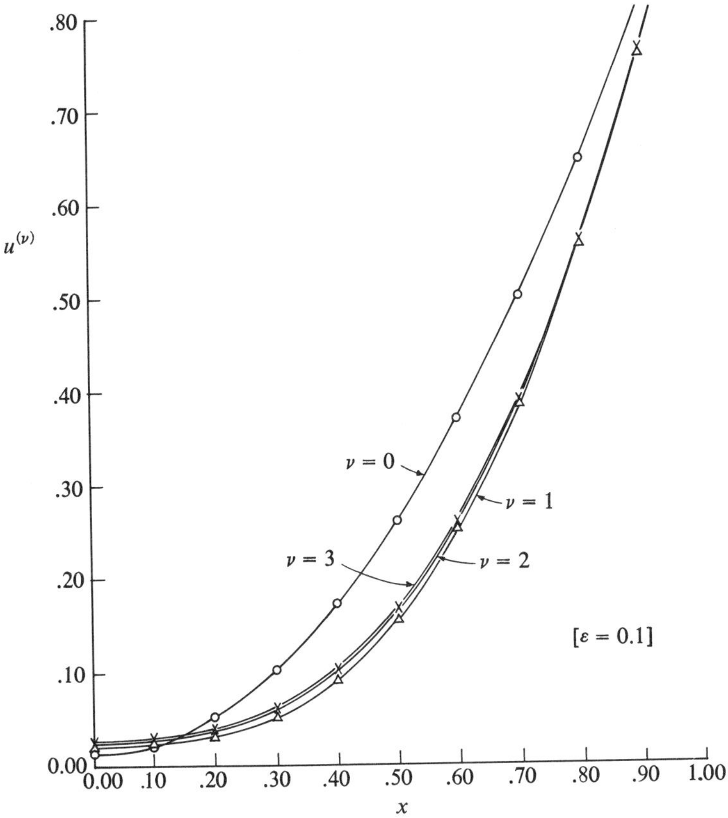

FIGURE 6.2.1

Using the same net, $k = 0.1$, and reducing ε to $\varepsilon = 0.05$, the initial iterate (6.2.10) again resulted in the convergence of Newton's method. Now, however, it required 15 iterations and the "solution" was not positive on $0 < x \leq 1$. These iterates are shown in Figure 6.2.2. It is clear that we have found a solution which is not physical. (Note that the slope of the solution is discontinuous at $u = -0.1$, where $f(u)$ has an infinite jump discontinuity. The corresponding solution of (6.2.1) does not have a second derivative at these points.) The desired positive solution for $\varepsilon = .05$, obtained as described below, is shown dashed in Figure 6.2.2.

The difficulty in the above case is of course due to the fact that the initial guess was too crude. But assuming (as we could in fact prove) that the

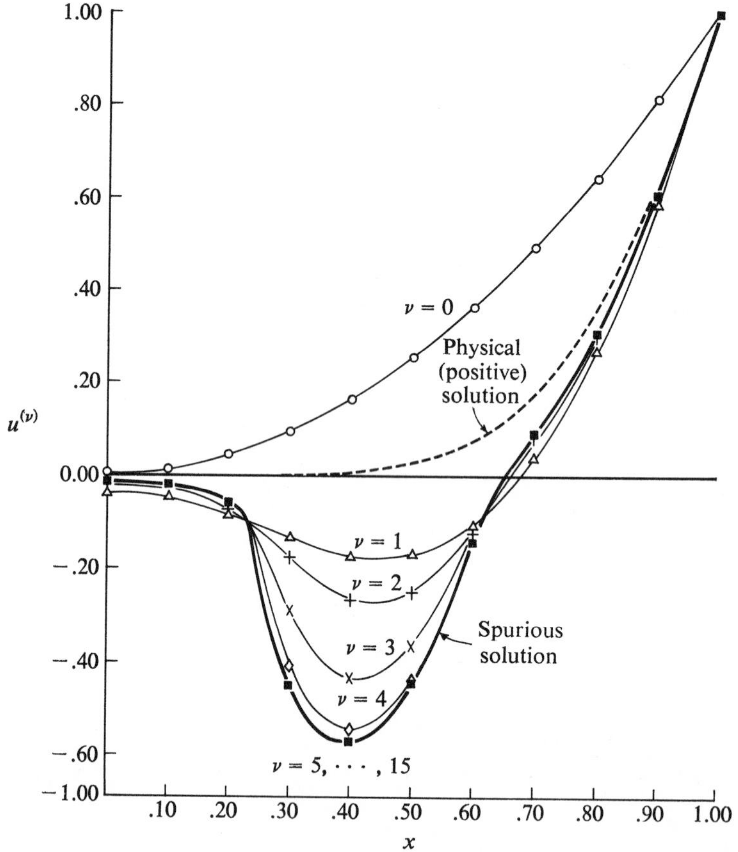

FIGURE 6.2.2

desired positive solution depends continuously on the parameter ε, we can employ a form of continuity method to get close initial estimates. Thus we use the converged solution for ε as the initial guess at the solution for $\varepsilon - \Delta\varepsilon$. In this way there was no difficulty starting with (6.2.10) for $\varepsilon = k = 0.1$ and continuing, with k fixed, as described for 19 values of ε in the decreasing sequence $\varepsilon = 10^{-1}(.01)10^{-2}(10^{-3})10^{-3}$. Some of these solutions are shown in Figure 6.2.3.

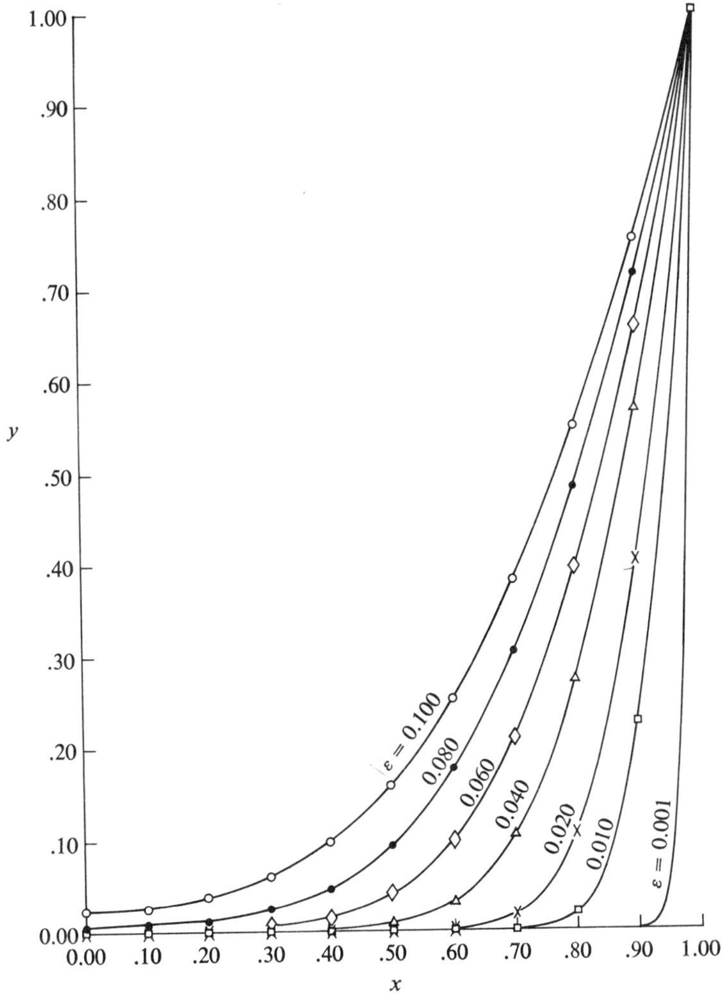

FIGURE 6.2.3

This sequence of calculations was done with the net spacing $h = 10^{-3}$, so that there were 1001 net points on $0 \le x \le 1$ [that is, $J + 1 = 1000$ nonlinear algebraic equations in the system (6.2.7)]. The convergence test for Newton's method required that

$$\delta_\nu \equiv h^{-2} \sum_{j=0}^{J} |\phi_j(\mathbf{u}^{(\nu)})| = \sum_{j=0}^{J} |L_h u_j^{(\nu)} - x_j^2 f(u_j^{(\nu)})| \le 10^{-6}. \tag{6.2.11}$$

A summary of the convergence properties is shown in Table 6.2.1. Since the factor h^{-2} was (inadvertently) included in this test we are assured that the final solution satisfies each difference equation to within 10^{-12} and on the average to within 10^{-15}, when $h = 10^{-3}$. The quantity

$$\Delta_\nu = \sum_{j=0}^{J} |\Delta u_j^{(\nu)}| \tag{6.2.12}$$

was also computed, and the final value is listed in the last column in Table 6.2.1. (This is a measure of the deviation in the numerical results from the exact solution of the difference equations.) The quadratic nature of the convergence of Newton's method in this case is clearly shown in the table. It is apparent that the last iterate did not significantly alter the results in at least five cases (that is, for $\varepsilon = .010, .009, .008, .007$, and $.002$). In fact our test was so severe when $h = 10^{-3}$ that there were not enough significant digits remaining to obtain the full benefit of the last iteration in almost all cases. This is also borne out by the entries in the last column, Δ_{final}.

6.2.1 *$h \to 0$ Extrapolation*

The machine code for the numerical scheme (6.2.7–9) was used to compute $h \to 0$ extrapolated solutions. This was done by using the Newton scheme to solve the difference equations for the two net spacings $h = \frac{1}{20}$ and $h = \frac{1}{40}$. Then at each net point of the cruder net we formed, in an obvious notation (see Section 3.1.1),

$$\tilde{u}(x_j) = \frac{4}{3} u_{2j}\left(\frac{h}{2}\right) - \frac{1}{3} u_j(h), \qquad x_j = jh, \qquad j = 0, 1, \ldots, 20. \tag{6.2.13}$$

Some results of these calculations are shown in Table 6.2.2. In the columns headed "$h = \frac{1}{20}$" and "$h = \frac{1}{1000}$" we list corresponding difference solutions, at the points $x_j = 0$ and $x_j = \frac{1}{2}$, for the indicated values of ε. The column headed "$\tilde{u}(x)$" contains the values computed by means of (6.2.13) for the appropriate values of x and ε.

The convergence properties of Newton's method for all three net spacings ($h = \frac{1}{20}, \frac{1}{40}, \frac{1}{1000}$) was the same. Thus the time to compute $\tilde{u}(x_j)$ as in (6.2.13) with $h = \frac{1}{20}$ would be about the same as that required to compute

TABLE 6.2.1

Convergence of Newton's method in solving the 1000 *difference equations* (6.2.7) *with* $h = 10^{-3}$ *to approximate solutions of the boundary-value problem* (6.2.1–2) *with* $k = 0.1$. *The quantity* δ_ν *measures the error in satisfying the difference equations by the* ν*th iterate* [*see* (6.2.11)] *and* Δ_{final} *is the last value of* Δ_ν *computed* [*see* (6.2.12)].

ε	δ_0	δ_1	δ_2	δ_3	δ_4	δ_5	Δ_{final}
.100	8.19×10^{2}	3.30×10	8.18×10^{-1}	1.08×10^{-3}	9.27×10^{-7}		3.75×10^{-9}
.090	2.91×10^{2}	9.89	9.39×10^{-2}	1.64×10^{-5}	9.80×10^{-7}		5.72×10^{-9}
.080	3.54×10^{2}	1.39×10	1.44×10^{-1}	2.64×10^{-5}	9.14×10^{-7}		8.95×10^{-10}
.070	4.41×10^{2}	1.99×10	2.29×10^{-1}	4.67×10^{-5}	8.64×10^{-7}		1.02×10^{-9}
.060	5.66×10^{2}	3.00×10	4.05×10^{-1}	1.04×10^{-4}	7.42×10^{-7}		4.37×10^{-10}
.050	7.54×10^{2}	4.86×10	8.20×10^{-1}	3.13×10^{-4}	6.91×10^{-7}		8.81×10^{-10}
.040	1.06×10^{3}	8.71×10	2.01	1.38×10^{-3}	6.05×10^{-7}		7.54×10^{-11}
.030	1.64×10^{3}	1.84×10^{2}	6.55	1.05×10^{-2}	5.71×10^{-7}		7.93×10^{-10}
.020	2.93×10^{3}	5.29×10^{2}	3.56×10	2.09×10^{-1}	7.90×10^{-6}	4.83×10^{-7}	1.48×10^{-10}
.010	7.46×10^{3}	3.47×10^{3}	6.68×10^{2}	3.87×10	1.47×10^{-1}	2.41×10^{-6} $(2.68 \times 10^{-7} = \delta_6)$	1.81×10^{-10}
.009	1.23×10^{3}	3.96×10	1.17×10^{-1}	1.49×10^{-6}	2.73×10^{-7}		1.08×10^{-10}
.008	1.47×10^{3}	5.33×10	1.95×10^{-1}	3.25×10^{-6}	2.62×10^{-7}		6.08×10^{-11}
.007	1.79×10^{3}	7.45×10	3.50×10^{-1}	9.02×10^{-6}	2.50×10^{-7}		3.25×10^{-11}
.006	2.24×10^{3}	1.10×10^{2}	6.86×10^{-1}	3.07×10^{-5}	2.58×10^{-7}		1.11×10^{-11}
.005	2.92×10^{3}	1.74×10^{2}	1.52	1.34×10^{-4}	1.94×10^{-7}		6.10×10^{-11}
.004	4.03×10^{3}	3.03×10^{2}	4.03	8.12×10^{-4}	2.25×10^{-7}		5.03×10^{-11}
.003	6.05×10^{3}	6.24×10^{2}	1.41×10	8.31×10^{-3}	1.74×10^{-7}		8.44×10^{-12}
.002	1.06×10^{4}	1.73×10^{3}	8.14×10	2.17×10^{-1}	1.70×10^{-6}	1.30×10^{-7}	4.89×10^{-12}
.001	2.61×10^{4}	1.05×10^{4}	1.59×10^{3}	5.24×10	6.15×10^{-2}	1.67×10^{-7}	1.91×10^{-11}

TABLE 6.2.2

Comparison of $h \to 0$ extrapolated solution (using $h = \frac{1}{20}$ and $h = \frac{1}{40}$) and difference solution for the net spacings $h = \frac{1}{20}$ and $h = \frac{1}{1000}$. The comparisons are of the computed approximations to $y(x)$ at $x = 0$ and $x = \frac{1}{2}$.

	$x_j = 0$			$x_j = \frac{1}{2}$		
ε	$h = \frac{1}{20}$	$\tilde{u}(0)$	$h = \frac{1}{1000}$	$h = \frac{1}{20}$	$\tilde{u}(\frac{1}{2})$	$h = \frac{1}{1000}$
.100	2.47141×10^{-2}	2.28706×10^{-2}	2.27927×10^{-2}	1.63074×10^{-1}	1.62561×10^{-1}	1.62562×10^{-1}
.090	1.37060×10^{-2}	1.24441×10^{-2}	1.23952×10^{-2}	1.28596×10^{-1}	1.28120×10^{-1}	1.28120×10^{-1}
.080	6.94581×10^{-3}	6.15949×10^{-3}	6.13261×10^{-3}	9.60633×10^{-2}	9.56128×10^{-2}	9.56133×10^{-2}
.070	3.14332×10^{-3}	2.70368×10^{-3}	2.69121×10^{-3}	6.66093×10^{-2}	6.61678×10^{-2}	6.61683×10^{-2}
.060	1.21684×10^{-3}	1.00338×10^{-3}	9.98898×10^{-4}	4.15850×10^{-2}	4.11370×10^{-2}	4.11375×10^{-2}
.050	3.73485×10^{-4}	2.89092×10^{-4}	2.88133×10^{-4}	2.23036×10^{-2}	2.18556×10^{-2}	2.18560×10^{-2}
.040	7.90739×10^{-5}	5.51098×10^{-5}	5.51652×10^{-5}	9.50409×10^{-3}	9.11223×10^{-3}	9.11220×10^{-3}
.030	8.68991×10^{-6}	4.94543×10^{-6}	5.03606×10^{-6}	2.76392×10^{-3}	2.51597×10^{-3}	2.51591×10^{-3}
.020	2.44811×10^{-7}	8.21931×10^{-8}	9.29283×10^{-8}	3.76343×10^{-4}	2.97021×10^{-4}	2.97518×10^{-4}
.010	1.35987×10^{-10}	-1.32918×10^{-11}	1.12845×10^{-11}	5.97694×10^{-6}	2.44531×10^{-6}	2.62288×10^{-6}
.009	3.70678×10^{-11}	-5.53370×10^{-12}	2.12627×10^{-12}	2.92525×10^{-6}	9.82325×10^{-7}	1.10258×10^{-6}
.008	8.24765×10^{-12}	-1.63822×10^{-12}	2.95151×10^{-13}	1.28161×10^{-6}	3.23194×10^{-7}	3.96498×10^{-7}
.007	1.40995×10^{-12}	-3.43248×10^{-13}	2.71460×10^{-14}	4.86301×10^{-7}	7.67188×10^{-8}	1.15575×10^{-7}
.006	1.69239×10^{-13}	-4.75580×10^{-14}	1.40153×10^{-15}	1.52150×10^{-7}	8.11170×10^{-9}	2.51026×10^{-8}
.005	1.23842×10^{-14}	-3.82330×10^{-15}	3.07870×10^{-17}	3.63431×10^{-8}	-2.10557×10^{-9}	3.52988×10^{-9}
.004	4.35024×10^{-16}	-1.41489×10^{-16}	1.73200×10^{-19}	5.82349×10^{-9}	-9.78488×10^{-10}	2.48493×10^{-10}
.003	4.65226×10^{-18}	-1.54438×10^{-18}	8.62714×10^{-23}	5.88562×10^{-10}	-1.27689×10^{-10}	5.11736×10^{-12}
.002	5.34157×10^{-21}	-1.78023×10^{-21}	2.41798×10^{-28}	1.21321×10^{-11}	-3.85515×10^{-12}	7.66428×10^{-15}
.001	2.52302×10^{-26}	-8.41008×10^{-27}	6.53728×10^{-41}	1.56115×10^{-14}	-5.19796×10^{-15}	3.30634×10^{-21}

u_j for a spacing of $h = \frac{1}{60}$ and so is *less than* $\frac{1}{16}$ *the time* required for the calculations with $h = 10^{-3}$. It was found, as expected, that the agreement of all the different calculations was best near $x = 1$ and worst near $x = 0$ (in fact the differences were monotonic in x). A glance at Table 6.2.2 reveals that the $h \to 0$ extrapolated solution, $\tilde{u}(x_j)$, has at least three significant digits in agreement with the difference solution $u_j(\frac{1}{1000})$ in $\frac{1}{2} \le x_j \le 1$ for all $\varepsilon \ge 0.20$. At least one significant digit agrees at $x_j = 0$ for the same ε range. While the relative error in $\tilde{u}(x_j)$ is rather large for x_j near zero and ε small (say $\varepsilon \le .03$), it should be noted that the absolute error is extremely small even down to $\varepsilon = .001$. Of course the negative values of $\tilde{u}(0)$ for $\varepsilon \le 0.01$ are nonsense and are caused by the coarseness of the net spacing used in the extrapolation method. In fact the values of u_0 computed with $h = \frac{1}{1000}$ may not have any significant digits for $\varepsilon \le .009$, say, but it is of interest to note that the difference scheme gives only positive values of u_j for all j and ε.

It is suggested, by the analysis in Section 3.1 and 3.2, that

$$|y(x_j) - \tilde{u}(x_j)| = \mathcal{O}(h^4) \qquad \text{and} \qquad |y(x_j) - u_j| = \mathcal{O}(h^2).$$

For $h = 10^{-3}$ we see that the error $\mathcal{O}(10^{-6})$ in the difference solution, u_j, is easily the order of magnitude of the solution for x_j near zero (if the higher derivatives of the solution are not very small). Thus, as seems to be the case, very accurate results are obtained over part of the interval [0, 1] and only order of magnitude results over the remainder (for small ε).

In conclusion it seems quite clear that we could easily obtain the same accuracy as in the $h = 10^{-3}$ case with much less computing effort (say using $h \to 0$ extrapolation with $h = 10^{-2}$ and $h = 10^{-2}/2$ in about one third the time).

6.2.2 *Computing Exercise; Nonlinear Diffusivity*

In many reactions the diffusion coefficient is a function of the substrate concentration. For the spherically-symmetric case treated above we need only replace (6.2.1) by

$$Ly \equiv (x^2 D(y)y')' = x^2 f(y), \qquad 0 < x < 1, \tag{6.2.14}$$

where $D(y)$ is the nonlinear diffusion coefficient. The difference operator (6.2.4) can now be replaced by

$$h^2 L_h v_j \equiv x_{j-1/2}^2 D_{j-1/2}(\mathbf{v}) v_{j-1} - [x_{j-1/2}^2 D_{j-1/2}(\mathbf{v}) + x_{j+1/2}^2 D_{j+1/2}(\mathbf{v})] v_j + x_{j+1/2}^2 D_{j+1/2}(\mathbf{v}) v_{j+1}, \tag{6.2.15a}$$

where

$$D_{j\pm 1/2}(\mathbf{v}) \equiv D\left(\frac{v_{j\pm 1} + v_j}{2}\right). \tag{6.2.15b}$$

The difference equations (6.2.5) and (6.2.6), using the above definition of L_h, can again be solved by Newton's method. In fact the linear system (6.2.9b) still has a tridiagonal coefficient matrix, but now the elements of this matrix involve derivatives of $D(\mathbf{u})$. For instance, we find from the derivative of the jth equation with respect to u_{j-1} that

$$A_j(\mathbf{u}) = -x_{j-1/2}^2 D_{j-1/2}(\mathbf{u}) + x_{j-1/2}^2 D'_{j-1/2}(\mathbf{u})\cdot\frac{(u_j - u_{j-1})}{2}, \qquad 1 \leq j \leq J, \tag{6.2.16a}$$

where

$$D'_{j\pm 1/2}(\mathbf{u}) \equiv \left.\frac{dD(\xi)}{d\xi}\right|_{\xi=(u_{j\pm 1}+u_j)/2} \tag{6.2.16b}$$

We leave the remaining details and the somewhat different treatment that results at x_0 to the reader.

A diffusion coefficient of particular interest is

$$D(y) \equiv 1 + \frac{\lambda}{(y + k_2)^2}. \tag{6.2.17}$$

Computations are of interest for various parameter values; try $\lambda = k_2 = 10^{-2}$ with the same ε and k values used above. Note that for $\lambda = 0$ the problem reduces to the one previously solved. Thus if there is difficulty in getting close initial guesses for the smaller ε values, the continuity procedure in λ can be employed, working in "small" steps from $\lambda = 0$ to $\lambda = 10^{-2}$. An alternative, of course, is to use continuity in ε as in the worked-out example, but with possibly smaller steps in ε. (It turns out that only one extra value, $\varepsilon = 0.015$, is required.)

Finally, it would seem to be quite simple to solve either the linear or nonlinear diffusivity case by the shooting method. Try shooting in both directions for small ε and observe the striking difference.

APPENDIX A

FUNCTION SPACE APPROXIMATION METHODS

We present here a very brief sketch of some approximation methods which are frequently as practical and effective as any of those methods previously studied. However, the theoretical justification of these methods is considerably more difficult and less well developed. These methods are expansion procedures, of which the power-series method in Section 2.4.1 is a (not so practical or typical) special case. More specifically, the solution is approximated by a linear combination of linearly-independent functions in an appropriate function space. The coefficients in the expansion are to be determined so that this combination minimizes some measure of the error in satisfying the boundary-value problem. There is tremendous variety in the choice of approximating functions and in the choice of "measure of error" in satisfying the problem. We proceed to indicate three procedures of this type.

The problems we consider are the most general nonlinear eigenvalue problems discussed in Section 5.5.

$$\mathbf{y}' = \mathbf{f}(x, \mathbf{y}; \boldsymbol{\lambda}, \boldsymbol{\sigma}), \qquad a < x < b; \tag{A.1}$$

$$\mathbf{g}(\mathbf{y}(a), \mathbf{y}(b); \boldsymbol{\lambda}, \boldsymbol{\sigma}) = 0. \tag{A.2}$$

Here $\mathbf{y}(x)$ and $\mathbf{f}$ are n-vectors, $\boldsymbol{\lambda}$ is an m-vector, $\boldsymbol{\sigma}$ is a q-vector, and $\mathbf{g}$ is an $(n + m)$-vector. For some fixed value of the parameter $\boldsymbol{\sigma}$ we seek a value of the "eigenvalue" $\boldsymbol{\lambda}$, and a function $\mathbf{y}(t) \equiv \mathbf{y}(t; \boldsymbol{\lambda}, \boldsymbol{\sigma})$ such that (A.1) and (A.2) are satisfied. These solutions are required for all $\boldsymbol{\sigma}$ in some parameter set. Of course, as special cases, the eigenvalue $\boldsymbol{\lambda}$ may be absent, in which case $m = 0$ and the parameter $\boldsymbol{\sigma}$ may not appear.

A.1 Galerkin's Method

The approximating functions are chosen from some Hilbert space $\mathscr{H}$ of functions that are, say, piecewise continuously differentiable on $[a, b]$.

Specifically let us assume that $\{\phi_j(x)\}$ is an orthonormal basis for $\mathscr{H}$, with

$$(\phi_j, \phi_k) \equiv \int_a^b \phi_j(x)\phi_k(x)w(x)\,dx = \delta_{ij}, \qquad i, j = 1, 2, \ldots. \tag{A.3}$$

Here $w(x)$ is the (positive) weight function defining the inner product on $\mathscr{H}$. Then we may define an Nth-order approximation to a solution of (A.1) and (A.2) as a combination of the form

$$\mathbf{u}_N(x) \equiv \sum_{j=1}^{N} \boldsymbol{\xi}_j\phi_j(x). \tag{A.4}$$

The N coefficient vectors $\boldsymbol{\xi}_j$ of dimension n are to be determined by requiring that

$$\boldsymbol{\gamma}_k(\boldsymbol{\xi}_1, \ldots, \boldsymbol{\xi}_N; \boldsymbol{\lambda}, \boldsymbol{\sigma}) \equiv \int_a^b [\mathbf{u}'_N(x) - \mathbf{f}(x, \mathbf{u}_N(x); \boldsymbol{\lambda}, \boldsymbol{\sigma})]\phi_k(x)w(x)\,dx = 0,$$
$$k = 1, 2, \ldots, N - 1, \tag{A.5a}$$

and

$$\boldsymbol{\gamma}_N(\boldsymbol{\xi}_1, \ldots, \boldsymbol{\xi}_N; \boldsymbol{\lambda}, \boldsymbol{\sigma}) \equiv \mathbf{g}(\mathbf{u}_N(a), \mathbf{u}_N(b); \boldsymbol{\lambda}, \boldsymbol{\sigma}) = 0. \tag{A.5b}$$

Thus in Galerkin's method to "approximately satisfy the boundary value problem" means to satisfy the boundary conditions exactly and for the error in satisfying the equation to be orthogonal to the first $N - 1$ basis functions. That is, the measure of error to be made zero is the projection of the error in satisfying the equation on the subspace spanned by $\phi_1, \ldots, \phi_{N-1}$. Of course any function orthogonal to all of the basis functions vanishes and so in the limit as $N \to \infty$ the functions $\mathbf{u}_N(x)$ may very well converge to a solution. This convergence question is quite open in the indicated general case, but see Urabe [2].

The system (A.5) contains $(nN + m)$ equations in the as many unknowns $\boldsymbol{\xi}_1, \ldots, \boldsymbol{\xi}_N, \boldsymbol{\lambda}$. In some problems it is possible to pick the basis functions $\phi_j(x)$ so that n of the boundary conditions in (A.2) are identically satisfied. Then an additional orthogonality condition, from (A.5a) with $k = N$, can be imposed. If the differential equation and the boundary conditions are all linear, then the system (A.5) is linear in the components of the $\boldsymbol{\xi}_j$. If the original linear problem is also homogeneous, then $\boldsymbol{\xi}_j \equiv 0$, $j = 1, 2, \ldots, N$ is clearly a solution, and we have in fact an algebraic eigenvalue problem in which $\boldsymbol{\lambda}$ is to be determined so that nontrivial solutions exist. In the nonlinear case, Newton's method is again an excellent scheme for seeking accurate approximations to the roots of (A.5). Continuity techniques employing the parameters $\boldsymbol{\sigma}$, or even introducing such parameters, are also clearly suggested as in Section 5.5.1. Another alternative iteration scheme is suggested in the next subsection.

A.2 Collocation Methods

Now let the functions $\{\psi_j(x)\}$ form a complete set on some Banach space, $\mathscr{B}$, which includes $C^1(a, b)$. Further, let interpolation by linear combinations of these functions converge, together with the first derivatives, for functions in $C^1(a, b)$. This means that for each finite set

$$\{\psi_1(x), \psi_2(x), \ldots, \psi_N(x)\}$$

there are corresponding points $x_{j,N} \in [a, b]$ such that for any $\phi(x) \in C^1(a, b)$ there exist (unique) constants $a_{k,N}$ such that

$$\phi(x_{j,N}) = \sum_{k=1}^{N} a_{k,N}\psi_k(x_{j,N}), \qquad j = 1, 2, \ldots, N.$$

Further, for these combinations, it is required that:

$$\lim_{N\to\infty} \sum_{k=1}^{N} a_{k,N}\psi_k(x) = \phi(x),$$

$$\lim_{N\to\infty} \sum_{k=1}^{N} a_{k,N}\psi_k'(x) = \phi'(x).$$

In particular cases these conditions may be varied or relaxed somewhat.

Approximate solutions of (A.1–2) are now sought, for a given value of N, in the form

$$\mathbf{v}_N(x) \equiv \sum_{j=1}^{N} \boldsymbol{\eta}_j\psi_j(x). \tag{A.6}$$

The conditions for determining the coefficients $\boldsymbol{\eta}_j$ and the eigenvalues $\boldsymbol{\lambda}$ are that $\mathbf{v}_N(x)$ should satisfy the differential equation at $N - 1$ distinct points $x_{j,N-1} \in [a, b]$ and should satisfy the boundary conditions. Thus we obtain the $(nN + m)$ equations in as many unknowns

$$\boldsymbol{\gamma}_k(\boldsymbol{\eta}_1, \ldots, \boldsymbol{\eta}_N; \boldsymbol{\lambda}, \boldsymbol{\sigma}) \equiv \mathbf{v}_N'(x_{k,N-1}) - \mathbf{f}(x_{k,N-1}, \mathbf{v}_N)(x_{k,N-1}; \boldsymbol{\lambda}, \boldsymbol{\sigma}) = 0,$$
$$k = 1, 2, \ldots, N - 1; \tag{A.7a}$$

$$\boldsymbol{\gamma}_N(\boldsymbol{\eta}_1, \ldots, \boldsymbol{\eta}_N; \boldsymbol{\lambda}, \boldsymbol{\sigma}) \equiv \mathbf{g}(\mathbf{v}_N(a), \mathbf{v}_N(b); \boldsymbol{\lambda}, \boldsymbol{\sigma}) = 0. \tag{A.7b}$$

We have indicated in (A.7a) the use of the interpolation points corresponding to the first $N - 1$ basis functions. This may of course be altered, as an optimum choice of points for interpolation of any $\phi(x) \in C^1(a, b)$ need not be near optimum for interpolating $[\mathbf{v}_N'(x) - \mathbf{f}(x, \mathbf{v}_N(x); \boldsymbol{\lambda}, \boldsymbol{\sigma})]$ which may only be in $C(a, b)$. Again, if the functions $\psi_j(x)$ can be chosen so that n of the conditions in (A.7b) are identically satisfied, the condition in (A.7a) can be imposed at an additional point; that is, we could use the set $x_{k,N}$, $k = 1, 2, \ldots, N$. For linear problems this scheme, which we have called collocation, is but a generalization of Lanczos's (1957) selected-points method (in which the $\psi_j(x) \equiv x^{j-1}$ and the $x_{k,N}$ are the zeros or extrema of Chebyshev, Legendre, or other polynomials).

To solve (A.7), we can try the usual suggestions of Newton plus continuity. Another proposed iteration is based on a special case of collocation employed by Clenshaw (1966) [in which $n = 2$, $m = q = 1$, so that all of the features of the system (A.1–2) are included in this example]. The estimate $\boldsymbol{\lambda} = \boldsymbol{\lambda}^{(\nu)}$, say, of the eigenvalue is kept fixed, and Newton's method is employed on the $n(N - 1)$ equations in (A.7a) and some selected n of the $(n + m)$ equations in (A.7b) to determine the (converged) vectors $\boldsymbol{\eta}_1^{(\nu)}, \ldots, \boldsymbol{\eta}_N^{(\nu)}$. This stage corresponds to what is frequently called an "inner iteration" in related contexts. Upon the completion (that is, convergence) of the inner iterations, the function

$$\mathbf{v}_N^{(\nu)}(x) = \sum_{j=1}^{N} \boldsymbol{\eta}_j^{(\nu)} \psi_j(x)$$

is presumably an (accurate) approximation to the solution of *some* boundary-value problem for (A.1) with $\boldsymbol{\lambda} = \boldsymbol{\lambda}^{(\nu)}$. But it only satisfies n of the required $n + m$ boundary conditions (A.2).

The "outer iterations" are designed to change $\boldsymbol{\lambda}^{(\nu)}$ so that the remaining m conditions in (A.7b) are more nearly satisfied. For $m = 1$ some interpolation and false position procedures are straightforward. But quite generally we could again apply Newton's method to these m equations in the unknowns $\boldsymbol{\lambda}$. To do this rigorously we would have to determine $\partial\boldsymbol{\eta}_j^{(\nu)}/\partial\boldsymbol{\lambda}$ for $j = 1, 2, \ldots, N$, which is equivalent to the original application of Newton's method to all of (A.7). Of course other procedures, based on contracting maps, can be employed for the outer iterations.

A.3 Generalized Ritz Methods

Many boundary-value problems (for the determination of some equilibrium state of a physical system, for example) can be formulated in terms of variational problems. That is, some functional (representing for example the energy of the system) is to be minimized over an appropriate space of admissible functions, say $\mathscr{H}$, as in Section A.1. Or we may easily form some functional whose minimum value is attained for the solution of the boundary-value problem. As an *example*, consider the expression

$$I\{\mathbf{w}, \boldsymbol{\lambda}, \boldsymbol{\sigma}\} \equiv \int_a^b \|\mathbf{w}'(x) - \mathbf{f}(x, \mathbf{w}(x); \boldsymbol{\lambda}, \boldsymbol{\sigma})\|_I^2 \, dx + \|\mathbf{g}(\mathbf{w}(a), \mathbf{w}(b); \boldsymbol{\lambda}, \boldsymbol{\sigma})\|_{II}^2,$$

where $\|\cdot\|_I$ and $\|\cdot\|_{II}$ are some vector norms for n- and $(n + m)$-dimensional real vector spaces, respectively. Clearly $I\{\mathbf{w}, \boldsymbol{\lambda}, \boldsymbol{\sigma}\} \geq 0$ for all $\mathbf{w}(x) \in C^1(a, b)$ and $I\{\cdot\} = 0$ for a solution of (A.1–2).

For any functional $I\{\cdot\}$, using an approximation of the form

$$\mathbf{w}_N(x) = \sum_{j=1}^{N} \boldsymbol{\zeta}_j \phi_j(x), \tag{A.8}$$

a scalar function of the N coefficient vectors, $\boldsymbol{\zeta}_j$, and the eigenvalue parameters $\boldsymbol{\lambda}$, as well as $\boldsymbol{\sigma}$ is defined by

$$\Phi(\boldsymbol{\zeta}_1, \boldsymbol{\zeta}_2, \ldots, \boldsymbol{\zeta}_N, \boldsymbol{\lambda}, \boldsymbol{\sigma}) \equiv I\{\mathbf{w}_N(x), \boldsymbol{\lambda}, \boldsymbol{\sigma}\}. \tag{A.9}$$

The Ritz procedure is to minimize Φ with respect to the components ζ_{ij} of $\boldsymbol{\zeta}_j$ and the components λ_k of $\boldsymbol{\lambda}$. In general this leads to the system of $(nN + m)$ equations

$$\frac{\partial \Phi}{\partial \zeta_{ij}}(\boldsymbol{\zeta}_1, \ldots, \boldsymbol{\zeta}_N, \boldsymbol{\lambda}, \boldsymbol{\sigma}) = 0, \qquad i = 1, 2, \ldots, n, \qquad j = 1, 2, \ldots, N; \tag{A.10a}$$

$$\frac{\partial \Phi}{\partial \lambda_k}(\boldsymbol{\zeta}_1, \ldots, \boldsymbol{\zeta}_N, \boldsymbol{\lambda}, \boldsymbol{\sigma}) = 0, \qquad k = 1, 2, \ldots, m. \tag{A.10b}$$

For many important special cases that frequently occur in physical applications, the systems (A.10) simplify in various ways. The most familiar example occurs when the integrand in the functional $I\{\cdot\}$ is a homogeneous quadratic form in $\mathbf{w}$ and $\mathbf{w}'$ and the space $\mathscr{H}$ is such that the boundary conditions are automatically satisfied. In fact, under fairly common circumstances, the Ritz and Galerkin procedures are identical [see Kantorovich and Krylov (1958)].

Convergence of the Ritz method has been studied extensively for variational problems leading to linear boundary-value problems. An interesting exception, given by Ciarlet, Schultz, and Varga (1967), leads to higher-order-accurate approximations to the solution of various nonlinear problems of second and higher order.

For linear boundary-value problems that come from variational problems, the Ritz procedure can lead to finite-difference methods. The basic idea here is to employ "basis" functions $\phi_j(x)$ which are piecewise-linear and continuous, and vanish at all but one point (at which it takes on the value unity) of some net. See, for example, Courant (1943) or Friedrichs and Keller (1966).

BIBLIOGRAPHY

ANTOSIEWICZ, H., and W. GAUTSCHI
"Numerical methods in ordinary differential equations." In: *Survey of Numerical Analysis*, J. Todd, Ed. New York: McGraw-Hill, 1962, pp. 314–346.

ATKINSON, F. V.
Discrete and Continuous Boundary Problems. New York: Academic Press, 1964.

BABUŠKA, I., M. PRÁGER and E. VITÁSEK
Numerical Processes in Differential Equations. New York: Interscience, 1966.

BELLMAN, R., and R. KALABA
Quasilinearization and Nonlinear Boundary Value Problems. New York: Amer. Elsevier Publ. Co., 1965.

BERS, L., F. JOHN and M. SCHECHTER
Partial Differential Equations. New York: Interscience, 1964.

BIRKHOFF, G., and G-C. ROTA
Ordinary Differential Equations. Boston: Ginn & Co., 1962.

BIRKHOFF, G., C. DE BOOR, B. SWARTZ and B. WENDROFF
"Rayleigh-Ritz approximation by piecewise cubic polynomials." *SIAM Jour. Numer. Anal. 3* (1966), pp. 188–203.

BROWN, R. R.
"Numerical solution of boundary value problems using nonuniform grids." *SIAM Journal* ***10*** (1962), pp. 475–495.

CIARLET, P., M. SCHULTZ and R. VARGA
"Numerical methods of high-order accuracy for nonlinear boundary value problems." *Numer. Mathe.*, *9* (1967), pp. 394–430.

CLENSHAW, C. W.
"The solution of van der Pol's equation in Chebyshev series." In: *Numerical Solutions of Nonlinear Differential Equations*, D. Greenspan, Ed. New York: John Wiley & Sons, 1966, pp. 55–63.

CODDINGTON, E. A., and N. LEVINSON
Theory of Ordinary Differential Equations. New York: McGraw-Hill, 1955.

COLE, J. D., H. B. KELLER and P. G. SAFFMAN
"The flow of a viscous compressible fluid through a very narrow gap." *SIAM Jour. Appl. Math. 15* (1967), pp. 605–617.

COLLATZ, L.
The Numerical Treatment of Differential Equations, 3rd ed. Berlin: Springer, 1960.

CONTI, R.
"Problemes lineaires pour les equations differentielles ordinaires." *Math. Nachrichten 23* (1961), pp. 161–178.

CONTI, S. D.
"The numerical solution of linear boundary value problems." *SIAM Review 8* (1966), pp. 309–321.

COURANT, R., and D. HILBERT
1. *Methoden der Mathematischen Physik*, Vol. II. New York: Interscience, 1937.
2. *Methods of Mathematical Physics*, Vol. I. New York: Interscience, 1953.

COURANT, R.
"Variational methods for the solution of problems of equilibrium and vibrations." *Bull. Amer. Math. Soc. 49* (1943), pp. 1–23.

DAHLQUIST, G.
1. "Convergence and stability in the numerical integration of ordinary differential equations." *Math. Scand. 4* (1956), pp. 33–53.
2. "Stability and error bounds in the numerical integration of ordinary differential equations." *Trans. Roy. Inst. Tech.*, Stockholm, No. 130 (1959).

DAVIS, P. J. and P. RABINOWITZ
Numerical Integration. Waltham, Mass.: Blaisdell Publishing Co., 1967.

FARRINGTON, C. C., R. T. GREGORY and A. H. TAUB
"On the numerical solution of Sturm-Liouville differential equations." *M. O. C. 11* (1957), pp. 131–150.

FOX, L.
1. *The Numerical Solution of Two-Point Boundary Problems in Ordinary Differential Equations.* Fairlawn, N. J.: Oxford University Press, 1957.
2. "Some numerical experiments with eigenvalue problems in ordinary differential equations." In: *Boundary Problems in Differential Equations*, R. E. Langer, Ed. Madison: Univ. of Wisconsin Press, 1960, pp. 243–255.

FRIEDRICHS, K. O., and H. B. KELLER
"A finite difference scheme for generalized Neumann problems." In: *Numerical Solution of Partial Differential Equations*, J. Bramble, Ed. New York: Academic Press, 1966, pp. 1–19.

GARY, J.
"Computing eigenvalues of ordinary differential equations by finite differences." *M. O. C. 19* (1965), pp. 365–379.

GODUNOV, S.
"On the numerical solution of boundary value problems for systems of linear ordinary differential equations." *Uspehi Mat. Nauk. 16* (1961), pp. 171–174.

GOODMAN, T. R., and G. N. Lance
"The numerical integration of two-point boundary value problems." *M. O. C. 10* (1956), pp. 82–86.

HARTMAN, P.
Ordinary Differential Equations. New York: John Wiley & Sons, 1964.

HENRICI, P.
1. *Discrete Variable Methods in Ordinary Differential Equations.* New York: John Wiley & Sons, 1962.
2. *Error Propagation for Difference Methods.* New York: John Wiley & Sons, 1963.

INCE, E. L.
Ordinary Differential Equations. New York: Dover, 1944.

ISAACSON, E., and H. B. KELLER
Analysis of Numerical Methods. New York: John Wiley & Sons, 1966.

KALNINS, A.
"Analysis of shells of revolution subjected to symmetrical and nonsymmetrical loads." *Trans. A.S.M.E., Ser. E., Jour. Appl. Mech. 31* (1964), pp. 467–476.

KALABA, R.
"On nonlinear differential equations, the maximum operation and monotone convergence." *Jour. Math. Mech. 8* (1959), pp. 519–574.

KANTOROVICH, L. V., and G. P. AKILOV
Functional Analysis in Normed Spaces. New York: Macmillan, 1964.

KANTOROVICH, L. V., and V. I. KRYLOV
Approximate Methods of Higher Analysis. New York: Interscience, 1958.

KELLER, H. B., and E. L. REISS
"Computers in solid mechanics—a case history." *Amer. Math. Monthly 72*, Part II (1965), pp. 92–98.

KELLER, H. B., and A. W. WOLFE
"On the nonunique equilibrium states and buckling mechanism of spherical shells." *SIAM Jour. Appl. Math. 13* (1965), pp. 674–705.

KELLER, H.
1. "On the accuracy of finite difference approximations to the eigenvalues of differential and integral operators." *Numer. Math. 7* (1965), pp. 412–419.
2. "Existence theory for two point boundary value problems." *Bulletin A. M. S. 72* (1966), pp. 728–731.

LANCASTER, P.
"Error analysis for the Newton-Raphson Method." *Numer. Math. 9* (1966), pp. 55–68.

LANCZOS, C.
Applied Analysis. London: Pitman, 1957.

LASOTA, A., and Z. OPIAL
"On the existence of solutions of linear problems for ordinary differential equations." *Bull. Acad. Polon. Sci. 14* (1966), pp. 371–376.

LEES, M.
1. "A boundary value problem for nonlinear ordinary differential equations." *J. Math. and Mech. 10* (1961), pp. 423–430.
2. "Discrete methods for nonlinear two-point boundary value problems." In: *Numerical Solution of Partial Differential Equations*, J. H. Bramble, Ed. New York: Academic Press, 1966, pp. 59–72.

MORRISON, D. D., J. D. RILEY and J. F. ZANCANARO
"Multiple shooting method for two-point boundary value problems." *Comm. ACM 5* (1962), pp. 613–614.

MURRAY, J. D.
"A simple method for obtaining approximate solutions for a large class of diffusion-kinetic enzyme problems." *Math. Biosciences 2* (1968), pp. 379–411.

ORTEGA, J., and W. RHEINBOLDT
"On discretization and differentiation of operators with application to Newton's method." *SIAM Jour. Numer. Anal. 3* (1966), pp. 143–156.

ORTEGA, J., and M. ROCKOFF
"Nonlinear difference equations and Gauss-Seidel type iterative methods." *SIAM Jour. Numer. Anal. 3* (1966), pp. 497–513.

OSTROWSKI, A.
Solution of Equations and Systems of Equations. New York: Academic Press, 1960.

PEREYRA, V.
"On improving an approximate solution of a functional equation by deferred corrections." *Numer. Math. 8* (1966), pp. 376–391.

REISS, E., H. GREENBERG and H. KELLER
"Nonlinear deflections of shallow spherical shells." *J. Aero./Space Sci. 24* (1957), pp. 533–543.

RILEY, J., M. BENNETT and E. MCCORMICK
"Numerical integration of variational equations." *M.O.C. 21* (1967), pp. 12–17.

SYLVESTER, R., and F. MEYER
"Two-point boundary problems by quasilinearization." *J. Appl. Math. 13* (1965), pp. 586–602.

TRAUB, J. F.
Iterative Methods for the Solution of Equations. Engelwood Cliffs, N. J.: Prentice-Hall, 1964.

TROESCH, B. A.
"Intrinsic difficulties in the numerical solution of a boundary value problem." Space Tech. Labs., Tech. Note NN-142 (1960).

URABE, M.
1. "Convergence of numerical integration in solution of equations." *J. Sci. Hiroshima Univ. 19* (1956), pp. 479–489.
2. "Galerkin's procedure for nonlinear periodic systems and its extension to multipoint boundary value problems for general nonlinear systems." In: *Numerical Solutions of Nonlinear Differential Equations,* D. Greenspan, Ed. New York: John Wiley & Sons, 1966, pp. 297–327.

VARGA, R. S.
1. *Matrix Iterative Analysis.* Engelwood Cliffs, N. J.; Prentice-Hall, 1962.
2. "Hermite interpolation-type ritz methods for two point boundary value problems." In: *Numerical Solution of Partial Differential Equations,* J. H. Bramble, Ed. New York: Academic Press, 1965, pp. 365–373.

WEINBERGER, H. F.
Variational Methods for Eigenvalue Problems, Lecture Notes, University of Minnesota, Inst. of Tech., Dept. of Math., 1962.

WEINITSCHKE, H. J.
"On the stability problem for shallow spherical shells." *Jour. Math. Phys. 38* (1960), pp. 209–231.

WENDROFF, B.
Theoretical Numerical Analysis. New York: Academic Press, 1966.

WIELANDT, H.
"Error bounds for eigenvalues of symmetric integral equations." In *Proceedings of Symposia in Applied Mathematics,* Vol. VI, *Numerical Analysis.* Providence, R. I: American Math. Socy., 1956, pp. 261–282.

WILKINSON, J. H.
The Algebraic Eigenvalue Problem. Fairlawn, N. J.: Oxford University Press. 1965.

INDEX

APPENDIX B

NUMERICAL SOLUTION OF TWO POINT BOUNDARY VALUE PROBLEMS

Contents

Preface

This monograph is an account of ten lectures I presented at the Regional Research Conference on Numerical Solution of Two-Point Boundary Value Problems. The Conference was held at Texas Tech. University on July 7–11, 1975, and was one of a series supported by the National Science Foundation and managed by the Conference Board of the Mathematical Sciences.

I have used this opportunity, in part, to augment and bring up to date portions of a previous monograph on this subject [24]. However, as the lectures could only be a partial survey of the field, many interesting topics had to be omitted. Indeed the omissions form the basis for another conference in this area. They include: automatic net selection, singular perturbation problems, bifurcation problems, optimal control problems, the completion of difference schemes to yield invariant embedding schemes, Galerkin methods, collocation (with other than C_0 piecewise polynomials), singular nonlinear problems and no doubt other significant topics. However, the main core of the subject is covered rather broadly and some proofs have been included when they are brief and contribute to a better understanding. Quite a bit of new material is covered and numerous areas in need of study are exposed.

It is a pleasure to thank Professor Paul Nelson and the rest of his committee for organizing this very well-run conference. The participants were a most stimulating group and any success of my lectures was due in no small part to the general spirit of the gathering. The National Science Foundation is performing a most useful and innovative service in supporting the Regional Conference Series and I was honored to have been the principal lecturer at this one. Much of my research in this area has been sponsored over the years by the U.S. Army Research Office (Durham) under Contract DAHC 04-68-C-0006 and by the Energy Research and Development Administration under Contract AT(04-3-767) Project Agreement 12. Finally I wish to thank Pam Hagan for her cheerful help in the arduous task of preparing this manuscript for publication.

Pasadena
August 1975

HERBERT B. KELLER

Introduction

The theory and practice of the numerical solution of two-point boundary value problems are currently very active areas. Indeed, the signs are rather clear that over the next five to ten years (or less) standard computer codes will be available to "solve" most such problems. This is already very much the case for systems of linear algebraic equations and it is very close to being true for initial value problems for ordinary differential equations. The general pattern seems to be that some years after a thorough theoretical understanding of the numerical methods is known, the more or less standard codes are developed. Thus it was not until after the definitive work of Wilkinson, summarized in [58], that the best linear algebra codes were developed. In ordinary differential equations the Dahlquist theory, presented and greatly extended by Henrici [21], has more recently been followed by serious attempts to produce standard initial value problem codes. We believe that two-point boundary value problem solvers are next in line. These lectures will attempt to supply a fairly unified theory of numerical methods along with some practical procedures for approximating the solutions of a very general class of linear and nonlinear two-point boundary value problems. Our hope is that this will supply a significant part of the background that seems necessary for the development of good general purpose codes.

We examine in some detail shooting, Chapter 1, and finite difference methods, Chapter 2, which, when broadly interpreted, include almost all of the currently used methods. Galerkin procedures are not included but collocation with C_0 piecewise polynomials is briefly treated in § 2.5. Although the developing general theory is presented in some detail we are most interested in practical methods of broad applicability.

The notion of partially separated endconditions is introduced in § 1.2 and used to clarify the roles of the method of complementary functions and the method of adjoints. Parallel shooting and the related stabilized-march technique are discussed in § 1.3 and § 1.5. Continuation or embedding for shooting methods is discussed in § 1.6 and most of these considerations hold also for solving nonlinear finite difference equations.

For linear boundary value problems with unique solutions general classes of difference schemes are shown, in § 2.1, to be stable and consistent if and only if, trivially modified, they are stable and consistent for the initial value problem. The well-developed theory for initial value problems is then available and one-step schemes seem particularly attractive. These are briefly considered in § 2.2 where efficient and stable solution procedures are presented for the special block tridiagonal linear systems that arise for separated endconditions. These are easily extended to partially separated endconditions. A theory for nonlinear problems is sketched in § 2.3 where Newton's method is shown to converge quadratically. Higher order accuracy is easily and efficiently obtained in § 2.4 by means of Richardson extrapolation or deferred corrections.

The linear problems considered are of the fairly general first order form:

$$\text{(0.1a)} \qquad \mathscr{L}\mathbf{y}(t) \equiv \mathbf{y}'(t) - A(t)\mathbf{y}(t) = \mathbf{f}(t), \qquad a \leqq t \leqq b,$$

$$\text{(0.1b)} \qquad \mathscr{B}\mathbf{y}(t) \equiv B_a\mathbf{y}(a) + B_b\mathbf{y}(b) = \boldsymbol{\beta}.$$

Here $A(t)$, B_a and B_b are $n \times n$ matrices while $\mathbf{y}(t)$, $\mathbf{f}(t)$ and $\boldsymbol{\beta}$ are n-vectors. Appropriate smoothness or piecewise smoothness of $A(t)$ and $\mathbf{f}(t)$ is specified as required in Chapters 1 and 2. In Chapter 3 we consider eigenvalue problems in which $\mathbf{f}(t) \equiv \boldsymbol{\beta} \equiv \mathbf{0}$ and $A(t)$, B_a and B_b are replaced by $A(t, \lambda)$, $B_a(\lambda)$ and $B_b(\lambda)$ depending analytically on λ. It is shown in § 3.2 and § 3.3, respectively, how all the shooting and finite difference methods go over quite generally for the eigenvalue problems. As a very practical matter, however, we show in § 3.4 how to reformulate eigenvalue problems for more efficient numerical solution.

The nonlinear problems considered are of the general form:

$$\text{(0.2a)} \qquad \mathscr{N}\mathbf{z}(t) \equiv \mathbf{z}'(t) - \mathbf{f}(t, \mathbf{z}(t)) = \mathbf{0}, \qquad a \leqq t \leqq b,$$

$$\text{(0.2b)} \qquad \mathbf{g}(\mathbf{z}(a), \mathbf{z}(b)) = \mathbf{0}.$$

We study here only the computation of isolated solutions of (0.2a,b). The important case of bifurcation along with many other interesting topics (listed in the Preface) could not be included in these lectures.

In Chapter 4 we briefly consider some singular linear problems. In particular with

$$A(t) \equiv \frac{1}{t-a}R + A_0(t)$$

problem (0.1a,b) has a regular singularity at one endpoint and we treat this case in § 4.1. We also allow $\mathbf{f}(t)$, $\boldsymbol{\beta}(t)$ and $B_a \equiv B_a(t)$ to have singularities at $t = a$. Finally, some linear problems over $[a, b] \equiv [a, \infty)$ are discussed in § 4.2. These problems have $\lim_{t\to\infty} A(t) = A_\infty \not\equiv 0$ so that the point at infinity is an irregular singular point. In particular we show, in each of the above cases, how to replace the singular problem by a smooth problem over a reduced interval. For regular singular endpoints we show how to estimate the error in this process. However, for infinite intervals we exhibit an important class of problems (with constant "tails" at ∞) for which *no error* is introduced by the truncation process. This also includes a broad class of eigenvalue problems over $[a, \infty)$.

CHAPTER 1

Shooting methods

1.1. Linear problems. A straightforward approach in attempting to solve (0.1a,b) consists in first solving the $n + 1$ initial value problems:

(1.1a) $$\mathscr{L}\mathbf{y}_0(t) = \mathbf{f}(t), \qquad \mathbf{y}_0(a) = \mathbf{y}_0,$$

(1.1b) $$\mathscr{L}\mathbf{y}_\nu(t) = \mathbf{0}, \qquad \mathbf{y}_\nu(a) = \mathbf{e}_\nu, \qquad 1 \leqq \nu \leqq n.$$

Here $\{\mathbf{e}_\nu\}$ are the unit vectors in $\mathbb{E}^n$; that is, $I = (\mathbf{e}_1, \cdots, \mathbf{e}_n)$. Then we seek a linear combination of the form:

(1.2) $$\begin{aligned} \mathbf{y}(t) &= \mathbf{y}_0(t) + \sum_{\nu=1}^{n} \xi_\nu \mathbf{y}_\nu(t), \\ &= \mathbf{y}_0(t) + Y(t)\boldsymbol{\xi}, \end{aligned}$$

to satisfy (0.1b). This yields the linear system

(1.3) $$Q\boldsymbol{\xi} = \boldsymbol{\beta} - \mathscr{B}\mathbf{y}_0(t) \equiv \boldsymbol{\beta} - B_a\mathbf{y}_0 - B_b\mathbf{y}_0(b).$$

Here we have introduced the fundamental solution matrix

(1.4a) $$Y(t) \equiv (\mathbf{y}_1(t), \cdots, \mathbf{y}_n(t)),$$

and the very important matrix

(1.4b) $$Q \equiv \mathscr{B}Y(t) = B_a + B_bY(b).$$

Indeed the basic existence theory for (0.1a,b) can be stated as follows.

THEOREM 1.5. *Let* $A(t)$, $\mathbf{f}(t) \in C^p[a, b]$. *Then* (0.1a,b) *has a unique solution* $\mathbf{y}(t) \in C^{p+1}[a, b]$ *if and only if* Q *is nonsingular.*

Proof. Since every C^{p+1} solution of (0.1a) has the representation (1.2) it clearly follows that the unique solvability of (1.3) is equivalent to that of (0.1a,b). ▨

The smoothness conditions can easily be altered in obvious ways. We assume that Q is nonsingular so that (0.1a,b) has a unique solution.

The above is the continuous or analytical shooting procedure. The analogous numerical shooting procedure is simply to carry out the "solution" of (1.1a,b) and (1.3) using appropriate numerical techniques. For simplicity of exposition let us assume that a net of points $\{t_j\}$, given by

$$t_0 = a, \quad t_j = t_{j-1} + h_j, \quad 1 \leqq j \leqq J, \quad t_J = b, \quad h = \max_j h_j \leqq \theta \min_j h_j$$

is placed on $[a, b]$. Then suppose some *stable, accurate of order m, initial value method* is used to compute approximate solutions to (1.1a,b) on this net (see

Henrici [21]). With $A(t)$ and $\mathbf{f}(t)$ sufficiently smooth these numerical solutions, denoted by $\{\mathbf{u}_{\nu,j}\}_0^J$, $0 \leqq \nu \leqq n$, satisfy

$$\|\mathbf{y}_\nu(t_j) - \mathbf{u}_{\nu,j}\| \leqq Mh^m, \qquad 0 \leqq j \leqq J. \tag{1.6}$$

Then with $U_j \equiv (\mathbf{u}_{1,j}, \cdots, \mathbf{u}_{n,j})$ and $Q_h \equiv [B_a + B_b U_J]$ we get, as the approximation to (1.3),

$$Q_h \xi_h = \boldsymbol{\beta} - B_a \mathbf{u}_{0,0} - B_b \mathbf{u}_{0,J}.$$

Since Q is nonsingular and $\|Q - Q_h\| = O(h^m)$ by (1.6) it follows that Q_h is nonsingular for h sufficiently small. Thus

$$\xi_h = Q_h^{-1}(\boldsymbol{\beta} - B_a \mathbf{u}_{0,0} - B_b \mathbf{u}_{0,J})$$

and the numerical approximation to the solution of (0.1a,b) is

$$\mathbf{u}_j = \mathbf{u}_{0,j} + U_j \xi_h, \qquad 0 \leqq j \leqq J. \tag{1.7}$$

Just as Q_h is shown to be nonsingular, using the Banach lemma, we also get that

$$\|Q^{-1} - Q_h^{-1}\| = O(h^m).$$

Then (1.6) and (1.7) yield with (1.2) that

$$\|\mathbf{y}(t_j) - \mathbf{u}_j\| = O(h^m).$$

Thus we conclude that $O(h^m)$ accurate approximations can be computed by the basic shooting method.

While this is true in principle it may be quite misleading in actual calculations. In particular if Q_h is not well-conditioned, then numerical errors may be magnified in solving for ξ_h. Also in forming the final linear combinations (1.7) cancellation errors frequently occur. The latter problem is easily overcome by solving an extra initial value problem[1] with initial data $\mathbf{y}_{n+1}(a) = \mathbf{y}_0 + \xi_h$. The former problem however is not so easy to cure. That is because its cause is usually due to some exponentially growing solutions in (1.1b). If they exist then when computing the other solutions of (1.1b) roundoff invariably introduces small components of the growing solutions which soon dominate. As a result the computed U_j for t_j near b are nearly singular. We return to some possible cures for this problem in § 1.3. The possible difficulties with shooting are discussed frequently in the literature, for example, Keller[24], Osborne [39], Conte [10], etc.

1.2. Separated and partially separated endconditions. We examine ways in which to reduce the number of initial value problems (1.1a,b) that must be solved. This can always be done if the boundary conditions are what we term *partially separated.* The "method of adjoints" and "method of complementary functions" are related to these considerations.

[1] This procedure also reduces the storage requirements dramatically since only U_J, $\mathbf{u}_{0,0}$ and $\mathbf{u}_{0,J}$ need be retained to compute ξ_h.

In general the boundary conditions (0.1b) must represent n independent linear constraints on $\mathbf{y}(a)$ and $\mathbf{y}(b)$. Thus it is required that

$$\text{rank}\,(B_a, B_b) = n, \tag{1.8}$$

and this is a necessary condition in order that Q of (1.4b) be nonsingular. However it frequently happens that rank $B_a < n$ or rank $B_b < n$ or both. If either holds we call the boundary conditions *partially separated.* Indeed the case in which rank B_a = rank $B_b = n$ must be considered rather rare, its most obvious occurrence being periodic boundary conditions.

Suppose that rank $B_b = q < n$. Then for some nonsingular $n \times n$ matrix R_b, representing appropriate linear combinations of the rows of B_b, we must have

$$R_b B_b = \begin{pmatrix} 0 \\ C_b \end{pmatrix} \begin{matrix} \}n-q \\ \}q \end{matrix}, \quad \text{rank } C_b = q.$$

We also introduce the partitions:

$$R_b \boldsymbol{\beta} = \begin{pmatrix} \boldsymbol{\beta}_a \\ \boldsymbol{\beta}_b \end{pmatrix} \begin{matrix} \}n-q \\ \}q \end{matrix}, \quad R_b B_a = \begin{pmatrix} C_a \\ C_{ba} \end{pmatrix} \begin{matrix} \}n-q \\ \}q \end{matrix}, \quad \text{where rank } C_a = n - q.$$

The latter rank condition follows from (1.8). Thus forming $R_b \mathscr{B}\mathbf{y}(t) = R_b \boldsymbol{\beta}$ we find that the boundary conditions (0.1b) are equivalent to

$$\begin{aligned} C_a \mathbf{y}(a) &= \boldsymbol{\beta}_a, \quad \boldsymbol{\beta}_a \in \mathbb{E}^{n-q}, \\ C_{ba}\mathbf{y}(a) + C_b \mathbf{y}(b) &= \boldsymbol{\beta}_b, \quad \boldsymbol{\beta}_b \in \mathbb{E}^q. \end{aligned} \tag{1.9a}$$

Obviously if rank $B_a = p < n$ we obtain by the analogous procedure, but with different matrices and vectors:

$$\begin{aligned} C_a \mathbf{y}(a) + C_{ab}\mathbf{y}(b) &= \boldsymbol{\beta}_a, \quad \boldsymbol{\beta}_a \in \mathbb{E}^p, \\ C_b \mathbf{y}(b) &= \boldsymbol{\beta}_b, \quad \boldsymbol{\beta}_b \in \mathbb{E}^{n-p}. \end{aligned} \tag{1.9b}$$

Either of the forms (1.9a,b) are *partially separated* boundary conditions. If $C_{ab} \equiv 0$ and $C_{ba} \equiv 0$, then they are *separated boundary* conditions which are perhaps the most commonly occurring forms in applications.

To solve (0.1a) subject to (1.9a), say, we augment C_a by adjoining q-linearly independent rows, D_a. Then

$$M_a \equiv \begin{pmatrix} C_a \\ D_a \end{pmatrix}$$

is nonsingular and we partition its inverse as

$$M_a^{-1} \equiv (E_a, F_a),$$

where F_a is $n \times q$. We solve the $q + 1$ initial value problems

$$\mathscr{L}\mathbf{y}_0(t) \equiv \mathbf{f}(t), \qquad M_a \mathbf{y}_0(a) = \begin{pmatrix} \boldsymbol{\beta}_a \\ \mathbf{0} \end{pmatrix}, \tag{1.10a}$$

$$\mathscr{L}V(t) = 0, \qquad V(a) = F_a. \tag{1.10b}$$

Here $V(t)$ is $n \times q$ and the solution of the boundary value problem is sought in the form:

$$\text{(1.10c)} \qquad \mathbf{y}(t) = \mathbf{y}_0(t) + V(t)\boldsymbol{\eta}.$$

Since $C_aF_a = 0$ this will be a solution provided $\boldsymbol{\eta} \in \mathbb{E}^q$ satisfies

$$\text{(1.11a)} \qquad [C_{ba}F_a + C_bV(b)]\boldsymbol{\eta} = \boldsymbol{\beta}_b - C_{ba}\mathbf{y}_0(a) - C_b\mathbf{y}_0(b).$$

Recall from (1.4a) that $V(t) = Y(t)F_a$ and so the above $q \times q$ coefficient matrix has the form

$$\text{(1.11b)} \qquad Q_b \equiv [C_{ba} - C_bY(b)]F_a.$$

This must be nonsingular if our problem has a unique solution. But the form (1.4b) with $B_a \equiv \begin{pmatrix} C_a \\ C_{ba} \end{pmatrix}$, $B_b \equiv \begin{pmatrix} 0 \\ C_b \end{pmatrix}$ to represent (1.9a) yields

$$QM_a^{-1} = \begin{pmatrix} I & 0 \\ [\cdot]E_a & Q_b \end{pmatrix}.$$

Thus Q_b is singular if and only if Q is singular and the so-called method of *complementary functions* is justified. This technique has been described by various authors in many modifications with a host of new names which we avoid listing here.

We have shown above how the problem with the partially separated end-condition (1.9a) can be solved in terms of only $q + 1$ initial value problems. By using the method of adjoints this problem can also be solved using $q + 2$ initial value problems. From (1.11b) we note that $C_bY(b)$, a $q \times n$ matrix, would suffice to determine Q_b and hence $\boldsymbol{\eta}$. It is well known that $Y(b) \equiv Y(b; a) = Z^T(a; b)$, where $Z(t; b)$ is the fundamental solution of the adjoint equation with initial point $t = b$. Thus we have

$$\begin{aligned} C_bY(b) &= (Y^T(b; a)C_b^T)^T \\ &= (Z(a; b)C_b^T)^T \equiv (\tilde{Z}(a; b))^T. \end{aligned}$$

So we need only compute the q-columns of the $n \times q$ matrix $\tilde{Z}(t; b)$ defined by

$$\text{(1.12)} \qquad \tilde{Z}' = -A^T(t)\tilde{Z}, \qquad \tilde{Z}(b; b) = C_b^T.$$

Using these columns we form Q_b, solve for $\boldsymbol{\eta}$, and note that $\mathbf{v}(t) \equiv V(t)\boldsymbol{\eta} = Y(t)F_a\boldsymbol{\eta}$ is the solution of

$$\text{(1.13)} \qquad \mathscr{L}\mathbf{v} = \mathbf{0}, \qquad \mathbf{v}(a) = F_a\boldsymbol{\eta}.$$

Then we form (1.10a,b,c), now as $\mathbf{y}(t) = \mathbf{y}_0(t) + \mathbf{v}(t)$, to obtain the solution. In this procedure we solve the $q + 2$ initial value problems (1.10a), (1.12) and (1.13). Note that q of them are integrated from b to a and two of them from a to b. Since equations must be integrated in both directions and one extra equation must be solved, we can see no reason to recommend the method of adjoints. We have in fact discussed it here merely to point out its inefficiency compared to the method of complementary functions.

It should be recalled that if both reductions (1.9a) and (1.9b) are valid and $p < q$, then the above procedures should be applied reversing the roles of a and b. Then only $p + 1$ or $p + 2$ initial value problems need be solved.

A slightly different view of the method of adjoints was first introduced by Goodman and Lance [18] and this is also described in Roberts and Shipman [45].

1.3. Parallel shooting for linear problems; stabilized march. This powerful method is not thoroughly understood as is evidenced by the fact that there is no generally accepted rule for picking the nodes. However, some theoretical results and strong heuristic motivations are available so we sketch them. Recall first the standard shooting procedure of (1.1a,b)–(1.4a,b). As previously mentioned growing solutions of (1.1b) can be expected to cause difficulties. So to reduce this effect we simply reduce the length of the intervals over which we solve initial value problems. Somehow nodes are chosen, say,

$$\tau_0 = a < \tau_1 < \cdots < \tau_J = b$$

and over the jth interval, $[\tau_{j-1}, \tau_j]$, we represent the solution of (0.1a,b), in analogy with (1.2) or (1.10a,b,c), in the form

$$\mathbf{y}(t) = \mathbf{y}_j(t) \equiv \mathbf{v}_j(t) + V_j(t)\boldsymbol{\xi}_j, \qquad \tau_{j-1} \leqq t \leqq \tau_j,\ 1 \leqq j \leqq J, \tag{1.14a}$$

where

$$\mathscr{L}\mathbf{v}_j(t) = \mathbf{f}(t), \quad \mathbf{v}_j(\tau_{j-1}) = \mathbf{v}_j^0, \qquad \tau_{j-1} \leqq t \leqq \tau_j,\ 1 \leqq j \leqq J, \tag{1.14b}$$

$$\mathscr{L}V_j(t) = 0, \quad V_j(\tau_{j-1}) = F_j, \qquad \tau_{j-1} \leqq t \leqq \tau_j,\ 1 \leqq j \leqq J. \tag{1.14c}$$

Here $V_j(t)$ and F_j are $n \times n$ matrices (or perhaps $n \times q$ matrices); the latter, along with $\mathbf{v}_j^0$, are not yet specified. Continuity of the solution at the interior nodes requires that $\mathbf{y}_j(\tau_{j-1}) = \mathbf{y}_{j-1}(\tau_{j-1})$, and hence

$$\mathbf{v}_j^0 + F_j\boldsymbol{\xi}_j = \mathbf{v}_{j-1}(\tau_{j-1}) + V_{j-1}(\tau_{j-1})\boldsymbol{\xi}_{j-1}, \qquad 2 \leqq j \leqq J. \tag{1.15a}$$

The boundary conditions (0.1b) require

$$B_a[\mathbf{v}_1^0 + F_1\boldsymbol{\xi}_1] + B_b[\mathbf{v}_J(b) + V_J(b)\boldsymbol{\xi}_J] = \boldsymbol{\beta}. \tag{1.15b}$$

In matrix form, if we write (1.15b) first, this system becomes

$$\begin{pmatrix} B_aF_1 & & & & B_bV_J(b) \\ -V_1(\tau_1) & F_2 & & & \\ & -V_2(\tau_2) & F_3 & & \\ & & \ddots & \ddots & \\ & & & -V_{J-1}(\tau_{J-1}) & F_J \end{pmatrix} \begin{pmatrix} \boldsymbol{\xi}_1 \\ \boldsymbol{\xi}_2 \\ \vdots \\ \boldsymbol{\xi}_J \end{pmatrix} = \begin{pmatrix} \boldsymbol{\beta} - [B_a\mathbf{v}_1^0 + B_b\mathbf{v}_J(b)] \\ \mathbf{v}_1(\tau_1) - \mathbf{v}_2^0 \\ \vdots \\ \mathbf{v}_{J-1}(\tau_{J-1}) - \mathbf{v}_J^0 \end{pmatrix}. \tag{1.15c}$$

The solution of this system, $\{\xi_j\}_1^J$, used in (1.14a,b,c) yields the solution of (0.1a,b).

If all of the F_j are nonsingular we can easily relate parallel shooting to ordinary shooting as follows. Call the coefficient matrix in (1.15c) $\mathbb{A}_J$ and factor it in the form:

$$\text{(1.16a)} \qquad \mathbb{A}_J = \begin{pmatrix} Q_1 & Q_2 & \cdots & Q_J \\ & I & & \\ & & \ddots & \\ & & & I \end{pmatrix} \begin{pmatrix} F_1 & & & \\ -V_1(\tau_1) & F_2 & & \\ & \ddots & \ddots & \\ & & -V_{J-1}(\tau_{J-1}) & F_J \end{pmatrix}.$$

We need only take

$$\begin{aligned} &Q_J = B_b V_J(b) F_J^{-1}, \\ \text{(1.16b)} \quad &Q_j = Q_{j+1} V_j(\tau_j) F_j^{-1}, \qquad\qquad j = J-1, J-2, \cdots, 2, \\ &Q_1 = B_a + Q_2 V_1(\tau_1) F_1^{-1} = B_a + B_b[V_J(\tau_J)F_J^{-1}] \cdots [V_1(\tau_1)F_1^{-1}] \end{aligned}$$

to justify (1.16a). Clearly the right-hand factor is nonsingular and the left-hand factor is nonsingular provided Q_1 is. However it easily follows from (1.14c) and (1.1b) that

$$[V_J(\tau_J)F_J^{-1}] \cdots [V_1(\tau_1)F_1^{-1}] = Y(b),$$

and hence $Q_1 = Q$ in (1.4b).

The above indicated factorization also yields a simple way to solve the system (1.15c) since it is a block UL-decomposition. However this "obvious" solution procedure should probably not be used in general as it comes close to being the original shooting scheme which we claim has difficulties. Rather some block or band elimination using some form of (partial) pivoting should be used. Indeed one of the hopes for parallel shooting is that with a proper choice of the nodes, τ_j, and appropriate pivoting the condition of the system actually solved, in place of $\mathbb{A}_J$, will be better than the condition of Q! Some work in this direction is reported by George and Gunderson [16], and Osborne [40].

An important case in which pivoting has been used with great success is that of separated or even partially separated endconditions. Thus suppose (0.1b) represents the boundary conditions (1.9a). Then simply writing the last q-conditions of (1.15b) last and the first $(n-q)$-conditions first we get for the new coefficient matrix the form

$$\text{(1.17)} \qquad \mathbb{A}_J \equiv \begin{pmatrix} C_a F_1 & & & \\ -V_1(\tau_1) & F_2 & & \\ & \ddots & \ddots & \\ & & -V_{J-1}(\tau_{J-1}) & F_J \\ C_{ba} F_1 & & & C_b V_J(b) \end{pmatrix}.$$

We study matrices of this form in § 2.2 when devising efficient stable schemes to solve one-step difference schemes. The techniques derived there can be used to good advantage in parallel shooting applications.

In the numerical implementation of parallel shooting we obtain, in place of the $\mathbf{v}_j(t)$ and $V_j(t)$ of (1.14b,c), some numerical approximations, say, $\mathbf{v}_{jk}$ and V_{jk} on some net $t = t_k$. If a stable one-step or multistep scheme is used, then as shown in Henrici [22], the errors have asymptotic expansions of the form:

$$V_j(t_k) - V_{jk} = h^\alpha Y(t_k, \tau_{j-1})M_j(t_k) + O(h^\alpha).$$

Here h^α is the leading order of the error in the scheme (including the generation of initial data), $Y(t, \tau)$ is the fundamental solution matrix for $\mathscr{L}$ with $Y(\tau, \tau) = I$, and $M_j(t)$ is either some fixed matrix determined by the initial data generator or else has the form $M_j(t) = \int_0^t Y(t, \tau)T_j(\tau)\,d\tau$, where $T_j(\tau)$ is the principal error function (i.e., leading term in the truncation error). Thus for ordinary shooting in place of the matrix $Q = B_a + B_b Y(b, a)$ of (1.4b) we obtain essentially, as a result of computations,

$$Q_{\text{sh}} = B_a + B_b\{Y(b, a)[I + h^\alpha M(b)]\}.$$

However in parallel shooting, from (1.16a,b), we are led to consider

$$Q_{\text{psh}} = B_a + B_b\{Y(\tau_J, \tau_{J-1})[I + h^\alpha M_J(\tau_\tau)] \cdots Y(\tau_1, a)[I + h^\alpha M_1(\tau_1)]\}.$$

Clearly as $h \to 0$ both schemes converge to the correct result. But for finite h there may be significant benefits even using the ill-advised factorization (1.16a) to solve (1.15c). This is suggested by examining the leading error terms:

$$Q_{\text{sh}} - Q = h^\alpha B_b Y(b, a)M(b),$$

$$Q_{\text{psh}} - Q = h^\alpha B_b \sum_{j=1}^{J} Y(b, \tau_{j-1})M_j(\tau_j)Y(\tau_{j-1}, a) + O(h^{2\alpha}).$$

Recall that if $\|A(t)\| \leqq K$ uniformly on $a \leqq t \leqq b$, then

$$\|Y(b, a)\| \leqq e^{K[b-a]}, \quad \|Y(\tau_j, \tau_{j-1})\| \leqq e^{K[\tau_j - \tau_{j-1}]}.$$

Estimates of $\|M(b)\|$ and $\|M_j(\tau_j)\|$ yield similar reductions in the exponential growth rate of the errors so that Q_{psh} may be a better approximation to Q than is Q_{sh}. However, the improvement in the condition of $\mathbb{A}_J$ properly factored compared to that of Q would seem to be the more significant if less well-understood feature of parallel shooting. Exponential error reductions of the above form have been noted and discussed by Keller [24] and Falkenberg [13]. They enter in a more dramatic fashion in the nonlinear case of § 1.5.

Of course the choices for the F_j and $\mathbf{v}_j^0$ are rather important and yield many variants of the method. "Standard" parallel shooting is, say, the choice: $F_j = I$ and $\mathbf{v}_j^0 = \mathbf{0}$. The Godunov–Conte or *stabilized march* technique can also be included. We consider the partially separated endpoint case since then a reduction in the order of the system (1.15c) (or matrix (1.17)) is obtained. Using the form of representation in (1.14a) with $V_j(t)$ an $n \times q$ matrix and ξ_j a q-vector we start as in (1.10a,b) with

$$F_1 = F_a, \qquad \mathbf{v}_1^0 = \mathbf{v}_0(a), \tag{1.18a}$$

so that the boundary conditions at $t = a$ are satisfied. Then having integrated up to the node τ_j we orthogonalize the q vectors of $V_j(\tau_j)$ and take them as F_{j+1}, that is,

$$F_{j+1} = V_j(\tau_j)P_j, \qquad 1 \leqq j \leqq J, \tag{1.18b}$$

for some $q \times q$ upper triangular matrix P_j. We pick $\mathbf{v}_{j+1}^0$ as the projection of $\mathbf{v}_j(\tau_j)$ orthogonal to the column vectors in F_{j+1}; thus

$$\mathbf{v}_{j+1}^0 = (I - F_{j+1}F_{j+1}^T)\mathbf{v}_j(\tau_j), \qquad 1 \leqq j \leqq J. \tag{1.18c}$$

The continuity conditions (1.15a) now give, *since* $V_j(\tau_j)$ *must have rank* q,

$$\xi_j = P_j[\xi_{j+1} - F_{j+1}^T\mathbf{v}_j(\tau_j)], \qquad j = J, J-1, \cdots, 1. \tag{1.19a}$$

The final coefficient ξ_{J+1}, which is introduced merely for consistent notation (and ease in writing machine codes), is obtained from the end condition at $t = \tau_J = b$ which from (1.9a) becomes

$$C_{ba}F_1\xi_1 + C_bF_{J+1}\xi_{J+1} = \boldsymbol{\beta}_b - C_{ba}\mathbf{v}_1^0 - C_b\mathbf{v}_{J+1}^0. \tag{1.19b}$$

The boundary conditions and continuity conditions (1.19a,b) can be written as the system

$$\begin{pmatrix} -I & P_1 & & & \\ & \ddots & \ddots & & \\ & & -I & P_J & \\ C_{ba}F_1 & & & C_bF_{J+1} \end{pmatrix} \begin{pmatrix} \xi_1 \\ \vdots \\ \\ \xi_{J+1} \end{pmatrix} = \begin{pmatrix} P_1F_2^T\mathbf{v}_1(\tau_1) \\ \vdots \\ P_JF_{J+1}^T\mathbf{v}_J(\tau_J) \\ \boldsymbol{\beta}_b - [C_{ba}\mathbf{v}_1^0 + C_b\mathbf{v}_{J+1}^0] \end{pmatrix}. \tag{1.19c}$$

The similarity of this system with that for parallel shooting is clear. However the blocks in the coefficient matrix of (1.19c) are $q \times q$ while those in (1.17) are $n \times n$. The above is essentially the procedure described by Conte [10] but generalized for the partially separated endpoint case. Again we stress that care should be taken solving (1.19c) and the procedures of § 2.2 can be used. Conte suggests two techniques for selecting the nodes $\{\tau_j\}$. The first is simply to monitor the growth of the columns of $V_j(t_i)$ and the second is essentially to monitor the Gramm matrix of the columns of $V_j(t_i)$. Notice that the entire formulation (1.19a,b,c) is retained for *any choice* of the $q \times q$ nonsingular matrices P_j in (1.18b). Indeed it seems likely that the singular value decomposition of $V_j(t_i)$, if they can be computed efficiently, might be preferable to the Gramm procedure usually advocated as it yields a better measure of rank of $V_j(t_i)$. Of course the full singular value decomposition is not required to compute the P_j but merely the RQ-transformation of $V_j(\tau_j)$. Then the P_j would be orthogonal matrices rather than upper triangular and so for fully separated endconditions the back substitution is easily accomplished. These speculations have not yet been completely justified or implemented.

1.4. Nonlinear problems; Newton's method. Shooting appears to be a straightforward way to solve nonlinear boundary value problems. In particular with

$$\text{(1.20a)} \qquad \mathbf{z}'(t) = \mathbf{f}(t, \mathbf{z}), \qquad a \leqq t \leqq b,$$

$$\text{(1.20b)} \qquad \mathbf{g}(\mathbf{z}(a), \mathbf{z}(b)) = \mathbf{0},$$

we associate the related initial value problem

$$\text{(1.21a)} \qquad \mathbf{u}'(t) = \mathbf{f}(t, \mathbf{u}),$$

$$\text{(1.21b)} \qquad \mathbf{u}(a) = \mathbf{s}.$$

Then, denoting the solution of (1.21a,b) by $\mathbf{u} \equiv \mathbf{u}(t; \mathbf{s})$, we seek $\mathbf{s}$ such that

$$\text{(1.22)} \qquad \boldsymbol{\phi}(\mathbf{s}) \equiv \mathbf{g}(\mathbf{s}, \mathbf{u}(b; \mathbf{s})) = \mathbf{0}.$$

If $\mathbf{s} = \mathbf{s}^*$ is a root of (1.22), then $\mathbf{z}(t) \equiv \mathbf{u}(t, \mathbf{s}^*)$ is a solution of (1.20a,b). Conversely for any solution $\mathbf{z}(t)$ of (1.20a,b) the value $\mathbf{s} = \mathbf{z}(a)$ is a root of (1.22). Thus our problem is "reduced" to that of computing the zeros of $\boldsymbol{\phi}(\mathbf{s})$. It is not difficult to devise existence theorems for (1.20a,b) in this way (see Keller [24]). However we can easily show that when (1.20a,b) has an *isolated solution* and some mild smoothness also holds, then (1.22) has a *simple* (*isolated*) *root*. Then, in principle, the shooting method and its numerical implementation can be used to compute this solution. Many practical difficulties arise in actually carrying out this procedure and we shall discuss some of them.

Recall that $\mathbf{z}(t)$ is an isolated solution of (1.20a,b) if the linearized problem

$$\text{(1.23a)} \qquad \mathscr{L}[\mathbf{z}]\mathbf{y}(t) \equiv \mathbf{y}'(t) - A(t, \mathbf{z})\mathbf{y}(t) = \mathbf{0}, \qquad a \leqq t \leqq b,$$

$$\text{(1.23b)} \qquad \mathscr{B}[\mathbf{z}]\mathbf{y}(t) \equiv B_a[\mathbf{z}]\mathbf{y}(a) + B_b[\mathbf{z}]\mathbf{y}(b) = \mathbf{0},$$

with

$$\text{(1.23c)} \qquad A(t, \mathbf{z}) \equiv \frac{\partial \mathbf{f}}{\partial \mathbf{z}}(t, \mathbf{z}(t)), \quad B_x[\mathbf{z}] \equiv \frac{\partial \mathbf{g}(\mathbf{z}(a), \mathbf{z}(b))}{\partial \mathbf{z}(x)}, \qquad x = a, b,$$

has only the trivial solution $\mathbf{y}(t) \equiv \mathbf{0}$. Smoothness on $\mathbf{f}(t, \mathbf{v})$ will be required in some tube about $\mathbf{z}(t)$, say in

$$\text{(1.24)} \qquad T_\delta\{\mathbf{z}\} \equiv \{(t, \mathbf{w}) : a \leqq t \leqq b, \|\mathbf{z}(t) - \mathbf{w}\| \leqq \delta\}.$$

Indeed in order that $\boldsymbol{\phi}(\mathbf{s})$ be well-defined by (1.22) it must be true that, for the $\mathbf{s}$ in question, the solution of (1.21a,b) exist on the entire interval $[a, b]$. This and some smoothness of $\mathbf{u}(t; \mathbf{s})$ are consequences of standard results (see Coddington and Levinson [9, Chap. 1, § 7]) which we state as the following lemma.

LEMMA 1.25. *Let* $\mathbf{z}(t)$ *be a solution of* (1.20a,b) *and for some* $K > 0$ *and* $\delta > 0$ *assume*:

$$\text{(1.25a)} \qquad \|\mathbf{f}(t, \mathbf{v}) - \mathbf{f}(t, \mathbf{w})\| \leqq K\|\mathbf{v} - \mathbf{w}\| \quad \textit{for all} \quad (t, \mathbf{v}), (t, \mathbf{w}) \in T_\delta\{\mathbf{z}\}.$$

Then for each $\mathbf{s} \in S_\rho(\mathbf{z}(a))$, *where*

$$\text{(1.25b)} \qquad \rho = \delta\, e^{-K(b-a)}, \quad S_\rho(z(a)) \equiv \{\mathbf{s} : \|\mathbf{z}(a) - \mathbf{s}\| \leqq \rho\},$$

there exists a unique solution $\mathbf{u}(t, \mathbf{s})$ *of* (1.21a,b) *on* $a \leqq t \leqq b$. *If in addition* $\mathbf{f}(t, \mathbf{v}) \in C_1(T_\rho\{\mathbf{z}\})$, *then* $\partial \mathbf{u}(t, \mathbf{s})/\partial \mathbf{s} \equiv U(t; \mathbf{s})$ *exists and is the fundamental solution which satisfies*

$$\mathscr{L}[\mathbf{u}]U(t, \mathbf{s}) = 0, \qquad U(a; \mathbf{s}) = I. \tag{1.25c}$$

To justify numerical methods for solving (1.20a,b) by shooting we can easily show that iterative schemes for finding the roots of (1.22) are available (see Keller [24]). In particular, contraction mappings and Newton's method are frequently used and we consider the latter here. The Newton iterates $\{\mathbf{s}_\nu\}$ are defined for any $\mathbf{s}_0$ by

$$Q(\mathbf{s}_\nu)[\mathbf{s}_{\nu+1} - \mathbf{s}_\nu] = -\boldsymbol{\phi}(\mathbf{s}_\nu), \qquad \nu = 0, 1, \cdots, \tag{1.26}$$

where

$$Q(\mathbf{s}_\nu) \equiv \frac{\partial \boldsymbol{\phi}(\mathbf{s}_\nu)}{\partial \mathbf{s}_\nu} = B_a[\mathbf{u}_\nu] + B_b[\mathbf{u}_\nu]U(b; \mathbf{s}_\nu), \qquad \nu = 0, 1, \cdots. \tag{1.27}$$

Here $\mathbf{u}_\nu \equiv \mathbf{u}(t, \mathbf{s}_\nu)$, $U(t; \mathbf{s}_\nu)$ is the fundamental solution defined in (1.25c) and $B_x[\mathbf{u}]$ are defined in (1.23c). Note that when (1.20a,b) has an isolated solution $\mathbf{z}(t)$, then $Q(\mathbf{s}^*)$, with $\mathbf{s}^* = \mathbf{z}(0)$, is nonsingular. The convergence properties of this scheme have been thoroughly studied (see, for example, Ortega and Rheinboldt [38], Keller [28], Rall [43]). For example, the well-known Kantorovich result can be stated as follows.

THEOREM 1.28. *Let* $\boldsymbol{\varphi}(\mathbf{s})$ *satisfy for some* $\mathbf{s}_0$ *and* ρ_0:

$$\|Q^{-1}(\mathbf{s}_0)\boldsymbol{\varphi}(\mathbf{s}_0)\| \leqq \alpha, \tag{1.28a}$$

$$\|Q^{-1}(\mathbf{s}_0)\| \leqq \beta, \tag{1.28b}$$

$$\|Q(\mathbf{s}) - Q(\mathbf{s}')\| \leqq \gamma\|\mathbf{s} - \mathbf{s}'\| \quad \textit{for all} \quad \mathbf{s}, \mathbf{s}' \in S_{\rho_0^+}(\mathbf{s}_0), \tag{1.28c}$$

$$\alpha\beta\gamma \leqq 1/2, \qquad \rho_0\beta\gamma \leqq [1 - \sqrt{1 - 2\alpha\beta\gamma}]. \tag{1.28d}$$

Then the iterates (1.26) *remain in* $S_{\rho_0}(\mathbf{s}_0)$ *and converge quadratically to the unique root* $\mathbf{s}^*$ *of* $\boldsymbol{\varphi}(\mathbf{s}) = \mathbf{0}$ *in* $S_{\rho_0^+}(\mathbf{s}_0)$, $\rho_0^+ = [1 + \sqrt{1 - 2\alpha\beta\gamma}]/\beta\gamma$.

However, this is not always the type of result we require. Rather we would like to know, given that a solution exists, what is the largest sphere about that solution such that every point in it can be used as a converging starting iterate, $\mathbf{s}_0$ (i.e., what is the domain of attraction about a root?) A simple result in this direction is easily proven.

THEOREM 1.29. *Let* $\boldsymbol{\varphi}(\mathbf{s}) = \mathbf{0}$ *have an isolated root* $\mathbf{s} = \mathbf{s}^*$ *and for some* $\rho_* > 0$:

$$\|Q^{-1}(\mathbf{s}^*)\| \leqq \beta, \tag{1.29a}$$

$$\|Q(\mathbf{s}) - Q(\mathbf{s}')\| \leqq \gamma\|\mathbf{s} - \mathbf{s}'\| \quad \textit{for all} \quad \mathbf{s}, \mathbf{s}' \in S_{\rho_*}(\mathbf{s}^*), \tag{1.29b}$$

$$\rho_*\beta\gamma < 2/3. \tag{1.29c}$$

Then for every $\mathbf{s}_0 \in S\rho_*(\mathbf{s}^*)$ *the iterates in* (1.26) *remain in* $S\rho_*(\mathbf{s}^*)$ *and converge quadratically to* $\mathbf{s}^*$; *in fact, for all* ν,

(1.29d) $$\|\mathbf{s}_{\nu+1} - \mathbf{s}^*\| \leqq a\|\mathbf{s}_\nu - \mathbf{s}^*\|^2, \quad a \equiv \frac{\beta\gamma}{2(1 - \rho_*\beta\gamma)} < 1/\rho_*.$$

Proof. For any $\mathbf{s}$ we have the identity:

$$Q(\mathbf{s}) = Q(\mathbf{s}^*)\{I + Q^{-1}(\mathbf{s}^*)[Q(\mathbf{s}) - Q(\mathbf{s}^*)]\}.$$

But for $\mathbf{s} \in S_{\rho_*}(\mathbf{s}^*)$, by (1.29a,b,c),

$$\|Q^{-1}(\mathbf{s}^*)[Q(\mathbf{s}) - Q(\mathbf{s}^*)]\| \leqq \beta\gamma\rho_* < 2/3.$$

Thus $Q(\mathbf{s})$ is nonsingular and in fact

$$\|Q^{-1}(\mathbf{s})\| \leqq \frac{\beta}{1 - \rho_*\beta\gamma}.$$

For an induction assume $\mathbf{s}_\nu \in S_{\rho_*}(\mathbf{s}^*)$. Then since $\boldsymbol{\varphi}(\mathbf{s}^*) = \mathbf{0}$ we get from (1.26)

$$\begin{aligned}\mathbf{s}_{\nu+1} - \mathbf{s}^* &= [\mathbf{s}_\nu - \mathbf{s}^*] + Q^{-1}(\mathbf{s}_\nu)[\boldsymbol{\phi}(\mathbf{s}^*) - \boldsymbol{\phi}(\mathbf{s}_\nu)] \\ &= Q^{-1}(\mathbf{s}_\nu)[Q(\mathbf{s}_\nu) - Q(\mathbf{s}^*, \mathbf{s}_\nu)][\mathbf{s}_\nu - \mathbf{s}^*],\end{aligned}$$

where we have introduced

$$Q(\mathbf{s}, \mathbf{s}') \equiv \int_0^1 Q(\theta\mathbf{s} + [1 - \theta]\mathbf{s}')\, d\theta.$$

By (1.29b) it follows that $\|Q(\mathbf{s}') - Q(\mathbf{s}, \mathbf{s}')\| \leqq \gamma\|\mathbf{s} - \mathbf{s}'\|/2$ and so we get (1.29d). However, for $\nu = 0$ we have that $\|\mathbf{s}_0 - \mathbf{s}^*\| \leqq \rho_*$ and $a\rho_* < 1$ so that our induction is concluded. ▧

We point out that if, in (1.28d), $\alpha\beta\gamma < 1/2.25$, then $\rho_0\beta\gamma < 2/3$. So our result in (1.29c) is not so different from the Kantorovich result in (1.28d). It is easy to see by Theorem 1.29 that the existence of a smooth isolated solution of (1.20a,b) implies that Newton's method can be used as in (1.26) and (1.27). The location of a suitable $\mathbf{s}_0$ is one of the big open questions in such applications and we return to it in § 1.6. We also must recall that the $\mathbf{s}_\nu$ cannot be evaluated exactly since, for each iteration, we must solve (1.21a,b) once and (1.23a,b) n times. Assuming these initial value problems are computed with a stable accurate of order m scheme on a net with maximum spacing h, it follows, for h sufficiently small, that we actually use in place of Newton's method an altered scheme of the form:

(1.30a) $$Q(\mathbf{s}_\nu^h)[\mathbf{s}_{\nu+1}^h - \mathbf{s}_\nu^h] = -\boldsymbol{\phi}(\mathbf{s}_\nu^h) + \boldsymbol{\Delta}_\nu(h), \qquad \nu = 0, 1, \cdots.$$

We may call this "noisy Newton" where the noise, $\boldsymbol{\Delta}_\nu(h)$, is supplied by the numerical errors in solving the initial value problems. Thus we have, by our assumptions

(1.30b) $$\|\boldsymbol{\Delta}_\nu(h)\| \leqq Mh^m.$$

We could also include roundoff errors here, and they are frequently very noticeable in actual computations. To examine this scheme we have the following result.

THEOREM 1.31. *Let the hypothesis of Theorem 1.29 hold with ρ_* replaced by $\rho = \rho_* + \delta\rho_*$ and with (1.29c) strengthened to*

(1.31a) $$\rho\beta\gamma < 1/2.$$

For some $\theta < 1$ let

(1.31b) $$\left(\frac{\beta\gamma}{1 - 2\rho\beta\gamma}\right)^2 \frac{2Mh^m}{\gamma} \leqq \theta$$

and

(1.31c) $$\delta \equiv \frac{1}{1 + \sqrt{1 - \theta}}\left(\frac{2\beta Mh^m}{1 - 2\rho\beta\gamma}\right) \leqq \delta\rho_*.$$

Then if $\mathbf{s}_0 \in S\rho_*(\mathbf{s}^*)$ is such that $\|\mathbf{s}_1 - \mathbf{s}_0\| \leqq \rho_*$, the iterates $\{\mathbf{s}_\nu\}$ of (1.26) and $\{\mathbf{s}_\nu^h\}$ of (1.30a,b) with $\mathbf{s}_0 = \mathbf{s}_0^h$ satisfy

(1.32a) $$\|\mathbf{s}_\nu^h - \mathbf{s}_\nu\| \leqq \delta$$

and

(1.32b) $$\|\mathbf{s}_\nu^h - \mathbf{s}^*\| \leqq \frac{1}{a}(a\|\mathbf{s}_0 - \mathbf{s}^*\|)^{2^\nu} + \delta.$$

Proof. As in the proof of Theorem 1.29 we have that, for any $\mathbf{s} \in S_\rho(\mathbf{s}^*)$, $Q(\mathbf{s})$ is nonsingular and $\|Q^{-1}(\mathbf{s})\| \leqq \beta/(1 - \rho\beta\gamma)$. We also easily show, using induction and (1.31a), that $\|\mathbf{s}_{\nu+1} - \mathbf{s}_\nu\| \leqq \rho_*$, $\nu = 0, 1, \cdots$. Now let $\mathbf{e}_\nu \equiv \mathbf{s}_\nu - \mathbf{s}_\nu^h$ and subtract (1.30a) from (1.26) to get

$$Q(\mathbf{s}_\nu^h)\mathbf{e}_{\nu+1} = [Q(\mathbf{s}_\nu^h) - Q(\mathbf{s}_\nu, \mathbf{s}_\nu^h)]\mathbf{e}_\nu + [Q(\mathbf{s}_\nu^h) - Q(\mathbf{s}_\nu)](\mathbf{s}_{\nu+1} - \mathbf{s}_\nu) - \mathbf{\Delta}_\nu(h).$$

For an induction assume $\mathbf{s}_\nu^h \in S_\rho(\mathbf{s}^*)$ and then, recalling the last line but one in the proof of Theorem 1.29 and (1.30b), we get

$$\|\mathbf{e}_{\nu+1}\| \leqq \frac{\beta}{1 - \rho\beta\gamma}\left[\frac{\gamma}{2}\|\mathbf{e}_\nu\|^2 + \gamma\rho_*\|\mathbf{e}_\nu\| + Mh^m\right] \equiv \frac{\beta}{1 - \rho\beta\gamma}p(\|\mathbf{e}_\nu\|) + \|\mathbf{e}_\nu\|.$$

Here we have introduced the quadratic

$$p(z) \equiv \frac{\gamma}{2}z^2 - \left(\frac{1 - 2\rho\beta\gamma}{\beta}\right)z + Mh^m,$$

which by (1.31b) has two distinct positive zeros, say, $z_+ > z_- > 0$. Since $\mathbf{e}_0 = \mathbf{0}$ it will easily follow by induction that $\|\mathbf{e}_\nu\| \leqq z_-$. However, since the roots must satisfy $z_+z_- = 2Mh^m/\gamma$ and (1.31b) yields $z_+ > (1 + \sqrt{1 - \theta})(1 - 2\rho\beta\gamma)/\beta\gamma$, we find that

$$z_- = \frac{2Mh^m}{\gamma z_+} \leqq \frac{2Mh^m}{1 + \sqrt{1 - \theta}}\left(\frac{\beta}{1 - 2\rho\beta\gamma}\right) \equiv \delta.$$

To conclude the inductive proof of (1.32a) we need only use (1.31c) to observe that

$\mathbf{s}_{v+1}^h \in S_\rho(\mathbf{s}^*)$ since $\|\mathbf{s}_{v+1}^h - \mathbf{s}^*\| \leqq \|\mathbf{s}_{v+1}^h - \mathbf{s}_v\| + \|\mathbf{s}_v - \mathbf{s}^*\| \leqq \delta + \rho_* \leqq \rho$. The estimate in (1.29d) used in the above triangle inequality gives (1.32b). ▧

According to the estimate in (1.32b) a level may be reached beyond which further iterations do not yield any improvement. This actually occurs in practice and indeed the roundoff tolerance should usually be added to the truncation term Mh^m. To illustrate the behavior predicted by (1.32b) suppose the exact iterates satisfy $a\|\mathbf{s}_0 - \mathbf{s}^*\| = 10^{-q}$ and the noise level is $Mh^m = 10^{-r}$, where, say, $r = 10q$. Then

$$\|\mathbf{s}_v^h - \mathbf{s}^*\| \doteq 10^{-2^v q} + 10^{-10q}$$

so the vth iterate contains approximately $2^v q$ correct digits provided $2^v \leqq 10$, i.e., for the first three iterates. Thereafter the number of correct digits remains fixed at about $10q$ in this example. Further iterations are useless without mesh refinement.

Note that the most important amplification of the noise level, i.e., δ in (1.31c) rather than simply Mh^m, is due to $\rho\beta\gamma$ being close to its limiting value 1/2. This is most important in the early iterates which are likely to be close to the boundary of the domain of attraction. As the iterates converge and are well within the domain of attraction the inequality (1.31a) is, with a new smaller ρ, easily satisfied and the relevant value for δ is much reduced. The limiting magnification is simply $\delta \approx \beta Mh^m$ which from (1.29a) and (1.30a,b) is always to be expected.

A more complete analysis of the effect of errors in Newton's method is given by Lancaster [36].

1.5. Parallel shooting for nonlinear problems. Parallel shooting applied to nonlinear problems has much more dramatic and easily demonstrated benefits. To treat problems of the form (1.20a,b) it again proceeds from some set of nodes, say,

$$a = \tau_0 < \tau_1 < \cdots < \tau_J = b.$$

On each interval $[\tau_{j-1}, \tau_j]$ we pose an initial value problem:

$$\text{(1.33a)} \qquad \mathbf{u}_j' = \mathbf{f}(t, \mathbf{u}_j), \quad \tau_{j-1} \leqq t \leqq \tau_j, \qquad 1 \leqq j \leqq J,$$

$$\text{(1.33b)} \qquad \mathbf{u}_j(\tau_{j-1}) = \mathbf{s}_j, \qquad 1 \leqq j \leqq J.$$

Then we seek to piece together these J solutions, by appropriate choice of the J initial n-vectors $\mathbf{s}_j$, to form a continuous solution of (1.20a) which also satisfies (1.20b). Writing the boundary conditions first and then the $J - 1$ continuity conditions at the interior nodes τ_j, $1 \leqq j \leqq J - 1$, we get the (transcendental) system of nJ equations, using the notation $\mathbf{u}_j \equiv \mathbf{u}_j(t, \mathbf{s}_j)$:

$$\text{(1.33c)} \qquad \boldsymbol{\Phi}(\hat{\mathbf{s}}) = \begin{pmatrix} \mathbf{g}(\mathbf{s}_1, \mathbf{u}_J(b, \mathbf{s}_J)) \\ \mathbf{s}_2 - \mathbf{u}_1(\tau_1, \mathbf{s}_1) \\ \vdots \\ \mathbf{s}_J - \mathbf{u}_{J-1}(\tau_{J-1}, \mathbf{s}_{J-1}) \end{pmatrix} = \mathbf{0}, \quad \hat{\mathbf{s}} = \begin{pmatrix} \mathbf{s}_1 \\ \mathbf{s}_2 \\ \vdots \\ \mathbf{s}_J \end{pmatrix}.$$

This parallel shooting procedure can also be viewed as ordinary shooting for an (artificially) enlarged two-point boundary value problem. To do this we reformulate (1.20a,b). A new independent variable $\hat{t}$ is introduced on each subinterval by

$$\hat{t} \equiv (t - \tau_{j-1})/\Delta_j, \quad \Delta_j \equiv \tau_j - \tau_{j-1}, \qquad j = 1, 2, \cdots, J. \tag{1.34a}$$

Then with $z_j(\hat{t}) \equiv z(\tau_{j-1} + \hat{t}\Delta_j)$ and

$$\mathbf{f}_j(\hat{t}, \mathbf{z}_j(\hat{t})) \equiv \Delta_j \mathbf{f}(\tau_{j-1} + \hat{t}\Delta_j, z(\tau_{j-1} + \hat{t}\Delta_j)), \qquad 1 \leqq j \leqq J, \tag{1.34b}$$

we can write (1.20a) as

$$\hat{\mathbf{z}}' = \hat{\mathbf{f}}(\hat{t}, \hat{\mathbf{z}}), \quad 0 \leqq \hat{t} \leqq 1, \qquad \hat{\mathbf{z}} \equiv \begin{pmatrix} \mathbf{z}_1 \\ \vdots \\ \mathbf{z}_J \end{pmatrix}, \quad \hat{\mathbf{f}} \equiv \begin{pmatrix} \mathbf{f}_1 \\ \vdots \\ \mathbf{f}_J \end{pmatrix}. \tag{1.34c}$$

The continuity and boundary conditions become

$$\hat{\mathbf{g}}(\hat{\mathbf{z}}(0), \hat{\mathbf{z}}(1)) \equiv \begin{pmatrix} g(\mathbf{z}_1(0), \mathbf{z}_J(1)) \\ \mathbf{z}_2(0) - \mathbf{z}_1(1) \\ \vdots \\ \mathbf{z}_J(0) - \mathbf{z}_{J-1}(1) \end{pmatrix} = \mathbf{0}. \tag{1.34d}$$

The problem (1.34a,b,c,d) is equivalent to (1.20a,b) and ordinary shooting for the former is equivalent to (1.33a,b,c). Indeed it is simpler to write now:

$$\hat{\mathbf{u}}' = \hat{\mathbf{f}}(\hat{t}, \hat{\mathbf{u}}), \qquad 0 \leqq \hat{t} \leqq 1, \tag{1.35a}$$

$$\hat{\mathbf{u}}(0) = \hat{\mathbf{s}}, \tag{1.35b}$$

and from (1.34d):

$$\hat{\mathbf{\Phi}}(\hat{\mathbf{s}}) \equiv \hat{\mathbf{g}}(\hat{\mathbf{s}}, \hat{\mathbf{u}}(1, \hat{\mathbf{s}})). \tag{1.35c}$$

An immediate advantage of our reformulation is obtained by examining how Lemma 1.25 goes over from (1.20a,b) to (1.34a,b,c,d). In particular consider, for example, the case in which $\Delta_j \equiv (b - a)/J$ for all $j = 1, 2, \cdots, J$. Then we have as in Weiss [55]:

LEMMA 1.36. *Let the hypothesis of Lemma 1.25 hold. Then for each* $\hat{\mathbf{s}} \in S_{\hat{\rho}}(\mathbf{z}(0))$, *where*

$$\hat{\rho} = \delta\, e^{-K(b-a)/J}, \tag{1.36a}$$

$$S_{\hat{\rho}}(\hat{\mathbf{z}}(0)) \equiv \{\hat{\mathbf{s}} : \|\hat{\mathbf{z}}(0) - \hat{\mathbf{s}}\| \leqq \hat{\rho}\}, \tag{1.36b}$$

there exists a unique solution $\hat{\mathbf{u}}(\hat{t}, \hat{\mathbf{s}})$ *of* (1.35a,b) *on* $0 \leqq \hat{t} \leqq 1$.

Proof. We need only observe, using the appropriate norm, that (1.25a) implies by (1.34b,c) and $\Delta_j \equiv (b - a)/J$ that

$$\|\hat{\mathbf{f}}(\hat{t}, \hat{\mathbf{v}}) - \hat{\mathbf{f}}(\hat{t}, \hat{\mathbf{w}})\| \leqq \frac{K(b-a)}{J}\|\hat{\mathbf{v}} - \hat{\mathbf{w}}\| \quad \text{for all} \quad (\hat{t}, \hat{\mathbf{v}}), (\hat{t}, \hat{\mathbf{w}}) \in T_\delta\{\hat{\mathbf{z}}\}. \qquad ▧$$

Comparing (1.25b) and (1.36a) we get

$$\hat{\rho}/\rho = e^{(1-1/J)K(b-a)},$$

for an exponential increase in the radius of the sphere whose initial data allows an integration over the total interval required ([0, 1] in the present case). In fact this goes over as we might suspect to the convergence of Newton's method. Indeed suppose (1.20a,b) has a smooth isolated solution and for some $\rho_* \leqq \rho$ of Lemma 1.25 the hypothesis of Theorem 1.29 holds. Then clearly (1.34a,b,c,d) has a smooth isolated solution and the relevant matrix of the linearized problem, which is the Jacobian of (1.33c) or (1.35c) with respect to $\hat{\mathbf{s}}$, has the form:

$$(1.37)\qquad \hat{Q}(\hat{\mathbf{s}}) \equiv \begin{pmatrix} B_a(\mathbf{s}_1, \hat{\mathbf{u}}_J(1, \mathbf{s}_J)) & & & B_b(\cdot, \cdot)U_J(1, \mathbf{s}_J) \\ -U_1(1, \mathbf{s}_1) & I & & \\ & \ddots & \ddots & \\ & & -U_{J-1}(1, \mathbf{s}_J) & I \end{pmatrix}.$$

This matrix for $\hat{\mathbf{s}} = \hat{\mathbf{s}}^* \equiv \hat{\mathbf{z}}(0)$ is nonsingular since $\hat{\mathbf{z}}(t)$ is an isolated solution of (1.34a,b,c,d) or by the argument used to study $\mathbb{A}_J$ in (1.16a,b). The verification of (1.29b) follows trivially from the Lipschitz continuity of the fundamental solutions $U_j(t, \mathbf{s}_j)$ with respect to the initial data, with the new constant $\hat{\gamma} = \gamma\, e^{-(1-1/J)K(b-a)}$ (assuming the first row of blocks supplies the max of the row norms). The bound on $\|\hat{Q}^{-1}\|$ is a bit messier but can be obtained from the factorization (1.16a); it is carried out in Keller and White [30] for other purposes (see also § 2.1 following (2.10e)). We find that, essentially,

$$\|\hat{Q}^{-1}(s)\| \leqq e^{2K(b-a)/J}\|Q^{-1}(s^*)\| = \beta\, e^{2K(b-a)/J}.$$

Thus the product $\rho_*\beta\gamma$ in (1.29c) is now replaced by $\hat{\rho}_*\beta\gamma\, e^{-(1-3/J)K(b-a)}$. Now to satisfy (1.29c) we can use

$$\hat{\rho}_* = \rho_*\, e^{(1-3/J)K(b-a)}.$$

These estimates imply that more than 3 shooting intervals must be employed before the enlarged sphere of attraction is noted. However we believe this to be caused by crude estimates.

In conclusion we advocate using Newton's method with parallel shooting to solve the continuity and boundary conditions as in (1.33c). The linear systems to be solved for each iterate have coefficient matrices of the above form $\hat{Q}(\hat{\mathbf{s}})$ or, for partially separated endpoints properly ordered, of the form of $\mathbb{A}_J$ in (1.17). Again these systems should be solved by one of the stable efficient factorization procedures discussed in § 2.2. Techniques for selecting the shooting nodes can be patterned after those for linear problems when Newton's method is used. Thus either the growth of the columns of the $U_j(t_i, \mathbf{s}_j)$ or the "condition" of these matrices (i.e., how nearly singular) can be monitored.

1.6. Continuation or embedding. In solving nonlinear problems by shooting or parallel shooting the major difficulty is in locating any point $\mathbf{s}_0$ or $\hat{\mathbf{s}}_0$ in the domain of attraction for, say, Newton's method. Of course if the problem contains some natural parameter such that for extreme values of the parameter the solution is known (or easily obtained), then it is clear that we may try to continue in this parameter. Thus suppose our problem can be formulated as:

$$\text{(1.38a)} \qquad \mathbf{z}' = \mathbf{f}(t, \mathbf{z}; \sigma), \qquad a \leqq t \leqq b,$$

$$\text{(1.38b)} \qquad \mathbf{g}(\mathbf{z}(a), \mathbf{z}(b); \sigma) = \mathbf{0}.$$

We assume that for each $\sigma \in [\sigma_0, \sigma_F]$ an isolated solution, $\mathbf{z} = \mathbf{z}(t, \sigma)$, exists and depends smoothly on σ. We call such a family of solutions a *continuation branch* on $[\sigma_0, \sigma_F]$.

If natural parameters do not occur, as in (1.20a,b) say, then we can introduce them *for example* by forming:

$$\mathbf{f}(t, \mathbf{z}; \sigma) \equiv \sigma\mathbf{f}(t; \mathbf{z}) + (1 - \sigma)[A(t)z + \mathbf{f}(z)],$$

$$\mathbf{g}(\mathbf{u}, \mathbf{v}; \sigma) \equiv \sigma\mathbf{g}(\mathbf{u}, \mathbf{v}) + (1 - \sigma)[B_a\mathbf{u} + B_b\mathbf{v} - \boldsymbol{\beta}].$$

For $\sigma = 0$ in this embedding our problem reduces to (0.1a,b) while for $\sigma = 1$ it is (1.20a,b). Of course even if (0.1a,b) has a known solution we do not know if a continuation branch exists on [0, 1]. This more basic problem of devising embeddings for which continuation branches exist receives very little attention while the rather straightforward problem of continuing along a branch is frequently rehashed as a new discovery. There are many other forms of embedding that can be and have been used. One favorite is to rescale the independent variable t so that the interval length $[b - a]$ can be the parameter (see Keller [29], Roberts and Shipman [46]).

To compute a continuation branch of (1.38a b) by shooting we consider the initial value problem:

$$\text{(1.39a)} \qquad \mathbf{u}' = \mathbf{f}(t, \mathbf{u}; \sigma),$$

$$\text{(1.39b)} \qquad \mathbf{u}(a) = \mathbf{s},$$

and the system of equations

$$\text{(1.39c)} \qquad \boldsymbol{\phi}(\mathbf{s}, \sigma) \equiv \mathbf{g}(\mathbf{s}, \mathbf{u}(b; \mathbf{s}, \sigma); \sigma) = \mathbf{0}.$$

Here we have denoted the solution of (1.39a,b) by $\mathbf{u}(t; \mathbf{s}, \sigma)$. A well-known result of Poincaré ensuring continuation over small intervals is now easily presented using (1.39a,b,c).

THEOREM 1.40. *Let* (1.38a,b) *have an isolated solution* $\mathbf{z} = \mathbf{z}(t, \sigma_1)$ *for* $\sigma = \sigma_1$. *Let* $\mathbf{f}(t, \mathbf{z}; \sigma)$ *and* $\mathbf{g}(\mathbf{z}(a), \mathbf{z}(b); \sigma)$ *be* C^1 *in some neighborhood of* $(t, \mathbf{z}(t; \sigma_1), \sigma_1)$. *Then* (1.38a,b) *has a continuation branch on* $[\sigma_1 - \delta\sigma, \sigma_1 + \delta\sigma]$ *for some* $\delta\sigma > 0$.

Proof. By the hypothesis, (1.39c) must have, for $\sigma = \sigma_1$, a root $\mathbf{s} = \mathbf{z}(a, \sigma_1) \equiv \mathbf{s}_1$. This is an isolated root since

$$\text{(1.41a)} \qquad Q(\mathbf{s}, \sigma) \equiv \frac{\partial\boldsymbol{\phi}(\mathbf{s}, \sigma)}{\partial\mathbf{s}} = B_a[\mathbf{u}, \sigma] + B_b[\mathbf{u}, \sigma]U(b; \mathbf{s}, \sigma)$$

is nonsingular at $(\mathbf{s}, \sigma) = (\mathbf{s}_1, \sigma_1)$. Here $Q(\mathbf{s}, \sigma)$ is clearly defined in analogy with

(1.23a,b) and (1.27). Now the implicit function theorem can be applied to (1.39c) at $(\mathbf{s}_1, \sigma_1)$ to yield a family of isolated roots $\mathbf{s} = \mathbf{s}(\sigma)$ for all σ in $|\sigma - \sigma_1| \leqq \delta\sigma$ and $\mathbf{s}(\sigma_1) = \mathbf{s}_1$. ▧

The hypothesis of Theorem 1.40 yields even more, namely, that $\mathbf{s}(\sigma)$ has a continuous derivative. From the identity $\boldsymbol{\phi}(\mathbf{s}(\sigma), \sigma) = \mathbf{0}$ we thus get that:

$$Q(\mathbf{s}(\sigma), \sigma)\frac{d\mathbf{s}(\sigma)}{d\sigma} + \frac{\partial\boldsymbol{\phi}(\mathbf{s}(\sigma), \sigma)}{\partial\sigma} = \mathbf{0}. \tag{1.41b}$$

This is a differential equation for $\mathbf{s}(\sigma)$ subject to the initial condition, say:

$$\mathbf{s}(\sigma_1) = \mathbf{s}_1. \tag{1.41c}$$

It is quite practicable to solve or rather approximate the solution $\mathbf{s}(\sigma)$ of (1.41b,c) numerically and this is frequently done in practice (see Wasserstrom [54]). In fact when Newton's method is being used to solve (1.39c) the matrix $Q(\sigma) \equiv Q(\mathbf{s}(\sigma), \sigma)$ is available for a given σ when the iterations converge. Then we note that

$$\frac{\partial\boldsymbol{\phi}(\mathbf{s}(\sigma), \sigma)}{\partial\sigma} = \mathbf{g}_\sigma(\mathbf{s}, \mathbf{u}(b; \mathbf{s}, \sigma); \sigma) + B_b[\mathbf{u}, \sigma]\mathbf{v}(b; \mathbf{s}, \sigma),$$

where $\mathbf{v}(t; \mathbf{s}, \sigma) = \partial\mathbf{u}(t; \mathbf{s}, \sigma)/\partial\sigma$ is the solution of the variational problem

$$\mathscr{L}[\mathbf{u}]\mathbf{v} = \mathbf{f}_\sigma(t, \mathbf{u}(t; \mathbf{s}, \sigma), \sigma), \quad \mathbf{v}(a) = \mathbf{0}. \tag{1.42}$$

The simplest way to use these results for continuation is to observe that:

$$\mathbf{s}(\sigma + \delta\sigma) = \mathbf{s}(\sigma) + \frac{d\mathbf{s}(\sigma)}{d\sigma}\delta\sigma + O(\delta\sigma^2).$$

Then we use as the initial guess in Newton's method, $\mathbf{s}_0(\sigma + \delta\sigma)$, the first two terms on the right-hand side above. The only extra calculations required in this process are those to solve (1.42). However this is simply one linear initial value problem while n of them in (1.25c) must be solved for each evaluation of $Q(\mathbf{s}, \sigma)$.

It is not difficult, using the above or similar techniques, to prove that one can continue on some existing branch over $[\sigma_0, \sigma_F]$ (see Bosarge [5]). Indeed one can study the effects of step length, $\delta\sigma$, and incomplete iteration or fixed number of Newton iterations per step that still yields $\mathbf{s}_0(\sigma_F)$ in the domain of attraction (see for example Avila [2]).

Note that the existence of a continuation branch on $[\sigma_0, \sigma_F]$ implies that the family of matrices $Q(\sigma) \equiv Q(\mathbf{s}(\sigma), \sigma)$ must be nonsingular for $\sigma_0 \leqq \sigma \leqq \sigma_F$. Thus $\det Q(\sigma)$ must be of one sign on the branch. For real problems with real parameters the question is thus raised of how many components there are to the sets $D_n^\pm$ of all $n \times n$ real matrices with positive (negative) determinant. (For the complex case we ask how many components there are to the set of all $n \times n$ nonsingular complex matrices.) For the scalar case $n = 1$, it is trivial that $D_1^\pm$ each have one component. Fortunately the same is true for all n as we proceed to show.

THEOREM 1.43. *Each set $D_n^\pm$ is connected. In fact for each $A \in D_n^\pm$ there exists $A(\theta)$ such that:* (i) $A(1) = A$, (ii) $A(\theta) \in C[0, 1]$, (iii) $\det A(\theta) \neq 0$, (iv) $A(0) = I$ *if* $A \in D_n^+$, $A(0) = ({}^{-1}I)$ *if* $A \in D_n^-$.

Proof. If $\det A \neq 0$, then for some permutation matrix P,

$$PA = LDU,$$

where $L(U)$ is unit-lower (-upper) triangular and D is diagonal. Let $L(\theta)$ and $U(\theta)$ be obtained by multiplying all off-diagonal terms by θ. Then $L(0) = U(0) = I$ while $L(1) = L$ and $U(1) = U$. Let $D(\theta)$ have the diagonal elements

$$d_i(\theta) = \theta\, d_i + (1 - \theta)\, d_i/|d_i|$$

so that $D(1) = D$ and $D(0) = (\operatorname{sign} d_i\, \delta_{ij})$. Then with

$$A_1(\theta) \equiv P^T L(\theta)D(\theta)U(\theta)$$

we have $A_1(1) = A$, $A_1(0) = P^T D(0)$, $\det A_1(\theta) \det A > 0$.

Any permutation matrix, P^T, is the product of simple row interchanges. Neglecting unaltered elements all such matrices are essentially $P_{12} = \begin{pmatrix} 0 & 1 \\ 1 & 0 \end{pmatrix}$.

But note that

$$P_{12}(\theta) = \begin{pmatrix} \theta & 1 - \theta \\ 1 - \theta & -\theta \end{pmatrix}$$

has $\det P_{12}(\theta) < 0$, $P_{12}(0) = P_{12}$ and $P_{12}(1) = \begin{pmatrix} 1 & 0 \\ 0 & -1 \end{pmatrix}$. So with several such matrices we can form $P^T(\theta)$ such that:

$$P^T(0) = P^T, \quad P^T(1) = \begin{pmatrix} \pm 1 & & 0 \\ & \ddots & \\ 0 & & \pm 1 \end{pmatrix}, \quad (\det P^T(\theta) \det P^T) > 0.$$

Now $A_2(\theta) = P^T(\theta)A_1(\theta)$ has $\det A_2(\theta) \det A > 0$ and

$$A_2(1) = \begin{pmatrix} \pm 1 & & 0 \\ & \ddots & \\ 0 & & \pm 1 \end{pmatrix}.$$

Next consider $M \equiv \begin{pmatrix} -1 & 0 \\ 0 & -1 \end{pmatrix}$ and $M(\theta) \equiv \begin{pmatrix} 1 - 2\theta & -\theta(1 - \theta) \\ \theta(1 - \theta) & 1 - 2\theta \end{pmatrix}$ which shows, since $\det M(\theta) > 0$, that all pairs of minus signs in $A_2(1)$ can be rotated to unity. Thus either one or no minus signs remain. In the latter case we are finished since $\det A > 0$ must have held. Otherwise $\det A < 0$ and the transformation $N(\theta) \equiv \begin{pmatrix} 1 - 2\theta & \theta(1 - \theta) \\ \theta(1 - \theta) & 2\theta - 1 \end{pmatrix}$ takes $\begin{pmatrix} 1 & 0 \\ 0 & -1 \end{pmatrix}$ to $\begin{pmatrix} -1 & 0 \\ 0 & 1 \end{pmatrix}$ while $\det N(\theta) < 0$. Thus we can bring A to the canonical form $\begin{pmatrix} -1 & 0 \\ 0 & I \end{pmatrix}$ in this case. ▧

As a result of the above some practical procedures are suggested for attempts at embedding in which the desired final state is never reached. Simply alter the initial state so that $\det Q(\sigma_0)$ has the opposite sign from the original choice.

A somewhat simpler proof of the connectedness of $D_n^\pm$ can be given by induction on n using minoring. This has been suggested independently by B. Fornberg and G. Latta. We have given the above version as it is constructive.

A survey of some general studies of continuation is contained in Rheinboldt [44].

CHAPTER 2

Finite Difference Methods

2.1. Difference methods for linear problems. A very general theory of difference methods is available for linear problems of the form (0.1a,b), which have unique solutions. The theory states in essence that a difference scheme is stable and consistent for the boundary value problem if and only if it is so for the initial value problem. For practical purposes it is only the sufficiency part that is important. But a simple result relating pairs of boundary value problems makes the necessity proof trivially follow from sufficiency. So we first present this useful theoretical result as:

THEOREM 2.1. *Consider the pair of boundary value problems $BV(\nu)$, $\nu = 0, 1$ given by*

$$\mathscr{L}\mathbf{y}(t) \equiv \mathbf{y}'(t) - A(t)\mathbf{y}(t) = \mathbf{f}(t), \qquad a \leqq t \leqq b, \tag{2.1a}$$

$$\mathscr{B}^{(\nu)}\mathbf{y}(t) \equiv B_a^{(\nu)}\mathbf{y}(a) + B_b^{(\nu)}\mathbf{y}(b) = \boldsymbol{\beta}. \tag{2.1b}$$

The corresponding fundamental solution matrix, $Y^{(\nu)}(t)$, defined by

$$\mathscr{L}Y^{(\nu)}(t) = 0, \tag{2.2a}$$

$$\mathscr{B}^{(\nu)}Y^{(\nu)}(t) = I \tag{2.2b}$$

exists if and only if $BV(\nu)$ has a unique solution.

If $BV(0)$ has a unique solution, then $BV(1)$ has a unique solution if and only if $\mathscr{B}^{(1)}Y^{(0)}(t)$ is nonsingular.

Proof. It is clear from Theorem 1.5 that $BV(\nu)$ has a unique solution if and only if

$$Q_\nu \equiv \mathscr{B}^{(\nu)}Y(t) \equiv B_a + B_bY(b)$$

is nonsingular. When this is so we can define $Y^{(\nu)}(t)$ as

$$Y^{(\nu)}(t) = Y(t)Q_\nu^{-1}.$$

On the other hand if $Y^{(\nu)}(t)$ exists, then it must have this representation.

Now suppose $BV(0)$ has a unique solution. Then $Y^{(0)}(t) = Y(t)Q_0^{-1}$ exists and $\mathscr{B}^{(1)}Y^{(0)}(t) = \mathscr{B}^{(1)}Y(t)Q_0^{-1} = Q_1Q_0^{-1}$. Thus $\mathscr{B}^{(1)}Y^{(0)}(t)$ is nonsingular if and only if Q_1 is nonsingular. ▧

The difference schemes employ a net of the form

$$t_0 = a, \quad t_j = t_{j-1} + h_j, \quad 1 \leqq j \leqq J, \quad t_J = b, \quad h \equiv \max_j h_j \leqq \theta \min_k h_k.$$

Then corresponding to each interval $[t_{j-1}, t_j]$ of this net we consider an approximation to (0.1a) in the form:

$$\mathscr{L}_h \mathbf{u}_j \equiv \sum_{k=0}^{J} C_{jk}(h)\mathbf{u}_k = \mathbf{F}_j(h, \mathbf{f}), \qquad 1 \leqq j \leqq J. \tag{2.3a}$$

The boundary conditions (0.1b) are simply taken over as

$$\mathscr{B}_h \mathbf{u}^h \equiv B_a \mathbf{u}_0 + B_b \mathbf{u}_J = \boldsymbol{\beta}. \tag{2.3b}$$

The net function $\mathbf{u}^h \equiv \{\mathbf{u}_j\}_0^J$ is to approximate $\mathbf{y}^h \equiv \{\mathbf{y}(t_j)\}_0^J$, the exact solution on the net. The $n \times n$ coefficient matrices $\{C_{jk}(h)\}$ and inhomogeneous terms $\{\mathbf{F}_j(h, \mathbf{f})\}$ define a very general family of difference schemes for which we develop a complete stability theory.

It will frequently be useful to formulate the difference equations (2.3a,b) in block matrix form as

$$\mathbb{A}_h \mathbf{u}^h = \mathbf{F}, \tag{2.4a}$$

where

$$\mathbb{A}_h \equiv \begin{pmatrix} B_a & \cdots & 0 & \cdots & B_b \\ C_{10}(h) & C_{11}(h) & & \cdots & C_{1J}(h) \\ \vdots & \vdots & & & \vdots \\ C_{J0}(h) & C_{J1}(h) & & \cdots & C_{JJ}(h) \end{pmatrix}, \quad \mathbf{u}^h \equiv \begin{pmatrix} \mathbf{u}_0 \\ \mathbf{u}_1 \\ \vdots \\ \mathbf{u}_J \end{pmatrix}, \quad \mathbf{F} \equiv \begin{pmatrix} \boldsymbol{\beta} \\ \mathbf{F}_1(h, \mathbf{f}) \\ \vdots \\ \mathbf{F}_J(h, \mathbf{f}) \end{pmatrix}. \tag{2.4b}$$

The truncation errors in (2.3a,b) as an approximation to (0.1a,b) are

$$\tau_0[\mathbf{y}] \equiv \mathscr{B}_h \mathbf{y}^h - \boldsymbol{\beta}, \quad \tau_j[\mathbf{y}] \equiv \mathscr{L}_h \mathbf{y}(t_j) - \mathbf{F}_j(h, \mathbf{f}), \qquad 1 \leqq j \leqq J, \tag{2.5a}$$

and the scheme is *consistent or accurate of order m* if for every (smooth) solution of (0.1a,b) there exist constants $K_0 > 0$, $h_0 > 0$ such that for all nets $\{t_j\}$ with $h \leqq h_0$,

$$\|\tau_j(\mathbf{y})\| \leqq K_0 h^m, \qquad 0 \leqq j \leqq J. \tag{2.5b}$$

The scheme is said to be *stable* provided there exist constants $K_1 > 0$ and $h_0 > 0$ such that for all nets $\{t_j\}$ with $h \leqq h_0$ and for all net functions $\mathbf{v}^h$,

$$\|\mathbf{v}_j\| \leqq K_1 \max\{\|\mathscr{B}_h \mathbf{v}^h\|, \max_{1 \leqq k \leqq J} \|\mathscr{L}_h \mathbf{v}_k\|\}, \qquad 0 \leqq j \leqq J. \tag{2.6a}$$

It is more or less well known, and easy to prove, that stability is equivalent to the existence and uniform boundedness of the inverses of the *family* of matrices $\mathbb{A}_h$; that is, for some $K_1 > 0$, $h_0 > 0$ and all $h \leqq h_0$,

$$\|\mathbb{A}_h^{-1}\| \leqq K_1. \tag{2.6b}$$

(Norms on block matrices, say, $B \equiv (B_{ij})$, are always to be taken as

$$\|B\| = \max_{0 \leqq i \leqq J} \sum_{j=0}^{J} \|B_{ij}\|_n,$$

where $\|\cdot\|_n$ is the matrix norm induced by the vector norm used on $\mathbb{E}^n$.) It is also well-known that consistency and stability imply convergence as in

(2.7) $$\|\mathbf{u}_j - \mathbf{y}(t_j)\| \leqq K_0 K_1 h^m.$$

Proofs of these facts are simple consequences of the definitions and can be found, for example, in Keller [24].

Now consider a pair of linear problems of the form (2.1a,b). The corresponding difference schemes resulting from applying (2.3a,b) to each of these problems differ only in the boundary conditions, say:

(2.8) $$\mathscr{B}_h^{(\nu)}\mathbf{u}^h \equiv B_a^{(\nu)}\mathbf{u}_0 + B_b^{(\nu)}\mathbf{u}_J = \boldsymbol{\beta}.$$

The deep and important connection between these schemes is contained in

THEOREM 2.9. *Let each problem $BV(\nu)$, $\nu = 0, 1$, have a unique solution. Then the corresponding difference schemes $BV_h(\nu)$, $\nu = 0, 1$, consisting of* (2.3a) *and* (2.8) *are such that $BV_h(0)$ is stable and consistent for $BV(0)$ if and only if $BV_h(1)$ is stable and consistent for $BV(1)$.*

Proof. The equivalence of consistency is trivial since the schemes differ only in the boundary conditions which, in each case, are exact.

For stability consider each matrix $\mathbb{A}_h^{(\nu)}$ in which B_a and B_b in (2.4b) are replaced by those of (2.8). We note that

$$\mathbb{D}_h \equiv \mathbb{A}_h^{(1)} - \mathbb{A}_h^{(0)} = \begin{pmatrix} [B_a^{(1)} - B_a^{(0)}] \cdots 0 \cdots [B_b^{(1)} - B_b^{(0)}] \\ 0 \end{pmatrix}.$$

Then, assuming $BV_h(0)$ is stable, $\mathbb{A}_h^{(0)}$ is nonsingular with

$$\|\mathbb{A}_h^{(0)^{-1}}\| \leqq K.$$

So we can write

(2.10a) $$\mathbb{A}_h^{(1)} = [\mathbb{I} + \mathbb{D}_h\mathbb{A}_h^{(0)^{-1}}]\mathbb{A}_h^{(0)}.$$

Denote the block structure of $\mathbb{A}_h^{(0)^{-1}}$ by

(2.10b) $$\mathbb{A}_h^{(0)^{-1}} \equiv (Z_{jk}^{(0)}),$$

where the $Z_{jk}^{(0)}$ are $n \times n$ matrices, $0 \leqq j, k \leqq J$. Now we have the form

(2.10c) $$\mathbb{A}_h^{(1)} = \begin{pmatrix} Q_{h0} & Q_{h1} & \cdots & Q_{hJ} \\ & I & & \\ & & \ddots & \\ & & & I \end{pmatrix}\mathbb{A}_h^{(0)},$$

where

(2.10d) $$Q_{hk} \equiv B_a^{(1)}Z_{0k}^{(0)} + B_b^{(1)}Z_{Jk}^{(k)}, \qquad 0 \leqq k \leqq J.$$

This follows from the fact that $\mathbb{A}_h^{(0)}(Z_{jk}^{(0)}) = \mathbb{I}$. In particular then, $B_a^{(0)}Z_{0k}^{(0)} + B_b^{(0)}Z_{Jk}^{(0)} = \delta_{0k}I$ and $\mathscr{L}_h Z_{jk}^{(0)} = 0$, $1 \leqq j \leqq J$, $0 \leqq k \leqq J$. Note that the column of blocks $\{Z_{j0}^{(0)}\}_0^J$ is just the difference approximation, using $BV_h(0)$, to $Y^{(0)}(t)$. Thus

$$\|Z_{jo}^{(0)} - Y^{(0)}(t_j)\| = O(h^m),$$

and hence

$$\|Q_{h0} - \mathscr{B}^{(1)}Y^{(0)}\| = O(h^m).$$

But $\mathscr{B}^{(1)}Y^{(0)}$ is nonsingular by Theorem 2.1. Thus Q_{h0} is nonsingular for h_0 sufficiently small and so is $\mathbb{A}_h^{(1)}$. In fact now

$$\text{(2.10e)} \qquad \mathbb{A}_h^{(1)^{-1}} = \mathbb{A}_h^{(0)^{-1}} \begin{pmatrix} Q_{h0}^{-1} & -Q_{h0}^{-1}Q_{h1} & \cdots & -Q_{h0}^{-1}Q_{hJ} \\ & I & & \\ & & \ddots & \\ & & & I \end{pmatrix}.$$

It easily follows, using

$$\sum_{j=1}^{J} \|Q_{hj}\| \leqq K(\|B_a^{(1)}\| + \|B_b^{(1)}\|), \quad \|Q_{h0}^{-1}\| \leqq C,$$

that, if $C > 1$,

$$\|\mathbb{A}_h^{(1)^{-1}}\| \leqq KC[1 + K(\|B_a^{(1)}\| + \|B_b^{(1)}\|)].$$

So $BV_h(1)$ is stable.

The converse follows by interchanging $\nu = 0$ and $\nu = 1$ in the above. ▧

As the most important application of Theorem 2.9 we have

COROLLARY 2.11. *Let problem* (0.1a,b) *have a unique solution. Then the difference scheme* (2.3a,b) *is stable and consistent for* (0.1a,b) *if and only if the scheme*

$$\text{(2.11a)} \qquad \mathbf{v}_0 = \boldsymbol{\alpha}, \qquad \mathscr{L}_h\mathbf{v}_j = \mathbf{F}_j(h, \mathbf{f}), \qquad 1 \leqq j \leqq J,$$

is stable and consistent for the initial value problem

$$\text{(2.11b)} \qquad \mathbf{v}(a) = \boldsymbol{\alpha}, \qquad \mathscr{L}\mathbf{v}(t) = \mathbf{f}(t), \qquad a \leqq t \leqq b.$$

Proof. We need only identify $BV(1)$ with (0.1a,b), $BV_h(1)$ with (2.3a,b), $BV(0)$ with (2.11b) (which has a unique solution) and $BV_h(0)$ with (2.11a). ▧

We can now devise many stable schemes for boundary value problems by simply using the well-developed theory for initial value problems. In addition asymptotic error expansions are easily obtained in the usual manner (see Stetter [51]). The required truncation expansions are given by Gragg [19] for a special midpoint scheme and by Henrici [22] and Engquist [12] for multistep schemes. We point out, however, that not all schemes are included in our theory—but we believe that all the useful ones are included.

The factorization in (2.10c) or inverse in (2.10e) yields specific procedures for solving the difference equations when $\mathbb{A}_h^{(0)}$ represents an explicit scheme (as would be the case in most initial value schemes). But we do not advocate using this obvious procedure. It would in effect be an initial value or shooting scheme with no special features to recommend it.

2.2. One-step schemes; solution procedures. A most important class of difference schemes of the form (2.3a,b) for problems of the form (0.1a,b) are one-step or two-point schemes. That is, schemes in which

$$\text{(2.12a)}\quad \begin{aligned} &C_{jk}(h) \equiv 0 \quad \text{for} \quad k \neq j-1, j, \\ &C_{j,j-1}(h) \equiv L_j(h) = -h_j^{-1}I + \tilde{C}_{j,j-1}(h), \qquad \|\tilde{C}_{j,j-1}\| \leqq M, \\ &C_{j,j}(h) \equiv R_j(h) = h_j I + \tilde{C}_{j,j}(h), \qquad \|\tilde{C}_{j,j}\| \leqq M. \end{aligned}$$

If such schemes are consistent, then they are stable as a result of Corollary 2.11 and a result for initial value problems (see Isaacson and Keller [23, p. 396]). If in addition the boundary conditions are separated, as in (1.9a) but with $C_{ba} \equiv 0$, and we write those conditions at $t = a$ first, then the difference equations (2.3a) using (2.12a), and then the boundary conditions at $t = b$ last, the matrix of the scheme becomes, in place of (2.4b),

$$\text{(2.12b)}\quad \mathbb{A}_h \equiv \begin{pmatrix} C_a & & & & \\ L_1 & R_1 & & & \\ & L_2 & R_2 & & \\ & & \ddots & \ddots & \\ & & & L_J & R_J \\ & & & & C_b \end{pmatrix}.$$

We recall that C_a is $p \times n$, C_b is $q \times n$, and all L_j, R_j are $n \times n$. Thus $\mathbb{A}_h$ can be written as a block tridiagonal matrix

$$\text{(2.13a)}\quad \mathbb{A}_h \equiv [B_j, A_j, C_j],$$

where the $n \times n$ bidiagonal matrices have the forms:

$$\text{(2.13b)}\quad B_j \equiv \begin{pmatrix} x & x \text{---} x \\ | & | \quad | \\ x & x \text{---} x \\ 0 & \text{------} 0 \\ | & \quad | \\ 0 & \text{------} 0 \end{pmatrix}\begin{matrix} \}p \\ \\ \}q \end{matrix}, \quad C_j \equiv \begin{pmatrix} 0 & \text{------} 0 \\ | & \quad | \\ 0 & \text{------} 0 \\ x & x \text{---} x \\ | & | \quad | \\ x & x \text{---} x \end{pmatrix}\begin{matrix} \}p \\ \\ \}q \end{matrix}, \quad p + q = n.$$

To solve linear systems

$$\text{(2.14)}\quad \mathbb{A}_h \mathbf{u}^h = \mathbf{F}$$

with $\mathbb{A}_h$ of the above form we consider first block-$\mathbb{L}\mathbb{U}$ decompositions or band-factorizations. A general class of such procedures can be studied by writing

$$\text{(2.15a)}\quad \mathbb{A}_h = \mathbb{L}_h \mathbb{U}_h, \quad \mathbb{L}_h \equiv [\beta_j, \delta_j, 0], \quad \mathbb{U}_h \equiv [0, \alpha_j, \gamma_j], \qquad 0 \leqq j \leqq J.$$

Here α_j, β_j, γ_j, δ_j and 0 are $n \times n$ matrices so that $\mathbb{L}_n$ and $\mathbb{U}_h$ are block lower- or upper-triangular. They may indeed be strictly triangular, as in band-factorization, if δ_j and α_j are, respectively lower- and upper-triangular. In order for this factorization to be valid we must have:

$$\text{(2.15b)}\quad \begin{aligned} &\delta_0 \alpha_0 = A_0, \quad \delta_j \alpha_j = A_j - \beta_j \gamma_{j-1}, \qquad 1 \leqq j \leqq J, \\ &\beta_j \alpha_{j-1} = B_j, \quad 1 \leqq j \leqq J, \qquad \delta_j \gamma_j = C_j, \quad 0 \leqq j \leqq J-1. \end{aligned}$$

Then (2.14) can be solved by writing

$$\mathbb{L}_h \mathbf{v}^h = \mathbf{F}, \qquad \mathbb{U}_h \mathbf{u}^h = \mathbf{v}^h, \tag{2.15c}$$

and solving, in order:

$$\begin{aligned} &\delta_0 \mathbf{v}_0 = \mathbf{F}_0, \quad \delta_j \mathbf{v}_j = \mathbf{F}_j - \beta_j \mathbf{v}_{j-1}, && 1 \leqq j \leqq J, \\ &\alpha_J \mathbf{u}_J = \mathbf{v}_J, \quad \alpha_j \mathbf{u}_j = \mathbf{v}_j - \gamma_j \mathbf{u}_{j+1}, && J-1 \geqq j \geqq 0. \end{aligned} \tag{2.15d}$$

The indicated factorizations are not uniquely determined and do not exist in general. However, with a simple form of pivoting that does not introduce new zero elements we can justify such schemes. They become unique if we specify, for example:

$$\begin{aligned} &\text{Case (i)}: \ \delta_j = I; \quad \text{Case (iii)}: \ \delta_j \equiv \begin{pmatrix} 1 & & 0 \\ |||| & \ddots & \\ |||| & & 1 \end{pmatrix}, \ \alpha_j \equiv \begin{pmatrix} x & & \equiv \\ & \ddots & \\ 0 & & x \end{pmatrix}; \\ &\text{Case (ii)}: \ \alpha_j = I; \quad \text{Case (iv)}: \ \delta_j \equiv \begin{pmatrix} x & & 0 \\ |||| & \ddots & \\ |||| & & x \end{pmatrix}, \ \alpha_j \equiv \begin{pmatrix} 1 & & \equiv \\ & \ddots & \\ 0 & & 1 \end{pmatrix}. \end{aligned} \tag{2.16}$$

Cases (i) and (ii) yield standard block-tridiagonal factorizations while (iii) and (iv) are standard band-factorization or Gauss eliminations accounting for zero elements. When these schemes are valid it turns out that the β_j have the same zero pattern as the B_j in (2.13b) while the δ_j have zeros as in the C_j of (2.13b) only in Cases (i), (iii) and (iv). Surprisingly Case (ii) may fill in the γ_j and so should not be used. Operational counts (see Keller [26]) assuming the indicated zero patterns, show that (i) is preferable to (iii) when $p/n < 0.38$ (approximate) (see Varah [52]); (iii) is preferable to (iv) when $p < q$ and (i) is always preferable to (ii). In production codes we have generally used (i) or (iii) with some form of pivoting. To justify these procedures we have

THEOREM 2.17. *Let* $\mathbb{A}_h$ *of* (2.13a) *be nonsingular. Then with appropriate row interchanges within each of the n rows with indices k in*: $jn + p < k < (j+1)n + p$ *for each* $j = 0, 1, \cdots, J-1$, *the block-factorization* $\mathbb{A}_h = [\beta_j, I, 0][0, \alpha_j, C_j]$ *is valid* (Case (i) of (2.16)).

The proof is given in White [57] or Keller [26], where Cases (iii) and (iv) are also justified with a mixed partial row and column pivoting. ▧

Alternative schemes using maximal partial row and column pivoting have been studied by Lam [35] and Varah [53] to reduce the growth of the pivot elements. These procedures are quite stable but a bit more involved to program. Their operational counts are quite similar to those above. The study and development of effective procedures for solving systems of the form (2.12a,b), (2.13a,b) is very important; extensions to partially separated endpoints are also of great interest and so we consider them briefly.

For partially separated endpoints we get, in place of (2.12b) and (2.13a,b) a coefficient matrix of the form

$$(2.18)\qquad \hat{\mathbb{A}}_h \equiv \begin{pmatrix} \hat{C}_a & & & \\ L_1 & R_1 & & \\ & \ddots & \ddots & \\ & & L_J & R_J \\ \hat{C}_{ba} & & & \hat{C}_b \end{pmatrix} \equiv \begin{pmatrix} \hat{A}_0 & C_0 & & \\ B_1 & A_1 & C_1 & \\ & \ddots & \ddots & C_{J-1} \\ \hat{C}_J & & B_J & \hat{A}_J \end{pmatrix}.$$

Here $\hat{C}_a$ is $p \times n$ while $\hat{C}_{ba}$ and $\hat{C}_b$ are $q \times n$. These matrices have been used to replace C_a, C_{ba} and C_b in the partially separated endconditions (1.9a). The zero patterns in (2.13a,b) are maintained and $\hat{C}_J$ has the same form as the other C_j's. In place of the factorizations (2.15a) we seek a "bordered" form, say,

$$(2.19a)\qquad \hat{\mathbb{A}}_h = \begin{pmatrix} I & & & \\ \beta_1 & \ddots & & \\ & \ddots & \ddots & \\ \lambda_0 \cdots \lambda_{J-2} & & \beta_J & I \end{pmatrix} \begin{pmatrix} \alpha_0 & C_0 & & \\ & \ddots & \ddots & \\ & & \ddots & C_{J-1} \\ & & & \alpha_J \end{pmatrix} \equiv \hat{\mathbb{L}}_h \hat{\mathbb{U}}_h.$$

Thus the α_j, β_j and λ_j must satisfy

$$(2.19b)\qquad \alpha_0 = \hat{A}_0, \qquad \alpha_j = A_j - \beta_j C_{j-1}, \quad 1 \leqq j \leqq J-1,$$
$$\beta_j \alpha_{j-1} = B_j, \quad 1 \leqq j \leqq J-1,$$

$$(2.19c)\qquad \lambda_0 \alpha_0 = \hat{C}_J, \qquad \lambda_j \alpha_j = -\lambda_{j-1} C_{j-1}, \quad 1 \leqq j \leqq J-2,$$
$$\beta_J \alpha_{J-1} = B_J - \lambda_{J-2} C_{J-2}, \qquad \alpha_J = \hat{A}_J - \beta_J C_{J-1}.$$

If this procedure can be carried out, then the λ_j have the same zero pattern as the C_j (i.e., the first p-rows null). Note that (2.19a,b,c) is the analogue of Case (i) in (2.16); we could just as easily consider the other cases. To solve $\hat{\mathbb{A}}_h \mathbf{u}^h = \mathbf{F}$ we again use

$$\hat{\mathbb{L}}_h \mathbf{v}^h = \mathbf{F}, \qquad \hat{\mathbb{U}}_h \mathbf{u}^h = \mathbf{v}^h.$$

The recursions are the same as those in (2.15d) using $\delta_j \equiv I$, with the single exception of $\mathbf{v}_J$ which is now given by

$$\mathbf{v}_J = \mathbf{F}_J - \beta_J \mathbf{v}_{J-1} - \sum_{k=0}^{J-2} \lambda_k \mathbf{v}_k.$$

The additional operations caused by partial separation are easily obtained by simply adding all operations involving the nonzero elements of the λ_j matrices.

To justify this procedure we have

THEOREM 2.20. *Let the partially separated endpoint problem* (0.1a) *subject to* (1.9a) (*with the C's replaced by $\hat{C}$'s*) *have a unique solution. Let the one-step scheme* (2.12a) *be consistent for the initial value problem for* (0.1a). *Then if h_0 is sufficiently small, for all $h \leqq h_0$ the factorization* (2.19a) *of $\hat{\mathbb{A}}_h$ in* (2.18) *is valid with the same restricted partial pivoting of Theorem* 2.17.

Proof. Our hypothesis and Corollary 2.11 ensure that $\hat{\mathbb{A}}_h$ is nonsingular. Pick some $q \times n$ matrix $\hat{D}_a$, say, such that

$$\hat{M}_a \equiv \begin{pmatrix} \hat{C}_a \\ \hat{D}_a \end{pmatrix} \quad \text{is nonsingular.}$$

Define $C_a \equiv \hat{C}_a$ and $C_b \equiv \hat{D}_a Y^{-1}(b)$ so that the separated endpoint problem (0.1a) subject to (1.9a) with $C_{ba} \equiv 0$ and C_a, C_b as defined has a unique solution. (This follows since for this problem $Q = \hat{M}_a$.) Now the hypothesis and Corollary 2.11 imply $\mathbb{A}_h$ of (2.12b) is nonsingular. Thus Theorem 2.17 implies that, with the indicated form of partial pivoting,

$$\mathbb{A}_h = \mathbb{L}_h \mathbb{U}_h,$$

as in (2.15a) (with all $\delta_j \equiv I$ and all $\gamma_j \equiv C_j$).

Define

$$\mathbb{D}_h \equiv \hat{\mathbb{A}}_h - \mathbb{A}_h \equiv \begin{pmatrix} 0 \\ \hat{C}_{ba} 0 — 0 [\hat{C}_b - C_b] \end{pmatrix}$$

and use the indicated factorization to get

$$\begin{aligned} \hat{\mathbb{A}}_h &= \mathbb{A}_h + \mathbb{D}_h \\ &= \mathbb{L}_h \mathbb{U}_h + \mathbb{D}_h \\ &= (\mathbb{L}_h + \mathbb{D}_h \mathbb{U}_h^{-1}) \mathbb{U}_h \equiv \hat{\mathbb{L}}_h \hat{\mathbb{U}}_h. \end{aligned}$$ ▧

Of course the factorization in the last line of the above proof does not in general give exactly that obtained in (2.19a,b,c). The difference can occur only in the last $n \times n$ blocks of $\hat{\mathbb{L}}_h$ and $\hat{\mathbb{U}}_h$. But since the choice of $\hat{D}_a$ was arbitrary and we can insert the identity in the form $\mathbb{P}^{-1}\mathbb{P}$, where

$$\mathbb{P} \equiv \begin{pmatrix} I & & \\ & \ddots & \\ & & I & \\ & & & M \end{pmatrix}, \quad M \text{ is } n \times n \text{ nonsingular},$$

to get $\hat{\mathbb{L}}_h \mathbb{P}^{-1}$ and $\mathbb{P}\hat{\mathbb{U}}_h$ in place of $\hat{\mathbb{L}}_h$ and $\hat{\mathbb{U}}_h$, the ambiguity is resolved. It is clear, or should be on reflection, that all of the proposed schemes for factoring $\mathbb{A}_h$, including mixed partial pivoting as in Keller [26] or Varah [53], go over unaltered for the first J blocks, to that of $\hat{\mathbb{A}}_h$. In fact even if $\hat{\mathbb{A}}_h$ is singular but rank $\hat{C}_a = p$, the factorization works but α_J turns out to be singular. Our statement of Theorem 2.20 merely stressed the most important case.

One-step difference schemes can be chosen or used to get high order accuracy. The two simplest and most popular forms are the Box-scheme (or centered Euler) with

(2.21a) $L_j \equiv -h_j^{-1} I - \tfrac{1}{2} A(t_{j-\frac{1}{2}}), \quad R_j \equiv h_j^{-1} I - \tfrac{1}{2} A(t_{j-\frac{1}{2}}), \quad \mathbf{F}_j(h, \mathbf{f}) \equiv \mathbf{f}(t_{j-\frac{1}{2}}),$

and the trapezoidal rule with

(2.21b) $$L_j \equiv -h_j^{-1} I - \tfrac{1}{2} A(t_{j-1}), \quad R_j \equiv h_j^{-1} I - \tfrac{1}{2} A(t_j),$$
$$\mathbf{F}_j(h, \mathbf{f}) \equiv \tfrac{1}{2}[\mathbf{f}(t_j) + \mathbf{f}(t_{j-1})].$$

The former can be used easily with piecewise smooth coefficients provided all jump discontinuities lie on points of the net $\{t_j\}$. In both cases the truncation error expansions (2.5a) are of the form

$$\tau_j[\mathbf{y}(t)] = \sum_{\nu=1}^{N} h^{2\nu}\mathbf{T}_\nu[\mathbf{y}(\zeta_j)] + O(h^{2N+2}). \tag{2.21c}$$

Then it easily follows (see Keller [25]) that the errors have asymptotic expansions of the corresponding form

$$\mathbf{u}_j - \mathbf{y}(t_j) = \sum_{\nu=1}^{N} h^{2\nu}\mathbf{e}_\nu(t_j) + O(h^{2N+2}), \tag{2.21d}$$

where the $\mathbf{e}_\nu(t)$ are independent of h. However these results are valid only on special families of nets; for example, any arbitrary initial net $\{t_j\}$ with h sufficiently small which is then uniformly subdivided in any manner (i.e., all subintervals are divided in the same ratios). Then Richardson extrapolation or deferred corrections can be used (see § 2.4) to get two orders of accuracy improvement per application.

Higher order one-step schemes than those in (2.21a,b) are easily formulated. An interesting class of them, called gap- or Hermite-schemes can be devised by writing (0.1a) as

$$\mathbf{y}(t_j) - \mathbf{y}(t_{j-1}) = \int_{t_{j-1}}^{t_j} \mathbf{y}'(t)\,dt = \int_{t_{j-1}}^{t_j} [A(t)\mathbf{y}(t) - \mathbf{f}(t)]\,dt.$$

Then using Hermite interpolation of degree $2m + 1$ for the integrand, evaluating the higher derivatives up to order m by differentiating in (0.1a), we obtain one-step schemes of accuracy h^{2m+3}. Furthermore, the truncation error expansions *start* with the $(2m + 3)$rd derivative of $\mathbf{y}(t)$, hence the name "gap-schemes". If the derivatives of $A(t)$ and $\mathbf{f}(t)$ are easily evaluated these schemes can be quite effective. More details are contained in Keller [31] and White [57].

2.3. Difference methods for nonlinear problems. The general theory of difference schemes for linear boundary value problems, as in § 2.1, can be used to develop a fairly complete theory of difference methods for isolated solutions of nonlinear problems. In particular to approximate (1.20a,b) on the net $\{t_j\}_0^J$ we consider

$$\mathbf{G}_j(\mathbf{u}^h) \equiv \mathbf{G}_j(\mathbf{u}_0, \mathbf{u}_1, \cdots, \mathbf{u}_J; h) = \mathbf{0}, \qquad 1 \leqq j \leqq J, \tag{2.22a}$$

$$\mathbf{g}(\mathbf{u}_0, \mathbf{u}_J) = \mathbf{0}. \tag{2.22b}$$

The truncation errors are now defined as

$$\tau_0[\mathbf{z}] \equiv \mathbf{g}(\mathbf{z}(a), \mathbf{z}(b)), \quad \tau_j[\mathbf{z}] \equiv \mathbf{G}_j(\mathbf{z}^h) = \mathbf{G}_j(\mathbf{z}(t_0), \cdots, \mathbf{z}(t_J); h), \qquad 1 \leqq j \leqq J, \tag{2.23a}$$

and (2.22a,b) is *consistent* (*accurate*) *for* (1.20a,b) *of order* m if for each solution $\mathbf{z}(t)$:

$$\|\tau_j[\mathbf{z}]\| \leqq K_0 h^m \tag{2.23b}$$

for some constants $K_0 > 0$, $h_0 > 0$ and all $h \leqq h_0$. The scheme *is stable for* $\mathbf{z}^h$

provided there exist positive constants K_ρ, ρ and h_0 such that for all net functions $\mathbf{v}^h$, $\mathbf{w}^h$ in

$$S_\rho(\mathbf{z}^h) \equiv \{\mathbf{v}^h : \|\mathbf{v}_j - \mathbf{z}(t_j)\| \leqq \rho, 0 \leqq j \leqq J\}$$

and on all nets $\{t_j\}_0^J$ with $h \leqq h_0$,

$$\text{(2.24)} \qquad \|\mathbf{v}_j - \mathbf{w}_j\| \leqq K_\rho \max\{\|\mathbf{g}(\mathbf{v}_0, \mathbf{v}_J) - \mathbf{g}(\mathbf{w}_0, \mathbf{w}_J)\|, \max_{1 \leqq k \leqq J} \|\mathbf{G}_k(\mathbf{v}^h) - \mathbf{G}_k(\mathbf{w}^h)\|\}, \quad 0 \leqq j \leqq J.$$

Again it easily follows from consistency and stability, if the appropriate solutions exist, that

$$\text{(2.25)} \qquad \|\mathbf{z}(t_j) - \mathbf{u}_j\| \leqq K_0 K_\rho h^m, \qquad 0 \leqq j \leqq J.$$

To demonstrate stability of (2.22a,b) and also to give a constructive existence proof of solutions of (2.22a,b) we require some properties of the *linearized difference equations*:

$$\text{(2.26a)} \qquad \mathscr{L}_h[\mathbf{u}^h]\mathbf{v}_j \equiv \sum_{k=0}^{J} C_{jk}(h, \mathbf{u}^h)\mathbf{v}_k = \mathbf{0}, \qquad 1 \leqq j \leqq J;$$

$$\text{(2.26b)} \qquad \mathscr{B}_h[\mathbf{u}^h]\mathbf{v}^h \equiv B_a[\mathbf{u}^h]\mathbf{v}_0 + B_b[\mathbf{u}^h]\mathbf{v}_J = \mathbf{0}.$$

Here the $n \times n$ coefficient matrices are defined by

$$\text{(2.26c)} \qquad \begin{gathered} C_{jk}(h, \mathbf{u}^h) \equiv \frac{\partial \mathbf{G}_j(\mathbf{u}^h)}{\partial \mathbf{u}_k}, \qquad 1 \leqq j \leqq J, 0 \leqq k \leqq J, \\ B_a[\mathbf{u}^h] \equiv \frac{\partial \mathbf{g}(\mathbf{u}_0, \mathbf{u}_J)}{\partial \mathbf{u}_0}, \quad B_b[\mathbf{u}^h] \equiv \frac{\partial \mathbf{g}(\mathbf{u}_0, \mathbf{u}_J)}{\partial \mathbf{u}_J}. \end{gathered}$$

Stability is covered by

THEOREM 2.27. *Let* $\mathbf{z}(t)$ *be an isolated solution of* (1.20a,b). *Assume that the linear difference scheme*

$$\text{(2.27a)} \qquad \mathscr{L}_h[\mathbf{z}^h]\mathbf{v}_j = \mathbf{0}, \quad 1 \leqq j \leqq J, \quad \mathbf{v}_0 = \boldsymbol{\alpha},$$

is stable and consistent for the initial value problem (recall (1.23a,b))

$$\text{(2.27b)} \qquad \mathscr{L}[\mathbf{z}]\mathbf{v}(t) = \mathbf{0}, \quad a \leqq t \leqq b, \quad \mathbf{v}(a) = \boldsymbol{\alpha}.$$

Let $\mathscr{L}_h[\mathbf{w}^h]$ *be Lipshitz continuous in* $\mathbf{w}^h$ *about* $\mathbf{z}^h$; *that is,*

$$\text{(2.27c)} \qquad \|\mathscr{L}_h[\mathbf{z}^h] - \mathscr{L}_h[\mathbf{w}^h]\| \leqq K_L\|\mathbf{z}^h - \mathbf{w}^h\|$$

for all $\mathbf{w}^h \in S_\rho(\mathbf{z}^h)$, $h \leqq h_0$ *and some* $K_L > 0$, $\rho > 0$, $h_0 > 0$. *Similarly let*

$$\text{(2.27d)} \qquad \|B_x[\mathbf{z}^h] - B_x[\mathbf{w}^h]\| \leqq K_L \max\{\|\mathbf{z}(t_0) - \mathbf{w}_0\|, \|\mathbf{z}(t_J) - \mathbf{w}_J\|\}, \qquad x = a, b.$$

Then if ρ *is sufficiently small, the schemes* $\{\mathscr{L}_h[\mathbf{v}^h], \mathscr{B}_h[\mathbf{v}^h]\}$ *are stable for all* $\mathbf{v}^h \in S_\rho(\mathbf{z}^h)$ (*that is,* $\mathbb{A}_h[\mathbf{v}^h]$ *has a uniformly bounded inverse for all* $\mathbf{v}^h \in S_\rho(\mathbf{z}^h)$), *and the nonlinear scheme* (2.22a,b) *is stable for* $\mathbf{z}^h$.

Proof. With $\mathbb{A}_h[\mathbf{u}^h]$ defined by (2.4b) using the $n \times n$ matrices (2.26c) it follows from (2.27a,b,c,d), the isolation of $\mathbf{z}(t)$ and Corollary 2.11 that $\|\mathbb{A}_h^{-1}[\mathbf{z}^h]\| \leqq K$.

Then (2.27c,d) and the Banach lemma imply the uniform boundedness of $\|\mathbb{A}_h^{-1}[\mathbf{v}^h]\|$ for all $\mathbf{v}^h \in S_\rho(\mathbf{z}^h)$, provided ρ is sufficiently small.

If we write (2.22a,b) as the system

(2.28) $$\mathbf{\Phi}(\mathbf{u}^h) \equiv \begin{pmatrix} \mathbf{g}(\mathbf{u}_0, \mathbf{u}_J) \\ \vdots \\ \mathbf{G}_j(\mathbf{u}^h) \\ \vdots \end{pmatrix} = \mathbf{0},$$

the identity

(2.29a) $$\mathbf{\Phi}(\mathbf{v}^h) - \mathbf{\Phi}(\mathbf{w}^h) = \hat{\mathbb{A}}_h[\mathbf{v}^h, \mathbf{w}^h](\mathbf{v}^h - \mathbf{w}^h)$$

with

(2.29b) $$\hat{\mathbb{A}}_h[\mathbf{v}^h, \mathbf{w}^h] \equiv \int_0^1 \mathbb{A}_h[s\mathbf{v}^h + (1-s)\mathbf{w}^h]\, ds$$

follows from the assumed continuous differentiability of the $\mathbf{G}_j(\cdot)$ and $\mathbf{g}(\cdot, \cdot)$. But $\|\hat{\mathbb{A}}_h[\mathbf{v}^h, \mathbf{w}^h] - \mathbb{A}_h[\mathbf{z}^h]\| \leqq K_L\rho$ for all $\mathbf{v}^h$, $\mathbf{w}^h$ in $S_\rho(\mathbf{z}^h)$. Then

(2.29c) $$\|\hat{\mathbb{A}}_h^{-1}[\mathbf{v}^h, \mathbf{w}^h]\| \leqq K_\rho \equiv K/[1 - \rho K_L K],$$

and the stability of (2.22a,b) follows from

$$(\mathbf{v}^h - \mathbf{w}^h) = \hat{\mathbb{A}}_h^{-1}[\mathbf{v}^h, \mathbf{w}^h](\mathbf{\Phi}(\mathbf{v}^h) - \mathbf{\Phi}(\mathbf{w}^h)). \quad ▨$$

Newton's method can be used to prove the existence of a unique solution of the nonlinear difference equations (2.22a,b) in $S_{\rho_0}(\mathbf{z}^h)$, for some $\rho_0 \leqq \rho$. The iterates $\{\mathbf{u}_\nu^h\}$ are defined, using (2.28), as

(2.30a) $$\mathbf{u}_0^h \in S_{\rho_0}(\mathbf{z}^h),$$

(2.30b) $$\mathbb{A}_h[\mathbf{u}_\nu^h](\mathbf{u}_{\nu+1}^h - \mathbf{u}_\nu^h) = -\mathbf{\Phi}(\mathbf{u}_\nu^h), \qquad \nu = 0, 1, 2, \cdots.$$

Equivalently the iterates can be defined by

(a) $$\mathbf{u}_{\nu+1}^h = \mathbf{u}_\nu^h + \delta\mathbf{u}_\nu^h, \qquad \nu = 0, 1, 2, \cdots,$$

where $\delta\mathbf{u}_\nu^h$ is the solution of the linearized difference equations:

(b) $$\mathscr{L}_h[\mathbf{u}_\nu^h]\, \delta\mathbf{u}_{\nu j} = -\mathbf{G}_j(\mathbf{u}_\nu^h), \qquad 1 \leqq j \leqq J,$$

(c) $$\mathscr{B}_h[\mathbf{u}_\nu^h]\, \delta\mathbf{u}_\nu^h = -\mathbf{g}(\mathbf{u}_{\nu 0}, \mathbf{u}_{\nu J}).$$

The actual computation of the $\delta\mathbf{u}_\nu^h$ can be carried out using the stable and efficient schemes discussed in § 2.2 when one-step schemes are used with separated endpoint conditions.

THEOREM 2.31. *Let the hypothesis of Theorem 2.27 hold. Let* (2.22a,b) *be accurate of order m for* (1.20a,b). *Then for some $\rho_0 \leqq \rho$ and h_0 sufficiently small* (2.22a,b) *has for all $h \leqq h_0$, a unique solution $\mathbf{u}^h \in S_\rho(\mathbf{z}^h)$. This solution can be computed by Newton's method* (2.30a,b) *which converges quadratically.*

Proof. We simply verify the hypothesis of the Newton–Kantorovich Theorem 1.28 using $\boldsymbol{\varphi} \equiv \boldsymbol{\Phi}(\mathbf{u}^h)$ of (2.28) and $Q \equiv \mathbb{A}_h(\mathbf{u}^h)$ with elements in (2.26c). From (2.27c,d) we can use $\gamma = K_L$ to satisfy (1.28c). Also (2.29c) implies (1.28b) with $\beta = K_{\rho_0}$. To estimate α as in (1.28a) we write, recalling (2.29a),

$$\mathbb{A}_h^{-1}[\mathbf{u}_0^h]\boldsymbol{\Phi}(\mathbf{u}_0^h) = \mathbb{A}_h^{-1}[\mathbf{u}_0^h][\boldsymbol{\Phi}(\mathbf{z}^h) + \hat{\mathbb{A}}_h[\mathbf{u}_0^h, \mathbf{z}^h](\mathbf{u}_0^h - \mathbf{z}^h)].$$

However,

$$\mathbb{A}_h^{-1}[\mathbf{u}_0^h]\hat{\mathbb{A}}_h[\mathbf{u}_0^h, \mathbf{z}^h] = \mathbb{I} + \mathbb{A}_h^{-1}[\mathbf{u}_0^h][\hat{\mathbb{A}}_h[\mathbf{u}_0^h, \mathbf{z}^h] - \mathbb{A}_h[\mathbf{u}_0^h]],$$

and so

$$\|\mathbb{A}_h^{-1}[\mathbf{u}_0^h]\hat{\mathbb{A}}_h[\mathbf{u}_0^h, \mathbf{z}^h]\| \leqq 1 + K_{\rho_0}K_L\rho_0/2 \equiv C.$$

Thus we get, recalling that $\boldsymbol{\Phi}(\mathbf{z}^h)$ is the vector of truncation errors,

$$\|\mathbb{A}_h^{-1}(\mathbf{u}_0^h)\boldsymbol{\Phi}(\mathbf{u}_0^h)\| \leqq K_{\rho_0}K_0h_0^m + C\rho_0 \equiv \alpha.$$

Now taking h_0 and ρ_0 sufficiently small we are assured that

$$\alpha\beta\gamma = (K_{\rho_0}K_0h_0^m + C\rho_0)K_{\rho_0}K_L \leqq \tfrac{1}{2}$$

and

$$\rho_0\beta\gamma = \rho_0K_{\rho_0}K_L \leqq 1 - \sqrt{1 - 2\alpha\beta\gamma}. \quad ▨$$

The difficulty in finding a suitable initial guess $\mathbf{u}_0^h$ in some sphere about the exact solution $\mathbf{z}^h$ is generally much less for finite difference methods than it is for shooting methods on the same problem. The reason is to be found in the close analogy between parallel shooting methods (see § 1.3, § 1.5) and finite difference methods. In fact, J, the number of mesh intervals for finite differences is usually much larger than the number of shooting intervals one would normally think of employing. However, a closer inspection of our finite difference schemes and parallel shooting analysis shows that the same procedures can frequently belong to both classifications!

2.4. Richardson extrapolation and deferred corrections. The accuracy of a given finite difference scheme may frequently be improved in a very efficient manner. So much so in fact that the most effective difference procedures are based on one-step schemes of second order accuracy enhanced by Richardson extrapolation or deferred corrections. The basis for these techniques is the existence of an asymptotic expansion of the *local truncation error*, say, of the form

(2.32a) $$\boldsymbol{\tau}_j[\mathbf{z}] = \sum_{\nu=1}^{N} h^{m_\nu}\mathbf{T}_\nu[\mathbf{z}(\zeta_j)] + O(h^{m_{N+1}}).$$

Of course $0 < m_1 < \cdots < m_{N+1}$ and we assume that for a subset of the subscripts $\{\nu_k\}$, say, with $\nu_1 = 1$,

(2.32b) $$m_{\nu_k} = km_1, \qquad k = 1, 2, \cdots, L.$$

For example, the Box-scheme and trapezoidal rule applied to nonlinear problems still yield truncation expansions of the form (2.21c). Here $m_1 = 2$ and $\nu_k = k$. In other schemes, the first gap-scheme beyond trapezoidal rule for instance, $m_1 = 4$ and $\nu_k = 2k - 1$.

When (2.32a) holds for a stable scheme on a uniform net it is not difficult to show that the errors in the numerical solution also have asymptotic error expansions of the form

$$\text{(2.32c)} \qquad \mathbf{u}_j - \mathbf{z}(t_j) = \sum_{\nu=1}^{N} h^{m_\nu}\mathbf{e}_\nu(t_j) + O(h^{m_{N+1}}).$$

Here the functions $\mathbf{e}_\nu(t)$ are independent of h and they are defined as the solutions of linear boundary value problems. For linear problems, for example, they satisfy

$$\text{(2.33a)} \qquad \mathscr{L}\mathbf{e}_\nu(t) = \mathbf{T}_\nu[\mathbf{z}(t)] + \sum_{k=1}^{\nu-1} \mathbf{T}_k[\mathbf{e}_{\nu-k}(t)], \qquad a \leqq t \leqq b,$$

$$\text{(2.33b)} \qquad \mathscr{B}\mathbf{e}_\nu(t) = \mathbf{0}.$$

In the nonlinear case, $\mathscr{L}$, $\mathscr{B}$ are replaced by $\mathscr{L}[\mathbf{z}]$, $\mathscr{B}[\mathbf{z}]$ and the inhomogeneous terms are slightly more complicated (as a result of the nonlinearity). Nonuniform nets of appropriate families are easily included as shown in Keller [25].

The knowledge that an expansion of the form (2.32c) exists enables us to apply Richardson extrapolation on a sequence of refined nets. Each successive refinement requires only the solution of the difference scheme (2.22a,b) on the latest net. At the kth refinement the error term of order h^{m_k} is removed; the details are rather well known being based on iterative extrapolation so we do not repeat them here. It should be stressed however that successive halvings of the net spacing is an unfortunate choice of refinement, assuming that some fixed maximum number of extrapolations, say, 10 or less, are to be done. Rather an arithmetic reduction with $h^{(k)} = (k + 1)^{-1}h^{(0)}$ is much more efficient. In anticipation of Richardson extrapolation we must use an appropriately severe convergence criterion for the iterative solution of the difference equations. That is, the iteration error (as well as the truncation error) must be reduced beyond what is anticipated of the final truncation error. The particular effectiveness of quadratic convergence, and hence Newton's method, in this regard is very important (see, for example, Keller [26], [31]). Specifically if the initial guess $\mathbf{u}_0^h$ is within $O(h)$ of the solution at the kth stage, then at most $\nu_k = \log_2 m_k$ iterations are required; that is, h^8 accuracy would require 3 Newton iterations.

The deferred correction technique need only be based on (2.32a) rather than on (2.32c), as is frequently stated. In effect this method determines successively higher order difference schemes by including sufficiently accurate approximations to the leading terms in (2.32a) in the latest scheme. In more detail, suppose (2.32a,b) hold. We seek a set of net functions $\{\mathbf{u}_j(k)\}_0^J \equiv \mathbf{u}^h(k)$ such that

$$\text{(2.34a)} \qquad \|\mathbf{z}(t_j) - \mathbf{u}_j(k)\| \leqq M_0 h^{m_{\nu_k}}, \qquad k = 1, 2, \cdots, N.$$

We assume that the net $\{t_j\}$ and operators $\mathbf{T}_\nu[\mathbf{z}(t)]$ are such that for some "difference" operators $\mathbf{T}_{\nu,j,k}[\,\cdot\,]$ and net functions $\mathbf{u}^h(k)$, satisfying (2.34a),

$$h^{m_\nu}\|\mathbf{T}_\nu[\mathbf{z}(t_j)] - \mathbf{T}_{\nu,j,k}[\mathbf{u}^h(k)]\| \leqq M_1 h^{m_{\nu_{k+1}}}, \quad 1 \leqq \nu \leqq \nu_{k+1}, \quad k = 1, 2, \cdots, N. \tag{2.34b}$$

Then the set of net functions can be determined recursively by solving, in order,

$$\mathbf{G}_j(\mathbf{u}^h(k+1); h) = \sum_{\nu=1}^{\nu_{k+1}} h^{m_\nu}\mathbf{T}_{\nu,j,k}[\mathbf{u}^h(k)], \qquad 1 \leqq j \leqq J, \quad k = 1, 2, \cdots, N, \tag{2.35a}$$

$$\mathbf{g}(\mathbf{u}_0^h(k+1), \mathbf{u}_J^h(k+1)) = \mathbf{0}, \qquad k = 1, 2, \cdots, N. \tag{2.35b}$$

To show that (2.34a) holds for the net functions defined in (2.35a,b) we use the stability of the scheme (2.22a,b), (2.32a) and (2.34a,b) to get

$$\begin{aligned} \|\mathbf{z}(t_j) - \mathbf{u}_j(k+1)\| &\leqq \max_i \|\mathbf{G}_i(\mathbf{z}^h) - \mathbf{G}_i(\mathbf{u}^h(k))\| \\ &\leqq \max_i \sum_{\nu=1}^{\nu_{k+1}} h^{m_\nu}\|\mathbf{T}_\nu[\mathbf{z}(t_i)] - \mathbf{T}_{\nu,i,k}[\mathbf{u}^h(k)]\| \\ &\quad + \sum_{\nu=\nu_{k+1}+1}^{N} h^{m_\nu}\|\mathbf{T}_\nu[\mathbf{z}(t_i)]\| + O(h^{m_{N+1}}) \\ &= O(h^{m_{\nu_{k+1}}}). \end{aligned}$$

Thus the result follows by induction.

The important practical problem is thus reduced to constructing the difference approximations to the truncation terms. This is done in some generality by Pereyra [42] where "centered" formulae are used at the interior netpoints and uncentered formulae are used near the endpoints. This causes considerable complications in coding the scheme as well as in justifying it rigorously. However, all these difficulties can be circumvented by simply extending the net $\{t_j\}$ to include points external to $[a, b]$ and computing the original solution on the complete extended net. Then on each successive improvement, $\mathbf{u}_j(k)$, the approximations need to be computed over fewer external points. However, this procedure assumes that the maximum intended order of correction is known in advance and that the correspondingly extended net is used.

Another theoretically simple way to determine appropriate operators $\mathbf{T}_{\nu,j,k}[\mathbf{u}^h(k)]$ is to define and compute approximations to the μth derivative, $\mathbf{z}^{(\mu)}(t_j)$, of the exact solution by differentiation of the equation (1.20a). Thus since

$$\mathbf{z}^{(\mu+1)}(t) \equiv \left(\frac{d}{dt}\right)^{\mu+1}\mathbf{z}(t) = \left(\frac{d}{dt}\right)^{\mu}\mathbf{f}(t, \mathbf{z}(t))$$

we define, starting with $\mathbf{u}_j^{(0)}(k) \equiv \mathbf{u}_j(k)$,

$$\mathbf{u}_j^{(\mu+1)}(k) \equiv \left[\left(\frac{d}{dt}\right)^{\mu}\mathbf{f}(t, \boldsymbol{\varphi}(t))\right]\Bigg|\, t = t_j;\ \boldsymbol{\varphi}^{(\nu)}(t_j) = \mathbf{u}_j^{(\nu)}(k), \qquad 0 \leqq \nu \leqq \mu. \tag{2.36}$$

It follows by induction from (2.34a,b) and the assumed smoothness of $\mathbf{f}(t, \mathbf{z})$ that

$$\|\mathbf{z}^{(\mu)}(t_j) - \mathbf{u}_j^{(\mu)}(k)\| \leqq O(h^{m_{\nu k}}), \qquad \mu = 0, 1, 2, \cdots.$$

The number of function evaluations in (2.36) grows extremely fast and so it is unlikely to be practical for a general purpose code.

Finally we point out that the first correction, of order h^{m_1} to get an $O(h^{m_2})$ accurate solution, can be done by solving only one *linear* difference system. In fact it is just equivalent to one more Newton iterate when the $\mathbf{T}_{\nu,j,1}[\mathbf{u}^h(1)]$ have been evaluated for $\nu = 1, 2, \cdots, \nu_2$. This idea was first used by Fox [14] and has also been used by Pereyra [42].

2.5. Collocation and implicit Runge–Kutta. Collocation has been studied and used increasingly for two-point problems. We shall show in some generality that it is identical to difference schemes based on implicit Runge–Kutta methods when $C^0[a, b]$ piecewise polynomials are used in the standard fashion. Then our theory of § 2.3 and the work of Axelsson [3] shows the stability of the method and its super-convergence properties. This work follows closely that of Weiss [56] who first carried out such an analysis.

To define the schemes a basic net $\{t_j\}_0^J$ with

$$t_0 = a; \quad t_j = t_{j-1} + h_j, \quad 1 \leqq j \leqq J, \quad t_J = b, \tag{2.37a}$$

is refined in each interval to get $\{t_{j\nu}\}_{0,0}^{J,N}$, where

$$t_{j\nu} = t_{j-1} + \theta_\nu h_j, \qquad 0 \leqq \nu \leqq N, 0 \leqq \theta_0 < \theta_1 < \cdots < \theta_N \leqq 1. \tag{2.37b}$$

Using the θ_ν as nodes on [0, 1] we define $N + 1$ interpolatory quadrature rules:

$$\int_0^{\theta_\nu} \psi(s)\, ds \approx \sum_{k=0}^{N} w_{\nu k}\psi(\theta_k), \qquad 0 \leqq \nu \leqq N. \tag{2.37c}$$

If $\theta_0 > 0$ and $\theta_N < 1$ we can define an $(N + 2)$nd rule $\{w_{N+1,k}\}$ over [0, 1] which has maximum degree of precision $2N + 1$, i.e., Gaussian quadrature. If $\theta_0 = 0$ and $\theta_N = 1$ the Nth rule can have degree of precision $2N - 1$, i.e., Lobatto quadrature. If $\theta_0 > 0$ and $\theta_n = 1$ (or conversely) the Nth rule can have degree of precision $2N$, i.e., Radau quadrature. In the Lobatto case all $w_{0k} \equiv 0$ and we only have N nontrivial quadrature formulas in (2.37c). We shall only employ the latter case, i.e., $\theta_0 = 0$, $\theta_N = 1$, in our present study. The others are easily treated in a similar manner (see Weiss [56]).

The difference scheme for (1.20a,b) based on the quadrature rules (2.37c) is formulated as:

$$\mathbf{u}_{j\nu} - \mathbf{u}_{j0} = h_j \sum_{k=0}^{N} w_{\nu k}\mathbf{f}(t_{jk}, \mathbf{u}_{jk}), \qquad 1 \leqq \nu \leqq N, 1 \leqq j \leqq J, \tag{2.38a}$$

$$\mathbf{g}(\mathbf{u}_{10}, \mathbf{u}_{JN}) = \mathbf{0}. \tag{2.38b}$$

If $\mathbf{u}_{j0}$ is known we see that (2.38a) for $1 \leqq \nu \leqq N$, j fixed, consists of a system of nN equations for the nN components of $\mathbf{u}_{j\nu}$, $1 \leqq \nu \leqq N$. This is an implicit Runge–Kutta scheme of the type studied by Axelsson [3]. If we imagine these equations solved for each $\mathbf{u}_{jN}$ in terms of $\mathbf{u}_{j0}$ the scheme can be viewed as a one-step scheme. More important it is stable for the initial value problem and the accuracy at the nodes $\{t_j\}_0^J$ is the same as that of the quadrature scheme over [0, 1]. In the Lobatto case we get $O(h^{2N+1})$ accuracy. The linearized version of (2.38a,b) clearly satisfies the conditions (2.27a,b,c,d) of Theorem 2.27. Then it follows from this result, the results of Axelsson and Theorem 2.31 that the scheme (2.38a,b) is stable and accurate to $O(h^{2N+1})$ on the net $\{t_j\}$ for isolated solutions of (1.20a,b). Similar results hold for the Radau and Gaussian cases. Now we turn to the relation with collocation.

To approximate the solution of (1.20a,b) by collocation we introduce a piecewise polynomial function, $\mathbf{v}(t)$, which on each interval $[t_{j-1}, t_j]$ of (2.37a) is a polynomial of degree $N + 1$, say:

$$\text{(2.39)} \qquad \mathbf{v}(t) = \mathbf{p}_j(t) \equiv \sum_{k=0}^{N+1} \mathbf{a}_{jk}(t - t_{j-1})^k, \qquad t_{j-1} \leqq t \leqq t_j, \quad 1 \leqq j \leqq J.$$

We seek to determine the coefficients $\{\mathbf{a}_{jk}\}$ by collocating (1.20a) with $z(t)$ replaced by $\mathbf{v}(t)$ at each of the $t_{j\nu}$, by imposing the boundary conditions (1.20b) on $\mathbf{v}(t)$ and by requiring continuity of $\mathbf{v}(t)$ at each internal netpoint t_j. This yields the system:

$$\text{(2.40a)} \qquad \mathbf{p}_j'(t_{j\nu}) = \mathbf{f}(t_{j\nu}, \mathbf{p}_j(t_{j\nu})), \qquad 0 \leqq \nu \leqq N, \quad 1 \leqq j \leqq J,$$

$$\text{(2.40b)} \qquad \mathbf{g}(\mathbf{p}_1(t_{10}), \mathbf{p}_J(t_{JN})) = \mathbf{0},$$

$$\text{(2.40c)} \qquad \mathbf{p}_j(t_j) = \mathbf{p}_{j-1}(t_j), \qquad 1 \leqq j \leqq J - 1.$$

In (2.39) there are $(N + 2)J$ unknown coefficient vectors $\mathbf{a}_{jk}$ and in (2.40a,b,c) there are the same number of vector equations. The basic relation between collocation and our implicit Runge–Kutta scheme is given in

THEOREM 2.41. (A) *Let* $\mathbf{p}_j(t)$ *satisfy* (2.40a,b,c). *Then the* $\mathbf{u}_{j\nu} \equiv \mathbf{p}_j(t_{j\nu})$ *satisfy* (2.38a,b).

(B) *Let* $\mathbf{u}_{j\nu}$ *satisfy* (2.38a,b). *Define* $\hat{\mathbf{p}}_j(t)$ *to be the unique polynomials of degree* $N + 1$ *which interpolate the* $\mathbf{u}_{j\nu}$, $0 \leqq \nu \leqq N$ (*i.e., satisfy* $\hat{\mathbf{p}}_j(t_{j\nu}) = \mathbf{u}_{j\nu}$) *and satisfy* (2.40a) *for* $\nu = 0$. *Then the* $\hat{\mathbf{p}}_j(t)$ *satisfy* (2.40a,b,c).

Proof. Since the $\mathbf{p}_j(t)$ are of degree $N + 1$ and the quadrature schemes (2.37c) being interpolatory have degrees of precision at least N, we must have that

$$\begin{aligned} \mathbf{p}_j(t_{j\nu}) - \mathbf{p}_j(t_{j0}) &= \int_{t_{j0}}^{t_{j\nu}} \mathbf{p}_j'(s)\, ds \\ &= h_j \sum_{k=0}^{N} w_{\nu k} \mathbf{p}_j'(t_{jk}). \end{aligned}$$

Using (2.40a) we get that (2.38a) is satisfied by $\mathbf{p}_j(t_{j\nu}) = \mathbf{u}_{j\nu}$. Trivially (2.40b) implies (2.38b) so our collocation solution does satisfy the difference scheme and part (A) is proven.

In a similar fashion, since the $\hat{\mathbf{p}}_j(t)$ interpolate the $\mathbf{u}_{j\nu}$ and are of degree $N+1$,

$$\mathbf{u}_{j\nu} - \mathbf{u}_{j0} = \hat{\mathbf{p}}_j(t_{j\nu}) - \hat{\mathbf{p}}_j(t_{j0}) = \int_{t_{j0}}^{t_{j\nu}} \hat{\mathbf{p}}_j'(s)\,ds$$

$$= h_j \sum_{k=0}^{N} w_{\nu k}\hat{\mathbf{p}}_j'(t_{jk}), \qquad 1 \leqq \nu \leqq N.$$

Now use (2.38a) to get that

$$\sum_{k=0}^{N} w_{\nu k}[\hat{\mathbf{p}}_j'(t_{jk}) - \mathbf{f}(t_{jk}, \hat{\mathbf{p}}_j(t_{jk}))] = \mathbf{0}, \qquad 1 \leqq \nu \leqq N.$$

However, the $n \times n$ matrix $\Omega = (w_{\nu k})$, $1 \leqq \nu, k \leqq N$ is nonsingular[1] so we get that (2.40a) is satisfied by the $\hat{\mathbf{p}}_j(t)$ for $1 \leqq \nu \leqq N$, $1 \leqq j \leqq J$. For $\nu = 0$ these relations were imposed in defining the $\hat{\mathbf{p}}_j(t)$. The continuity (2.40c) follows since $\hat{\mathbf{p}}_j(t_{j0}) = \mathbf{u}_{j0} \equiv \mathbf{u}_{j-1N} = \hat{\mathbf{p}}_{j-1}(t_{j-1N})$. The boundary conditions are similarly satisfied as a result of interpolating the difference solution. ▧

We do not believe that implicit Runge–Kutta schemes are as efficient as the same order one-step schemes using the techniques of § 2.4 (see Keller [31]). Thus we do not advocate $C^0[a, b]$ collocation either. However, Russell [48], [49] has shown that collocation with smoother piecewise polynomials is much more efficient, so collocation cannot be ruled out in general.

[1] To see this, apply the rules (2.37c) to $\psi(s) \equiv s^m$, $m = 1, \cdots, N$, recalling $\theta_0 = 0$ and degree of precision at least N. Then Ω times a nonsingular Vandermonde matrix equals another.

CHAPTER 3

Eigenvalue Problems

3.1. Analytic eigenvalue problems. We consider very general eigenvalue-eigenfunction problems of the form

$$\mathscr{L}[\lambda]\mathbf{y}(t) \equiv \mathbf{y}' - A(t, \lambda)\mathbf{y} = \mathbf{0}, \qquad a \leqq t \leqq b, \tag{3.1a}$$

$$\mathscr{B}[\lambda]\mathbf{y}(t) \equiv B_a(\lambda)\mathbf{y}(a) + B_b(\lambda)\mathbf{y}(b) = \mathbf{0}. \tag{3.1b}$$

We shall frequently consider the special case of partially separated endconditions, say, in which

$$B_a(\lambda) \equiv \begin{pmatrix} C_a(\lambda) \\ C_{ba}(\lambda) \end{pmatrix}, \quad B_b(\lambda) \equiv \begin{pmatrix} 0 \\ C_b(\lambda) \end{pmatrix}, \quad \begin{cases} C_a \text{ is } p \times n \text{ of rank } p, \\ C_b \text{ is } q \times n \text{ of rank } q. \end{cases} \tag{3.1c}$$

Here $p + q = n$ and we assume $p \geqq q$; otherwise we use the complementary form of (1.9b). In all cases we assume that:

(i) $\qquad \operatorname{rank}(B_a(\lambda), B_b(\lambda)) = n$ for all λ;

(ii) $\qquad A(t, \lambda), B_a(\lambda), B_b(\lambda)$ are analytic in λ uniformly on $a \leqq t \leqq b$.

In § 4.2 the case of $b \to \infty$ is briefly considered for separated endpoints.

The fundamental solution matrix $Y(t, \lambda)$ defined by:

$$\mathscr{L}[\lambda]Y(t; \lambda) = 0, \tag{3.2a}$$

$$Y(a; \lambda) = I, \tag{3.2b}$$

is known to be analytic in λ uniformly in t on $[a, b]$. Since for any fixed λ, every solution of (3.1a) has the form

$$\mathbf{y}(t; \lambda) = Y(t; \lambda)\boldsymbol{\xi}, \tag{3.3a}$$

it will satisfy the boundary conditions (3.1b) if and only if

$$Q(\lambda)\boldsymbol{\xi} \equiv [B_a(\lambda) + B_b(\lambda)Y(b; \lambda)]\boldsymbol{\xi} = \mathbf{0}. \tag{3.3b}$$

Since $Q(\lambda)$ is analytic in λ we deduce the basic result:

THEOREM 3.4. *Either every λ is an eigenvalue of* (3.1a,b) *or else there are at most a denumerable set of distinct eigenvalues $\{\lambda_k\}$ with no finite accumulation point. In the latter case each λ_k has geometric multiplicity (i.e., number of linearly independent eigenfunctions)*:

$$r_k = \dim \mathscr{N}[Q(\lambda_k)] \leqq n. \tag{3.4a}$$

In the partially separated endpoint case, with $p \geqq q$,

(3.4b) $$r_k \leqq q.$$

Proof. Since (3.3b) has nontrivial solutions if and only if

(3.4c) $$\Delta(\lambda) \equiv \det Q(\lambda) = 0,$$

the first part of the theorem follows from the analyticity of $\Delta(\lambda)$. The number of independent solutions $\boldsymbol{\xi} = \boldsymbol{\xi}(\lambda_k)$ of (3.3b) is the geometric multiplicity of λ_k and hence (3.4a) follows. For separated endpoints as in (3.1c) we observe that (3.3b) takes the form

$$Q(\lambda)\boldsymbol{\xi} = \begin{pmatrix} C_a(\lambda) \\ [C_{ba}(\lambda) + C_b(\lambda)Y(b, \lambda)] \end{pmatrix} \boldsymbol{\xi} = \mathbf{0}.$$

Since $C_a(\lambda)$ has rank p it follows that $\dim \mathcal{N}[Q(\lambda)] \leqq n - p = q$. ▨

As a very practical matter in the separated endpoint case we note that for some $n \times n$ nonsingular analytic matrix

$$R(\lambda) \equiv (R_p(\lambda), R_q(\lambda)), \quad R_p \text{ is } n \times p, \quad R_q \text{ is } n \times q,$$

we have

$$C_a(\lambda)R(\lambda) = (I_{p \times p}, 0_{p \times q}).$$

Then writing

$$Q(\lambda)\boldsymbol{\xi} \equiv [Q(\lambda)R(\lambda)][R^{-1}(\lambda)\boldsymbol{\xi}] = Q(\lambda)R(\lambda)\boldsymbol{\eta},$$

we find that (3.3b) is equivalent to

(3.5a) $$Q_q(\lambda)\mathbf{s} \equiv [C_{ba}(\lambda) + C_b(\lambda)Y(b, \lambda)]R_q(\lambda)\mathbf{s} = \mathbf{0}.$$

Here we have set

(3.5b) $$\boldsymbol{\eta} \equiv R^{-1}(\lambda)\boldsymbol{\xi} = \begin{pmatrix} \mathbf{0} \\ \mathbf{s} \end{pmatrix}, \qquad \mathbf{s} \in \mathbb{E}^q.$$

The eigenvalue equation, replacing (3.4c), is now

(3.5c) $$\Delta_q(\lambda) \equiv \det Q_q(\lambda) = 0.$$

We assume that the algebraic and geometric multiplicity of each eigenvalue are the same. Our numerical methods for such eigenvalue problems will lead to a study of the roots of perturbations of (3.4c) or (3.5c), say, in the form

(3.6a) $$\Delta(\lambda) + \sum_{\nu=1}^{N} h^{\nu m}\delta_\nu(\lambda) + O(h^{(N+1)m}) = 0.$$

Here $m > 0$, the $\delta_\nu(\lambda)$ are also analytic in λ and we are interested in the behavior of the roots of (3.6a) as $h \to 0$. It is well known that for each simple root of (3.4c) there is a corresponding simple root of (3.6a) with the same asymptotic h dependence as in (3.6a). This is not true of multiple roots however. For example if λ_0 is an r-fold root of (3.4c), then it easily follows that there are r roots of (3.6a), say, $\lambda_{0,k}(h)$, $k = 1, 2, \cdots, r$, with the forms

$$\lambda_{0,k}(h) = \lambda_0 + a_{k1}h^{m/p} + a_{k2}h^{2m/p} + \cdots + O(h^{(N+1)m/p}). \tag{3.6b}$$

However, we shall show that in fact

$$\bar{\lambda}_0(h) \equiv \frac{1}{r}\sum_{k=1}^{r} \lambda_{0,k}(h) = \lambda_0 + \sum_{\nu=1}^{N} a_\nu h^{\nu m} + O(h^{(N+1)m}), \tag{3.6c}$$

that is, the original asymptotic form is recovered by the average of the roots. Similar results, but for more specialized eigenvalue problems, were first obtained by Kreiss [33]. Subsequently Atkinson [1] and Bramble and Osborne [8] have generalized such estimates to other nonself-adjoint linear operator equations. In all of these results, however, the eigenvalue parameter enters in the standard linear fashion. Our proof is but a simple combination of the Residue Theorem and the Principal of the Argument.

Suppose $f(w, z)$ is analytic in w and z on $|w - w_0| \leqq a$, $|z| \leqq a$. Let w_0 be an r-fold root of $f(w, 0) = 0$. Then by Rouche's theorem, for some $\delta > 0$, the equation $f(w, z) = 0$ has for each z in $|z| < \delta$, r roots $w_k(z)$ that satisfy $w_k(0) = w_0$, $k = 1, 2, \cdots, r$.

THEOREM 3.7. *The average of the roots $w_k(z)$ has the form*

$$\bar{w}(z) \equiv \frac{1}{r}\sum_{k=1}^{r} w_k(z) = w_0 + g(z), \tag{3.7a}$$

where $g(z) = g_0 z + g_1 z^2 + \cdots$ is analytic about $z = 0$.

Proof. Pick δ_0 and δ_1 so small that for $|z| \leqq \delta_0$ all the roots $w_\nu(z)$ are in $|w - w_0| < \delta_1$. Further no other roots of $f(w, z) = 0$ are in this δ_1-disk about w_0. Then for $z \neq 0$,

$$r\bar{w}(z) \equiv \sum_{k=1}^{r} w_k(z) = \frac{1}{2\pi i}\oint_{|w-w_0|=\delta_1} w\left[\frac{d}{dw}\log f(w, z)\right] dw. \tag{3.7b}$$

For $z = 0$, since w_0 is an r-fold root,

$$rw_0 = \frac{1}{2\pi i}\oint_{|w-w_0|=\delta_1} w\left[\frac{d}{dw}\log f(w, 0)\right] dw. \tag{3.7c}$$

However,

$$\log f(w, z) = \log f(w, 0) + \phi(w, z),$$

where on $|w - w_0| = \delta_1$,

$$\begin{aligned}\phi(w, z) &= \log\left(\frac{f(w, z)}{f(w, 0)}\right) = \log\left(\frac{f(w, 0) + zf_z(w, 0) + \cdots}{f(w, 0)}\right)\\ &= z\varphi_1(w) + z^2\varphi_2(w) + \cdots.\end{aligned} \tag{3.7d}$$

The result (3.7a) clearly follows from (3.7c,d) in (3.7b). ▨

To use this result for our numerical work we must first generalize the analytic dependence on z to an asymptotic dependence as in

$$f(w, z) \sim f(w, 0) + zf_1(w) + z^2 f_2(w) + \cdots .$$

But if all dependence on w is analytic, then our arguments go over essentially unchanged and $g(z)$ in (3.7a) can be replaced by a corresponding asymptotic power series in z. Now (3.6c) follows with $z \equiv h^m$, $w \equiv \lambda$, etc.

3.2. Shooting for eigenvalues. We examine shooting methods for approximating the eigenvalues and eigenfunctions of (3.1a,b,c). (If the boundary conditions are *not* partially separated, then simply taking $q = n$, $R(\lambda) \equiv I$, and setting $C_a \equiv 0$ yields the appropriate results.) Recalling (3.5a,b,c) we consider the initial value problem

$$\mathscr{L}[\lambda]\mathbf{u} = \mathbf{0}, \tag{3.8a}$$

$$\mathbf{u}(a) = R(\lambda)\begin{pmatrix} \mathbf{0} \\ \mathbf{s} \end{pmatrix} = R_q(\lambda)\mathbf{s}, \qquad \mathbf{u}(t) \equiv \mathbf{u}(t; \lambda, \mathbf{s}), \mathbf{s} \in \mathbb{E}^q. \tag{3.8b}$$

Clearly $C_a(\lambda)\mathbf{u}(a) = \mathbf{0}$ since $C_a(\lambda)R_q(\lambda) = 0_{p \times q}$ by construction of $R(\lambda)$. Thus $\mathbf{u}(t; \lambda, \mathbf{s})$ will be an eigenvector and λ an eigenvalue if $\mathbf{s} \not\equiv \mathbf{0}$ and

$$\boldsymbol{\phi}(\lambda, \mathbf{s}) \equiv C_{ba}(\lambda)R_q(\lambda)\mathbf{s} + C_b(\lambda)\mathbf{u}(b; \lambda, \mathbf{s}) = \mathbf{0}. \tag{3.8c}$$

This is a system of q equations in the $q + 1$ unknowns $(\lambda, \mathbf{s})$. However, we can easily normalize $\mathbf{s}$, say, by taking

$$s_q = \sqrt{1 - (s_1^2 + \cdots + s_{q-1}^2)} \equiv \sqrt{1 - \|\hat{\mathbf{s}}\|_2^2}. \tag{3.8d}$$

In terms of the fundamental solution $Y(t; \lambda)$ of (3.2a,b),

$$\mathbf{u}(t; \lambda, \mathbf{s}) = Y(t; \lambda)R_q(\lambda)\mathbf{s},$$

and so (3.8c) is just another way of writing (3.5a). To evaluate $Q_q(\lambda)$ in (3.5a) we could solve for the q-columns of $V(t, \lambda) \equiv Y(t, \lambda)R_q(\lambda)$ as in (1.10b) which is discussed in the method of complementary functions, or for the q-columns of $\tilde{Z}(a, \lambda) = Y^T(b, \lambda)C_b^T(\lambda)$ as in (1.12) which is discussed in the method of adjoints. Either of these evaluation methods could be used to seek the eigenvalues as the roots of (3.5c). However by solving one additional linear system Newton's method can be used to solve (3.8c,d). This gives simultaneously an eigenvalue and corresponding eigenfunction.

The Jacobian of $\boldsymbol{\phi}(\lambda, \mathbf{s})$ subject to (3.8d) is, from (3.8c), using the notation $\mathbf{s} = \begin{pmatrix} \hat{\mathbf{s}} \\ s_q \end{pmatrix}$ with $\hat{\mathbf{s}} \in \mathbb{E}^{q-1}$:

$$\frac{\partial \boldsymbol{\phi}(\lambda, \mathbf{s})}{\partial(\lambda, \hat{\mathbf{s}})} = (\dot{\boldsymbol{\phi}}, \boldsymbol{\phi}_{\hat{\mathbf{s}}}), \tag{3.9a}$$

$$\dot{\boldsymbol{\phi}}(\lambda, \mathbf{s}) = [\dot{C}_{ba}R_q + C_{ba}\dot{R}_q]\mathbf{s} + \dot{C}_b\mathbf{u}(b; \lambda, \mathbf{s}) + C_b\dot{\mathbf{u}}(b; \lambda, \mathbf{s}), \tag{3.9b}$$

$$\boldsymbol{\phi}_{\hat{\mathbf{s}}}(\lambda, \mathbf{s}) = C_{ba}R_q\begin{pmatrix} I \\ -\hat{\mathbf{s}}^T/s_q \end{pmatrix} + C_bV(b; \lambda, \mathbf{s}). \tag{3.9c}$$

Here the dot represents differentiation with respect to λ and $V(t;\lambda,\mathbf{s})$ is an $n \times (q-1)$ matrix with columns $(\mathbf{v}_1, \cdots, \mathbf{v}_{q-1})$, where $\mathbf{v}_k = \partial\mathbf{u}/\partial s_k$. Thus from (3.8a,b,d), with $\hat{\mathbf{e}}_k$ unit vectors in $\mathbb{E}^{q-1}$,

$$\text{(3.10a)} \qquad \mathscr{L}[\lambda]\dot{\mathbf{u}} = \dot{A}(t,\lambda)\mathbf{u}, \quad \dot{\mathbf{u}}(a) = \dot{R}_q(\lambda)\mathbf{s},$$

$$\text{(3.10b)} \qquad \mathscr{L}[\lambda]\mathbf{v}_k = 0, \quad \mathbf{v}_k(a) = R_q(\lambda)\begin{pmatrix} \mathbf{0} \\ \mathbf{e}_k \\ -s_k/s_q \end{pmatrix}, \qquad k = 1, 2, \cdots, q-1.$$

In summary, for one Newton iterate, given $(\lambda, \hat{\mathbf{s}})$, we must solve (3.8a,b), (3.10a) and (3.10b), a total of $q+1$ initial value problems. This is perhaps the most efficient direct shooting procedure for the eigenvalue problem (see, however, § 3.4 for even more efficient procedures).

To assess the errors in carrying out the above scheme numerically or directly computing roots of (3.5c) we assume that some stable accurate of order h^m initial value method is used to solve the appropriate $q+1$ or q initial value problems. The method must also retain analyticity of the numerical solution in λ; all standard schemes we know of do this. (But the current set of adaptive codes, say, due to Gear [15], Krogh [34], Shampine et al. [50], which switch from one scheme to another, might destroy analyticity. This has not been examined seriously; it should be done.) Then assuming convergence of our iterations the computed eigenvalue $\lambda(h)$ must be a root of $\Delta_q^h(\lambda) = 0$, where $\Delta_q^h(\lambda)$ is some analytic approximation to $\Delta_q(\lambda)$ in (3.5c). Indeed if in integrating an initial value problem, say (3.8a,b), our numerical scheme has an asymptotic error expansion of the form

$$\mathbf{u}_j(\lambda) = \mathbf{u}(t_j,\lambda) + \sum_{\nu=1}^{N} h^{\nu m}\mathbf{e}_\nu(t_j,\lambda) + O(h^{(N+1)m}),$$

then we easily find that

$$\text{(3.11)} \qquad \Delta_q^h(\lambda) = \Delta_q(\lambda) + \sum_{\nu=1}^{N} h^{\nu m}\delta_\nu(\lambda) + O(h^{(N+1)m}).$$

Now the theory indicated in (3.6a,b,c) is applicable and so Richardson extrapolation is valid for the average of the r_k approximations, $\lambda_{k\nu}(h)$, $\nu = 1, 2, \cdots, r_k$ to an eigenvalue λ_k of multiplicity r_k.

3.3. Finite differences for eigenvalues. Using any of the stable accurate of order m finite difference schemes of § 2.1 on an appropriate net $\{t_j\}_0^J$ we approximate (3.1a,b,c) by the finite problem

$$\text{(3.12a)} \qquad \mathscr{L}_h[\lambda]\mathbf{u}_j \equiv \sum_{k=0}^{J} C_{jk}(\lambda,h)\mathbf{u}_k = \mathbf{0}, \qquad 1 \leqq j \leqq J,$$

$$\text{(3.12b)} \qquad \mathscr{B}_h[\lambda]\mathbf{u}^h \equiv B_a(\lambda)\mathbf{u}_0 + B_b(\lambda)\mathbf{u}_J = \mathbf{0}.$$

However, the coefficients $C_{jk}(\lambda, h)$ are determined from the matrix $A(t;\lambda)$ and the net $\{t_j\}$; we assume analytic dependence on λ. In block-matrix form the difference equations can be written

$$
\text{(3.12c)}\qquad \mathbb{A}_h(\lambda)\mathbf{u}^h \equiv \begin{pmatrix} B_a(\lambda) & \cdots & 0 & \cdots & B_b(\lambda) \\ C_{10}(\lambda) & C_{11}(\lambda) & \cdots & & C_{1J}(\lambda) \\ \vdots & \vdots & & & \vdots \\ C_{J0}(\lambda) & C_{J1}(\lambda) & \cdots & & C_{JJ}(\lambda) \end{pmatrix} \mathbf{u}^h = \mathbf{0}.
$$

To show how this finite procedure relates to the continuous problem we have

THEOREM 3.13. *Let the scheme* (3.12a) *with* $\mathbf{u}_0 = \boldsymbol{\alpha}$ *be stable and consistent of order* h^m *for the initial value problem* $\mathscr{L}[\lambda]\mathbf{u}(t) = \mathbf{0}$, $\mathbf{u}(a) = \boldsymbol{\alpha}$. *Then*

$$
\text{(3.13a)}\qquad \det \mathbb{A}_h(\lambda) = [\det Q(\lambda) + O(h^m)]\Theta(\lambda, h),
$$

where for some $K(\lambda) > 0$,

$$
\text{(3.13b)}\qquad |\Theta(\lambda, h)| > K(\lambda)^{-(J+1)n}.
$$

If the scheme has an asymptotic error expansion, then the $O(h^m)$ *term can be replaced by such an expansion.*

Proof. Just as in the proof of Theorem 2.9 we factor $\mathbb{A}_h(\lambda)$ in the form

$$
\mathbb{A}_h(\lambda) = \begin{pmatrix} Q_{h,0}(\lambda) & Q_{h,1}(\lambda) & \cdots & Q_{h,J}(\lambda) \\ & I & & \\ & & \ddots & \\ & & & I \end{pmatrix} \mathbb{A}_h^0(\lambda),
$$

where $\mathbb{A}_h^0(\lambda)$ is obtained from $\mathbb{A}_h(\lambda)$ by replacing $B_a(\lambda)$ by I, $B_b(\lambda)$ by 0, $\mathbb{A}_h^0(\lambda)$ has a bounded inverse say in block form $(\mathbb{A}_h^0(\lambda))^{-1} = (Z_{ij}(\lambda))$ with

$$
\|(\mathbb{A}_h^0(\lambda))^{-1}\| \leqq K(\lambda),
$$

and

$$
Q_{h,j}(\lambda) = B_a(\lambda)Z_{0j}(\lambda) + B_b(\lambda)Z_{Jj}(\lambda), \qquad 0 \leqq j \leqq J.
$$

The bounded inverse follows from the assumed stability for the initial value problem. We also note that now

$$
Z_{i0}(\lambda) = Y(t_i, \lambda) + O(h^m)
$$

since $\mathbb{A}_h^0(\lambda)\begin{pmatrix} \vdots \\ Z_{i0} \\ \vdots \end{pmatrix} = \begin{pmatrix} I \\ 0 \\ \vdots \\ 0 \end{pmatrix}$ is the numerical approximation to (3.2a,b). Thus using (3.3a,b),

$$
Q_{h,0}(\lambda) = Q(\lambda) + O(h^m),
$$

and (3.13a) follows with

$$
\Theta(\lambda, h) = \det \mathbb{A}_h^0(\lambda).
$$

Since we use only induced matrix norms it follows that

$$\rho(\mathbb{A}_h^{0^{-1}}(\lambda)) \leqq K(\lambda).$$

Then using $\mu_k(A)$ to represent the kth eigenvalue of A,

$$\left|\det \mathbb{A}_h^0(\lambda)\right| = \left|\prod_{k=1}^{n(J+1)} \mu_k(\mathbb{A}_h^0(\lambda))\right| = \left|\prod_{k=1}^{n(J+1)} \mu_k(\mathbb{A}_h^0(\lambda)^{-1})\right|^{-1} \geqq K(\lambda)^{-(J+1)n},$$

and (3.13b) follows. ▧

Clearly we can now apply root-finding techniques to solve

$$\Delta_h(\lambda) \equiv \det \mathbb{A}_h(\lambda) = 0,$$

and if they converge to a root λ_h, our previous error estimates as in (3.6a,b,c) apply. The annoying term $\Theta(\lambda, h)$ may get quite small as J gets large and this signals possible difficulties with this direct approach. Indeed most reasonable determinant evaluations will use a form of pivoting with possible scaling that could reduce these effects. Of course when one-step schemes are used the special factorization procedures of § 2.2 should be employed. It is tempting to form the $n \times n$ matrix $Q_{h,0}(\lambda)$. This is not a wise procedure for difference methods. It is precisely equivalent to using shooting with the scheme $\mathscr{L}_h[\lambda]$ to approximate $Y(b, \lambda)$. But then we might as well use the more efficient schemes implied in § 3.2.

In many eigenvalue problems the parameter λ enters into the coefficients of the differential equations and of the boundary conditions linearly. Then in all reasonable difference approximations of such linear equations the parameter λ will also enter the coefficients of the difference equations linearly. The coefficient matrix in (3.12c) then has the form:

$$\mathbb{A}_h(\lambda) \equiv \hat{\mathbb{A}}_h + \lambda \mathbb{B}_h,$$

where $\hat{\mathbb{A}}_h$ and $\mathbb{B}_h$ are independent of λ. In this form it seems natural to employ some "standard" eigenvalue-eigenvector code for the general eigenproblem $(A + \lambda B)\mathbf{x} = \mathbf{0}$, and this is frequently done in practice. We do not advocate such procedures in general as they seem much less efficient than using our standard difference schemes on appropriate reformulations of the eigenvalue problem presented in the next section.

3.4. Reformulations of eigenvalue problems. We present here two very powerful ways in which the general eigenvalue problem can be reformulated so that more or less standard shooting or finite difference techniques can be applied. In particular they both avoid determinant evaluations and eigenvalue calculations for matrices while yielding simultaneously approximations to the eigenvalue and eigenfunction.

The first idea is essentially to drop one of the original boundary conditions and impose a new independent inhomogeneous boundary condition. Then this new *inhomogeneous* problem is solved and the eigenvalue parameter λ is "adjusted" until the solution satisfies the dropped boundary constraint. The adjusting can readily be done by Newton's method and we note that both an eigenvalue and eigenfunction approximation are simultaneously obtained.

To illustrate these ideas in more detail, suppose that the original eigenvalue problem had separated endpoint conditions, say, for (3.1b):

$$C_a(\lambda)\mathbf{y}(a) = 0, \quad C_b(\lambda)\mathbf{y}(b) = 0, \tag{3.14a}$$

where

$$C_a(\lambda) \equiv \begin{pmatrix} \mathbf{l}_1^T(\lambda) \\ \vdots \\ \mathbf{l}_p^T(\lambda) \end{pmatrix}, \quad C_b(\lambda) \equiv \begin{pmatrix} \mathbf{r}_1^T(\lambda) \\ \vdots \\ \mathbf{r}_q^T(\lambda) \end{pmatrix}, \quad p + q = n. \tag{3.14b}$$

The p vectors $\mathbf{l}_\nu(\lambda) \in \mathbb{E}^n$ are assumed linearly independent as are the q vectors $\mathbf{r}_\nu(\lambda) \in \mathbb{E}^n$. Thus rank $C_a = p$ and rank $C_b = q$.

Now pick some n-vector $\mathbf{l}_{p+1}(\lambda)$ which is independent of the set $\{\mathbf{l}_\nu(\lambda)\}_1^p$ for all λ and replace the conditions (3.14a) by

$$\hat{C}_a(\lambda)\mathbf{u}(a) = \mathbf{e}_{p+1}, \quad \hat{C}_b(\lambda)\mathbf{u}(b) = 0, \tag{3.15a}$$

where $\mathbf{e}_{p+1}$ is the $(p+1)$st unit vector in $\mathbb{E}^{p+1}$ and

$$\hat{C}_a(\lambda) \equiv \begin{pmatrix} C_a(\lambda) \\ \mathbf{l}_{p+1}^T(\lambda) \end{pmatrix}, \quad \hat{C}_b(\lambda) \equiv \begin{pmatrix} \mathbf{r}_1^T(\lambda) \\ \vdots \\ \mathbf{r}_{q-1}^T(\lambda) \end{pmatrix}. \tag{3.15b}$$

Here we have dropped the last or qth condition at $t = b$; any other choice would have done at this stage, say, drop the vector $\mathbf{r}_\nu(\lambda)$ and retain all the others. The solution of

$$\mathscr{L}[\lambda]\mathbf{u}(t) = \mathbf{0}, \qquad a \leqq t \leqq b, \tag{3.15c}$$

subject to (3.15a) is denoted by $\mathbf{u}(t) = \mathbf{u}(t;\lambda)$. We now seek λ such that

$$\phi(\lambda) \equiv \mathbf{r}_q^T(\lambda)\mathbf{u}(b, \lambda) = 0. \tag{3.15d}$$

This is the missing boundary condition at $t = b$. Note that we could just as well use shooting or finite differences to solve (3.15a,b,c,d). Each iteration, say with Newton's method on (3.15d), requires the solution of only two linear two-point boundary value problems; thus it yields a very effective technique.

If we are employing a one-step difference scheme, then our approximation to (3.15a,b,c) yields a system of the form

$$\hat{\mathbb{A}}_h(\lambda)\mathbf{u}^h \equiv \begin{pmatrix} \hat{C}_a(\lambda) & & & \\ L_1(h,\lambda) & R_1(h,\lambda) & & \\ & \ddots & \ddots & \\ & & L_J(h,\lambda) & R_J(h,\lambda) \\ & & & \hat{C}_b(\lambda) \end{pmatrix} \begin{pmatrix} \mathbf{u}_0 \\ \mathbf{u}_1 \\ \vdots \\ \mathbf{u}_J \end{pmatrix} = \begin{pmatrix} \mathbf{e}_{p+1} \\ \mathbf{0} \\ \vdots \\ \mathbf{0} \end{pmatrix}.$$

To solve this system by one of the efficient block- or band-factorization schemes of § 2.2 we first consider it in the block tridiagonal form

$$\hat{\mathbb{A}}_h(\lambda) \equiv [\hat{B}_j, \hat{A}_j, \hat{C}_j].$$

Note that if the original boundary conditions (3.14a) had been used with our difference scheme the corresponding block form, say,

$$\mathbb{A}_h(\lambda) \equiv [B_j, A_j, C_j]$$

would have had a somewhat different zero pattern. That is, each $\hat{B}_j$ has one *less* row of zeros than the corresponding B_j while each $\hat{C}_j$ has one *more* row of zeros than the C_j.

Of course the basic theoretical question is how to pick the new $\mathbf{l}_{p+1}(\lambda)$ and which $\mathbf{r}_\nu(\lambda)$ to drop in order that the altered problem has a unique solution, at least for the λ-values in some neighborhood of the desired eigenvalues of (3.1a,b). We know of no studies of this problem. It is clear that our procedure easily goes over to partially separated endconditions with no difficulties. Indeed if R is some $q \times q$ nonsingular matrix, then the more general form of our scheme is to drop one of the elements of $R[C_{ba}(\lambda)\mathbf{y}(a) + C_b(\lambda)\mathbf{y}(b)] = 0$ and then to pose this dropped condition in place of (3.15d). The $(q - 1)$ retained conditions take the place of those in $\hat{C}_b(\lambda)\mathbf{u}(b) = \mathbf{0}$ of (3.15b).

The implementation of the above ideas require special procedures to account for the unknown parameter λ (when employing Newton's method). We show another reformulation which yields the general form (1.20a,b). Then any standard code prepared to solve this general nonlinear two-point problem can be used to solve the eigenvalue problem. It also works very well in practice.

To make the eigenfunctions of (3.1a,b,c) unique, to within a sign (for simple eigenvalues), we normalize them, say, by

$$\int_a^b \mathbf{y}^T(t)\mathbf{y}(t)\,dt = 1.$$

This condition is equivalent to:

$$z'_{n+1}(t) = \mathbf{y}^T(t)\mathbf{y}(t), \quad a \leqq t \leqq b, \quad z_{n+1}(a) = 0, \quad z_{n+1}(b) = 1.$$

The fact that the eigenvalue is a constant, say, $z_0(t) \equiv \lambda$, can be written as

$$z'_0(t) = 0.$$

No boundary conditions can be imposed here since the value of λ is unknown. Combining these relations with (3.1a,b), using the notation $z_\nu(t) \equiv y_\nu(t)$, $1 \leqq \nu \leqq n$, we get the system

$$\text{(3.16a)} \qquad \mathbf{z}' = \mathbf{f}(t, \mathbf{z}), \qquad a \leqq t \leqq b,$$

$$\text{(3.16b)} \qquad \mathbf{g}(\mathbf{z}(a), \mathbf{z}(b)) = \mathbf{0}.$$

Here $\mathbf{z}^T(t) = (z_0(t), z_1(t), \cdots, z_{n+1}(t))$,

$$\text{(3.17a)} \qquad f_0(t, \mathbf{z}) \equiv 0,$$

$$\text{(3.17b)} \qquad f_i(t, \mathbf{z}) \equiv \sum_{j=1}^{n} a_{ij}(t, z_0(t))z_j(t), \qquad 1 \leqq i \leqq n,$$

$$\text{(3.17c)} \qquad f_{n+1}(t, \mathbf{z}) \equiv \sum_{k=1}^{n} z_k^2(t),$$

and with $\hat{\mathbf{z}}^T(t) = (z_1(t), \cdots, z_n(t))$,

$$\mathbf{g}(\mathbf{z}(a), \mathbf{z}(b)) \equiv \begin{pmatrix} B_a(\dot{z}_0(a))\hat{\mathbf{z}}(a) + B_b(z_0(b))\hat{\mathbf{z}}(b) \\ z_{n+1}(a) \\ z_{n+1}(b) - 1 \end{pmatrix}.$$

Note that the added boundary conditions are separated. Thus we have not destroyed this important feature if it was present initially. Although the order of the system (3.1a,b) is increased by two in going to (3.17a,b,c) there are numerous advantages. At a multiple eigenvalue the solution of (3.16a,b) is not isolated. This has not been treated extensively and is one of the important areas in need of study. Of course many successful calculations have been done on such problems, in particular for bifurcation phenomena (see, for example, Keller and Wolfe [32], Bauer, Keller and Reiss [4]).

It should be noted that many other normalizations of the eigenfunction can be used which avoid the need for introducing $z_{n+1}(t)$. The most obvious choice, for separated endconditions, would be to use $\hat{C}_a(\lambda)$ as in (3.15b) and $C_b(\lambda)$ as in (3.14a,b). That is, a new linearly independent inhomogeneous linear constraint is simply adjoined at $t = a$. Then we have $n + 1$ boundary conditions for (3.16a) with the last components of $\mathbf{z}$ and $\mathbf{f}$ eliminated.

CHAPTER 4

Singular Problems

4.1. Regular singular points. We study linear boundary value problems with a regular singularity at one endpoint, say,

$$\mathscr{L}\mathbf{y} \equiv \mathbf{y}' - A(t)\mathbf{y} = \mathbf{f}(t), \qquad 0 \leqq t \leqq 1, \tag{4.1a}$$

where $A(t)$ has a simple pole at $t = 0$:

$$A(t) \equiv \frac{1}{t}R + A_0(t). \tag{4.1b}$$

The boundary conditions are posed as

$$\mathscr{B}\mathbf{y}(t) \equiv \lim_{t \downarrow 0}\,[B_0(t)\mathbf{y}(t) + B_1\mathbf{y}(1) - \mathbf{b}(t)] = \mathbf{0}. \tag{4.1c}$$

This allows the typical incident wave conditions or those resulting from singular coordinate systems. With $B_0(t) = B_0 t^{-R}$ conditions can be imposed on the regular part of the solution, as studied by Natterer [37]. In (4.1a,b,c): $\mathbf{y}(t)$, $\mathbf{f}(t)$, $\mathbf{b}(t)$ are n-vectors, R, $A_0(t)$, $B_0(t)$ and B_1 are $n \times n$ matrices. We assume $A_0(t)$ analytic on $[0, 1]$, while $B_0(t)$, $\mathbf{b}(t)$ and $\mathbf{f}(t)$ may be singular at $t = 0$ but sufficiently smooth on $(0, 1]$.

We examine and justify the more or less standard procedure of expanding about the singular point, solving a regular problem over a reduced interval, $[\delta, 1]$, say, and then matching to the expansion. The extension of these techniques to problems with two singular endpoints, an interior regular singular point and even unbounded intervals, if the point at ∞ is regular, are indicated in Brabston [6]. The first serious study of this device was by Gustafsson [20] who applied it to scalar problems. More detailed accounts of our present work are given in Brabston [6] and Brabston and Keller [7].

Let the fundamental solution matrix, $Y(t)$, for (4.1a) be defined by

$$\mathscr{L}Y(t) = 0, \qquad 0 < t < 1, \tag{4.2a}$$

$$Y(\delta_0) = I. \tag{4.2b}$$

It is well known that we may take

$$Y(t) = P(t)t^R, \qquad 0 < t \leqq \delta_0, \tag{4.2c}$$

where $P(t)$ is analytic on $0 \leqq t \leqq \delta_0$. Then every solution of (4.1a) can be written as

$$\mathbf{y}(t) = Y(t)\mathbf{c} + \mathbf{y}_p(t), \qquad 0 < t \leqq 1, \tag{4.3a}$$

where the particular solution $\mathbf{y}_p(t)$ satisfies

(4.3b) $$\mathscr{L}\mathbf{y}_p(t) = \mathbf{f}(t), \quad 0 < t \leqq 1, \quad \mathbf{y}_p(\delta_0) = 0.$$

Thus we can use the representation

(4.3c) $$\mathbf{y}_p(t) = Y(t) \int_{\delta_0}^{t} Y^{-1}(\xi)\mathbf{f}(\xi)\, d\xi.$$

If $\mathbf{y}(t)$ given by (4.3a) is to satisfy the boundary conditions (4.1c), then we must have

(4.4a) $$\lim_{t \downarrow 0} \{[B(t) + B_1 Y(1)]\mathbf{c} - [\mathbf{g}(t) - B_1 \mathbf{y}_p(1)]\} = \mathbf{0},$$

where

(4.4b) $$B(t) \equiv B_0(t) Y(t),$$

(4.4c) $$\mathbf{g}(t) \equiv \mathbf{b}(t) - B_0(t)\mathbf{y}_p(t).$$

Since $B_0(t)$, $\mathbf{b}(t)$, $A(t)$ and hence $Y(t)$ have a finite number of singularities only at $t = 0$ we can write, without loss of generality, that for some integer s and for some scalar functions $\{\varphi_v(t)\}_1^s$,

(4.5a) $$B(t) = M_0(t) + \sum_{v=1}^{s} \varphi_v(t) M_v,$$

(4.5b) $$\mathbf{g}(t) = \mathbf{g}_0(t) + \sum_{v=1}^{s} \varphi_v(t)\mathbf{g}_v,$$

where for some $N > 0$,

(4.5c) $$\lim_{t \downarrow 0} |\varphi_k(t)| = \infty, \quad \lim_{t \downarrow 0} \frac{\varphi_k(t)}{\varphi_{k+1}(t)} = 0, \quad \lim_{t \downarrow 0} t^N \varphi_s(t) = 0.$$

Here $M_0(t)$ and $\mathbf{g}_0(t)$ are analytic on $[0, \delta_0]$ while M_v and $\mathbf{g}_v$ are constant. From these assumptions we easily obtain the following theorem.

THEOREM 4.6. *Let* (4.5a,b,c) *hold. Then* (4.1a,b,c) *has a solution if and only if*

(4.6a) $$\operatorname{rank} M = \operatorname{rank}(M, G),$$
$$M \equiv \begin{pmatrix} M_0(0) + B_1 Y(1) \\ M_1 \\ \vdots \\ M_s \end{pmatrix}, \quad G \equiv \begin{pmatrix} \mathbf{g}_0(0) - B_1 \mathbf{y}_p(1) \\ \mathbf{g}_1 \\ \vdots \\ \mathbf{g}_s \end{pmatrix}.$$

The solution is unique if and only if

(4.6b) $$\operatorname{rank} M = n.$$

For practical applications it should be noted that the matrices M_k and the vectors $\mathbf{g}_k$ can be determined explicitly if $B_0(t)$, $\mathbf{b}(t)$, $A(t)$ and hence $B(t)$ and $\mathbf{g}(t)$ are analytic in $0 < |t| \leqq \delta_0$. We shall employ the representations (4.3a,b,c) and (4.5a,b,c) in devising numerical procedures.

The numerical scheme is to employ an *approximation* to the solution (4.3a,b,c) on some interval $0 < t \leqq \delta$, with $\delta \leqq \delta_0$, and by means of this to define a *regular* boundary value problem on $[\delta, 1]$. The justification of this procedure which is also used for obtaining error estimates is based on an exact such reduction to a regular boundary value problem. This procedure has also been used by Gustafsson [20] in treating a second order scalar problem. In particular from (4.3a) it follows, assuming (4.6a) holds, that $\mathbf{y}(\delta) = Y(\delta)\mathbf{c} + \mathbf{y}_p(\delta)$. Using this in (4.4a) to eliminate $\mathbf{c}$ in the term $B(t)\mathbf{c}$, and using $Y(1)\mathbf{c} = \mathbf{y}(1) - \mathbf{y}_p(1)$ to eliminate $Y(1)\mathbf{c}$, we get the boundary conditions satisfied on $[\delta, 1]$ by any solution of (4.1a,b,c) to be:

(4.7a) $$B_{0\delta}\mathbf{y}(\delta) + B_{1\delta}\mathbf{y}(1) = \mathbf{b}_\delta.$$

Here $B_{0\delta}$ and $B_{1\delta}$ are $n(s+1) \times n$ matrices while $\mathbf{b}_\delta$ is an $n(s+1)$-vector:

(4.7b) $$B_{0\delta} \equiv \begin{pmatrix} M_0(0) \\ M_1 \\ \cdot \\ \vdots \\ M_s \end{pmatrix} Y^{-1}(\delta), \quad B_{1\delta} \equiv \begin{pmatrix} B_1 \\ 0 \\ \cdot \\ \vdots \\ 0 \end{pmatrix}, \quad \mathbf{b}_\delta \equiv \begin{pmatrix} \mathbf{g}_0(0) \\ \mathbf{g}_1 \\ \cdot \\ \vdots \\ \mathbf{g}_s \end{pmatrix} + B_{0\delta}\mathbf{y}_p(\delta).$$

With $\mathbf{y}_p(t)$ uniquely defined by (4.3b) it is not difficult to show that: *the boundary value problem* (4.1a,b) *has a (unique) solution if and only if the boundary value problem* (4.1a), (4.7a,b) *has a (unique) solution. The solution of* (4.1a,b,c) *on* $[\delta, 1]$ *satisfies* (4.7a,b). This result is formally established in Brabston and Keller [7].

To determine the boundary conditions (4.7a) we need only $Y(t)$ and $\mathbf{y}_p(t)$ on $(0, \delta]$ and then the matrices and vectors in (4.7b) can be evaluated. By the assumed analyticity of $A_0(x)$ on $0 \leqq x \leqq \delta_0$ we can obtain as many terms as desired in the expansion of $P(x)$ in (4.2c). Thus for some integer N we define the approximate or "truncated" fundamental solution matrix:

(4.8a) $$Y^N(t) \equiv \left(\sum_{k=0}^{N} P_k t^k\right) t^R.$$

We also define the truncated particular solution:

(4.8b) $$\mathbf{y}_p^N(t) \equiv Y^N(t) \int_{\delta_0}^{t} [Y^N(\xi)]^{-1}\mathbf{f}(\xi)\, d\xi.$$

If N is sufficiently large we can use (4.8a,b) in (4.4b,c) to determine the exact values of M_ν and $\mathbf{g}_\nu$, $\nu = 1, 2, \cdots, s$. The largest value of N required for this is $N_{\max} = \max_\nu \operatorname{Re}[-\lambda_\nu(R)] + 1$, where $\lambda_\nu(R)$ are the eigenvalues of R. This of course assumes that the "highest order" singularity occurs in $Y(t)$. If singularities of $B_0(t)$ or of $\mathbf{f}(t)$ increase the order, then a larger N may be required.

Employing (4.8a,b) we define the truncated quantities

(4.9a) $$B^N(t) \equiv B_0(t) Y^N(t),$$

(4.9b) $$\mathbf{g}^N(t) \equiv \mathbf{b}(t) - B_0(t)\mathbf{y}_p^N(t).$$

Then expansions analogous to (4.5a,b,c) yield the M_ν and g_ν for $\nu = 1, 2, \cdots, s$ and the truncated quantities $M_0^N(t)$ and $g_0^N(t)$. Using the appropriate truncated quantities in (4.7b) we define

$$(4.9c)\qquad B_{0\delta}^N \equiv \begin{pmatrix} M_0^N(0) \\ M_1 \\ \vdots \\ M_s \end{pmatrix} [Y^N(\delta)]^{-1}, \quad B_{1\delta}^N \equiv B_{1\delta}, \quad \mathbf{b}_\delta^N \equiv \begin{pmatrix} \mathbf{g}_0^N(0) \\ \mathbf{g}_1 \\ \vdots \\ \mathbf{g}_s \end{pmatrix} + B_{0\delta}^N \mathbf{y}_p^N(\delta).$$

The exact boundary conditions (4.7a) are now replaced by the truncated boundary condition

$$(4.10a)\qquad B_{0\delta}^N \mathbf{y}^N(\delta) + B_{1\delta}^N \mathbf{y}^N(1) = \mathbf{b}_\delta^N,$$

where $\mathbf{y}^N(t)$ is a solution of

$$(4.10b)\qquad \mathscr{L}\mathbf{y}^N(t) = \mathbf{f}(t), \qquad \delta \leqq t \leqq 1.$$

If the original boundary value problem (4.1a,b,c) has a unique solution, then it is established in [7] that for sufficiently large N the truncated regular boundary value problem (4.10a,b) also has a unique solution. The truncated solution $\mathbf{y}^N(t)$ defined on $[\delta, 1]$ is extended to $(0, \delta]$ by using

$$(4.11a)\qquad \mathbf{C}^N \equiv [Y^N(\delta)]^{-1}[\mathbf{y}^N(\delta) - \mathbf{y}_p^N(\delta)],$$

$$(4.11b)\qquad \mathbf{y}^N(t) \equiv Y^N(t)\mathbf{C}^N + \mathbf{y}_p^N(t), \qquad 0 < t \leqq \delta.$$

Our numerical procedure is to solve the linear boundary value problem (4.10a,b) by finite differences, say, with an accuracy $O(h^r)$ on some net $\{t_j\}_0^J$ with $t_0 = \delta$, $t_J = 1$. Denoting this numerical solution by $\{\mathbf{u}_j\}_0^J$ we replace $\mathbf{y}^N(\delta)$ in (4.11a) by $\mathbf{u}_0$ to define $\mathbf{C}_h^N$ and then use this in (4.11b) to define $\mathbf{y}_h^N(t)$ on $0 < t \leqq \delta$.

The error in our numerical solution is estimated by using

$$(4.12a)\qquad \|\mathbf{y}(t_j) - \mathbf{u}_j\| \leqq \|\mathbf{y}(t_j) - \mathbf{y}^N(t_j)\| + \|\mathbf{y}^N(t_j) - \mathbf{u}_j\|, \qquad \delta \leqq t_j \leqq 1,$$

$$(4.12b)\qquad \|\mathbf{y}(t) - \mathbf{y}_h^N(t)\| \leqq \|\mathbf{y}(t) - \mathbf{y}^N(t)\| + \|\mathbf{y}^N(t) - \mathbf{y}_h^N(t)\|, \qquad 0 < t \leqq \delta.$$

For fixed N and δ we obviously have

$$(4.13)\qquad \|\mathbf{y}^N(t_j) - \mathbf{u}_j\| \leqq K_0(N, \delta)h^m, \qquad 0 \leqq j \leqq J,$$

assuming a stable, consistent of order m scheme has been used to solve 4.10a,b). The error in the truncated solution $\mathbf{y}^N(t)$ can be estimated in terms of

$$(4.14a)\qquad \Delta_N(\delta) \equiv \max\{\sup_{0<t\leqq\delta} \|Y(t) - Y^N(t)\|, \sup_{0<t\leqq\delta} \|\mathbf{y}_p(t) - \mathbf{y}_p^N(t)\|\}.$$

It is clear for any fixed δ in $0 < \delta \leqq \delta_0$ that $\lim_{N\to\infty} \Delta_N(\delta) = 0$. In particular cases it is easy to establish more precise estimates of the form

$$(4.14b)\qquad \Delta_N(\delta) \leqq K_1\, \delta^{N-\rho}$$

for some fixed K_1 and ρ independent of δ and N. It is now easy to deduce that, for N sufficiently large,

(4.15) $$\|\mathbf{y}(t) - \mathbf{y}^N(t)\| \leqq K_2(\delta)\Delta_N(\delta), \qquad \delta \leqq t \leqq 1.$$

Combining (4.14a,b) and (4.15) in (4.12a) we get error estimates on $\delta \leqq t \leqq 1$.

On the interval $0 < t \leqq \delta$ we obtain, in a similar fashion, estimates of the form

(4.16) $$\|\mathbf{y}(t) - \mathbf{y}_h^N(t)\| \leqq \|Y(t)\| \cdot [K_4(N, \delta)h^m + K_5(\delta)\Delta_N(\delta)] + K_6(\delta)\Delta_N(\delta).$$

Since $\|Y(t)\|$ may be unbounded as $t \downarrow 0$, the absolute accuracy may degrade as $t \downarrow 0$. This is unavoidable if, as may be the case, the exact solution is singular at $t = 0$. It would seem that relative error estimates are desirable in such cases, but they are unknown at present. On the other hand, if any row of the matrix $Y(t)$ is nonsingular, then a more careful analysis reveals that the corresponding component of the error is bounded as in (4.16) with $\|Y(t)\|$ replaced by the norm of the appropriate row. For such components we can thus obtain small absolute errors on the entire closed interval $0 \leqq t \leqq 1$.

Numerical examples of this method, showing also that Richardson extrapolation can be employed, are contained in [6] and [7]. Very recent work of de Hoog and Weiss [11] shows that some one-step difference schemes can be used directly on (4.1a,b,c) provided the solutions are bounded. They also examine eigenvalue problems and some nonlinear singular equations.

The techniques employed here can also be used to develop and justify numerical methods for problems with irregular singular points. The convergent power series expansions are essentially replaced by asymptotic expansions and somewhat more work must be done to determine appropriate values for the integer N, if it exists in a given case. The use of one-step difference schemes up to the singular point, as in de Hoog and Weiss [11], may also be used in many cases. Calculations on such problems are performed very often but theoretical studies along the above indicated lines are just now in progress. Another aspect of frequently occurring irregular singular point problems is considered in § 4.2.

4.2. ∞-intervals; irregular singular points. There is at present no theoretical work justifying numerical methods for solving problems with irregular singular points. The main practical occurrence of such problems seems to be those formulated on infinite intervals and we examine some simple examples here.

We consider first problems of the form

(4.17a) $$\mathscr{L}\mathbf{y}(t) \equiv \mathbf{y}'(t) - A(t)\mathbf{y} = \mathbf{0}, \qquad t \geqq a,$$

(4.17b) $$C_a\mathbf{y}(a) = \boldsymbol{\alpha},$$

(4.17c) $$\lim_{t \to \infty} \mathbf{y}(t) = \mathbf{0}.$$

It is assumed that

(4.18) $$\lim_{t \to \infty} A(t) = A_\infty \not\equiv 0$$

so that the point at ∞ is an irregular singular point. The matrix C_a is $p \times n$ of rank $p < n$ and $\alpha \in E^p$. If we let $A(t)$ and C_a depend analytically on λ and set $\boldsymbol{\alpha} \equiv \mathbf{0}$ our theory includes an important class of eigenvalue problems. Most numerical work on such problems proceeds by replacing the infinite interval by a finite one, say, $[a, \infty] \to [a, b]$. However, the boundary conditions to be imposed at $t = b$ are not always chosen correctly. To do this correctly we examine the eigenvalues and eigenvectors of A_∞.

Suppose the eigenvalues, λ_j, of A_∞ satisfy

$$\text{(4.19a)} \qquad \operatorname{Re} \lambda_j \geqq 0, \quad 1 \leqq j \leqq q, \quad \operatorname{Re} \lambda_j < 0, \quad q + 1 \leqq j \leqq n.$$

Further assume $\boldsymbol{\xi}_j$, $j = 1, 2, \cdots, q$, are q independent *left* eigenvectors of A_∞ belonging to the corresponding λ_j. Then with the $q \times n$ matrix

$$\text{(4.19b)} \qquad C_\infty^+ \equiv \begin{pmatrix} \boldsymbol{\xi}_1^T \\ \vdots \\ \boldsymbol{\xi}_q^T \end{pmatrix},$$

we can replace (4.17c) by

$$\text{(4.17d)} \qquad \lim_{t \to \infty} C_\infty^+ \mathbf{y}(t) = \mathbf{0}.$$

To see that these are the correct conditions assume $\boldsymbol{\zeta}_j$, $1 \leqq j \leqq n$, are the right eigenvectors of A_∞. Then every solution of $\mathbf{y}' = A_\infty \mathbf{y}$ has the form

$$\mathbf{y}(t) = \sum_{j=1}^{n} a_j \boldsymbol{\zeta}_j e^{\lambda_j t}.$$

Since the $\boldsymbol{\xi}_j$ and $\boldsymbol{\zeta}_j$ are bi-orthogonal, condition (4.17d) simply ensures that $a_j = 0$ if $\operatorname{Re} \lambda_j \geqq 0$. More briefly it projects the solution into the subspace of decaying solutions at ∞. It is assumed that $p + q = n$ so (4.17b) and (4.17d) represent n constraints. The appropriate replacement of (4.17a,b,c) by a finite problem is now:

$$\text{(4.20a)} \qquad \mathscr{L}\mathbf{u}(t) = \mathbf{0}, \qquad a \leqq t \leqq b,$$

$$\text{(4.20b)} \qquad C_a \mathbf{u}(a) = \boldsymbol{\alpha},$$

$$\text{(4.20c)} \qquad C_\infty^+ \mathbf{u}(b) = \mathbf{0}.$$

To see how well the above procedure works compared to frequently used alternatives consider the trivial example

$$\text{(4.21)} \qquad \varphi'' = \sigma^2 \varphi, \quad \varphi(0) = \alpha, \quad \varphi(\infty) = 0, \quad \sigma > 0,$$

whose solution is

$$\varphi(t) = \alpha e^{-\sigma t}.$$

A common approach to (4.21) is to replace the condition at $t = \infty$ by $\varphi(L) = 0$ for some finite $t = L$. Calling the solution of the altered problem $\varphi_0(t)$ we have

$$\varphi_0(t) = \alpha \frac{\sinh \sigma(L - t)}{\sinh \sigma L}.$$

To get the equivalent of (4.20c) for (4.21) we note that as a system with $\mathbf{y}^T(t) = (\varphi(t), \varphi'(t))$ the latter yields $A(t) \equiv \begin{pmatrix} 0 & 1 \\ \sigma^2 & 0 \end{pmatrix} = A_\infty$. Hence $p = q = 1$ with $\xi_1^T = (\sigma, 1) \equiv C_\infty^+$. Thus the finite condition should be $\sigma\varphi(L) + \varphi'(L) = 0$. The altered problem now yields the *exact solution*. Of course this example seems extreme but it illustrates the important feature that using (4.20c) yields better approximations over any finite interval $[a, b]$ than does frequently used alternatives. We shall see in fact that it is not such an extreme example.

The general formulation of (4.20c) does not require the assumptions inherent in (4.19a,b). Rather we need only assume that, for some nonsingular $n \times n$ matrix P,

(4.22a) $$PA_\infty P^{-1} = \begin{pmatrix} A^+ & 0 \\ 0 & A^- \end{pmatrix},$$

where A^+ is $q \times q$, A^- is $p \times p$, $p + q = n$ and

(4.22b) $$\operatorname{Re} \lambda(A^+) \geqq 0, \quad \operatorname{Re} \lambda(A^-) < 0.$$

Then the first q-rows of P suffice for C_∞^+. In fact we partition both P and P^{-1} as follows:

(4.22c) $$P \equiv \begin{pmatrix} C_\infty^+ \\ C_\infty^- \end{pmatrix} \begin{matrix} \}q \\ \}p \end{matrix}, \quad P^{-1} \equiv (\underbrace{D_x^+}_{q} \; \underbrace{D_x^-}_{p}).$$

Now a rather striking generalization of the treatment of (4.21) can be given.

Consider (4.17a,b,d) where $A(t) \in C[a, \infty]$ and has a constant "tail", that is,

(4.23) $$A(t) \equiv A_\infty, \quad t \geqq t_0.$$

We replace this problem by the finite problem (4.20a,b,c) with $b \geqq t_0$. By our previous analysis we know that this finite problem has a unique solution if and only if

(4.24) $$Q(b) \equiv \begin{pmatrix} C_a \\ C_\infty^+ Y(b, a) \end{pmatrix}$$

is nonsingular. We first examine the nonsingularity question in

LEMMA 4.25. *Let $Q(b)$ be nonsingular for some $b \geqq t_0$. Then $Q(t)$ is nonsingular for all $t \geqq t_0$.*

Proof. The fundamental solution $Y(t, a)$ has the property that

$$Y(t, a) = Y(t, t_0)Y(t_0, a).$$

In particular by (4.23) and (4.22a,b,c), for $t \geqq t_0$,

$$Y(t, t_0) = e^{A_\infty[t - t_0]} = P^{-1}\begin{pmatrix} e^{A^+[t - t_0]} & 0 \\ 0 & e^{A^-[t - t_0]} \end{pmatrix}P.$$

Further recall that $C_\infty^+ P^{-1} = (I_{q \times q}, 0_{q \times p})$ since $PP^{-1} = I$. Thus we see that for $t \geqq t_0$,

$$C_\infty^+ Y(t, a) = e^{A^+[t - t_0]}C_\infty^+ Y(t_0, a).$$

Now (4.24) with $b \equiv t \geqq t_0$ yields

(4.25a)
$$Q(t) = \begin{pmatrix} I_{p\times p} & 0 \\ 0 & e^{A^+[t-t_0]} \end{pmatrix} Q(t_0), \qquad t \geqq t_0.$$

So if $Q(t)$ is nonsingular for any $t \geqq t_0$, then the same is true for all $t \geqq t_0$. ▧

Now we have the more dramatic result in

THEOREM 4.26. *Let $A(t)$ be as in* (4.23). *Let* (4.20a,b,c) *have the unique solution* $\mathbf{u}(t) \equiv \mathbf{u}(t, b)$ *for some $b \geqq t_0$. Then $\mathbf{u}(t, b)$ is independent of b for $b \geqq t_0$, i.e., $\mathbf{u}(t, t_1) = \mathbf{u}(t, t_0)$ for all $t_1 \geqq t_0$ and all $t \geqq a$.*

Proof. If $Q(t_1)$ is nonsingular, then the solution of (4.20a,b,c) for $b = t_1 > t_0$ is simply

$$\mathbf{u}(t, t_1) = Y(t, a)Q^{-1}(t_1)\begin{pmatrix} \boldsymbol{\alpha} \\ \mathbf{0} \end{pmatrix}.$$

However, from (4.25a) it follows that

$$Q^{-1}(t_1) = Q^{-1}(t_0)\begin{pmatrix} I_{p\times p} & 0 \\ 0 & e^{-A^+[t-t_0]} \end{pmatrix}.$$

Now note that

$$\mathbf{u}(t, t_0) - \mathbf{u}(t, t_1) = Y(t, a)Q^{-1}(t_0)\begin{pmatrix} 0 & 0 \\ 0 & I_{q\times q} - e^{-A^+[t-t_0]} \end{pmatrix}\begin{pmatrix} \boldsymbol{\alpha} \\ \mathbf{0} \end{pmatrix} = \mathbf{0}. \quad ▧$$

The result above shows how effective our procedure can be for problems with constant tails in the coefficients. Similar but not quite so dramatic results hold if the approach of $A(t)$ to A_∞ is sufficiently rapid. Such studies are in progress. It should also be stressed that for eigenvalue problems on infinite domains the same technique can be used. The analysis for the constant tail case is quite similar and indeed *the exact eigenvalues and eigenfunctions can be found in terms of a finite interval eigenvalue problem.* The key element here is again the identity (4.25a) where now $A_\infty(\lambda)$ and $Q(t_0, \lambda)$ depend analytically on the eigenvalue parameter λ.

In problems of physical significance over semi-infinite intervals the proper condition at infinity is usually the "outgoing wave" condition (i.e., Sommerfield radiation condition). That is what our above analysis yields in all such cases. Indeed from the physical point of view the results for constant tail at infinity are not surprising at all. If at any point in a homogeneous tail there are no incoming waves present, then none can be generated in the entire uniform tail. However, it is surprising that so many quantum mechanical eigenvalue problems and fluid mechanical stability problems are treated numerically without employing the outgoing wave condition.

In particular it is of interest to consider the Orr–Sommerfield equation for external flow problems (see, for example, Rosenhead [47]):

(4.27a)
$$\phi'''' - 2\alpha^2\phi'' + \alpha^4\phi = i\alpha R[(U(\eta) - c)(\phi'' - \alpha^2\phi) - U''(\eta)\phi].$$

Here $U(\eta)$ is the "external" velocity profile normalized such that $\lim_{\eta\to\infty} U(\eta) = 1$.

Indeed it is usually more realistic to assume that $U(\eta) = 1$ for all $\eta \geqq \eta_\infty$; this results in the constant tail case treated above. The boundary conditions at $\eta = 0$ are

$$\varphi(0) = \varphi'(0) = 0. \tag{4.27b}$$

Conditions at $\eta = \infty$ are specified in various ways (see, for instance, Gill and Davey [17] or Rosenhead [47]). The simplest statement seems to be $\varphi(\infty) = \varphi'(\infty) = 0$, which, for numerical purposes, is then imposed at a finite point.

We shall derive the proper "outgoing wave" conditions here by first reformulating (4.27a) as a first order system with

$$\mathbf{y}^T(\eta) \equiv (\psi(\eta), \psi'(\eta), \varphi(\eta), \varphi'(\eta)), \quad \psi(\eta) \equiv \varphi''(\eta) - \alpha^2\phi(\eta). \tag{4.28a}$$

Then we get, with $\boldsymbol{\sigma} \equiv (\alpha, R, c)$,

$$\mathbf{y}'(\eta) = A(\eta; \boldsymbol{\sigma})\mathbf{y}(\eta), \tag{4.28b}$$

$$A(\eta; \boldsymbol{\sigma}) \equiv \begin{pmatrix} 0 & 1 & 0 & 0 \\ [\alpha^2 + i\alpha R(U(\eta) - c)] & 0 & -i\alpha R U''(\eta) & 0 \\ 0 & 0 & 0 & 1 \\ 1 & 0 & \alpha^2 & 0 \end{pmatrix}. \tag{4.28c}$$

In the tail $\eta \geqq \eta_\infty$ we have

$$A(\eta, \boldsymbol{\sigma}) = \lim_{\eta \to \infty} A(\eta; \boldsymbol{\sigma}) \equiv A_\infty(\boldsymbol{\sigma}) \equiv \begin{pmatrix} 0 & 1 & 0 & 0 \\ \beta^2 & 0 & 0 & 0 \\ 0 & 0 & 0 & 1 \\ 1 & 0 & \alpha^2 & 0 \end{pmatrix},$$

$$\beta^2 \equiv \alpha^2 + iR(1 - c). \tag{4.29}$$

It easily follows that the eigenvalues of $A_\infty(\boldsymbol{\sigma})$ are $\pm\alpha$ and $\pm\beta$. Suppose the roots are taken such that $\lambda_1 = \alpha$ and $\lambda_2 = \beta$ have $\operatorname{Re}\lambda_i > 0$. The corresponding left eigenvectors are found to be

$$\boldsymbol{\xi}_1^T = (\beta, 1, 0, 0), \quad \boldsymbol{\xi}_2^T = (\alpha, 1, \alpha[\alpha^2 - \beta^2], [\alpha^2 - \beta^2]).$$

Then the boundary conditions are from (4.27b) and the above in (4.19b):

$$C_0\mathbf{y}(0) = \mathbf{0}, \quad C_0 \equiv \begin{pmatrix} 0 & 0 & 1 & 0 \\ 0 & 0 & 0 & 1 \end{pmatrix}, \tag{4.30a}$$

$$C_\infty^+\mathbf{y}(\eta_\infty) = \mathbf{0}, \quad C_\infty^+ \equiv \begin{pmatrix} \beta & 1 & 0 & 0 \\ \alpha & 1 & i\alpha R[c - 1] & iR[c - 1] \end{pmatrix}. \tag{4.30b}$$

Note that all of the parameters $\boldsymbol{\sigma} = (\alpha, R, c)$ in the Orr–Sommerfield equation also enter into the boundary conditions at $\eta = \eta_\infty$. Our methods for eigenvalue problems in Chapter 3 can now be applied to (4.28b) subject to (4.30a,b). In fact for such problems solutions are frequently desired for large ranges of some of the parameters and hence continuation procedures are also quite relevant.

It is of interest to note that the conditions (4.30b) are, in terms of $\phi(\eta)$ and with $D \equiv d/d\eta$, simply

$$\left.\begin{aligned}(D^2 - \alpha^2)(D + \beta)\phi(\eta) = 0\\ (D^2 - \beta^2)(D + \alpha)\phi(\eta) = 0\end{aligned}\right\} \eta = \eta_\infty.$$

We have never seen these conditions imposed on the Orr–Sommerfield equation although related forms, such as

$$(D + \beta)\phi(\eta) = 0, \quad (D + \alpha)\phi(\eta) = 0$$

have been used. The latter are incorrect attempts to eliminate the growing components of $\varphi(\eta)$ while the former clearly do the job.

References

[1] K. W. Atkinson, *Convergence rates for approximate eigenvalues of compact integral operators*, SIAM J. Numer. Anal., 12 (1975), pp. 213–222.

[2] J. Avila, *The feasibility of continuation methods for nonlinear equations*, Ibid., 11 (1974), pp. 102–122.

[3] O. Axelsson, *A class of A-stable methods*, BIT, 9 (1969), pp. 185–199.

[4] L. Bauer, E. L. Reiss and H. B. Keller, *Axisymmetric buckling of hollow spheres and hemispheres*, Comm. Pure Appl. Math., 23 (1970), pp. 529–568.

[5] W. Bosarge, *Iterative continuation and the solution of nonlinear two-point boundary value problems*, Numer. Math., 17 (1971), pp. 268–283.

[6] D. C. Brabston, Jr., *Numerical solution of singular endpoint boundary value problems*, Doctoral thesis, Part II, California Institute of Technology, Pasadena, Calif., 1974.

[7] D. C. Brabston and H. B. Keller, *Numerical solution of singular endpoint boundary value problems*, submitted to SIAM J. Numer. Anal.

[8] J. H. Bramble and J. E. Osborn, *Rate of convergence estimates for nonselfadjoint eigenvalue approximations*, Math. Comp., 27 (1973), pp. 525–549.

[9] E. A. Coddington and N. Levinson, *Theory of Ordinary Differential Equations*, McGraw-Hill, New York, 1955.

[10] S. D. Conte, *The numerical solution of linear boundary value problems*, SIAM Rev., 8 (1966), pp. 309–321.

[11] F. R. de Hoog and R. Weiss, *Difference methods for boundary value problems with a singularity of the first kind*, Tech. Summary Rep. 1536, Mathematics Research Center, University of Wisconsin, Madison, 1975.

[12] B. Engquist, *Asymptotic error expansions for multistep methods*, Reas. Rep., Computer Sci. Dep., Uppsala Univ., 1969.

[13] J. C. Falkenberg, *A method for integration of unstable systems of ordinary differential equations subject to two-point boundary conditions*, BIT, 8 (1968), pp. 86–103.

[14] L. Fox, *The Numerical Solution of Two-Point Boundary Problems in Ordinary Differential Equations*, Oxford University Press, Fairlawn, N.J., 1957.

[15] C. W. Gear, *Numerical Initial Value Problems in Ordinary Differential Equations*, Prentice-Hall, Englewood Cliffs, N.J., 1971.

[16] J. H. George and R. W. Gunderson, *Conditioning of linear boundary value problems*, BIT, 12 (1972), pp. 172–181.

[17] A. E. Gill and A. Davey, *Instabilities of a buoyancy-driven system*, J. Fluid Mech., 35 (1969), pp. 775–798.

[18] T. R. Goodman and G. N. Lance, *The numerical solution of two-point boundary value problems*, Math. Tables Aid. Comput., 10 (1956), pp. 82–86.

[19] W. B. Gragg, *On extrapolation algorithms for ordinary initial value problems*, SIAM J. Numer. Anal., 2 (1965), pp. 384–403.

[20] B. Gustafsson, *A numerical method for solving singular boundary value problems*, Numer. Math., 21 (1973), pp. 328–344.

[21] P. Henrici, *Discrete Variable Methods in Ordinary Differential Equations*, John Wiley, New York, 1962.

[22] ———, *Error Propagation for Difference Methods*, John Wiley, New York, 1963.

[23] E. ISAACSON AND H. B. KELLER, *Analysis of Numerical Methods*, John Wiley, New York, 1966.
[24] H. B. KELLER, *Numerical Methods for Two-Point Boundary Value Problems*, Ginn-Blaisdell, Waltham, Mass., 1968.
[25] ———, *Accurate difference methods for linear ordinary differential systems subject to linear constraints*, SIAM J. Numer. Anal., 6 (1969), pp. 8–30.
[26] ———, *Accurate difference methods for nonlinear two-point boundary value problems*, Ibid., 11 (1974), pp. 305–320.
[27] ———, *Approximation methods for nonlinear problems with application to two-point boundary value problems*, Math. Comp., 29 (1975), pp. 464–474.
[28] ———, *Newton's method under mild differentiability conditions*, J. Comput. System Sci., 4 (1970), pp. 15–28.
[29] ———, *Shooting and embedding for two-point boundary value problems*, J. Math. Anal. Appl., 36 (1971), pp. 598–610.
[30] H. B. KELLER AND A. B. WHITE, JR., *Difference methods for boundary value problems in ordinary differential equations*, SIAM J. Numer. Anal., 12 (1975), pp. 791–802.
[31] H. B. KELLER, *Numerical solution of boundary value problems for ordinary differential equations: Survey and some recent results on difference methods*, Numerical Solutions of Boundary Value Problems for Ordinary Differential Equations, A. K. Aziz, ed., Academic Press, New York, 1975, pp. 27–88.
[32] H. B. KELLER AND A. W. WOLFE, *On the nonunique equilibrium states and buckling mechanism of spherical shells*, SIAM J. Appl. Math., 13 (1965), pp. 674–705.
[33] H.-O. KREISS, *Difference approximations for boundary and eigenvalue problems for ordinary differential equations*, Math. Comp., 10 (1972), pp. 605–624.
[34] F. T. KROGH, *Variable order integrators for the numerical solution of ordinary differential equations*, J. P. L. Tech. Memo, California Institute of Technology, Pasadena, Calif., 1969.
[35] D. C. L. LAM, *Implementation of the Box scheme and modal analysis of diffusion–convection equations*, Doctoral thesis, Univ. of Waterloo, Ontario, 1974.
[36] P. LANCASTER, *Error analysis for the Newton–Raphson method*, Numer. Math., 9 (1966), pp. 55–68.
[37] F. NATTERER, *A generalized spline method for singular boundary value problems of ordinary differential equations*, Lin. Alg. & Appl., 7 (1973), pp. 189–216.
[38] J. ORTEGA AND W. RHEINBOLDT, *Iterative Solution of Nonlinear Equations in Several Variables*, Academic Press, New York, 1970.
[39] M R. OSBORNE, *On shooting methods for boundary value problems*, J. Math. Anal. Appl., 27 (1969), pp. 417–433.
[40] ———, *On the numerical solution of boundary value problems for ordinary differential equations*, Information Processing '74, North-Holland, Amsterdam, 1974, pp. 673–677.
[41] V. PEREYRA, *Iterated deferred corrections for nonlinear boundary value problems*, Numer. Math., 11 (1968), pp. 111–125.
[42] ———, *High order finite difference solution of differential equations*, Comp. Sci. Rep. STAN-CS-73-348, Stanford University, Stanford, Calif., 1973.
[43] L. RALL, *Computational Solution of Nonlinear Operator Equations*, John Wiley, New York, 1969.
[44] W. C. RHEINBOLDT, *Methods for Solving Systems of Nonlinear Equations*, Regional Conference Series in Applied Mathematics, no. 14, SIAM, Philadelphia, Pa., 1974.
[45] S. M. ROBERTS AND J. S. SHIPMAN, *Two-Point Boundary Value Problems: Shooting Methods*, American Elsevier, New York, 1972.
[46] ———, *Continuation in shooting methods for two-point boundary problems*, J. Math. Anal. Appl., 18 (1967), pp. 45–58.
[47] L. ROSENHEAD, *Laminar Boundary Layers*, Oxford University Press, Fairlawn, N.J., 1963.
[48] R. D. RUSSELL, *Collocation for systems of boundary value problems*, Numer. Math., 23 (1974), pp. 119–133.
[49] ———, *Application of B-splines for solving differential equations*, preprint, 1975.
[50] L. SHAMPINE AND M. GORDON, *Computer Solution of Ordinary Differential Equations*, W. H. Freeman, San Francisco, Calif., 1974.

[51] H. J. STETTER, *Asymptotic expansions for the error of discretization algorithms for non-linear functional equations*, Numer. Math., 7 (1965), pp. 18–31.

[52] J. M. VARAH, *On the solution of block-tridiagonal systems arising from certain finite-difference equations*, Math. Comp., 26 (1972), pp. 859–868.

[53] ———, *Alternate row and column elimination for solving certain linear systems*, preprint, 1975.

[54] E. WASSERSTROM, *Numerical solutions by the continuation method*, SIAM Rev., 15 (1973), pp. 89–119.

[55] R. WEISS, *The convergence of shooting methods*, BIT, 13 (1973), pp. 470–475.

[56] ———, *The application of implicit Runge–Kutta and collocation methods to boundary value problems*, Math. Comp., 28 (1974), pp. 449–464.

[57] A. B. WHITE, *Numerical Solution of Two-Point Boundary Value Problems*, Doctoral thesis, California Institute of Technology, Pasadena, Calif., 1974.

[58] J. H. WILKINSON, *The Algebraic Eigenvalue Problem*, Oxford University Press, Fairlawn, N.J., 1965.

APPENDIX C

SOME FURTHER RESULTS

CONTENTS

Reprinted from *Journal of Computer and Systems Sciences*, Vol. 4, No. 1, February, 1970 by permission of Academic Press.

Newton's Method under Mild Differentiability Conditions

HERBERT B. KELLER*

California Institute of Technology, Pasadena, California 94309

Received July 1, 1968

We study Newton's method for determining the solution of $f(x) = 0$ when $f(x)$ is required only to be continuous and piecewise continuously differentiable in some sphere about the initial iterate, $x^{(0)}$. First an existence, uniqueness and convergence theorem is obtained employing the modulus of continuity of the first derivative, $f_x(x)$. Under the more explicit assumption of Hölder continuity several other such results are obtained, some of which extend results of Kantorovich and Akilov [*1*] and Ostrowski [*5*]. Of course, when Newton's method converges, it is now of order $(1 + \alpha)$, where α is the Hölder exponent. Other results on Newton's method without second derivatives are given by Goldstein [*2*], Schroeder [*3*], Rheinboldt [*6*], and Antosiewicz [*7*], to mention a few. It seems clear that the error analysis for Newton's method given by Lancaster [*4*] can be extended to the present case.

Our results are applicable in the general setting of an operator $f(x)$ on some Banach space, B, into itself. The Frechet derivative, $f_x(x)$, is a linear operator on B whose dependence on x is as specified in Lemma 1. The operator norm, $\| \cdot \|$, is that induced by the norm, $| \cdot |$, of B. We employ throughout the notation $f_x(x) \equiv J(x)$ and $N_\rho(x^{(0)}) \equiv \{x \mid x \in B, \mid x - x^{(0)} \mid \leqslant \rho\}$. By $PC^1(D)$ for some convex set D in B, we denote the class of operators f, with Frechet derivative, J, which is bounded and continuous almost everywhere on the line segment joining any two points in D.

Some simple basic observations that will be required are collected as

LEMMA 1. *Let D be a convex set in B and for some fixed $x^{(0)} \in D$ assume:*

$$f(x) \in PC^1(D), \tag{1a}$$

$$J(x^{(0)}) \text{ is nonsingular,} \tag{1b}$$

$$\| J^{-1}(x^{(0)})[J(x) - J(y)]\| \leqslant \omega(| x - y |), \qquad \text{for all } x \text{ and } y \in D. \tag{1c}$$

* This work was supported by the U.S. Army Research Office, Durham, N.C., under contract DAHC 04-68-C-0006.

The function $\omega(\delta)$ *is continuous and nondecreasing for* $\delta > 0$ *with* $\omega(0) \geqslant 0$. *Then for all* x *and* $y \in D$,

$$(2) \qquad J(x, y) \equiv \int_0^1 J(x + t[y - x])\, dt = J(y, x),$$

$$(3) \qquad f(x) - f(y) = J(x, y)[x - y].$$

If in addition $z \in D$, $|x - z| \leqslant \delta_1$ *and* $|x - y| \leqslant \delta_2$, *then*

$$(4) \qquad \| J^{-1}(x^{(0)})[J(z) - J(x, y)]\| \leqslant \omega(\delta_1 + \delta_2).$$

Finally, if for some $K > 0$

$$(1d) \qquad \omega(\delta) \equiv K\delta^\alpha, \qquad \alpha \in [0, 1],$$

then

$$(5) \qquad \| J^{-1}(x^{(0)})[J(z) - J(x, y)]\| \leqslant \frac{K}{1 + \alpha} \frac{(\delta_1 + \delta_2)^{1+\alpha} - \delta_1^{1+\alpha}}{\delta_2}.$$

Proof. Replacing the integration variable in (2) by $\tau \equiv (1 - t)$, it easily follows that $J(x, y) = J(y, x)$. Since D is convex, it is clear that $(x + t[y - x]) \in D$ for all $t \in [0, 1]$, provided x and $y \in D$. Then, since $f(x) \in PC^1(D)$, we have in the finite-dimensional case

$$\begin{aligned} f(x) - f(y) &= \int_0^1 \frac{d}{dt} f(y + t[x - y])\, dt \\ &= \int_0^1 J(y + t[x - y])\, dt[x - y] \\ &= J(x, y)[x - y]. \end{aligned}$$

This is also justified in the Banach space setting by Graves [8]. Next, using (1c) and the fact that $\omega(\delta)$ is nondecreasing,

$$\begin{aligned} \| J^{-1}(x^{(0)})[J(z) - J(x, y)]\| &= \left\| \int_0^1 J^{-1}(x^{(0)})[J(z) - J(x + t[y - x])]\, dt \right\| \\ &\leqslant \int_0^1 \omega(\| z - x - t[y - x]\|)\, dt \\ &\leqslant \int_0^1 \omega(\delta_1 + t\delta_2)\, dt \\ &\leqslant \omega(\delta_1 + \delta_2). \end{aligned}$$

Finally with (1d) in the next to last inequality above we obtain (5). ∎

Our first result is an existence, uniqueness, and convergence proof for the chord method and Newton's method.

THEOREM 1. *For some* $x^{(0)} \in B$ *and* $\rho > 0$ *let* $f(x)$ *satisfy* (1a–c) *with* $D \equiv N_\rho(x^{(0)})$. *In addition assume that*

$$\text{(6a)} \qquad \omega(\rho) < 1,$$

$$\text{(6b)} \qquad \eta_0 \equiv |\, J^{-1}(x^{(0)}) f(x^{(0)})| \leqslant [1 - \omega(\rho)]\rho.$$

Then $f(x) = 0$ *has a solution, say* x^*, *which is unique in* $N_\rho(x^{(0)})$. *This solution is the limit of the iterates* $\{\xi^{(\nu)}\}$ *defined by the chord method:*

$$\text{(7a)} \qquad \xi^{(0)} = x^{(0)},$$

$$\text{(7b)} \qquad \xi^{(\nu+1)} = \xi^{(\nu)} - J^{-1}(x^{(0)}) f(\xi^{(\nu)}), \qquad \nu = 0, 1, \ldots.$$

If we have the stronger conditions

$$\text{(6c)} \qquad \omega(\rho) \leqslant \tfrac{1}{2}, \qquad \eta_0 \leqslant [1 - 2\omega(\rho)]\rho;$$

then the solution, x^*, *is also the limit of the Newton iterates,* $\{x^{(\nu)}\}$, *defined by*

$$\text{(8)} \qquad x^{(\nu+1)} = x^{(\nu)} - J^{-1}(x^{(\nu)}) f(x^{(\nu)}), \qquad \nu = 0, 1, \ldots.$$

The errors are bounded by

$$\text{(9a)} \qquad |\, \xi^{(\nu)} - x^* \,| \leqslant \omega^\nu(\rho)\rho, \qquad \nu = 0, 1, \ldots;$$

$$\text{(9b)} \qquad |\, x^{(\nu+1)} - x^* \,| \leqslant \left(\frac{\omega(\rho)}{1 - \omega(\rho)}\right)^\nu \left(\prod_{\mu=1}^{\nu} \frac{\omega(|\, x^{(\mu)} - x^* \,|)}{\omega(\rho)}\right) \frac{\rho}{2}, \qquad \nu = 1, 2, \ldots.$$

Proof. To show existence, uniqueness, and convergence of the chord method (7), we need only observe that

$$S(x) \equiv x - J^{-1}(x^{(0)}) f(x)$$

is a contracting map on $N_\rho(x^{(0)})$ into itself. In fact, for any $x \in N_\rho(x^{(0)})$, by (1c) and (6a),

$$\text{(10)} \qquad \begin{aligned} \| S_x(x) \| &= \| I - J^{-1}(x^{(0)})\, J(x) \| \\ &= \| J^{-1}(x^{(0)}) [J(x^{(0)}) - J(x)] \| \\ &\leqslant \omega(|\, x - x^{(0)} \,|) \leqslant \omega(\rho) < 1. \end{aligned}$$

Further, by (6b) and (6a),

$$\begin{aligned} |\, S(x^{(0)}) - x^{(0)} \,| &= |\, J^{-1}(x^{(0)}) f(x^{(0)}) | = \eta_0 \\ &\leqslant [1 - \omega(\rho)]\rho. \end{aligned}$$

Standard arguments now yield the estimates in (9a).

Writing

$$J^{-1}(x^{(0)})\, J(x) = I - S_x(x),$$

we find using (10) and the Banach lemma, that $J(x)$ is nonsingular for all $x \in N_\rho(x^{(0)})$ and

$$\| J^{-1}(x)\, J(x^{(0)}) \| \leqslant \frac{1}{1 - \omega(| x - x^{(0)} |)}. \tag{11}$$

From (6c) and (8) with $\nu = 0$, it follows that $x^{(1)} \in N_\rho(x^{(0)})$. For an induction assume $x^{(k)} \in N_\rho(x^{(0)})$ for $k = 1, 2,\ldots, n$. Then from (8) and (3) we write

$$\begin{aligned}(x^{(n+1)} - x^{(0)}) &= (x^{(n)} - x^{(0)}) - J^{-1}(x^{(n)}) f(x^{(n)})\\ &= (x^{(n)} - x^{(0)}) - J^{-1}(x^{(n)})[f(x^{(0)}) + J(x^{(n)}, x^{(0)})(x^{(n)} - x^{(0)})]\\ &= J^{-1}(x^{(n)})[J(x^{(n)}) - J(x^{(n)}, x^{(0)})](x^{(n)} - x^{(0)}) - J^{-1}(x^{(n)}) f(x^{(0)}).\end{aligned}$$

Taking norms and using (11), (4), and (6c) yields

$$\begin{aligned}| x^{(n+1)} - x^{(0)} | &\leqslant \frac{\omega(| x^{(n)} - x^{(0)} |)}{1 - \omega(| x^{(n)} - x^{(0)} |)} | x^{(n)} - x^{(0)} | + \frac{1 - 2\omega(\rho)}{1 - \omega(| x^{(n)} - x^{(0)} |)} \rho\\ &\leqslant \rho,\end{aligned}$$

since $| x^{(n)} - x^{(0)} | \leqslant \rho$. Thus, all the Newton iterates $x^{(\nu)}$ are in $N_\rho(x^{(0)})$, and hence the procedure is properly defined by (8); i.e., all required inverses exist.

Since $f(x) = 0$ has a unique solution $x^* \in N_\rho(x^{(0)})$, it is trivial that

$$| x^{(0)} - x^* | \leqslant \rho.$$

Now we write, using (8), (3), and $f(x^*) = 0$,

$$\begin{aligned}(x^{(\nu+1)} - x^*) &= (x^{(\nu)} - x^*) - J^{-1}(x^{(\nu)})[f(x^{(\nu)}) - f(x^*)]\\ &= J^{-1}(x^{(\nu)}[J(x^{(\nu)}) - J(x^{(\nu)}, x^*)](x^{(\nu)} - x^*), \qquad \nu = 0, 1,\ldots.\end{aligned}$$

From (11) and (4) it follows that

$$| x^{(1)} - x^* | \leqslant \omega(| x^{(0)} - x^* |)\, | x^{(0)} - x^* |$$

and

$$\begin{aligned}| x^{(\nu+1)} - x^* | &\leqslant \frac{\omega(| x^{(\nu)} - x^* |)}{1 - \omega(\rho)} | x^{(\nu)} - x^* |\\ &\leqslant \left(\frac{\omega(\rho)}{1 - \omega(\rho)}\right)\left(\frac{\omega(| x^{(\nu)} - x^* |)}{\omega(\rho)}\right) | x^{(\nu)} - x^* |, \qquad \nu = 1, 2,\ldots.\end{aligned}$$

The bound in (9b) follows by induction, recalling finally that

$$\omega(| x^{(0)} - x^* |) \leqslant \omega(\rho) \leqslant \tfrac{1}{2}. \qquad \blacksquare$$

We note that (9b) shows convergence of Newton's method with *arbitrarily* small convergence factor if $f_x(x)$ is continuous[1] at the solution, $x = x^*$. But from these error bounds, assuming they give the correct order of magnitude as $\nu \to \infty$,

$$\lim_{\nu\to\infty} \frac{\log | x^{(\nu+1)} - x^* |}{\log | x^{(\nu)} - x^* |} = 1,$$

so the iteration scheme is not shown to be more than first order. The theorem also shows that Newton's method can converge if $f_x(x)$ is only piecewise continuous at x^* with "small" jumps.

If we impose stronger smoothness conditions on $f_x(x)$, say Hölder continuity implied by (1c,d), then it easily follows from (9b) that the convergence is at least of order $1 + \alpha$. However, in this case we can weaken the conditions (6) considerably. A very simple such extension is stated as

THEOREM 2. *For some $x^{(0)} \in B$ and $\rho > 0$, let $f(x)$ satisfy* (1a–d) *with $D \equiv N_\rho(x^{(0)})$. In addition assume that, with the α and K of* (1d),

$$K\rho^\alpha < \frac{1+\alpha}{2+\alpha}, \tag{12a}$$

$$\eta_0 \equiv | J^{-1}(x^{(0)})\, f(x^{(0)})| \leqslant \left[1 - \frac{2+\alpha}{1+\alpha} K\rho^\alpha\right] \rho. \tag{12b}$$

Then Newton's method (8) *and the chord method* (7) *converge to a unique solution, x^*, of $f(x) = 0$ in $N_\rho(x^{(0)})$. The error in Newton's method is bounded, if $0 \leqslant \alpha \leqslant 1$, by*

$$| x^{(\nu+1)} - x^* | \leqslant \lambda^\nu \left(\prod_{\mu=1}^{\nu} \left| \frac{x^{(\mu)} - x^*}{\rho} \right|^\alpha\right) \frac{\rho}{2+\alpha}, \qquad \nu = 1, 2, \ldots, \tag{13a}$$

and if $0 < \alpha \leqslant 1$, by

$$| x^{(\nu+1)} - x^* | \leqslant \left(\frac{\lambda^{1/\alpha}}{2+\alpha}\right)^{(1+\alpha)^\nu} \frac{\rho}{\lambda^{1/\alpha}}, \qquad \nu = 1, 2, \ldots, \tag{13b}$$

where

$$\lambda \equiv \left(\frac{K\rho^\alpha}{1 - K\rho^\alpha}\right) \frac{1}{1+\alpha} < 1. \tag{13c}$$

Proof. The proof is very similar to that of Theorem 1, so we do not include it. In the two places where (4) had been used in the previous proof, we now employ (5), which accounts for all of the improvements. ∎

[1] In this case $\omega(| x^{(\nu)} - x^* |) \to 0$ as $\nu \to \infty$.

However, using condition (1d) properly, we can obtain results which are much sharper than the above. First we have an existence and uniqueness theorem involving only the chord method:

THEOREM 3. *For some* $x^{(0)} \in B$ *and* $\rho > 0$, *let* $f(x)$ *satisfy* (1a–d) *with* $D \equiv N_\rho(x^{(0)})$. *Define*

(14a)
$$\eta_0 \equiv | J^{-1}(x^{(0)}) f(x^{(0)})|,$$

and assume that, with the α *and* K *of* (1d),

(14b)
$$K\eta_0{}^\alpha < \left(\frac{\alpha}{1+\alpha}\right)^\alpha.$$

Let $r_0(\alpha)$ *and* $r_1(\alpha)$ *be, respectively, the minimum and maximum positive roots, with the convention* $r_1(0) = \infty$, *of*

(15)
$$\psi(t) \equiv Kt^{1+\alpha} - (1+\alpha)(t - \eta_0) = 0, \qquad \alpha \in [0, 1].$$

(A) *If* $\rho \geqslant r_0(\alpha)$ *then* $f(x) = 0$ *has a solution,* x^*, *in* $N_\rho(x^{(0)})$. *This solution is the limit of the sequence* $\{\xi^{(\nu)}\}$ *determined by the chord method* (7).

(B) *If* $r_0(\alpha) \leqslant \rho < r_1(\alpha)$, *then* x^* *is the only solution of* $f(x) = 0$ *in* $N_\rho(x^{(0)})$ *and the iterates in* (7b) *converge to this solution for any* $\xi^{(0)} \in N_\rho(x^{(0)})$.

Proof. The simplest proof is to define

(16a)
$$S(x) \equiv x - J^{-1}(x^{(0)}) f(x),$$

(16b)
$$\varphi(t) \equiv \frac{K}{1+\alpha} t^{1+\alpha} + \eta_0,$$

(16c)
$$[t_0, t'] \equiv [0, \rho],$$

and then apply Theorem 1, p. 697, in Kantorovich and Akilov [1], for part A) and Theorem 2, p. 700, *loc. cit.*, for part B). In particular from (1c,d), (14a) and, (16):

$$\begin{aligned} | S(x^{(0)}) - x^{(0)} | &= | J^{-1}(x^{(0)}) f(x^{(0)})| \\ &= \eta_0 \\ &= \varphi(t_0) - t_0 . \end{aligned}$$

For any $x \in N_\rho(x^{(0)})$, using (10) and (1c–d),

$$\begin{aligned} \| S_x(x)\| &\leqslant K | x - x^{(0)} |^\alpha, \\ &= \varphi'(x - x^{(0)}). \end{aligned}$$

Thus the equation $t = \varphi(t)$ "majorizes" the equation $x = S(x)$. But from (14b), (15), and (16b), it follows that $t = r_0(\alpha)$ is a root of $t = \varphi(t)$ in $[0, \rho]$ if $\rho \geqslant r_0(\alpha)$. Finally, for t' in $r_0(\alpha) \leqslant t' < r_1(\alpha)$, we see that $\varphi(t') \leqslant t'$ and $r_0(\alpha)$ is the only root of $t = \varphi(t)$ in $[0, t']$. Thus all the conditions of the cited theorems are satisfied. ∎

It should be observed that by setting $\alpha = 1$ in Theorem 3, we obtain a result formally similar to that of Theorem 6, p. 708, *loc. cit.* However, here we require only Lipschitz continuous first derivatives, while the Kantorovich and Akilov result employs continuous second derivatives. Our Lipschitz constant, of course, replaces their bound on the second derivative. It is also of interest to observe that Newton's method with initial iterate $t_0 = 0$ applied to (15) will converge monotonically to the root $r_0(\alpha)$ for all $\alpha \in [0, 1]$. However, for $\alpha \in (0, 1)$ neither the Kantorovich nor Ostrowski [5] proofs hold (since $\psi''(t_0)$ does not exist).

In order to prove the convergence of Newton's method we strengthen the condition (14) somewhat. Specifically we have

THEOREM 4. *For some $x^{(0)} \in B$ and $\rho > 0$ let $f(x)$ satisfy* (1a–d) *with $D \equiv N_\rho(x^{(0)})$. Define*

(17a) $$\eta_0 \equiv | J^{-1}(x^{(0)}) f(x^{(0)})|,$$

and assume that, with the α and K of (1d),

(17b) $$(2 + \alpha) K\eta_0{}^\alpha \leqslant \left(\frac{\alpha}{1 + \alpha}\right)^\alpha.$$

Let $R_0(\alpha)$ be the minimum positive root of

(18) $$\Psi(t) \equiv (2 + \alpha) Kt^{1+\alpha} - (1 + \alpha)(t - \eta_0) = 0.$$

If $\rho \geqslant R_0(\alpha)$, then $f(x) = 0$ has a solution $x^* \in N_\rho(x^{(0)})$ and the Newton iterates (8), *with initial point $x^{(0)}$, converge to this solution. The errors are bounded, for α in $0 \leqslant \alpha \leqslant 1$, by*

(19a) $$| x^{(\nu)} - x^* | \leqslant \frac{\Lambda^\nu}{1 - \Lambda} \eta_0, \qquad \nu = 0, 1, \ldots,$$

and for α in $0 < \alpha \leqslant 1$, by

(19b) $$| x^{(\nu)} - x^* | \leqslant \left(\frac{\Lambda^{1/\alpha}}{1 - \Lambda}\right)^{(1+\alpha)^\nu} \frac{\eta_0}{\Lambda^{1/\alpha}}, \qquad \nu = 0, 1, \ldots,$$

where

(19c) $$\Lambda \equiv \left(\frac{KR_0{}^\alpha(\alpha)}{1 - KR_0{}^\alpha(\alpha)}\right)\left(\frac{\eta_0}{R_0(\alpha)}\right)^\alpha \frac{1}{1 + \alpha} < 1.$$

Proof. Define the scalar function

(20) $$\Phi(t) \equiv \frac{Kt^{1+\alpha} + (1 + \alpha)\eta_0}{(1 + \alpha)(1 - Kt^\alpha)},$$

and consider the sequence $\{t_\nu\}$ defined by

$$\text{(21a)} \qquad t_0 = 0, \quad t_{\nu+1} = \Phi(t_\nu), \qquad \nu = 0, 1, \ldots .$$

Using condition (17b), it can be shown that

$$\text{(21b)} \qquad \eta_0 = t_1 < t_2 < \cdots < t_\nu < t_{\nu+1} < \cdots < R_0(\alpha),$$

and

$$\text{(21c)} \qquad \lim_{\nu\to\infty} t_\nu = R_0(\alpha).$$

To sketch the proof of these facts, we first observe that (17b) implies $\Psi(t) = 0$ has at least one and at most two real roots (since $\Psi(t_*) \leqslant 0$ at t_*, the minimum of $\Psi(t)$ for $t > 0$). The roots of $\Psi(t) = 0$ are also roots of $t = \Phi(t)$, and $\Phi(t)$ is a continuous monotone increasing function of t on $0 \leqslant t < K^{-1/\alpha}$. Clearly $\Phi(0) = \eta_0$, $\Phi(R_0(\alpha)) = R_0(\alpha) < K^{-1/\alpha}$ and $\Phi(t) > t$ on $0 \leqslant t < R_0(\alpha)$. The results in (21b,c) easily follow.

We shall now prove, by induction, that

$$\text{(22a)} \qquad | x^{(\nu)} - x^{(0)} | \leqslant t_\nu ,$$

$$\text{(22b)} \qquad \| J^{-1}(x^{(\nu)})\, J(x^{(0)})\| \leqslant \frac{1}{1 - Kt_\nu{}^\alpha}, \qquad \nu = 1, 2, \ldots .$$

For $\nu = 1$ using (17a), (8), and (21b), we get

$$| x^{(1)} - x^{(0)} | = | J^{-1}(x^{(0)}) f(x^{(0)})| = \eta_0 = t_1 .$$

Thus, $x^{(1)} \in N_\rho(x^{(0)})$, and from (1c–d) and the above:

$$\begin{aligned} \| I - J^{-1}(x^{(0)})\, J(x^{(1)})\| &= \| J^{-1}(x^{(0)})[J(x^{(0)}) - J(x^{(1)})]\| \\ &\leqslant Kt_1{}^\alpha = K\eta_0{}^\alpha \leqslant \left(\frac{\alpha}{1+\alpha}\right)^\alpha \frac{1}{2+\alpha} < 1. \end{aligned}$$

Hence, $J^{-1}(x^{(0)})\, J(x^{(1)})$ is nonsingular, and

$$\| J^{-1}(x^{(1)})\, J(x^{(0)})\| \leqslant \frac{1}{1 - Kt_1{}^\alpha} .$$

Now assume (22a,b) valid for $\nu = 1, 2, \ldots, n$. From (8), (3), (5), and (17a), we get, since $x^{(n)} \in N_\rho(x^{(0)})$,

$$\begin{aligned} | x^{(n+1)} - x^{(0)} | &= |[x^{(n)} - x^{(0)}] - J^{-1}(x^{(n)})\, f(x^{(n)})| \\ &= |[x^{(n)} - x^{(0)}] - J^{-1}(x^{(n)})(f(x^{(0)}) + J(x^{(n)}, x^{(0)})[x^{(n)} - x^{(0)}])| \\ &\leqslant | J^{-1}(x^{(n)})[J(x^{(n)}) - J(x^{(n)}, x^{(0)})][x^{(n)} - x^{(0)}]| + | J^{-1}(x^{(n)})\, f(x^{(0)})| \\ &\leqslant \| J^{-1}(x^{(n)})\, J(x^{(0)})\| \left(\frac{K}{1+\alpha} | x^{(n)} - x^{(0)} |^{1+\alpha} + \eta_0\right). \end{aligned}$$

Using (22a,b) with $\nu = n$, the above yields, from (20) and (21a),

$$| x^{(n+1)} - x^{(0)} | \leqslant \Phi(t_n) = t_{n+1} .$$

Thus, $x^{(n+1)} \in N_\rho(x^{(0)})$, since $t_{n+1} < R_0(\alpha) \leqslant \rho$, and again using (1c,d) with $\Psi(R_0) = 0$,

$$\begin{aligned} \| I - J^{-1}(x^{(0)})\, J(x^{(n+1)})\| &\leqslant K t_{n+1}^{\alpha} \\ &\leqslant K R_0{}^{\alpha}(\alpha) = \frac{1+\alpha}{2+\alpha}\left(1 - \frac{\eta_0}{R_0(\alpha)}\right) \\ &< 1. \end{aligned}$$

Now (22b) follows for $\nu = n + 1$, and (22a,b) is established. This result implies that $x^{(\nu)} \in N_\rho(x^{(0)})$ and $J(x^{(\nu)})$ is nonsingular for $\nu = 0, 1, \ldots$, and so Newton's method is well defined.

Next we shall verify, by induction, that

$$| x^{(\nu+1)} - x^{(\nu)} | \leqslant \Lambda^{\nu} \eta_0 , \qquad \nu = 0, 1, \ldots. \tag{23}$$

For $\nu = 0$ this follows trivially from (8) and (17a). Assuming (23) valid for $\nu = 1, 2, \ldots, n - 1$, we obtain from (8), (3), (22b) and (5):

$$\begin{aligned} | x^{(n+1)} - x^{(n)} | &= | J^{-1}(x^{(n)})(f(x^{(n-1)}) + J(x^{(n)}, x^{(n-1)})[x^{(n)} - x^{(n-1)}])| \\ &= | J^{-1}(x^{(n)})[J(x^{(n-1)}) - J(x^{(n)}, x^{(n-1)})][x^{(n)} - x^{(n-1)}]| \\ &\leqslant \| J^{-1}(x^{(n)})\, J(x^{(0)})\| \cdot \| J^{-1}(x^{(0)})[J(x^{(n-1)}) - J(x^{(n)}, x^{(n-1)})]\| \cdot \Lambda^{n-1}\eta_0 \\ &\leqslant \frac{K\, | x^{(n)} - x^{(n-1)} |^{\alpha}}{(1+\alpha)(1 - K t_n{}^{\alpha})} \Lambda^{n-1} \eta_0 . \end{aligned} \tag{24}$$

However, using

$$K R_0{}^{\alpha} = \frac{1+\alpha}{2+\alpha}\left(1 - \frac{\eta_0}{R_0{}^{\alpha}}\right),$$

in the definition (19c) of Λ, we get

$$\Lambda = \frac{\left(1 - \dfrac{\eta_0}{R_0}\right)}{\left(1 + [1+\alpha]\dfrac{\eta_0}{R_0}\right)} \left(\frac{\eta_0}{R_0}\right)^{\alpha}.$$

But $\eta_0 < R_0(\alpha)$ from (21b), and so $\Lambda < 1$ as asserted in (19c). Then from (23) with $\nu = n - 1$, we get $| x^{(n)} - x^{(n-1)} | \leqslant \eta_0$, and (24) yields, recalling (21b),

$$\begin{aligned} | x^{(n+1)} - x^{(n)} | &\leqslant \frac{K \eta_0{}^{\alpha}}{(1+\alpha)(1 - K t_0{}^{\alpha})} \Lambda^{n-1} \eta_0 \\ &\leqslant \frac{K R_0{}^{\alpha}}{(1+\alpha)(1 - K R_0{}^{\alpha})} \Lambda^{n-1} \eta_0 = \Lambda^n \eta. \end{aligned}$$

Thus (23) is demonstrated.

Since $\Lambda < 1$, it clearly follows from (23) that $\{x^{(\nu)}\}$ is a Cauchy sequence with limit $x^* \in N_\rho(x^{(0)})$. From (1c,d) we have that

$$\| J(x^{(\nu)})\| \leqslant (1 + K\rho)\| J(x^{(0)})\|, \qquad \nu = 0, 1, \ldots,$$

and using this and (23) in (8), we get

$$\begin{aligned}\|f(x^{(\nu)})\| &= \| J(x^{(\nu)})[x^{(\nu+1)} - x^{(\nu)}]\| \\ &\leqslant (1 + K\rho)\| J(x^{(0)})\| \Lambda^{\nu}\eta_0, \qquad \nu = 0, 1, \ldots.\end{aligned}$$

But $f(x)$ is continuous on $N_\rho(x^{(0)})$, so letting $\nu \to \infty$, we obtain $f(x^*) = 0$. Finally, from (23) it follows that for any $\nu \geqslant 0$ and $\mu \geqslant 0$,

$$| x^{(\nu+\mu)} - x^{(\nu)} | \leqslant \Lambda^{\nu} \frac{1 - \Lambda^{\mu}}{1 - \Lambda} \eta_0,$$

and (19a) results from $\mu \to \infty$.

Using the solution, x^*, we obtain from (8), (3), (22b), and (5)

$$\begin{aligned}| x^{(\nu+1)} - x^* | &= |[x^{(\nu)} - x^*] - J^{-1}(x^{(\nu)})[f(x^{(\nu)}) - f(x^*)]| \\ &\leqslant \| J^{-1}(x^{(\nu)}) J(x^{(0)})\| \cdot \| J^{-1}(x^{(0)})[J(x^{(\nu)}) - J(x^{(\nu)}, x^*)]\| \cdot | x^{(\nu)} - x^* | \\ &\leqslant \frac{K | x^{(\nu)} - x^* |^{1+\alpha}}{(1 + \alpha)(1 - Kt_\nu^{\alpha})}, \qquad \nu = 0, 1, \ldots .\end{aligned}$$

Using $t_\nu < R_0(\alpha)$ from (21b) and dividing by η_0, we get now

$$\frac{| x^{(\nu+1)} - x^* |}{\eta_0} \leqslant \Lambda \left(\frac{| x^{(\nu)} - x^* |}{\eta_0}\right)^{1+\alpha}, \qquad \nu = 0, 1, \ldots .$$

With $\alpha \in (0, 1]$ an induction yields (19b), since $| x^{(0)} - x^* | \leqslant \eta_0/(1 - \Lambda)$ by (19a). ∎

This result in particular shows the convergence of Newton's method to the root $R_0(\alpha)$ of the scalar equation (18) with $t_0 = 0$, while again the Kantorovich and Ostrowski results are not applicable in this case if $0 < \alpha < 1$.

We have not been able to prove the convergence of Newton's method under the conditions of Theorem 3. Thus, although Theorem 4 is an extension of the Newton–Kantorovich theorem, it does not formally reduce to that case when $\alpha = 1$. That is, in place of $K\eta_0 < \frac{1}{2}$, as in (14b) with $\alpha = 1$, we get $K\eta_0 < \frac{1}{6}$ from (17b). [Again we must recall that here K is the Lipschitz constant and not a bound on the second derivatives.] However, we can obtain a result which does formally reduce to the Newton–Kantorovich case when $\alpha = 1$, but this result does not yield conditions for convergence in the limit case $\alpha \to 0$.

THEOREM 5. *For some $x^{(0)} \in B$ and $\rho > 0$, let $f(x)$ satisfy* (1a–d) *with $D \equiv N_\rho(x^{(0)})$. Define*

$$\eta_0 \equiv | J^{-1}(x^{(0)}) f(x^{(0)})|,$$

and assume that, with $\alpha \in (0, 1]$ *and* K *of* (1d),

$$K\eta_0^{\alpha} \leqslant 1 - \left(\frac{1}{1+\alpha}\right)^{\alpha}, \tag{25a}$$

$$\rho \geqslant \frac{(1+\alpha)}{(2+\alpha)-(1+\alpha)^{\alpha}}\,\eta_0 . \tag{25b}$$

Then $f(x) = 0$ *has a root* $x^* \in N_\rho(x^{(0)})$ *and Newton's iterates* (8) *converge to* x^* *with*

$$|\, x^{(n)} - x^* \,| \leqslant \left[\frac{1}{1+\alpha}\right]^n [(1+\alpha)\,K^{1/\alpha}\eta_0]^{(1+\alpha)^n}\,K^{-1/\alpha}, \qquad n = 0, 1, 2, \ldots . \tag{25c}$$

Proof. If the iterates $\{x^{(\nu)}\}$ are well defined by (8), we introduce the notation $J_\nu \equiv J(x^{(\nu)})$ and

$$\text{(26)}\quad \left.\begin{aligned} &\text{(a)} && K_\nu \equiv K \,\|\, J_\nu^{-1} J_0 \,\| \\ &\text{(b)} && \eta_\nu \equiv |\, x^{(\nu+1)} - x^{(\nu)} \,| = |\, J_\nu^{-1} f(x^{(\nu)})| \\ &\text{(c)} && L_\nu \equiv (1+\alpha)^{\alpha}\, K_\nu \eta_\nu^{\alpha} \\ &\text{(d)} && A_{\nu+1} \equiv I - J_\nu^{-1} J_{\nu+1} \end{aligned}\right\} \quad \nu = 0, 1, \ldots .$$

Here we have employed the iteration scheme (8) in (26b). We will show by induction that

$$\text{(27)}\quad \left.\begin{aligned} &\text{(a)} && x^{(\nu)} \in N_\rho(x^{(0)}),\ J_\nu^{-1} \text{ exists} \\ &\text{(b)} && \eta_\nu \leqslant \frac{L_{\nu-1}}{1+\alpha}\,\eta_{\nu-1} \leqslant \left(\frac{L_0}{1+\alpha}\right)^{\nu} \eta_0 \\ &\text{(c)} && L_\nu \leqslant L_0^{(1+\alpha)^\nu} \end{aligned}\right\} \quad \nu = 1, 2, \ldots .$$

Since $|\, x^{(1)} - x^{(0)} \,| = \eta_0 \leqslant \rho$ by (25b), we have that $x^{(1)} \in N_\rho(x^{(0)})$. Using this, (1c,d), and (25a) yields

$$\|\, A_1 \,\| = \|\, J_0^{-1}[J_0 - J_1] \| \leqslant K\eta_0^{\alpha} < 1.$$

Thus, by the Banach lemma $J_1 = J_0[I - A_1]$ is nonsingular and (27a) is established for $\nu = 1$. In addition, the above gives

$$\|\, J_1^{-1} J_0 \,\| \leqslant \frac{1}{1 - K_0\eta_0^{\alpha}} ;$$

and so from (26a)

$$K_1 \leqslant \frac{K_0}{1 - K_0\eta_0^{\alpha}} . \tag{28}$$

However, using (26b) with $\nu = 0, 1$ and Lemma 1,

$$\begin{aligned}\eta_1 &= | J_1^{-1} f(x^{(1)})| = | J_1^{-1}[f(x^{(0)}) + J(x^{(0)}, x^{(1)})(x^{(1)} - x^{(0)})]| \\ &= | J_1^{-1}[J_0 - J(x^{(0)}, x^{(1)})](x^{(1)} - x^{(0)})| \\ &\leqslant \| J_1^{-1} J_0 \| \cdot \frac{K}{1+\alpha} \eta_0^{1+\alpha} = K_1 \frac{\eta_0^{1+\alpha}}{1+\alpha}.\end{aligned}$$

With (28) this yields

$$\eta_1 \leqslant \frac{K_0 \eta_0{}^\alpha}{1 - K_0 \eta_0{}^\alpha} \frac{\eta_0}{1+\alpha},$$

and so by (25a) and (26c) we deduce (27b) for $\nu = 1$. Raising this inequality to the α-power and multiplying by (28) gives with the aid of (25a)

$$(1+\alpha)^\alpha K_1 \eta_1{}^\alpha \leqslant \left(\frac{K_0 \eta_0{}^\alpha}{1 - K_0 \eta_0{}^\alpha}\right)^{1+\alpha} \leqslant [(1+\alpha)^\alpha K_0 \eta_0{}^\alpha]^{1+\alpha},$$

or equivalently

$$L_1 \leqslant L_0^{(1+\alpha)}.$$

Thus (27a–c) are established for $\nu = 1$.

We make the inductive assumption that (27a–c) hold for all $\nu \leqslant n$. Then, since $L_0 \leqslant (1+\alpha)^\alpha - 1 \leqslant 1$, by (25a), we obtain from (26b) and (27b) for $\nu \leqslant n$

$$\begin{aligned}| x^{(n+1)} - x^{(0)} | &\leqslant \eta_0 + \eta_1 + \cdots + \eta_n \\ &\leqslant \eta_0 \left[1 + \left(\frac{L_0}{1+\alpha}\right) + \cdots + \left(\frac{L_0}{1+\alpha}\right)^n\right] \\ &\leqslant \frac{\eta_0}{1 - \left(\dfrac{L_0}{1+\alpha}\right)} \leqslant \frac{(1+\alpha)\,\eta_0}{(2+\alpha) - (1+\alpha)^\alpha}.\end{aligned}$$

Now (25b) implies that $x^{(n+1)} \in N_\rho(x^{(0)})$. Since J_n is assumed nonsingular, we have by Lemma 1 and (26a)

$$\begin{aligned}\| A_{n+1} \| &= \| J_n^{-1}(J_n - J_{n+1})\| \\ &\leqslant \| J_n^{-1} J_0 \| \cdot \| J_0^{-1}(J_n - J_{n+1})\| \\ &\leqslant K_n \eta_n{}^\alpha = L_n/(1+\alpha)^\alpha < 1,\end{aligned}$$

since $L_n \leqslant 1$ by (27c) and $L_0 \leqslant 1$. Thus, $J_{n+1} = J_n(I - A_{n+1})$ is nonsingular by the Banach lemma and (27a) holds for $\nu = n + 1$. In addition, we now have

$$K_{n+1} = K \|(I - A_{n+1})^{-1} J_n^{-1} J_0\| \qquad (29)$$
$$\leqslant \frac{K_n}{1 - K_n \eta_n^{\alpha}}.$$

From (26a,b) and Lemma 1 we get, using (29),

$$\eta_{n+1} = | J_{n+1}^{-1}[f(x^{(n)}) + J(x^{(n)}, x^{(n+1)})(x^{(n+1)} - x^{(n)})]|$$
$$\leqslant \| J_{n+1}^{-1} J_0 \| \cdot \| J_0^{-1}[J_n - J(x^{(n)}, x^{(n+1)})]\| \cdot \eta_n$$
$$\leqslant K_{n+1} \frac{\eta_n^{1+\alpha}}{1+\alpha}$$
$$\leqslant \frac{K_n \eta_n^{\alpha}}{1 - K\eta_n^{\alpha}} \frac{\eta_n}{1+\alpha}. \qquad (30)$$

This implies

$$\eta_{n+1} \leqslant \frac{L_n}{(1+\alpha)^{\alpha} - L_n} \frac{\eta_n}{1+\alpha},$$

and since $L_n \leqslant L_0 \leqslant (1 + \alpha)^{\alpha} - 1$ by (27c) for $\nu = n$, we immediately get (27b) for $\nu = n + 1$. From (30) to the αth power multiplied by (29), we get

$$L_{n+1} = (1+\alpha)^{\alpha} K_{n+1} \eta_{n+1}^{\alpha} \leqslant \left(\frac{K_n \eta_n^{\alpha}}{1 - K_n \eta_n^{\alpha}}\right)^{1+\alpha},$$
$$= \left(\frac{L_n}{(1+\alpha)^{\alpha} - L_n}\right)^{1+\alpha},$$
$$\leqslant L_n^{1+\alpha}.$$

Thus the inductive proof of (27a–c) is concluded.

Since $L_0/(1 + \alpha) < 1$, it clearly follows from (27b) and (26b) that the Newton iterates $\{x^{(\nu)}\}$ form a Cauchy sequence. The limit $x^{(\nu)} \to x^* \in N_\rho(x^{(0)})$ is a solution of $f(x) = 0$ by the same reasoning as in the proof of Theorem 4. We use the bound

$$| x^{(n+m)} - x^{(n)} | \leqslant \eta_n + \eta_{n+1} + \cdots + \eta_{n+m-1},$$

and let $m \to \infty$ to get from (27b) and $L_n \leqslant L_0$

$$| x^* - x^{(n)} | \leqslant \frac{(1+\alpha)\,\eta_n}{(1+\alpha) - L_0}.$$

However, from the first inequality in (27b) used with (27c), we have that

$$\eta_n \leqslant \frac{L_{n-1}L_{n-2}\cdots L_0}{(1+\alpha)^n}\eta_0 \leqslant \frac{L_0^{[(1+\alpha)^n-1]/\alpha}}{(1+\alpha)^n}\eta_0 .$$

Thus, recalling that $1 + \alpha - L_0 > 1$, we obtain (25c). ∎

References

1. L. V. Kantorovich and G. P. Akilov. "Functional Analysis in Normed Spaces." Macmillan, New York, 1964.
2. A. A. Goldstein. "Constructive Real Analysis." Harper & Row, New York, 1967.
3. J. Schroeder. "Über das Newtonsche Verfahren." *Arch. Ratl. Mech. Anal.*, **1**, 154 (1957).
4. P. Lancaster. Error analysis for the Newton-Raphson method. *Numer. Math.* **9**, 55–68 (1966).
5. A. Ostrowski. "Solution of Equations and Systems of Equations." Academic Press, New York, 1960.
6. H. Antosiewicz. Newton's method and boundary-value problems. *J. Comp. and Sys. Sci.* **2**, 177–202 (1968).
7. W. Rheinboldt. A unified convergence theory for a class of iterative processes. *SIAM J. Num. Anal.* **5**, 42–63 (1968).
8. L. M. Graves. Riemann integration and Taylor's theorem in general analysis. *Trans. Am. Math. Soc.* **29**, 168–177 (1927).

Reprinted from *Mathematics of Computation,* Volume 29, Number 130, pp. 464–474, by permission of the American Mathematical Society.

Approximation Methods for Nonlinear Problems with Application to Two-Point Boundary Value Problems

By H. B. Keller*

Abstract. General nonlinear problems in the abstract form $F(x) = 0$ and corresponding families of approximating problems in the form $F_h(x_h) = 0$ are considered (in an appropriate Banach space setting). The relation between "isolation" and "stability" of solutions is briefly studied. The main result shows, essentially, that, if the nonlinear problem has an isolated solution and the approximating family has stable Lipschitz continuous linearizations, then the approximating problem has a stable solution which is close to the exact solution. Error estimates are obtained and Newton's method is shown to converge quadratically. These results are then used to justify a broad class of difference schemes (resembling linear multistep methods) for general nonlinear two-point boundary value problems.

1. Introduction. We present a general abstract study of methods for approximating the solution of nonlinear problems formulated in a Banach space setting. Our basic results are of the following kind: *If the nonlinear problem has a solution, and a consistent approximating problem has a stable Lipschitz continuous linearization* (*i.e.*, *Fréchet derivative*), *then the approximating problem has a stable solution which is close to the exact solution.* Estimates of the error are given in terms of the order of consistency, and Newton's method is shown to converge quadratically for computing the approximate solution. Asymptotic error expansions can also be derived under appropriate assumptions. We illustrate the theory by studying difference methods for approximating the *isolated* solutions of nonlinear two-point boundary value problems. Of course, all of these results are local in that they are confined to some sphere about the exact solution. The phenomenon of nonlinear instability does not occur here since, as has been shown by Stetter [10], this requires departing from the sphere. Indeed, parts of our theory are closely related to that of Stetter whose interest, however, was confined to the question of nonlinear instabilities; he therefore assumed the existence of solutions of the approximating problems. Similar existence results have been obtained by Pereyra [13] but his proofs are not constructive.

Received by the editors March 18, 1974.

AMS (MOS) subject classifications (1970). Primary 65J05, 65L10.

Key words and phrases. Nonlinear stability, Newton's method, two point boundary problems, finite difference schemes.

*This work was supported under A.E.C. Contract AT(04–3)–767.

For general nonlinear problems, the isolated nature of the solution replaces or is equivalent to the well posedness required of Cauchy problems in the Lax theory [7]. In particular, we show in Section 2 that "stability" and "isolation" are essentially equivalent. In Section 3, the family of approximating problems are introduced, and the main theorem is proven. This shows that a stable family of approximate solutions exists, that they can be obtained by Newton's method, and error estimates are given. Asymptotic error expansions are not discussed in Section 3 as they are easily obtained from our results by employing the techniques indicated in [3] or [9].

The present basic theory was initially developed and applied to study a specific difference method for nonlinear two-point boundary value problems [3]. However, the general simplicity and applicability of the theory to a variety of approximation problems prompted the more general treatment. In addition to the above cited use, the theory has already been applied by R. K. Weiss in [11] to study implicit Runge-Kutta and collocation schemes for nonlinear two-point boundary value problems. It can also be used for nonlinear Fredholm problems, for mildly nonlinear elliptic problems, and to justify the Box scheme applied to nonlinear parabolic problems, etc. Such applications will be presented elsewhere. However, we do show, in Section 4, how the theory can be applied to justify very general difference schemes for approximating isolated solutions of very general nonlinear two-point boundary value problems. A crucial step in this demonstration is supplied by a powerful stability result in [4] for linear problems. Indeed, we essentially show that any difference scheme which is stable and consistent for the initial-value problem is so for isolated solutions of the boundary value problem. Our theory is also used in [4] to study the most general difference methods for nonlinear boundary value problems in ordinary differential equations.

2. Stable and Isolated Solutions. For a mapping **$F\colon\ \mathcal{B}_1 \to \mathcal{B}_2$, where the $\mathcal{B}_\nu$ are appropriate Banach spaces, we consider the problem

$$F(x) = 0. \tag{2.1}$$

With the sphere $S_\rho(u) \equiv \{x\colon\ x \in \mathcal{B}_1,\ \|x - u\| \leqslant \rho\}$, we introduce the

(2.2) *Definitions.* (a) *The mapping $F(\cdot)$ is stable on $S_\rho(u)$ iff there exists a constant $K_\rho > 0$ such that****

$$\|v - w\| \leqslant K_\rho \|F(v) - F(w)\|$$

for all v, $w \in S_\rho(u)$.

(b) *A solution $x = u$ of (2.1) is stable iff $F(\cdot)$ is stable on $S_\rho(u)$ for some $\rho > 0$.*

Trivially, we note that a stable solution is also unique in $S_\rho(u)$. If $F(\cdot)$ is linear and stable, then our definition implies (for any $\rho > 0$) Lipschitz continuous dependence

---

**To simplify matters, we assume that the domain of F is $\mathcal{B}_1$; the restriction to a proper subset offers no difficulty.

***We do not distinguish notationally between norms on $\mathcal{B}_1$ and $\mathcal{B}_2$. Rather, we adopt the convention that $\|x\| \equiv \|x\|_{\mathcal{B}_\nu}$ if $x \in \mathcal{B}_\nu$.

of the solution of $F(x) = g$ on the inhomogeneous data g (which is the usual definition for linear problems).

The Fréchet derivative of F at x will be denoted by $L(x)$. This is a linear operator $L(x)$: $\mathcal{B}_1 \longrightarrow \mathcal{B}_2$ which is such that

$$(2.3) \qquad r(x, y) \equiv \frac{\|F(x + y) - [F(x) + L(x)y]\|}{\|y\|} \longrightarrow 0 \quad \text{as } \|y\| \longrightarrow 0.$$

In terms of the Fréchet derivative, we introduce the

(2.4) *Definition. A solution u of* (2.1) *is isolated iff $L(u)$ exists and is nonsingular; that is: if $L(u)y = 0$, then $y = 0$.*

Now we show that stability implies isolation as in

(2.5) THEOREM. *Let u be a stable solution of* (2.1). *Then, if $L(u)$ exists, u is an isolated solution.*

Proof. Suppose $L(u)y = 0$ and $\|y\| \neq 0$. Then, for all positive scalars $\alpha < \rho/\|y\|$, it follows that $v(\alpha) \equiv u + \alpha y \in S_\rho(u)$. By the stability of $F(\cdot)$ on $S_\rho(u)$, it follows that

$$\begin{aligned} \|u - v(\alpha)\| &\leqslant K_\rho \|F(u) - F(v)\| \\ &\leqslant K_\rho \{\|L(u)\alpha y\| + \|\alpha y\| r(u, \alpha y)\}. \end{aligned}$$

Thus, $\alpha\|y\| \leqslant K_\rho r(u, \alpha y)\alpha\|y\|$ and, if $\alpha > 0$ is chosen so small that $K_\rho r(u, \alpha y) < 1$, we must have $\|y\| = 0$.

Stability is such a strong condition that it implies $L(x)$ nonsingular wherever it exists in the *interior* of $S_\rho(u)$. The proof is identical to that of Theorem (2.5) since the latter does not employ $F(u) = 0$. Thus, it is not surprising that a form of converse to Theorem (2.5) requires the existence of $L(x)$ in some sphere about u. Indeed, we require even more in stating

(2.6) THEOREM. *Let $L(u)$ be nonsingular for some $u \in \mathcal{B}_1$. Let $L(x)$ exist and be Lipschitz continuous on $S_{\rho_0}(u)$ for some $\rho_0 > 0$; that is: for some constant $K_L > 0$, $\|L(x) - L(y)\| \leqslant K_L\|x - y\|$ for all $x, y \in S_{\rho_0}(u)$. Then, $F(\cdot)$ is stable on $S_\rho(u)$ for any $\rho < (K_L\|L^{-1}(u)\|)^{-1}$, and the stability constant is*

$$K_\rho = \|L^{-1}(u)\|(1 - \rho K_L\|L^{-1}(u)\|)^{-1}.$$

Proof. For any $x, y \in S_\rho(u)$ with $\rho \leqslant \rho_0$, we have $F(x) - F(y) = \hat{L}(x, y)(x - y)$, where

$$\hat{L}(x, y) \equiv \int_0^1 L(tx + [1 - t]y)\,dt.$$

Write $\hat{L}(x, y) = L(u) + [\hat{L}(x, y) - L(u)]$ and note that

$$\begin{aligned} \|\hat{L}(x, y) - L(u)\| &\leqslant \int_0^1 \|L(tx + [1 - t]y) - L(tu + [1 - t]u)\| \\ &\leqslant K_L \int_0^1 \|t(x - u) + [1 - t](y - u)\|\,dt \leqslant \rho K_L. \end{aligned}$$

Thus, if ρ is so small that $\rho K_L \|L^{-1}(u)\| < 1$, the Banach lemma implies that $\hat{L}(x, y)$ is nonsingular and

$$\|\hat{L}^{-1}(x, y)\| \leq \|L^{-1}(u)\|/(1 - \rho K_L \|L^{-1}(u)\|).$$

Now stability follows from $(x - y) = \hat{L}^{-1}(x, y)[F(x) - F(y)]$ with the indicated K_ρ. □

The hypothesis in Theorem (2.6) can be weakened to require only local Hölder continuity in the restricted form, with some $0 < \alpha \leq 1$:

$$\|L(x) - L(u)\| \leq K_L \|x - u\|^\alpha \quad \text{for all } x \in S_{\rho_0}(u).$$

In this case, ρ must be restricted by $\rho^\alpha < (K_L \|L^{-1}(u)\|)^{-1}$, and the stability constant must be suitably altered.

Finally, we note that if u is an isolated solution of (2.1), then $L(u)$ is nonsingular. So if, in addition, $L(x)$ exists and is Lipschitz continuous in some $S_{\rho_0}(u)$, then u is a stable solution, by Theorem (2.6). This is the essential converse of Theorem (2.5).

3. Approximation Problems. On a family of Banach spaces, $\{B_1^h, B_2^h\}$, we consider the family of approximating problems, for $0 < h \leq h_0$:

(3.1) $$F_h(x_h) = 0,$$

where F_h: $B_1^h \longrightarrow B_2^h$. To relate problems (2.1) and (3.1), we require that there exist a family of linear mappings $\{P_1^h, P_2^h\}$ where†

(3.2) $$\text{(a)}\ \ P_\nu^h: B_\nu \longrightarrow B_\nu^h, \quad \text{(b)}\ \lim_{h\to 0} \|P_\nu^h x\| = \|x\| \quad \forall x \in B_\nu.$$

We find it convenient to use the notation

$$P_\nu^h x \equiv [x]_h, \qquad \nu = 1, 2,$$

where, of course, $[x]_h \in B_\nu^h$ if $x \in B_\nu$. The Fréchet derivative of F_h at x_h is denoted by $L_h(x_h)$ and $S_\rho(x_h)$ is the sphere in B_1^h of radius ρ about x_h. We introduce several concepts.

(3.2) *Definition. The family* $\{F_h(\cdot)\}$ *is stable for* $u \in B_1$ *iff for some* $h_0 > 0$, $\rho > 0$ *and some constant* M_ρ, *independent of h.*

$$\|x_h - y_h\| \leq M_\rho \ \|F_h(x_h) - F_h(y_h)\|$$

for all $x_h, y_h \in S_\rho([u]_h)$ *and all* $h \in (0, h_0]$.

(3.3) *Definition. The family* $\{F_h(\cdot)\}$ *is consistent of order p with* $F(\cdot)$ *on* $S_\rho(u)$ *if and only if*

$$\|F_h([x]_h) - [F(x)]_h\| \equiv \|\tau_h(x)\| \leq M(x)h^p,$$

for all $x \in S_\rho(u)$ *and some bounded functional* $M(x) \geq 0$ *independent of h.*

---

†Again norms are those of the relevant spaces; $\|x_h\| \equiv \|x_h\|_{B_\nu^h}$ if $x_h \in B_\nu^h$.

The significance of these definitions is well known and, indeed, best summarized in the

(3.4) THEOREM. *Let $F(u) = 0$ and $F_h(v_h) = 0$ for some $v_h \in S_\rho([u]_h)$, $\rho > 0$ and all $h \in (0, h_0]$. Let $\{F_h(\cdot)\}$ be stable for u and consistent of order p on $S_0(u)$. Then*

$$\|[u]_h - v_h\| \leqslant M_\rho M(u) h^p.$$

Proof. By (3.2) with $x_h = [u]_h$ and $y_h = v_h$,

$$\|[u]_h - v_h\| \leqslant M_\rho \|F_h([u]_h) - F_h(v_h)\|.$$

Using $F_h(v_h) = 0$ and $[F(u)]_h = 0$ in the above and recalling (3.3) with $\rho = 0$ yields the result.□

We are thus faced with the basic problems: (i) to be insured that the approximating problems actually have solutions in some sphere about $[u]_h$; (ii) to be able to verify stability; and (iii) to determine the order of consistency. For many explicit difference schemes, it is trivial to verify (i), but, for implicit schemes and projection methods, this is frequently quite difficult. Again, for most difference approximation schemes, the order of consistency is determined by simple Taylor expansions. However, for projection or expansion methods, this is by no means a trivial task. The stability verification for nonlinear problems of great generality is also not a standard procedure. It is usually reduced to a study of the linearized problems. We present such a result as

(3.5) LEMMA. *Let the family of mappings $\{F_h(\cdot)\}$ have Fréchet derivatives (i.e., linearizations) $\{L_h(x_h)\}$ on some family of spheres $S_{\rho_0}(z_h)$ and satisfy for all $h \in (0, h_0]$:*

(a) *$\{L_h(z_h)\}$ have uniformly bounded inverses at the centers of the spheres; that is, for some constant $K_0 > 0$, $\|L_h^{-1}(z_h)\| \leqslant K_0$.*

(b) *$\{L_h(x_h)\}$ are uniformly Lipschitz continuous on $S_{\rho_0}(z_h)$; that is, for some constant $K_L > 0$,*

$$\|L_h(x_h) - L_h(y_h)\| \leqslant K_L \|x_h - y_h\|$$

for all $x_h, y_h \in S_{\rho_0}\{z_h\}$.

If $z_h = [u]_h$ for some $u \in B_1$, then the family $\{F_h(\cdot)\}$ is stable for u.

Proof. The proof is essentially identical to that of Theorem 2.6.□

To insure the existence of a family of solutions $\{v_h\}$ approximating a solution u of (2.1), we need only adjoin consistency to the above. More precisely, we have

(3.6) THEOREM. *Let $x = u$ be a solution of $F(x) = 0$. Let the family $\{F_h(\cdot)\}$ be consistent of order p with $F(\cdot)$ on $S_0(u)$. Let the hypothesis* (a) *and* (b) *of Lemma* (3.5) *hold with $z_h = [u]_h$. Then, for ρ_0 and h_0 sufficiently small and for each $h \in (0, h_0]$, the problem $F_h(x_h) = 0$ has a unique solution $x_h = v_h \in S_{\rho_0}([u]_h)$. These solutions satisfy*

$$\|[u]_h - v_h\| \leqslant M_{\rho_0} M(u) h^p.$$

Proof. We define the family of mappings $\{G_h(x_h)\}$ by

$$G_h(x_h) \equiv x_h - L_h^{-1}([u]_h)F_h(x_h)$$

and shall show that they are uniformly contracting on $S_{\rho_0}([u]_h)$, provided ρ_0 and h_0 are sufficiently small. Since the sphere is convex, we have, for any $x_h, y_h \in S_{\rho_0}([u]_h)$:

$$G_h(x_h) - G_h(y_h) = L_h^{-1}([u]_h)\{L_h([u]_h)(x_h - y_h) - (F(x_h) - F(y_h))\}$$

$$= L_h^{-1}([u]_h)\{L_h([u]_h) - \hat{L}_h(x_h, y_h)\}(x_h - y_h).$$

Here, as in Theorem (2.6), we used the generalized mean value theorem and

$$\hat{L}_h(x_h, y_h) \equiv \int_0^1 L_h(tx_h + [1-t]y_h)dt.$$

From (3.5b), it follows that

$$\|L_h([u]_h) - \hat{L}_h(x_h, y_h)\| \leq K_L\rho_0,$$

and thus, by (3.5a),

$$\|G_h(x_h) - G_h(y_h)\| \leq \alpha\|x_h - y_h\|, \qquad \alpha = K_0K_L\rho_0.$$

At the center of the sphere, $x_h = [u]_h$, we have, by consistency (3.3) and since $F(u) = 0$:

$$\|[u]_h - G_h([u]_h)\| \leq K_0\|F_h([u]_h) - [F(u)]_h\| \leq K_0M(u)h^p.$$

Now if $\alpha < 1$ and $K_0M(u)h^p \leq (1-\alpha)\rho_0$, the Contraction Mapping Theorem applied to $x_h = G(x_h)$ implies the existence of a unique solution in $S_{\rho_0}([u]_h)$.

The error estimate follows from Theorem (3.4), which is now applicable.□

Obviously, the iteration scheme implied in the proof of Theorem (3.6) cannot be used to compute the approximate solutions since $[u]_h$ is not known. However, Newton's method is frequently applicable for this purpose as we show in

(3.7) THEOREM. *Let the hypothesis of Theorem* (3.6) *hold. Then, for any* $h \in (0, h_0]$, *if* ρ_0, h_0 *and* $\rho_1 \leq \rho_0$ *are sufficiently small, the Newton iterates* $\{v_h^{(\nu)}\}$ *defined by*:

(a) $v_h^{(0)} \in S_{\rho_1}([u]_h)$,

(b) $L_h(v_h^{(\nu)})\,[v_h^{(\nu+1)} - v_h^{(\nu)}] = -F_h(v_h^{(\nu)}), \nu = 0, 1, 2, \ldots,$

converge quadratically to the unique solution of $F_h(x_h) = 0$ *in* $S_{\rho_0}([u]_h)$.

Proof. By writing

$$L_h(x_h) = L_h([u]_h) + [L_h(x_h) - L_h([u]_h)]$$

$$= L_h([u]_h)\{I + L_h^{-1}([u]_h)[L_h(x_h) - L_h([u_h])]\}$$

and using (3.5a, b) with $K_0K_L\rho_0 < 1$, the Banach lemma yields that $L_h(x_h)$ is nonsingular and, in fact,

(a) $\quad \|L_h^{-1}(x_h)\| \leq K_0/(1 - K_0 K_{L\rho_0}) = K_{\rho_0} \quad$ for all $x_h \in S_{\rho_0}([u]_h)$.

For any $v_h^{(0)} \in S_{\rho_1}([u]_h)$, we have from (3.7b) with $\nu = 0$, as in the proof of Theorem (3.6):

$$v_h^{(1)} - v_h^{(0)} = -L_h^{-1}(v_h^{(0)})F_h([u]_h) + L_h^{-1}(v_h^{(0)})\hat{L}_h(v_h^{(0)}, [u]_h)([u]_h - v_h^{(0)}).$$

However, the identity

$$L_h^{-1}(v_h^{(0)})\hat{L}_h(v_h^{(0)}, [u]_h) = I + L_h^{-1}(v_h^{(0)})(\hat{L}_h(v_h^{(0)}, [u]_h) - L_h(v_h^{(0)}))$$

implies by (a) above and (3.5b) that for some constant $C > 0$:

$$\|L_h^{-1}(v_h^{(0)})\hat{L}_h(v_h^{(0)}, [u]_h)\| \leq C.$$

Now, consistency in (3.3), recalling $[F(u)]_h = 0$, and the above yield

(b) $$\|v_h^{(1)} - v_h^{(0)}\| \leq K_{\rho_0} M(u)h^p + C_{\rho_1}.$$

Together with (3.5b), conditions (a) and (b) above are sufficient for the quadratic convergence of Newton's method, provided h_0 and ρ_1 are sufficiently small (see for instance [2] or [6]).□

The basic difficulty in applying the above theory is to verify (3.5a); that is, the establishment of the stability of the family of linear operators $\{L_h([u]_h)\}$. Frequently, this can be done by showing that there are some close approximations to $L_h([u]_h)$, say $\mathsf{L}_h([u]_h)$, which have uniformly bounded inverses. That is, if

$$\|L_h([u]_h) - \mathsf{L}_h([u]_h)\| = O(h)$$

and $\|\mathsf{L}_h^{-1}([u]_h)\| \leq K_0$ for all $h \leq h_0$, then, by the Banach lemma, (3.5a) holds with some modified constant, K_0^1. This technique is illustrated for two-point boundary value problems in Section 4; but, of course, the problem is just modified to show the stability of $\mathsf{L}_h([u]_h)$. If the problem $F(x) = 0$ has an isolated solution $x = u$, then we know that $L(u)$ has a bounded inverse. In Section 4, this fact and some additional assumptions on related (initial-value) problems are used to show that the consistent approximation $\mathsf{L}_h([u]_h)$ is stable. In another important class of problems, $L(u)$ is selfadjoint and, say positive definite. Then if the same is true, uniformly in h, of $\mathsf{L}_h([u]_h)$, the stability may easily follow. This technique is very close to that used by R. B. Simpson in [8]. Finally, the technique devised by H.-O. Kreiss is perhaps most powerful; see, for example, [5]. Here, since $L(u)$ does not have the eigenvalue zero, Kreiss shows that $\{\mathsf{L}_h([u]_h)\}$ must have eigenvalues bounded away from zero. This assumes the consistency of $\mathsf{L}_h([u]_h)$ with $L(u)$ and employs a contradiction obtained by using an appropriate map from B_h to B of the normalized solutions of $\mathsf{L}_h([u]_h)\phi_h = 0$ which can be shown to converge to ϕ, a nontrivial solution of $L(u)\phi = 0$.

4. Nonlinear Two-Point Boundary Value Problems. We assume the nonlinear two-point boundary value problem

(4.1)
$$\text{(a)}\quad N\mathbf{y}(t) \equiv \mathbf{y}'(t) - \mathbf{f}(t, \mathbf{y}(t)) = 0, \qquad a < t < b,$$
$$\text{(b)}\qquad \mathbf{g}(\mathbf{y}(a), \mathbf{y}(b)) = 0,$$

has an isolated solution, $\mathbf{y}(t)$. That is, the linear problem

(4.2)
$$\text{(a)}\quad L[\mathbf{y}]\phi(t) \equiv \phi'(t) - A(t)\phi(t) = 0, \qquad a < t < b,$$
$$\text{(b)}\quad B[\mathbf{y}]\phi \equiv B_a\phi(a) + B_b\phi(b) = 0,$$

where

(4.2)
$$\text{(c)}\quad A(t) \equiv \frac{\partial \mathbf{f}(t, \mathbf{y}(t))}{\partial \mathbf{y}}; \quad B_z \equiv \frac{\partial \mathbf{g}(\mathbf{y}(a), \mathbf{y}(b))}{\partial \mathbf{y}(z)}, \quad z = a, b;$$

has only the trivial solution, $\phi(t) \equiv 0$. We shall apply the previous theory to justify some fairly general difference schemes for approximating this solution of (4.1).

A family of nets is considered of which the general one is

(4.3)
$$\text{(a)}\quad t_0 = a: \ t_j = t_{j-1} + h_j, \qquad 1 \leqslant j \leqslant J; \quad t_J = b,$$
$$\text{(b)}\quad h \equiv \max_j h_j \leqslant \lambda \min_j h_j,$$

where λ is a fixed constant and on which $h \longrightarrow 0$ in some manner. For each such net, a difference scheme, determined by the coefficients $\{\alpha_{jk}(h), \beta_{jk}(h)\}$, is defined by:

(4.4)
$$\text{(a)}\quad N_h\mathbf{v}_j \equiv \sum_{k=0}^{J} \{\alpha_{jk}\mathbf{v}_k - \beta_{jk}\mathbf{f}(t_k, \mathbf{v}_k)\} = 0, \qquad 1 \leqslant j \leqslant J;$$
$$\text{(b)}\qquad \mathbf{g}(\mathbf{v}_0, \mathbf{v}_J) = 0.$$

(4.5) Our main assumption on the numerical method is that: *The family of schemes* (4.4a), *with* $\mathbf{v}_0 = \mathbf{u}_0$, *is consistent of order p and stable for all sufficiently smooth initial-value problems of the form*:

(4.6)
$$\mathbf{u}' = \mathbf{F}(t, \mathbf{u}), \qquad a < t < b; \quad \mathbf{u}(a) = \mathbf{u}_0.$$

For example, to satisfy (4.5), the scheme (4.4a) could be a one-step scheme such as Euler's method, centered Euler or the trapezoidal rule. It could equally well be some standard multistep scheme on a uniform net including a prescribed starting scheme on a refined net (to maintain uniform accuracy). Our present formulation does not include Runge-Kutta or implicit Runge-Kutta schemes, but this is essentially a notational simplification as we shall show later.

To apply the theory of Section 3, we introduce the family of Banach spaces $\mathcal{B}_1^h \equiv \mathcal{B}_2^h \equiv \mathbf{E}^{n(J+1)}$ and, say, $\mathcal{B}_1 \equiv C_{p+1}[a, b]$, $\mathcal{B}_2 \equiv C_p[a, b]$. Then we require that $\mathbf{f}(t, \mathbf{z}) \in C_p[[a, b] \times \mathbf{E}^n]$ and $F(x) = 0$ represents the boundary value problem (4.1a, b) with $x \equiv \mathbf{y}(t)$. For $x_h \equiv \{\mathbf{v}_j\}_0^J$, the family of problems $F_h(x_h) = 0$ represents the family of difference equations (4.4a, b). The mapping $[\ ]_h$ on $\mathcal{B}_\nu$ into $\mathcal{B}_\nu^h$ is defined by $[\mathbf{y}]_h = \{\mathbf{y}(t_j)\}_0^J$.

We now easily get the basic

(4.7) THEOREM. *Let* (4.1) *have an isolated solution* $\mathbf{y}(t) \in C_{p+1}[a, b]$. *Let* $\mathbf{f}(t, \mathbf{z}) \in C_p([a, b] \times E^n)$ *and* $\mathbf{g}(\mathbf{z}, \mathbf{w}) \in C_{1,\,\mathrm{Lip}}(E^n \times E^n)$. *If the scheme* (4.3)–(4.4) *satisfies* (4.5), *then for some* $\rho_0 > 0$, $h_0 > 0$ *sufficiently small and all* $h \leqslant h_0$:

(i) *The difference equations* (4.4) *have a unique solution* $\{\mathbf{v}_j\}_0^J$ *in* $\|\mathbf{v}_j - \mathbf{y}(t_j)\| \leqslant \rho_0$.

(ii) $\|\mathbf{v}_j - \mathbf{y}(t_j)\| \leqslant Mh^p$, $0 \leqslant j \leqslant J$.

(iii) *The difference solution can be computed by Newton's method which converges quadratically for any initial iterate* $\{\mathbf{v}_j^0\}_0^J$ *in* $\|\mathbf{v}_j^0 - \mathbf{y}(t_j)\| \leqslant \rho_1$, *provided* $\rho_1 \leqslant \rho_0$ *is sufficiently small.*

Proof. For (i) and (ii), we need only verify the hypothesis of Theorem (3.6). Clearly, by (4.5), the scheme (4.4) is consistent of order p with (4.1). It remains to verify (a) and (b) of Lemma (3.5) with $z_h = [\mathbf{y}]_h$. The linearized difference operators obtained from the nonlinear difference operators in (4.4) are, say, applied to $\phi_h \equiv \{\phi_j\}_0^J$:

$$
(4.8)\qquad
\begin{aligned}
&\text{(a)}\quad L_h(v_h)\phi_j \equiv \sum_{k=0}^{J}\left[\alpha_{jk}I - \beta_{jk}\frac{\partial \mathbf{f}}{\partial \mathbf{y}}(t_k, \mathbf{v}_k)\right]\phi_k, \qquad 1 \leqslant j \leqslant J;\\
&\text{(b)}\quad B_h(v_h)\phi_h \equiv \frac{\partial \mathbf{g}}{\partial \mathbf{v}_0}(\mathbf{v}_0, \mathbf{v}_J)\phi_0 + \frac{\partial \mathbf{g}}{\partial \mathbf{v}_J}(\mathbf{v}_0, \mathbf{v}_J)\phi_J.
\end{aligned}
$$

Now let us apply the difference scheme of (4.4) to the linear problem (4.2). This yields

$$
(4.9)\qquad
\begin{aligned}
&\text{(a)}\quad \sum_{k=0}^{J}[\alpha_{jk}I - \beta_{jk}A(t_k)]\phi_k = 0, \qquad 1 \leqslant j \leqslant J;\\
&\text{(b)}\qquad\qquad B_a\phi_0 + B_b\phi_J = 0.
\end{aligned}
$$

Recalling (4.2), it follows from (4.8) that (4.9) is just

$$
(4.10)\qquad \text{(a)}\ L_h([y]_h)\phi_j = 0, \quad 1 \leqslant j \leqslant J; \quad \text{(b)}\ B_h([y]_h)\phi_h = 0.
$$

However, Corollary (3.13) of [4] states essentially that: *if* (4.2) *has a unique solution, then the* (*linear*) *difference scheme* (4.10) *is stable and consistent for* (4.2) *if and only if* (4.10a) *with* $\phi_0 = \mathbf{c}$ *is stable and consistent with* (4.2a) *and* $\phi(a) = \mathbf{c}$. But the latter part follows from the assumption (4.5). Thus, since $\mathbf{y}(t)$ is isolated, and hence (4.2) has a unique solution, the scheme in (4.10) is stable. Hence, (a) of Lemma (3.5) is established. (We point out that the operator corresponding to $L_h([\mathbf{y}]_h)$ of (3.5) is just that represented by the coefficient matrix of the difference equations in (4.10a, b). Stability of this scheme is shown in [4] to be equivalent to the uniform boundedness of the inverses of these coefficient matrices for all $h \leqslant h_0$.)

From the Lipschitz continuity of the first derivatives of $\mathbf{f}(t, y)$ and $\mathbf{g}(\mathbf{y}(a), \mathbf{y}(b))$, it follows, using (4.6b), that (b) of Lemma (3.5) with $z_h = [\mathbf{y}]_h$ holds for the scheme in (4.10). Thus, (i) and (ii) of our theorem are established by an application of Theorem (3.6).

To establish (iii), we need only apply Theorem (3.7).□

We now show how to extend the above result to more general difference schemes, say including all Runge-Kutta and implicit Runge-Kutta types. Thus, in place of (4.4), we consider:

$$\text{(4.11)} \quad \begin{aligned} &\text{(a)} \quad N_h \mathbf{v}_j \equiv \sum_{k=0}^{J} \left\{ \alpha_{jk} \mathbf{v}_k - \beta_{jk} \mathbf{f}\left(t_{jk}, \sum_{l=0}^{J} \gamma_{jkl} \mathbf{v}_l \right) \right\} = 0, \qquad 1 \leqslant j \leqslant J; \\ &\text{(b)} \qquad\qquad \mathbf{g}(\mathbf{v}_0, \mathbf{v}_J) = 0. \end{aligned}$$

We again impose (4.5), now with (4.4a) replaced by (4.11a). The consistency requirement in (4.5) implies that for all sufficiently smooth functions $\mathbf{u}(t)$:

$$\text{(4.12)} \qquad \left\| \mathbf{u}(t_{jk}) - \sum_{l=0}^{J} \gamma_{jkl} \mathbf{u}(t_l) \right\| \leqslant C(u)h, \qquad 1 \leqslant j \leqslant J,\ 0 \leqslant k \leqslant J.$$

Note that this is also a restriction on the choice of the points t_{jk}.

The linearized difference equations obtained from (4.11) are:

$$\text{(4.13)} \quad \begin{aligned} &\text{(a)} \quad L_h(v_h)\phi_j \equiv \sum_{k=0}^{J} \left\{ \alpha_{jk} I - \sum_{s=0}^{J} \beta_{js} \frac{\partial \mathbf{f}}{\partial \mathbf{y}}\left(t_{js}, \sum_{l=0}^{J} \gamma_{jsl} \mathbf{v}_l \right) \gamma_{jsk} \right\} \phi_k = 0, \\ &\text{(b)} \qquad\qquad B_h(v_h)\phi_h = 0. \end{aligned}$$

However, when applied to the linear problem (4.2), our more general difference scheme (4.11a) yields

$$\text{(4.14)} \quad \begin{aligned} &\text{(a)} \quad L_h([\mathbf{y}]_h)\phi_j \equiv \sum_{k=0}^{J} \left\{ \alpha_{jk} I - \sum_{s=0}^{J} \beta_{js} A(t_{js}) \gamma_{jsk} \right\} \phi_k = 0, \\ &\text{(b)} \qquad\qquad B_h([\mathbf{y}]_h)\phi_h = 0. \end{aligned}$$

Since $\mathbf{y}(t)$ is isolated and (4.11a) is stable and consistent for initial-value problems, it follows, by the above cited Corollary (3.13) of [4], that the linear difference scheme in (4.14) is stable. Now use (4.12) to observe that $\|L_h([\mathbf{y}]_h) - L_h([y]_h)\| = O(h)$, provided $\partial\mathbf{f}/\partial\mathbf{y}$ and $\mathbf{y}(t)$ are sufficiently smooth. From the Banach lemma, it easily follows that the linear difference scheme in (4.13) is also stable. Thus, with no difficulty, we see that Theorem (4.7) goes over for difference schemes of the form (4.11).

The analog of Theorem (4.7) for implicit Runge-Kutta schemes has previously been demonstrated in [11] by R. K. Weiss. In place of $L_h([y]_h)$ given by (4.14a), Weiss employs the centered Euler (Box scheme) whose stability was demonstrated in [1]. It is a bit more involved to show the "consistency" of the Box scheme with $L_h([y]_h)$.

The analog of Theorem (4.7) for "gap-schemes" has been illustrated in [12] by A. B. White. These are high-order accurate two-point difference schemes in which the local truncation error has leading term $O(h^{2m})$, and the higher-order terms proceed in higher powers of h^2. The lowest-order part of these gap-difference schemes is just that obtained from the trapezoidal rule, and the corrections are bounded perturbations of order h^2. Thus, the stability proof could easily be obtained as above by showing consistency with the (stable) trapezoidal scheme.

Applied Mathematics
California Institute of Technology
Pasadena, California 91109

1. H. B. KELLER, "Accurate difference methods for linear ordinary differential systems subject to linear constraints," *SIAM J. Numer. Anal.*, v. 6, 1969, pp. 8–30. MR 40 #6776.

2. H. B. KELLER, "Newton's method under mild differentiability conditions," *J. Comput. System Sci.*, v. 4, 1970, pp. 15–28. MR 40 #3710.

3. H. B. KELLER, "Accurate difference methods for nonlinear two point boundary value problems," *SIAM J. Numer. Anal.*, v. 11, 1974, pp. 305–320.

4. H. B. KELLER & A. B. WHITE, "Difference methods for boundary value problems in ordinary differential equations," *SIAM J. Numer. Anal.*, v. 12, 1975. (To appear.)

5. H.-O. KREISS, "Difference approximations for boundary and eigenvalue problems for ordinary differential equations," *Math. Comp.*, v. 26, 1972, pp. 605–624.

6. J. M. ORTEGA & W. C. RHEINBOLDT, *Iterative Solution of Nonlinear Equations in Several Variables*, Academic Press, New York, 1970. MR 42 #8686.

7. R. D. RICHTMYER, *Difference Methods for Initial-Value Problems*, Interscience Tracts in Pure and Appl. Math., no. 4, Interscience, New York, 1957. MR 20 #438.

8. R. B. SIMPSON, "Existence and error estimates for solutions of a discrete analog of nonlinear eigenvalue problems," *Math. Comp.*, v. 26, 1972, pp. 359–375. MR 47 #4466.

9. H. J. STETTER, "Asymptotic expansions for the error of discretization algorithms for non-linear functional equations," *Numer. Math.*, v. 7, 1965, pp. 18–31. MR 30 #5505.

10. H. J. STETTER, "Stability of nonlinear discretization algorithms," *Numerical Solution of Partial Differential Equations* (Proc. Sympos. Univ. Maryland, 1965), Academic Press, New York, 1966, pp. 111–123. MR 34 #5322.

11. R. K. WEISS, "The application of implicit Runge-Kutta and collocation methods to boundary value problems," *Math. Comp.*, v. 28, 1974, pp. 449–464.

12. A. B. WHITE, *Numerical Solution of Two Point Boundary Value Problems*, Ph. D. Thesis, Calif. Inst. of Technology, Pasadena, 1974.

13. V. PEREYRA, "Iterated deferred corrections for nonlinear operator equations," *Numer. Math.*, v. 10, 1967, pp. 316–323.

Reprinted with permission from *SIAM Journal On Numerical Analysis*, Vol. 6, No. 1 (1969), pp. 8–30.

ACCURATE DIFFERENCE METHODS FOR LINEAR ORDINARY DIFFERENTIAL SYSTEMS SUBJECT TO LINEAR CONSTRAINTS*

HERBERT B. KELLER†

1. Introduction. We consider the general system of n first order linear ordinary differential equations

$$\mathbf{y}'(t) = A(t)\mathbf{y}(t) + \mathbf{g}(t), \qquad a < t < b, \tag{1.1}$$

subject to n "boundary" conditions, or rather n linear constraints, of the form

$$\sum_{\nu=1}^{N} B_\nu \mathbf{y}(\tau_\nu) = \boldsymbol{\beta}. \tag{1.2}$$

Here $\mathbf{y}(t)$, $\mathbf{g}(t)$ and $\boldsymbol{\beta}$ are n-vectors and $A(t), B_1, \cdots, B_N$ are $n \times n$ matrices. The N distinct points $\{\tau_\nu\}$ lie in $[a, b]$ and we only require $N \geqq 1$. Thus as special cases initial value problems, $N = 1$, are included as well as the general 2-point boundary value problem, $N = 2$, with $\tau_1 = a, \tau_2 = b$. (More general linear constraints are also studied, see (5.1) and (5.17).)

To include many other important practical applications we allow the coefficients $A(t)$ and inhomogeneous data $\mathbf{g}(t)$ to be piecewise smooth. More precisely we say that a function $\phi(t) \in PC_m[a, b]$ if and only if $\phi(t), \phi^{(1)}(t), \cdots, \phi^{(m)}(t)$ are piecewise continuous on $[a, b]$ with at most a finite number of jump discontinuities there. Thus left and right limits exist at the points of discontinuity. In any particular problem the fixed finite set of points of discontinuity will be assumed to be a subset of the points $\{\tau_\nu\}$ employed in (1.2). Of course if such a point is not intended to enter as a "boundary" point we simply set the corresponding $B_\nu \equiv 0$. This is strictly a device of notational convenience. We shall say that a matrix $A(t) \in PC_m[a, b]$ if and only if this is true for each of its components.

In § 2 we show that a unique continuous solution, with piecewise smooth derivatives, of the problem (1.1)–(1.2) exists if and only if a specific matrix is nonsingular. Then we present some sufficient conditions, which may be of practical utility, to insure the nonsingularity of this matrix. In § 3 a very simple difference scheme approximating (1.1), (1.2) on a nonuniform net is shown to have a solution, given explicitly, whenever the boundary value problem has a unique solution and the mesh is sufficiently fine. This scheme furnishes $O(h^2)$ accurate numerical solutions if $A(t)$ and $\mathbf{g}(t) \in PC_2[a, b]$. We show, in § 4, that $O(h^{2M+1})$ accuracy can be obtained by $h \to 0$ extrapolation if $A(t)$ and $\mathbf{g}(t) \in PC_{2M+1}[a, b]$.

* Received by the editors August 12, 1968.

† Applied Mathematics, Firestone Laboratories, California Institute of Technology, Pasadena, California 91109. This work was supported by the U.S. Army Research Office, Durham, under Contract DAHC 04-68-C-0006.

In § 5 we show how all of the above results apply to more general (integral) boundary conditions or constraints. Most of our results go over without difficulty to nonlinear systems of differential equations and even nonlinear boundary conditions. However we only present the linear theory here as it forms the basis for these other applications and is of independent interest. An account of some of the current work is contained in [7].

High order accurate approximations to solutions of boundary value problems with only piecewise smooth data and solutions do not seem to have been previously considered. However such problems occur frequently, and so we briefly discuss their formulation in the form (1.1), (1.2). The simplest and most familiar example consists in a self-adjoint equation, say

$$(p(x)u_x)_x - q(x)u = f(x),$$

where $p(x)$ has a point of discontinuity at which it is required that $p(x)u'(x)$ be continuous. To replace this equation by a first order system we introduce the variables

$$\mathbf{y}(x) \equiv \begin{pmatrix} y_1(x) \\ y_2(x) \end{pmatrix} \equiv \begin{pmatrix} u(x) \\ p(x)u'(x) \end{pmatrix}$$

to obtain

$$\mathbf{y}' = \begin{pmatrix} 0 & 1/p(x) \\ q(x) & 0 \end{pmatrix} \mathbf{y} + \begin{pmatrix} 0 \\ f(x) \end{pmatrix}.$$

The jump condition is now insured by simply requiring $\mathbf{y}(x)$ to be continuous.

Unfortunately the above device is not always applicable. To illustrate the general case we consider a 2-point boundary value problem for a first order system of n-equations:

$$\text{(1.3a)} \qquad \mathbf{u}' = P(t)\mathbf{u} + \mathbf{q}(t), \quad D_1\mathbf{u}(a) + D_2\mathbf{u}(b) = \boldsymbol{\alpha},$$

subject to the jump condition, at some point $c \in (a, b)$,

$$\text{(1.3b)} \qquad E_1\mathbf{u}(c - 0) + E_2\mathbf{u}(c + 0) = \boldsymbol{\gamma}.$$

Here $D_\nu, E_\nu, P(t)$ are $n \times n$ matrices, $\mathbf{u}(t), \mathbf{q}(t), \boldsymbol{\alpha}, \boldsymbol{\gamma}$ are n-vectors. The idea now is to consider two systems of differential equations of the above form, say one for $\mathbf{v}(t)$ on $[a, c]$, the other for $\mathbf{w}(t)$ on $(c, b]$, and to determine conditions such that with $\mathbf{u}(t) \equiv \mathbf{v}(t)$ on $[a, c)$ and $\mathbf{u}(t) \equiv \mathbf{w}(t)$ on $(c, b]$ the function $\mathbf{u}(t)$ satisfies (1.3). The simplest way to formulate such a procedure is to extend each subinterval to $[a, b]$ and introduce the $2n$-dimensional vectors and $2n \times 2n$ order matrices

$$\text{(1.4)} \qquad \begin{aligned} &\mathbf{y}(t) \equiv \begin{pmatrix} \mathbf{v}(t) \\ \mathbf{w}(t) \end{pmatrix}, \quad \mathbf{g}(t) \equiv \begin{pmatrix} \mathbf{q}(t) \\ \mathbf{q}(t) \end{pmatrix}, \quad \boldsymbol{\beta} \equiv \begin{pmatrix} \boldsymbol{\alpha} \\ \boldsymbol{\gamma} \end{pmatrix}, \quad A(t) \equiv \begin{pmatrix} P(t) & 0 \\ 0 & P(t) \end{pmatrix}, \\ &B_1 \equiv \begin{pmatrix} D_1 & 0 \\ 0 & 0 \end{pmatrix}, \quad B_2 \equiv \begin{pmatrix} 0 & 0 \\ E_1 & E_2 \end{pmatrix}, \quad B_3 \equiv \begin{pmatrix} 0 & D_2 \\ 0 & 0 \end{pmatrix}. \end{aligned}$$

Then (1.3) is equivalent to (1.1)–(1.2) with $N = 3$, $\tau_1 = a$, $\tau_2 = c$, $\tau_3 = b$. Of course the values of $\mathbf{v}(t)$ ($\mathbf{w}(t)$) are not required on $(c, b]$ ($[a, c)$) and this can simplify some of the work in carrying out the numerical solution.

Another, perhaps preferable, procedure is to introduce a new independent variable s on each subinterval by

$$s = \frac{t-a}{c-a}, \quad t \in [a, c], \qquad s = \frac{t-c}{b-c}, \quad t \in [c, b].$$

Now for $s \in [0, 1]$ we introduce the vectors and matrices

$$\mathbf{y}(s) \equiv \begin{pmatrix} \mathbf{v}(s) \\ \mathbf{w}(s) \end{pmatrix}, \quad \mathbf{g}(s) \equiv \begin{pmatrix} \mathbf{q}([c-a]s + a) \\ \mathbf{q}([b-c]s + c) \end{pmatrix}, \quad \boldsymbol{\beta} \equiv \begin{pmatrix} \boldsymbol{\alpha} \\ \boldsymbol{\gamma} \end{pmatrix},$$

$$A(s) = \begin{pmatrix} P([c-a]s + a) & 0 \\ 0 & P([b-c]s + c) \end{pmatrix}, \quad B_1 \equiv \begin{pmatrix} D_1 & 0 \\ 0 & E_1 \end{pmatrix},$$

$$B_2 \equiv \begin{pmatrix} 0 & D_2 \\ E_2 & 0 \end{pmatrix}.$$

Then problem (1.1)–(1.2) with $t \equiv s$ on $[a, b] \equiv [0, 1]$ and $N = 2$, $\tau_1 = 0$, $\tau_2 = 1$ is equivalent to (1.3). The solution of (1.3) is given in terms of $\mathbf{y}(s)$, the solution of (1.1)–(1.2) as

$$\mathbf{u}(t) = \mathbf{v}\left(\frac{t-a}{c-a}\right), \quad t \in [a, c]; \qquad \mathbf{u}(t) = \mathbf{w}\left(\frac{t-c}{b-c}\right), \quad t \in [c, b].$$

Our rescaling procedure does not introduce any new "boundary" points, but it again doubles the order of the system to be solved. However, since $A(s)$ is again block diagonal, the numerical computations simplify, for the most part, as in the treatment of systems of only n equations. The generalization to more than one point of discontinuity is clear. It should be mentioned that this device is quite useful in treating delay-differential equation problems, but we do not go into the details of such applications here.

2. Existence theory. The existence and uniqueness theory for linear multipoint boundary value problems of the form (1.1)–(1.2) is more or less standard. We first present the basic result in the form most useful for our work. Then we present a sequence of sufficient conditions which are more practical for verification in particular cases.

THEOREM 2.1. *Let $A(t) \in PC_m[a, b]$ for some integer $m \geqq 0$. For any $\tau_0 \in [a, b]$ define the (n-th order matrix) fundamental solution $Y(t, \tau_0)$ by*

$$(2.1) \qquad Y' = A(t)Y, \qquad Y(\tau_0, \tau_0) = I.$$

Then for each $\mathbf{g}(t) \in PC_m[a, b]$ *and* $\boldsymbol{\beta}$ *the problem* (1.1)–(1.2) *has a unique continuous solution* $y(t) \in PC_{m+1}[a, b]$ *if and only if the n-th order matrix* Q, *defined by*

$$Q \equiv \sum_{\nu=1}^{N} B_\nu Y(\tau_\nu, \tau_0), \tag{2.2}$$

is nonsingular.

Proof. By the standard uniqueness theorem for linear initial value problems and the variation of parameters formula it follows that a vector $\mathbf{y}(t) \in C[a, b]$, which takes on the value $\mathbf{y}(\tau_0) = \boldsymbol{\xi}$, is a solution of the differential equation (1.1) if and only if

$$\mathbf{y}(t) = Y(t, \tau_0)\boldsymbol{\xi} + \int_{\tau_0}^{t} Y(t, \tau)\mathbf{g}(\tau)\, d\tau. \tag{2.3}$$

However this function satisfies the boundary conditions (1.2) if and only if

$$Q\boldsymbol{\xi} = \boldsymbol{\beta} - \sum_{\nu=1}^{N} B_\nu \int_{\tau_0}^{\tau_\nu} Y(\tau_\nu, \tau)\mathbf{g}(\tau)\, d\tau. \tag{2.4}$$

Thus it follows that problem (1.1)–(1.2) has a unique continuous solution for each $\boldsymbol{\beta}$ and $\mathbf{g}(t)$ if and only if the linear system (2.4) has a unique solution $\boldsymbol{\xi}$ for each $\boldsymbol{\beta}$ and $\mathbf{g}(t)$. But this is just equivalent to the nonsingularity of Q.

Finally we observe that if (1.1) has a continuous solution, then, by differentiation of the equation, this solution has one more piecewise continuous derivative on $[a, b]$ than do the data $A(t)$ and $\mathbf{g}(t)$. This completes the proof.

An equivalent, perhaps more familiar, formulation of the above necessary and sufficient condition is that the homogeneous problem corresponding to (1.1)–(1.2) has only the trivial solution $\mathbf{y}(t) \equiv 0$. Clearly this yields, from (2.4) with $\boldsymbol{\beta} \equiv \mathbf{g}(t) \equiv 0$, the homogeneous linear system $Q\boldsymbol{\xi} = 0$, and the same result follows. In this form the result is well known for the case $N = 2$; see [1, pp. 295].

The statement in Theorem 2.1 yields, in principle, a specific test for the unique solvability of the boundary value problem. However it is usually not practical to determine the nonsingularity of the matrix Q directly. Thus we turn to the consideration of sufficient conditions which are more easily applied. For example, if the matrix $Q_0 \equiv \sum_{\nu=1}^{N} B_\nu$ is nonsingular, then we claim that Q is nonsingular provided $|b - a|$ is sufficiently small. In fact we define a sequence of matrices $\{Q_m\}$ which are such that if Q_M is nonsingular and $|b - a|$ is sufficiently small then so are Q_m for all $m > M$ and $Q = \lim_{m\to\infty} Q_m$ is also nonsingular. The allowed interval lengths $|b - a|$ increase with m. Specifically we have the following theorem.

THEOREM 2.2. *Let* $A(t) \in PC_0[a, b]$ *and define, for each* $m = 0, 1, 2, \cdots$, *the matrix*

$$Q_m \equiv \sum_{\nu=1}^{N} B_\nu Y_m(\tau_\nu, \tau_0), \tag{2.5a}$$

$$Y_0(\tau, \tau_0) \equiv I, \quad Y_m(\tau, \tau_0) = I + \int_{\tau_0}^{\tau} A(s)Y_{m-1}(s, \tau_0)\, ds, \quad m = 1, 2, \cdots. \tag{2.5b}$$

Define $k(t) \equiv \|A(t)\|$. *For some* λ *in* $0 \leqq \lambda < 1$ *and integer* $M \geqq 0$ *let* Q_M *be nonsingular and satisfy*

$$\sum_{\nu=1}^{N} \|Q_M^{-1}B_\nu\| \left\{ \exp\left(\left| \int_{\tau_0}^{\tau_\nu} k(s)\,ds \right| \right) - \left[1 + \left| \int_{\tau_0}^{\tau_\nu} k(s)\,ds \right| + \cdots + \frac{1}{M!} \left| \int_{\tau_0}^{\tau_\nu} k(s)\,ds \right|^M \right] \right\} \leqq \lambda. \tag{2.6}$$

Then the matrices Q_m *for all* $m > M$ *and the matrix* Q *defined by* (2.1)–(2.2) *are nonsingular.*

Let $\tau_0 = (a+b)/2$, $K \equiv \max_{a \leqq t \leqq b} k(t)$ *and suppose* Q_M *satisfies*

$$\frac{1}{(M+1)!} \left(\frac{K|b-a|}{2} \right)^{M+1} e^{K|b-a|/2} \leqq \frac{\lambda}{\sum_{\nu=1}^{N} \|Q_M^{-1}B_\nu\|}. \tag{2.7}$$

Then (2.6) *is satisfied.*

Proof. Since Q_M is nonsingular, by hypothesis, we can write Q_{M+r} for any $r = 1, 2, \cdots$ as

$$Q_{M+r} = Q_M \left\{ I + \sum_{\nu=1}^{N} Q_M^{-1} B_\nu [Y_{M+r}(\tau_\nu, \tau_0) - Y_M(\tau_\nu, \tau_0)] \right\}. \tag{2.8}$$

From the definition (2.5b) we get

$$\begin{aligned} \|Y_{M+r}(\tau, t) - Y_M(\tau, t)\| &\leqq \sum_{\mu=M+1}^{M+r} \left| \int_t^\tau k(t_1) \int_t^{t_1} k(t_2) \cdots \int_t^{t_{\mu-1}} k(t_\mu)\,dt_\mu \cdots dt_1 \right| \\ &\leqq \sum_{\mu=M+1}^{\infty} \frac{1}{\mu!} \left| \int_t^\tau k(\xi)\,d\xi \right|^\mu \\ &= \exp\left(\left| \int_t^\tau k(\xi)\,d\xi \right| \right) - \sum_{\mu=0}^{M} \frac{1}{\mu!} \left| \int_t^\tau k(\xi)\,d\xi \right|^\mu. \end{aligned} \tag{2.9a}$$

The final right-hand side above is independent of r and since the fundamental solution $Y(\tau, t)$ has the convergent series representation (i.e., the Matrizant, see [5, pp. 408–411])

$$Y(\tau, t) = \lim_{m \to \infty} Y_m(\tau, t),$$

we obtain, on letting $r \to \infty$ in (2.9a),

$$\|Y(\tau, t) - Y_M(\tau, t)\| \leqq \exp\left(\left| \int_t^\tau k(\xi)\,d\xi \right| \right) - \sum_{\mu=0}^{M} \frac{1}{\mu!} \left| \int_t^\tau k(\xi)\,d\xi \right|^\mu. \tag{2.9b}$$

Using (2.9) and (2.6) we obtain

$$\left\| \sum_{\nu=1}^{N} Q_M^{-1} B_\nu [Y_{M+r}(\tau_\nu, \tau_0) - Y_M(\tau_\nu, \tau_0)] \right\| \leqq \lambda$$

and

$$\left\| \sum_{\nu=1}^{N} Q_M^{-1} B_\nu [Y(\tau_\nu, \tau_0) - Y_M(\tau_\nu, \tau_0)] \right\| \leqq \lambda.$$

Since $\lambda < 1$ and Q_M is nonsingular we conclude from (2.8) and the Banach lemma that Q_{M+r} is nonsingular for all $r = 1, 2, \cdots$. Similarly from (2.2) and the above it follows that Q is nonsingular.

Finally the inequality (2.7) implies (2.6) since $|\tau_\nu - \tau_0| \leqq |b - a|/2$ when $\tau_0 = (a + b)/2$,

$$\left| \int_{\tau_0}^{\tau_\nu} k(s)\, ds \right| \leqq \frac{K|b - a|}{2}$$

and for any $z \geqq 0$,

$$e^z - \left[1 + z + \cdots + \frac{z^M}{M!} \right] \leqq \frac{z^{M+1}}{(M+1)!} e^z.$$

This completes the proof.

It should be observed that the choice $\tau_0 = \tau_\nu$ for some particular $\nu > 0$ results in the elimination of the term containing the factor $\|Q_M^{-1} B_\nu\|$ from the sum in (2.8). Thus, in particular, for two-point boundary value problems, $N = 2$, we find that (2.6) is satisfied if

$$\text{(2.10)} \qquad \frac{1}{(M+1)!}(K|b - a|)^{M+1} e^{K|b-a|} \leqq \frac{\lambda}{\theta}, \qquad \theta \equiv \min\left(\|Q_M^{-1} B_1\|, \|Q_M^{-1} B_2\|\right),$$

and τ_0 has the appropriate value $\tau_0 = a$ or $\tau_0 = b$. For pure initial value problems, $N = 1$, the condition is automatically satisfied by the obvious choice $\tau_0 = \tau_1$.

As an example of the use and limitations of Theorem 2.2 we consider briefly the general second order two-point boundary value problem with separated (or "unmixed") end conditions:

$$\text{(2.11)} \qquad \begin{gathered} \phi'' = p(t)\phi' + q(t)\phi + r(t), \\ a_0\phi(a) + a_1\phi'(a) = a_2, \quad b_0\phi(b) + b_1\phi'(b) = b_2. \end{gathered}$$

An equivalent first order system of the form (1.1)–(1.2) is obtained by means of the definitions:

$$\text{(2.12a)} \qquad \mathbf{y} \equiv \begin{pmatrix} \phi \\ \phi' \end{pmatrix}, \quad A(t) \equiv \begin{pmatrix} 0 & 1 \\ p(t) & q(t) \end{pmatrix}, \quad \mathbf{g}(t) \equiv \begin{pmatrix} 0 \\ r(t) \end{pmatrix},$$

$$\text{(2.12b)} \qquad B_1 \equiv \begin{pmatrix} a_0 & a_1 \\ 0 & 0 \end{pmatrix}, \quad B_2 \equiv \begin{pmatrix} 0 & 0 \\ b_0 & b_1 \end{pmatrix}, \quad \boldsymbol{\beta} \equiv \begin{pmatrix} a_2 \\ b_2 \end{pmatrix}.$$

From (2.5) we find that in this case

$$Q_0 = \begin{pmatrix} a_0 & a_1 \\ b_0 & b_1 \end{pmatrix}$$

and hence it is nonsingular if and only if $a_0 b_1 \neq a_1 b_0$. Thus, for instance, if ϕ (or ϕ') is specified at both endpoints, Q_0 is singular. (In general Q_0 is nonsingular if the boundary conditions are linearly independent when applied at the same point.) In the next order approximation we find with $\tau_0 = \tau_1 = a$ and $\tau_2 = b$ that

$$Q_1 = \begin{pmatrix} a_0 & a_1 \\ b_0 + b_1 \int_a^b p(t)\,dt & b_0(b - a) + b_1\left[1 + \int_a^b q(t)\,dt\right] \end{pmatrix}.$$

Thus

$$\det Q_1 = \det Q_0 + a_0 b_0 (b - a) + a_0 b_1 \int_a^b q(t)\,dt - a_1 b_1 \int_a^b p(t)\,dt,$$

and clearly Q_1 may be nonsingular when $\det Q_0 = 0$. In fact with $a_1 = b_1 = 0$, $a_0 b_0 \neq 0$, which corresponds to specifying ϕ at the endpoints, Q_1 is nonsingular. Then condition (2.10) becomes

$$\frac{K^2|b - a|}{2} e^{K|b-a|} \leqq \lambda < 1,$$

where $K \equiv \max_{a \leqq t \leqq b} (|p(t)| + |q(t)|, 1)$. Of course the problem (2.11) has a unique solution under much weaker conditions than the above.

Finally we point out that the *least* restrictions we can impose on the matrices B_ν in order to maintain uniqueness of solutions of (1.1)–(1.2) is that *the n boundary conditions in* (1.2) *should be linearly independent.* Their independence is easily shown to be equivalent to the assertion that

$$\operatorname{rank}(B_1, B_2, \cdots, B_N) = n,$$

where $(B_1, B_2, \cdots, B_N)$ is the $n \times nN$ order matrix whose columns are those of the B_ν. However a matrix of this form has rank n if and only if there exist nth order matrices C_ν such that $\sum_{\nu=1}^N B_\nu C_\nu$ is nonsingular. Thus from Theorem 2.1 it follows that the above independence condition is *necessary* for uniqueness.

3. Finite difference approximations. On the interval $[a, b]$ we place a net of points $\{t_j\}$ with

(3.1a) $$t_0 = a, \qquad t_j = t_{j-1} + h_j, \quad j = 1, 2, \cdots, J, \qquad t_J = b.$$

The mesh widths h_j are to be chosen so that all the special points τ_ν, $\nu = 1, 2, \cdots, N$, at which the data $A(t)$ and $\mathbf{g}(t)$ or their derivatives have jump discontinuities or at which "boundary" conditions are imposed are points of the net. Thus $J \geqq N$ and in particular for any such net we denote these special net points as

(3.1b) $$t_{j_\nu} = \tau_\nu, \qquad \nu = 1, 2, \cdots, N.$$

Our analysis usually refers to a family of such nets in which $J \to \infty$ while $\max_j h_j \to 0$. For each net in any such family we further require that for some fixed positive number $\lambda \geqq 1$,

$$h \equiv \max_j h_j \leqq \lambda \min_k h_k. \tag{3.1c}$$

At each point t_j of a fixed net we seek a vector $\mathbf{u}_j$ which is to approximate $\mathbf{y}(t_j)$, the solution of (1.1), (1.2). The approximating net function $\{\mathbf{u}_j\}$ is defined by the finite difference equations

$$L_h\mathbf{u}_j \equiv h_j^{-1}(\mathbf{u}_j - \mathbf{u}_{j-1}) - \tfrac{1}{2}A(t_{j-1/2})(\mathbf{u}_j + \mathbf{u}_{j-1}) = \mathbf{g}(t_{j-1/2}), \quad j = 1, 2, \cdots, J, \tag{3.2}$$

and the boundary conditions

$$\sum_{\nu=1}^{N} B_\nu \mathbf{u}_{j_\nu} = \boldsymbol{\beta}. \tag{3.3}$$

Here we have used the notation

$$t_{j-1/2} \equiv t_j - \tfrac{1}{2}h_j = t_{j-1} + \tfrac{1}{2}h_j, \qquad j = 1, 2, \cdots, J.$$

The difference scheme in (3.2) is frequently called the "centered difference" method or, in analogy with quadrature schemes, we shall call it the "midpoint rule." An obvious alternative would seem to be the "centered Euler" scheme, analogous to the trapezoidal rule, in which the zero order term is approximated by

$$\tfrac{1}{2}[A(t_j)\mathbf{u}_j + A(t_{j-1})\mathbf{u}_{j-1}].$$

However we *do not* employ this choice, as the treatment of discontinuous coefficients then requires special care.

The stability of the midpoint rule for *initial value* problems is straightforward. But since this result is basic for our work we present it in the following lemma.

LEMMA 3.1. *Let* $K \equiv \max_{a \leqq t \leqq b} \|A(t)\|$. *Then for any net* (3.1) *with* $h \leqq h_0 < 2/K$ *and for any net function* $\{\phi_j\}$ *on this net,*

$$\|\boldsymbol{\phi}_j\| \leqq e^{\lambda K^*|t_j - t_0|}\left\{\|\boldsymbol{\phi}_0\| + \frac{2}{K}\max_{1 \leqq i \leqq J} \|L_h\boldsymbol{\phi}_i\|\right\}, \quad j = 0, 1, \cdots, J, \tag{3.4}$$

where

$$K^* \equiv \frac{K}{1 - h_0K/2}.$$

Proof. It follows from the hypothesis and the Banach lemma that the matrices

$$R_j \equiv I - \tfrac{1}{2}h_jA(t_{j-1/2}), \qquad j = 1, 2, \cdots, J, \tag{3.5a}$$

are nonsingular. Then from the definition (3.2) of L_h and with

$$P_j \equiv [I - \tfrac{1}{2}h_jA(t_{j-1/2})]^{-1}[I + \tfrac{1}{2}h_jA(t_{j-1/2})], \quad j = 1, 2, \cdots, J, \tag{3.5b}$$

we have the identities

$$\phi_j \equiv P_j\phi_{j-1} + h_j R_j^{-1} L_h \phi_j, \qquad j = 1, 2, \cdots, J. \tag{3.6a}$$

By induction this yields

$$\begin{aligned}\phi_j &\equiv (P_j P_{j-1} \cdots P_1)\phi_0 \\ &+ \left\{ h_j R_j^{-1} L_h \phi_j + \sum_{i=1}^{j-1} (P_j P_{j-1} \cdots P_{i+1}) h_i R_i^{-1} L_h \phi_i \right\}, \quad j = 1, 2, \cdots, J.\end{aligned} \tag{3.6b}$$

Taking norms above and employing the inequalities, valid for $h \leqq h_0$,

$$\|R_j^{-1}\| \leqq \frac{1}{1 - hK/2}, \quad \|P_j\| \leqq \frac{1 + hK/2}{1 - hK/2} \leqq e^{hK^*}, \tag{3.6c}$$

we sum the resulting geometric progression and obtain (3.4) since

$$jh \leqq \lambda \sum_{i=1}^{j} h_i = \lambda |t_j - t_0|.$$

This completes the proof.

The solution to the finite difference boundary value problem (3.2)–(3.3) can be given explicitly for sufficiently small net spacing. This is most easily done by factoring the coefficient matrix into a product of block upper triangular by lower triangular matrices. The details and justification are contained in the proof of the following lemma.

LEMMA 3.2. *Define the n-th order matrices C_j and S_j by:*

$$C_j \equiv \begin{cases} 0, & j \neq j_\nu, \quad \nu = 1, 2, \cdots, N, \\ B_\nu, & j = j_\nu, \quad \nu = 1, 2, \cdots, N; \end{cases} \tag{3.7a}$$

$$S_J \equiv C_J, \qquad S_{j-1} = C_{j-1} + S_j P_j, \qquad j = J, J-1, \cdots, 1. \tag{3.7b}$$

Then if $A(t) \in PC_0[a, b]$ and the boundary value problem (1.1)–(1.2) *has a unique solution, the matrix S_0 is nonsingular for h sufficiently small. In this case the finite difference boundary value problem* (3.2)–(3.3) *also has a unique solution and it is given by*

$$\mathbf{u}_0 = S_0^{-1}\left[\boldsymbol{\beta} - \sum_{j=1}^{J} h_j S_j R_j^{-1} \mathbf{g}(t_{j-1/2})\right], \tag{3.8a}$$

$$\mathbf{u}_j = P_j \mathbf{u}_{j-1} + h_j R_j^{-1} \mathbf{g}(t_{j-1/2}), \qquad j = 1, 2, \cdots, J. \tag{3.8b}$$

Proof. Using (3.5) we can write (3.2) in an equivalent form which is just (3.8b). Then the coefficient matrix of the system consisting of (3.3) and (3.8b) can be written in block form and factored, using the definitions (3.7), as follows:

$$
\begin{pmatrix}
C_0 & C_1 & \cdot & \cdot & \cdot & \cdot & C_J \\
-P_1 & I & 0 & \cdot & \cdot & \cdot & 0 \\
 & -P_2 & I & & & & \\
 & & \cdot & \cdot & 0 & & \\
 & & & \cdot & \cdot & & \\
 & 0 & & & \cdot & \cdot & \\
 & & & & & -P_J & I
\end{pmatrix}
\tag{3.9}
$$

$$
= \begin{pmatrix}
S_0 & S_1 & \cdot & \cdot & \cdot & \cdot & S_J \\
 & I & & & & & \\
 & & I & & 0 & & \\
 & & & \cdot & & & \\
 & & & & \cdot & & \\
 & 0 & & & & \cdot & \\
 & & & & & & I
\end{pmatrix}
\begin{pmatrix}
I & & & & & \\
-P_1 & I & & & 0 & \\
 & -P_2 & I & & & \\
 & & \cdot & \cdot & & \\
 & & & \cdot & \cdot & \\
 & 0 & & & \cdot & \cdot \\
 & & & & -P_J & I
\end{pmatrix}.
$$

The solvability of the system (3.2)–(3.3) and the explicit form of the solution given in (3.8) clearly follow from the above factorization if the matrix S_0 is nonsingular.

We can write the matrix S_0 in the form

$$S_0 = \sum_{j=0}^{J} C_j Z_j, \tag{3.10a}$$

where the matrices Z_j are defined by the recursions

$$Z_0 = I, \qquad Z_j = P_j Z_{j-1}, \qquad j = 1, 2, \cdots, J. \tag{3.10b}$$

Note, by (3.5), that Z_j is the solution of the difference problem

$$L_h Z_j = 0, \quad j = 1, 2, \cdots, J, \qquad Z_0 = I. \tag{3.11a}$$

Thus the Z_j may be approximations on the net to the fundamental solution

$$Z'(t) = A(t)Z(t), \qquad Z(a) = I. \tag{3.11b}$$

Since $A(t) \in PC_0[a, b]$ it follows that $Z(t) \in PC_1[a, b]$ and satisfies

$$L_h Z(t_j) = \sigma_j, \; j = 1, 2, \cdots, J, \qquad Z(t_0) = I. \tag{3.11c}$$

Here σ_j are the local truncation errors and $\|\sigma_j\| = O(\omega_1(h/2))$, where $\omega_1(\delta)$ is the modulus of continuity of $Z'(t)$ over any interval of continuity on $[a, b]$. From (3.11a), (3.11c) and Lemma 3.1, assuming $h < 2/K$, we obtain

$$\|Z(t_j) - Z_j\| \leqq O(\|\sigma_j\|), \qquad j = 1, \cdots, J.$$

Since $Z(t)$ and $Y(t, \tau_0)$ defined in (2.1) are fundamental solutions of the same system, it follows that

$$Z(t) = Y(t, \tau_0)Y^{-1}(t_0, \tau_0),$$

and thus from the above estimate

$$Z_j = Y(t_j, \tau_0)Y^{-1}(t_0, \tau_0) + O(\sigma_j).$$

Recalling the definitions in (3.7) and (2.2) we find that

$$S_0 = QY^{-1}(t_0, \tau_0) + O\left(\sum_{j=1}^{J} C_j\sigma_j\right). \tag{3.12}$$

But by Theorem 2.1 the matrix Q is nonsingular since (1.1)–(1.2) is assumed to have a unique solution. Further since at most N (independent of J) of the matrices C_j are nonzero and the local truncation errors σ_j can be made arbitrarily small as $h \to 0$, it follows from the Banach lemma that S_0 is nonsingular for sufficiently small h. This completes the proof.

It should be noted that the terms $O(\omega_1(h/2))$ in the above proof can be replaced by terms which are $O(h^2)$ if $A(t) \in PC_2[a, b]$. The formulas (3.8) for the solution of the difference problem may not be the most appropriate to employ in actual calculations. Specifically suppose $N = 2$ and the two-point conditions are not mixed, a very common case. Then, by placing those conditions in (3.3) that occur at $\tau_1 = a$ first and those at $\tau_2 = b$ last, the coefficient matrix of the system replacing (3.2)–(3.3) becomes a band matrix (that is with all nonzero elements clustered about the diagonal). The resulting system can frequently be solved by the familiar band-matrix factoring procedure which is actually Gauss elimination accounting for zero elements, or by block tridiagonal factorization. Note that this is not the procedure employed in (3.9). In fact, when applicable, the present procedure may be somewhat more accurate than that in (3.9) as regards the growth of roundoff and possible loss in accuracy due to cancellation of leading digits. Of course, for exact arithmetic there is no difference between the two procedures. (In evaluating S_0 and $\mathbf{u}_0$ from (3.7) and (3.8) the universes R_j^{-1} should usually be avoided and the corresponding systems of order n solved by elimination instead.)

We may now employ the solution algorithm in Lemma 3.2 to demonstrate the stability of the centered difference boundary value scheme (3.2)–(3.3). This is to be contrasted to the stability result in Lemma 3.1 for the centered difference initial value scheme (3.2) with $\mathbf{u}_0$ given.

LEMMA 3.3. *Let $A(t) \in PC_0[a, b]$, define $K \equiv \max_{a \leq t \leq b} \|A(t)\|$ and let h be so small that Lemmas 3.1–3.2 hold. Then there exist constants h_0, K_1 and K_2 independent of J such that on all nets (3.1) with $h \leq h_0$ any net function $\{\boldsymbol{\phi}_j\}$ satisfies*

$$\|\boldsymbol{\phi}_j\| \leq K_1 \left\| \sum_{\nu=1}^{N} B_\nu \boldsymbol{\phi}_\nu \right\| + K_2 \max_{1 \leq i \leq J} \|L_h \boldsymbol{\phi}_i\|, \qquad j = 0, 1, \cdots, J. \tag{3.13}$$

Proof. From the result in Lemma 3.2 and with the identity (3.6a) we have the identity

$$\phi_0 = S_0^{-1}\left[\sum_{\nu=1}^{N} B_\nu \phi_\nu - \sum_{j=1}^{J} h_j S_j R_j^{-1} L_h \phi_j\right],$$

and thus

$$\|\phi_0\| \leqq \|S_0^{-1}\|\left[\left\|\sum_{\nu=1}^{N} B_\nu \phi_\nu\right\| + \frac{b-a}{1-hK/2}\max_j \|S_j\|\cdot\|L_h\phi_j\|\right]. \tag{3.14a}$$

It follows from (3.7b) that

$$S_j = C_j + C_{j+1}P_{j+1} + \cdots + C_J(P_J P_{J-1}\cdots P_{j+1}), \qquad j = 0, 1, \cdots, J-1.$$

Then using (3.6c) and (3.7a), we have

$$\begin{aligned} \|S_j\| &\leqq \sum_{k=j}^{J} \|C_k\|\, e^{K^*|t_k - t_j|} \\ &\leqq e^{K^*|b-a|}\sum_{\nu=1}^{N} \|B_\nu\|. \end{aligned} \tag{3.14b}$$

From (3.12) we get, since h is sufficiently small and $\|Y(t_0, \tau_0)\| \leqq e^{K|t_0 - \tau_0|}$,

$$\|S_0^{-1}\| \leqq \frac{\|Q^{-1}\|\, e^{K|t_0-\tau_0|}}{1 - O(\omega_1(h/2))}. \tag{3.14c}$$

Combining (3.14a)–(3.14c) with (3.4) we obtain (3.13) with

$$K_1 = \|Q^{-1}\|\frac{e^{K^*(|b-a|+|t_0-\tau_0|)}}{1-\delta_0},$$

$$K_2 = e^{\lambda K^*|b-a|}\left[\frac{2}{K} + \frac{K_1|b-a|}{1-\delta_1}\sum_{\nu=1}^{N}\|B_\nu\|\right],$$

where $\delta_0 = 1 - h_0K/2$, $K^* = K/\delta_0$ and $\delta_1 = O(\omega_1(h_0/2))$. This completes the proof.

Now we present the basic convergence theorem and error estimate for the finite difference approximation to the solution of the boundary value problem.

THEOREM 3.1. *Let $A(t) \in PC_p[a, b]$ for some integer $p \geqq 0$ and be such that the multipoint boundary value problem (1.1)–(1.2) with $\mathbf{g}(t) \in C_p[a, b]$ has a unique solution $\mathbf{y}(t)$. Then with the constants h_0, K and K_2 of Lemma 3.3, the solution $\{\mathbf{u}_j\}$ of the centered difference multipoint boundary value problem (3.2)–(3.3) satisfies, on all nets (3.1) with $h \leqq h_0$:*

$$\|\mathbf{y}(t_j) - \mathbf{u}_j\| \leqq \begin{cases} K_2\left[\dfrac{M_3}{24} + K\dfrac{M_2}{8}\right]h^2 & \text{if } p \geqq 2, \\[2ex] K_2\left[\dfrac{\omega_2(h)}{8} + K\dfrac{M_2}{8}h\right]h & \text{if } p = 1, \\[2ex] K_2\left[\omega_1\left(\dfrac{h}{2}\right) + K\dfrac{\omega_1(h)}{4}h\right] & \text{if } p = 0. \end{cases} \tag{3.15}$$

Here $M_\nu = \max_{a \leqq t \leqq b} \|\mathbf{y}^{(\nu)}(t)\|$ *for* $\nu = 2, 3$ *and* $\omega_\nu(\delta)$ *is the modulus of continuity of* $\mathbf{y}^{(\nu)}(t)$ *on every interval of continuity in* $[a, b]$.

Proof. Since $\mathbf{y}(t)$ and $\{\mathbf{u}_j\}$ satisfy the same boundary conditions (1.2) and (3.3), we have, on applying Lemma 3.3 to the net function $\boldsymbol{\phi}_j \equiv \mathbf{y}(t_j) - \mathbf{u}_j$ and using (3.2) and (1.1),

$$\begin{aligned}\|\mathbf{y}(t_j) - \mathbf{u}_j\| &\leqq K_2 \max_{1 \leqq i \leqq J} \|L_h\mathbf{y}(t_i) - \mathbf{g}(t_{i-1/2})\| \\ &\leqq K_2 \max_{1 \leqq i \leqq J} \|L_h\mathbf{y}(t_i) - L\mathbf{y}(t_{i-1/2})\| \\ &\leqq K_2 \max_{1 \leqq i \leqq J} \left\{ \left\| \frac{\mathbf{y}(t_i) - \mathbf{y}(t_{i-1})}{h_i} - \mathbf{y}'(t_{i-1/2}) \right\| \right. \\ &\qquad \left. + K \left\| \frac{\mathbf{y}(t_i) + \mathbf{y}(t_{i-1})}{2} - \mathbf{y}(t_{i-1/2}) \right\| \right\}.\end{aligned}$$

To estimate the right-hand side above we apply Taylor's theorem on each interval $[t_{i-1}, t_i]$, $i = 1, 2, \cdots, J$. But by the requirement (3.1b) and Theorem 2.1 it follows that $\mathbf{y}(t) \in C_{p+1}[t_{i-1}, t_i]$ for each $i = 1, 2, \cdots, J$. Thus (3.15) easily follows.

4. Higher order accurate approximations. Perhaps the simplest method for obtaining high order accurate numerical approximations is to apply Richardson's deferred approach to the limit or as it is also termed: $h \to 0$ extrapolation. Current work on such methods includes the extension of Romberg integration for nonlinear initial value problems [3] and generalizations of Fox's difference correction technique [9] for boundary value problems. The treatments to date have not noted the applicability to problems with only piecewise smooth solutions[1] and have usually required uniform nets. We shall first develop the theory for uniform spacing

$$h_j \equiv h \equiv \frac{b - a}{J}, \qquad j = 1, 2, \cdots, J. \tag{4.1}$$

This places an obvious restriction on the allowed spacing of points of discontinuity and boundary points $\{\tau_\nu\}$ which we assume to hold. Then in the next subsection we show that the severe restriction (4.1) can be removed and of course the spacing of the τ_ν is then unrestricted.

First we present the theoretical basis for higher order accurate approximations, which is an asymptotic expansion for the error $\mathbf{u}_j - \mathbf{y}(t_j)$ in powers of h^2. Then we discuss the computation of the improved or higher order accurate solution.

THEOREM 4.1. *For some integer* $M \geqq 1$ *let* $A(t)$, $\mathbf{g}(t) \in PC_{2M+1}[a, b]$ *and problem* (1.1)–(1.2) *have a unique solution. Let the boundary and discontinuity points* $\{\tau_\nu\}$ *satisfy* (3.1) *and* (4.1) *for some sequence of uniform nets with* $J = J_\mu \to \infty$. *For* $\boldsymbol{\phi}(t) \in PC_{2m+1}[a, b]$ *define the differential operator* $T_m\{\boldsymbol{\phi}(t)\} \in PC_0[a, b)$ *by*:

$$T_m\{\boldsymbol{\phi}(t)\} \equiv \frac{-1}{(2m)!}\left[\frac{1}{2m+1}\boldsymbol{\phi}^{(2m+1)}(t) - A(t)\boldsymbol{\phi}^{(2m)}(t)\right], \quad m = 1, 2, \cdots. \tag{4.2}$$

[1] For quadrature the integrand may be piecewise smooth. See § 5 where this is employed.

With $\mathbf{y}(t)$ *the solution of* (1.1)–(1.2) *define the principal error functions* $\mathbf{e}_m(t)$ *as the solutions of the multipoint boundary value problems:*

$$L\mathbf{e}_1(t) = T_1\{\mathbf{y}(t)\}, \qquad \sum_{\nu=1}^{N} B_\nu \mathbf{e}_1(\tau_\nu) = 0; \tag{4.3a}$$

$$L\mathbf{e}_m(t) = T_m\{\mathbf{y}(t)\} + \sum_{k=1}^{m-1} T_k\{\mathbf{e}_{m-k}(t)\}, \qquad \sum_{\nu=1}^{N} B_\nu \mathbf{e}_m(\tau_\nu) = 0, \tag{4.3b}$$

$$m = 2, 3, \cdots, M.$$

Then, if $\{\mathbf{u}_j\}$ *is the solution of the finite difference boundary value problem* (3.2)–(3.3) *on the net* (3.1), (4.1),

$$\mathbf{u}_j - \mathbf{y}(t_j) = \sum_{m=1}^{M} \left(\frac{h}{2}\right)^{2m} \mathbf{e}_m(t_j) + O(h^{2M+1}), \qquad j = 0, 1, \cdots, J. \tag{4.4}$$

Proof. By Theorem 2.1 it follows that $\mathbf{y}(t) \in PC_{2M+2}[a, b]$. Then applying Taylor's theorem to $\mathbf{y}(t_{j-1/2} \pm h/2)$ on $[t_{j-1}, t_j]$ we obtain, using (1.1), (3.2) and (4.2):

$$\begin{aligned} L_h[\mathbf{u}_j - \mathbf{y}(t_j)] &= L\mathbf{y}(t_{j-1/2}) - L_h\mathbf{y}(t_j) \\ &= \sum_{m=1}^{M} \left(\frac{h}{2}\right)^{2m} T_m\{\mathbf{y}(t_{j-1/2})\} + O(h^{2M+1}), \qquad j = 1, 2, \cdots, J. \end{aligned} \tag{4.5}$$

If $\boldsymbol{\phi}(t) \in PC_{2r+1}[a, b]$ and $s \leqq r$, then $T_s\{\boldsymbol{\phi}(t)\} \in PC_{2(r-s)}[a, b]$. Thus an induction employing Theorem 2.1 yields the fact that the $\mathbf{e}_m(t)$ defined in (4.3) satisfy $\mathbf{e}_m(t) \in PC_{2(M-m)+2}[a, b]$ for $m = 1, 2, \cdots, M$. Again applying Taylor's theorem on $[t_{j-1}, t_j]$ yields as above:

$$\begin{aligned} L_h\mathbf{e}_m(t_j) &= L\mathbf{e}_m(t_{j-1/2}) + [L_h\mathbf{e}_m(t_j) - L\mathbf{e}_m(t_{j-1/2})] \\ &= L\mathbf{e}_m(t_{j-1/2}) - \sum_{s=1}^{M-m} \left(\frac{h}{2}\right)^{2s} T_s\{\mathbf{e}_m(t_{j-1/2})\} + O(h^{2(M-m)+1}), \end{aligned} \tag{4.6a}$$

$$m = 1, 2, \cdots, M - 1,$$

and

$$L_h\mathbf{e}_M(t_j) = L\mathbf{e}_M(t_{j-1/2}) + O(h). \tag{4.6b}$$

Using the differential equations in (4.3) and (4.6a), (4.6b), we obtain

$$\begin{aligned} \sum_{m=1}^{M} \left(\frac{h}{2}\right)^{2m} L_h\mathbf{e}_m(t_j) &= \sum_{m=1}^{M} \left(\frac{h}{2}\right)^{2m} T_m\{\mathbf{y}(t_{j-1/2})\} + \sum_{m=2}^{M} \left(\frac{h}{2}\right)^{2m} \sum_{k=1}^{m-1} T_k\{\mathbf{e}_{m-k}(t_{j-1/2})\} \\ &\quad - \sum_{m=1}^{M-1} \left(\frac{h}{2}\right)^{2m} \sum_{s=1}^{M-m} \left(\frac{h}{2}\right)^{2s} T_s\{\mathbf{e}_m(t_{j-1/2})\} + O(h^{2M+1}) \\ &= \sum_{m=1}^{M} \left(\frac{h}{2}\right)^{2m} T_m\{\mathbf{y}(t_{j-1/2})\} + O(h^{2M+1}), \end{aligned} \tag{4.6c}$$

since the last two sums are found to be identical when like powers of h are compared. Thus (4.5) and (4.6c) yield

$$L_h\left[\mathbf{u}_j - \mathbf{y}(t_j) - \sum_{m=1}^{M}\left(\frac{h}{2}\right)^{2m}\mathbf{e}_m(t_j)\right] = O(h^{2M+1}), \quad j = 1, 2, \cdots, J. \tag{4.7a}$$

The boundary conditions in (4.3) and (3.3) yield

$$\sum_{\nu=1}^{N} B_\nu\left[\mathbf{u}_{j_\nu} - \mathbf{y}(t_{j_\nu}) - \sum_{m=1}^{M}\left(\frac{h}{2}\right)^{2m}\mathbf{e}_m(t_{j_\nu})\right] = 0. \tag{4.7b}$$

Now apply the stability result of Lemma 3.3, with

$$\boldsymbol{\phi}_j \equiv \left[\mathbf{u}_j - \mathbf{y}(t_j) - \sum_{m=1}^{M}\left(\frac{h}{2}\right)^{2m}\mathbf{e}_m(t_j)\right],$$

and (4.4) follows. This completes the proof.

It should be pointed out that the residual term in (4.4) is $O(h^{2M+2})$ if the data $A(t)$ and $\mathbf{g}(t)$ are in $PC_{2M+2}[a, b]$.

There are two general ways in which the asymptotic expansions (4.4) can be employed to justify computing higher order accuracy approximations. First suppose that we can determine net functions $\{\mathbf{E}_{m,j}\}$ such that on the net

$$\|\mathbf{e}_m(t_j) - \mathbf{E}_{m,j}\| = O(h^{2(M-m)+1}), \qquad m = 1, 2, \cdots, M.$$

Then it clearly follows from (4.4) that

$$\mathbf{U}_j = \mathbf{u}_j - \sum_{m=1}^{M}\left(\frac{h}{2}\right)^{2m}\mathbf{E}_{m,j}$$

satisfies

$$\|\mathbf{U}_j - \mathbf{y}(t_j)\| = O(h^{2M+1}).$$

This is the basis for Fox's difference correction method [2] in which the inhomogeneous terms in (4.2) and (4.3) are approximated by difference quotients (of decreasing order of accuracy as m increases). Then corresponding difference problems are solved for the $\{\mathbf{E}_{m,j}\}$. There are special difficulties near the endpoints where either external points must be introduced or noncentered differences must be used. A number of recent studies have justified these procedures in special cases [4], [8] and generalized them in various ways [9]. However if, as in the present case, we allow discontinuous data, the endpoint difficulties occur also in the interior. We are not aware of any justification for difference corrections in such cases but it seems clear that this could easily be done.

The other alternative is $h \to 0$ extrapolation in which there is no difficulty caused by nonsmooth data when the present difference scheme is employed. Specifically we consider a sequence of nets with uniform spacings $h \equiv h_\mu > 0$ satisfying (3.1), (4.1) and

$$h_{\mu+1} < h_\mu, \qquad \mu = 0, 1, 2, \cdots. \tag{4.8}$$

Let t be any fixed point common to all the nets specified above. Then on the μth net we introduce the integer

$$j(t, \mu) \equiv \frac{t - t_0}{h_\mu} \tag{4.9a}$$

so that $t_{j(t,\mu)} = t$. The corresponding net function satisfying (3.2)–(3.3) on the μth net has a value at this net point which we denote by

$$\mathbf{u}_{j(t,\mu)} \equiv \mathbf{v}(t, \mu). \tag{4.9b}$$

In terms of this notation the asymptotic error estimate (4.4) applied for $j = j(t, \mu)$ on the μth net yields

$$\mathbf{v}(t, \mu) = \mathbf{y}(t) + \sum_{m=1}^{M} \left(\frac{h_\mu}{2}\right)^{2m} \mathbf{e}_m(t) + O(h_\mu^{2M+1}), \qquad \mu = 0, 1, 2, \cdots. \tag{4.10}$$

With calculations performed on the nets corresponding to $\mu = 0, 1, 2, \cdots, M$ we can obtain approximations which are $O(h_0^{2M+1})$ accurate by using iterative interpolation. That is we define

$$\Delta_\mu \equiv \left(\frac{h_\mu}{2}\right)^2, \qquad \mu = 0, 1, \cdots, M,$$

$$\mathbf{v}(t; \mu, \mu + 1) \equiv \frac{\Delta_\mu \mathbf{v}(t; \mu + 1) - \Delta_{\mu+1}\mathbf{v}(t; \mu)}{\Delta_\mu - \Delta_{\mu+1}}, \quad \mu = 0, 1, \cdots, M - 1, \tag{4.11}$$

$$\mathbf{v}(t; \mu, \mu + 1, \mu + 2) \equiv \frac{\Delta_\mu \mathbf{v}(t; \mu + 1, \mu + 2) - \Delta_{\mu+2}\mathbf{v}(t; \mu, \mu + 1)}{\Delta_\mu - \Delta_{\mu+2}},$$

$$\vdots \qquad \mu = 0, 1, \cdots, M - 2.$$

It follows from (4.10) that

$$\begin{aligned} \mathbf{v}(t; \mu, \mu + 1) - \mathbf{y}(t) &= O(h_\mu^4), \qquad \mu = 0, 1, \cdots, M - 1, \\ \mathbf{v}(t; \mu, \mu + 1, \mu + 2) - \mathbf{y}(t) &= O(h_\mu^6), \qquad \mu = 0, 1, \cdots, M - 2, \\ &\vdots \\ \mathbf{v}(t; 0, 1, \cdots, M) - \mathbf{y}(t) &= O(h_0^{2M+1}). \end{aligned} \tag{4.12}$$

The final error will be $O(h_0^{2M+2})$, as expected, if the data $A(t)$ and $\mathbf{g}(t)$ are in $PC_{2M+2}[a, b]$. The scheme (4.11) is just Neville's iterated interpolation; see [3] or [6, pp. 258–260]. In actually performing the calculations in (4.11) there may be less danger of cancellation of leading digits and hence less loss in significance if we actually employ the forms

$$\mathbf{v}(t; \mu, \mu + 1) = \mathbf{v}(t; \mu + 1) + \frac{\Delta_{\mu+1}}{\Delta_\mu - \Delta_{\mu+1}}[\mathbf{v}(t; \mu + 1) - \mathbf{v}(t; \mu)],$$

and so on. Of course the results (4.12) apply for each point t common to the $M + 1$ nets employed. A rather obvious choice for the spacings h_μ is to take $h_\mu = (\frac{1}{2})^\mu h_0$. Then at all points on the initial net, with spacing h_0, the high order accuracy approximations can be determined. More general refinements of the spacing are briefly considered in [3] and in references given there, as are alternative rational fraction extrapolations. (The successive halving indicated above is not the most efficient way to reduce the net spacing.)

4.1. $h \to 0$ extrapolation with nonuniform nets. In place of the family of uniform nets defined by (4.1) we shall now consider specific sequences of nonuniform nets for which (3.1) is satisfied with no restriction on the points τ_ν. Let $\{t_j\}$ be any net with, say, $J + 1$ points on $[a, b]$ satisfying

$$t_0 = a, \qquad t_J = b, \qquad t_j > t_{j-1}, \qquad j = 1, 2, \cdots, J, \tag{4.13a}$$

$$\{\tau_\nu\} \subset \{t_j\}. \tag{4.13b}$$

We call $\{t_j\}$ a basic net and in terms of it define:

$$\begin{aligned} h_0 &\equiv \max_{1 \leqq j \leqq J} (t_j - t_{j-1}); \\ \theta(t) &\equiv \frac{t_j - t_{j-1}}{h_0}, \qquad t_{j-1} \leqq t < t_j, \qquad j = 1, 2, \cdots, J. \end{aligned} \tag{4.14}$$

Now we define a sequence of nets $\{t_{\mu,j}\}$, $\mu = 0, 1, 2, \cdots$, by

$$h_\mu = 2^{-\mu} h_0, \ \left(\text{or, say, } h_\mu = \frac{h_0}{1+\mu}\right), \ t_{\mu,0} = a; \tag{4.15a}$$

$$t_{\mu,j} = t_{\mu,j-1} + h_{\mu,j}, \qquad h_{\mu,j} \equiv \theta(t_{\mu,j-1} + 0)h_\mu, \qquad j = 1, 2, \cdots, J_\mu \equiv 2^\mu J \tag{4.15b}$$

$$j = 1, 2, \cdots, J_\mu \equiv 2^\mu J \ (\text{or } J_\mu = [1+\mu] J).$$

It easily follows that $t_{\mu,J_\mu} = b$ and $\{\tau_\nu\} \subset \{t_{\mu,j}\}$ for each $\mu = 1, 2, \cdots$. Note that $\theta(t)$ defined in (4.14) is piecewise constant and satisfies

$$0 < \lambda^{-1} \leqq \theta(t) \leqq 1, \qquad t \in [a, b], \tag{4.16}$$

where $\lambda \equiv \max_{j,k} (t_j - t_{j-1})/(t_k - t_{k-1})$. Thus each net (4.15) satisfies the conditions (3.1). In addition, after the $h_{\mu,j}$ have been defined as in (4.15) we can write

$$h_{\mu,j} = \theta(t_{\mu,j-1} + \tfrac{1}{2} h_{\mu,j}) h_\mu. \tag{4.17}$$

Finally we point out that the possible points of (jump) discontinuity of $\theta(t)$ are in the basic net $\{t_j\}$ and hence are included in all of the nets (4.15).

Now define in place of (4.3) the principal error functions $\boldsymbol{\varepsilon}_m(t)$ as solutions of the multipoint boundary value problems

$$L\boldsymbol{\varepsilon}_1(t) = \theta^2(t) T_1\{\mathbf{y}(t)\}, \qquad \sum_{\nu=1}^{N} B_\nu \boldsymbol{\varepsilon}_1(\tau_\nu) = 0, \tag{4.18a}$$

$$\begin{aligned} L\boldsymbol{\varepsilon}_m(t) &= \theta^{2m}(t) T_m\{\mathbf{y}(t)\} + \sum_{k=1}^{m-1} \theta^{2k}(t) T_k\{\boldsymbol{\varepsilon}_{m-k}(t)\}, \\ &\sum_{\nu=1}^{N} B_\nu \boldsymbol{\varepsilon}_m(\tau_\nu) = 0, \qquad m = 2, 3, \cdots, M. \end{aligned} \tag{4.18b}$$

If $\{\mathbf{u}_{\mu,j}\}$ is the solution of the finite difference boundary value problem (3.2)–(3.3) on the net $\{t_{\mu,j}\}$ in (4.15), then

$$\mathbf{u}_{\mu,j} - \mathbf{y}(t_{\mu,j}) = \sum_{m=1}^{M} \left(\frac{h_\mu}{2}\right)^{2m} \boldsymbol{\varepsilon}_m(t_{\mu,j}) + O(h_\mu^{2M+1}), \quad j = 0, 1, \cdots, J_\mu. \tag{4.19}$$

We assume here that $A(t)$, $\mathbf{g}(t) \in PC_{2M+1}[a, b]$ and problem (1.1)–(1.2) has a unique solution. The proof of the above result is almost exactly that of Theorem 4.1 with appropriate replacements of h by $h_{\mu,j}$ as represented in (4.17), t_j by $t_{\mu,j}$ and $t_{j-1/2}$ by $t_{\mu,j} - \frac{1}{2}h_{\mu,j}$. However the left-hand side of (4.6c) is now formed as

$$\sum_{m=1}^{M} \left(\frac{h_\mu}{2}\right)^{2m} L_h \boldsymbol{\varepsilon}_m(t_{\mu,j}),$$

and evaluated using the analogue of (4.6) with (4.18) in place of (4.3).

Since (4.19) is established, as above, for all $\mu = 0, 1, \cdots$ we can apply iterative interpolation as in (4.11) to obtain higher order accuracy solutions on the basic net $\{t_j\}$, which has arbitrary spacing. Our procedure for nonuniform nets can also be applied to the more general boundary conditions or integral constraints of § 5. The details should be fairly clear and hence are omitted. But we point out that these applications show that $h \to 0$ extrapolation or Romberg quadrature can be employed to get high order accuracy integral approximations with nonuniform nets and piecewise smooth integrands.

5. Integral constraints. In place of the multipoint boundary conditions (1.2) we consider (1.1) subject to integral constraints for the form

$$\int_a^b B(t)\mathbf{y}(t)\,dt = \boldsymbol{\beta}. \tag{5.1}$$

Here the rows of the $n \times n$ matrix $B(t)$ are assumed to be linearly independent vectors in $PC_q[a, b]$. (We discuss later the more general case of Stieltjes integrals with piecewise continuously differentiable integrators.)

The existence and uniqueness Theorem 2.1 applies to the problem (1.1), (5.1) if, in place of (2.2), we now define

$$Q \equiv \int_a^b B(\tau)Y(\tau, \tau_0)\,d\tau. \tag{5.2}$$

The sufficiency conditions of Theorem 2.2 employ (2.5b) and

$$Q_m \equiv \int_a^b B(\tau)Y_m(\tau, \tau_0)\,d\tau, \qquad m = 0, 1, \cdots. \tag{5.3}$$

Then (5.2) is nonsingular and the Q_m for all $m > M$ are nonsingular if Q_M is nonsingular and, for some $\lambda < 1$,

$$\int_a^b \|Q_M^{-1}B(\tau)\| \left\{\exp\left(\left|\int_{\tau_0}^{\tau} k(s)\,ds\right|\right) - \left[1 + \left|\int_{\tau_0}^{\tau} k(s)\,ds\right| + \cdots + \frac{1}{M!}\left|\int_{\tau_0}^{\tau} k(s)\,ds\right|^M\right]\right\} d\tau \leqq \lambda. \tag{5.4}$$

To approximate the solution of (1.1), (5.1) we use the net (3.1), where now the τ_ν include all discontinuity points of $A(t)$, $B(t)$ and $\mathbf{g}(t)$. The difference equations (3.2) are retained and in place of (3.3) we use a trapezoidal rule approximation to (5.1):

$$\sum_{j=0}^{J} C_j\mathbf{u}_j = \boldsymbol{\beta}, \tag{5.5a}$$

with

$$\begin{aligned} C_0 &\equiv \tfrac{1}{2}h_1B(t_0), \qquad C_J \equiv \tfrac{1}{2}h_JB(t_J), \\ C_j &\equiv \tfrac{1}{2}[h_jB(t_j - 0) + h_{j+1}B(t_j + 0)], \quad j = 1, 2 \cdots, J - 1. \end{aligned} \tag{5.5b}$$

See (5.16) for alternative coefficients which are simpler in applications.

If we define S_0 by (3.7b) using the C_j above, rather than (3.7a), then as in the proof of Lemma 3.2 we find

$$S_0 = \left[\sum_{j=0}^{J} C_jY(t_j, \tau_0)\right]Y^{-1}(t_0, \tau_0) + O\left(\sum_{j=0}^{J} C_j\sigma_j\right).$$

But since the trapezoidal rule converges for continuous integrands (see proof of Theorem 5.1 below) it follows that, with Q defined in (5.2),

$$\left\|Q - \sum_{j=0}^{J} C_jY(t_j, \tau_0)\right\|$$

can be made arbitrarily small as $h \to 0$. Thus, by the Banach lemma, S_0 is nonsingular provided Q is nonsingular and h is sufficiently small. Lemma 3.2 now gives, in (3.8), the solution of the difference system (3.2), (5.5) provided the modified S_j and C_j are used and Q in (5.2) is nonsingular.

The stability result (3.13) becomes, under the obvious modified hypothesis of Lemma 3.3,

$$\|\boldsymbol{\phi}_j\| \leqq K_1\left\|\sum_{j=0}^{J} C_j\boldsymbol{\phi}_j\right\| + K_3 \max_{1 \leqq i \leqq J} \|L_h\boldsymbol{\phi}_i\|, \quad j = 0, 1, \cdots, J, \tag{5.6}$$

where in place of K_2 we employ

$$K_3 = e^{\lambda K^*|b-a|}\left[\frac{2}{K} + \frac{K_1|b - a|^2}{1 - \delta_1} \max_{a \leqq t \leqq b} \|B(t)\|\right].$$

Now we have the following theorem.

THEOREM 5.1. *Let* $A(t)$ *and* $\mathbf{g}(t) \in PC_p[a, b]$ and $B(t) \in PC_q[a, b]$ *for some non-negative integers* p, q. *Let the generalized boundary value problem* (1.1), (5.1) *have a unique solution* $\mathbf{y}(t)$. *Then the difference equations* (3.2), (5.5) *have a unique solution* $\{\mathbf{u}_j\}$ *for all nets* (3.1) *with* h *sufficiently small and*

$$(5.7) \qquad \|\mathbf{y}(t_j) - \mathbf{u}_j\| = \begin{cases} O(h^2) & \text{if } p \geqq 2, \quad q \geqq 2, \\ O(h\omega_2(h)) & \text{if } p = 1, \quad q \geqq 2, \\ O(h) & \text{if } p \geqq 1, \quad q = 1, \\ O(\omega_1(h) + \Omega(h)) & \text{if } p = q = 0. \end{cases}$$

Here $\omega_\nu(h)$ *is the modulus of continuity of* $\mathbf{y}^{(\nu)}(t)$ *and* $\Omega(h)$ *that of* $B(t)\mathbf{y}(t)$ *over any interval of continuity on* $[a, b]$.

Proof. From (5.6) with $\boldsymbol{\phi}_j \equiv \mathbf{y}(t_j) - \mathbf{u}_j$, (5.1) and (5.5) we obtain

$$\|\mathbf{y}(t_k) - \mathbf{u}_k\| \leqq K_1 \left\| \sum_{j=1}^{J} \frac{h_j}{2}[B(t_{j-1} + 0)\mathbf{y}(t_{j-1}) + B(t_j - 0)\mathbf{y}(t_j)] - \int_{t_{j-1}}^{t_j} B(t)\mathbf{y}(t)\, dt \right\|$$
$$+ K_3 \max_{1 \leqq i \leqq J} \|L_h\mathbf{y}(t_i) - L\mathbf{y}(t_{i-1/2})\|, \qquad k = 0, 1, \cdots, J.$$

By Theorem 2.1, or rather its extension to the present case, and the choice of the net it follows that the integrand $B(y)\mathbf{y}(t)$ above has $r \equiv \min(p + 1, q)$ continuous derivatives on each interval $[t_{j-1}, t_j]$. Thus if $r \geqq 1$ we can use partial integration (Taylor's theorem) on each such interval to get

$$E_j \equiv \frac{h_j}{2}[B(t_{j-1} + 0)\mathbf{y}(t_{j-1}) + B(t_j - 0)\mathbf{y}(t_j)] - \int_{t_{j-1}}^{t_j} B(t)\mathbf{y}(t)\, dt$$
$$= \int_{-h_j/2}^{h_j/2} [B(t_{j-1/2} + \theta)\mathbf{y}(t_{j-1/2} + \theta)]'\theta\, d\theta.$$

Clearly $\|E_j\| \leqq M_1(h_j/2)^2$, where $M_1 = \frac{1}{2}\max_{[t_{j-1}, t_j]} \|[B(t)\mathbf{y}(t)]'\|$ in the above case. If $r \geqq 2$, so that another integration by parts is valid, we find that $\|E_j\| = O(h_j^3)$. If $r = 0$, so that the integrand is only continuous, we easily get $\|E_j\| = O(h_j\Omega(h_j))$, where $\Omega(\delta)$ is the modulus of continuity of $B(t)\mathbf{y}(t)$ on $[t_{j-1}, t_j]$. From these results and the expansions in the proof of Theorem 3.1 the estimates in (5.7) follow. This completes the proof.

To obtain higher order accuracy approximations by $h \to 0$ extrapolation we again specialize to uniformly spaced nets (for simplicity of presentation only). In analogy with Theorem 4.1 we now have the following theorem.

THEOREM 5.2. *For some integer* $M \geqq 1$ *let* $A(t), B(t), \mathbf{g}(t) \in PC_{2M+1}[a, b]$ *and problem* (1.1), (5.1) *have a unique solution. Let the possible discontinuity points* $\{\tau_\nu\}$ *satisfy* (3.1) *and* (4.1) *for some sequence of nets with* $J = J_\mu \to \infty$. *Define* $T_m\{\boldsymbol{\phi}(t)\}$ *as in* (4.2) *and for* $\boldsymbol{\phi}(t) \in PC_{2M-1}[a, b]$ *define the (sum) operator*:

$$(5.8) \qquad S_m\{\boldsymbol{\phi}(t)\} \equiv \frac{-2^{2m}}{(2m)!} B_{2m} \int_a^b [B(t)\boldsymbol{\phi}(t)]^{(2m-1)}\, dt, \quad m = 1, 2, \cdots, M,$$

where B_{2m} is the 2m-th Bernoulli number. With $\mathbf{y}(t)$ *the solution of* (1.1), (5.1) *define the principal error functions* $\mathbf{f}_m(t)$ *as the solutions of*

$$L\mathbf{f}_1(t) = T_1\{\mathbf{y}(t)\},$$

$$\int_a^b B(t)\mathbf{f}_1(t)\,dt = S_1\{\mathbf{y}(t)\}; \tag{5.9a}$$

$$L\mathbf{f}_m(t) = T_m\{\mathbf{y}(t)\} + \sum_{k=1}^{m-1} T_k\{\mathbf{f}_{m-k}(t)\}, \qquad m = 1, 2, \cdots, M,$$

$$\int_a^b B(t)\mathbf{f}_m(t)\,dt = S_m\{\mathbf{y}(t)\} + \sum_{k=1}^{m-1} S_k\{\mathbf{f}_{m-k}(t)\}, \quad m = 1, 2, \cdots, M. \tag{5.9b}$$

Then $\{\mathbf{u}_j\}$, *the solution of the finite difference problem* (3.2), (5.5) *on the net* (3.1), (4.1), *satisfies*:

$$\mathbf{u}_j - \mathbf{y}(t_j) = \sum_{m=1}^{M} \left(\frac{h}{2}\right)^{2m} \mathbf{f}_m(t_j) + O(h^{2M+1}), \qquad j = 0, 1, \cdots, J. \tag{5.10}$$

Proof. From the modified form of Theorem 2.1 which applies to problem (1.1), (5.1) it follows that $\mathbf{y}(t) \in PC_{2M+2}[a, b]$. Thus the development in (4.5) also holds for the present $\mathbf{u}_j$ and $\mathbf{y}(t_j)$. We also see that the relations (4.6a) and (4.6b) apply if the $\mathbf{e}_m(t)$ are replaced by the $\mathbf{f}_m(t)$ defined in (5.9) since only appropriate smoothness properties of these solutions are required and, clearly, $\mathbf{f}_m(t) \in PC_{2(M-m)+2}[a, b]$. The result in (4.6c) is also valid with the $\mathbf{e}_m(t)$ replaced by the $\mathbf{f}_m(t)$ since the differential equations in (4.3) and (5.9) are formally identical. Thus we have, exactly as in the derivation of (4.7a), the relations

$$L_h\left[\mathbf{u}_j - \mathbf{y}(t_j) - \sum_{m=1}^{M} \left(\frac{h}{2}\right)^{2m} \mathbf{f}_m(t_j)\right] = O(h^{2M+1}), \quad j = 1, 2, \cdots, J. \tag{5.11}$$

From (5.1) and (5.5) we have, employing the Euler–Maclaurin sum formula on $[t_{j-1}, t_j]$, where $B(t)\mathbf{y}(t) \in C_{2M+1}[t_{j-1}, t_j]$ for $j = 1, 2, \cdots, J$:

$$\begin{aligned}
\sum_{j=0}^{J} C_j[\mathbf{y}(t_j) - \mathbf{u}_j] &= \sum_{j=1}^{J} \left\{\frac{h}{2}[B(t_{j-1} + 0)\mathbf{y}(t_{j-1}) + B(t_j - 0)\mathbf{y}(t_j)] - \int_{t_{j-1}}^{t_j} B(t)\mathbf{y}(t)\,dt\right\} \\
&= \sum_{j=1}^{J} \left\{\sum_{m=1}^{M} \left(\frac{h}{2}\right)^{2m} \frac{2^{2m}}{(2m)!} B_{2m}[B(t)\mathbf{y}(t)]^{(2m-1)}\Bigg|_{t_{j-1}}^{t_j} + O(h^{2M+2})\right\} \\
&= -\sum_{m=1}^{M} \left(\frac{h}{2}\right)^{2m} S_m\{\mathbf{y}(t)\} + O(h^{2M+1}).
\end{aligned} \tag{5.12}$$

In the final step above we have interchanged orders of summation and used the fact that jump discontinuities of $[B(t)\mathbf{y}(t)]^{(m)}$ occur, if at all, at points $\tau_\nu = t_{j_\nu}$ of the net. Similarly from (5.5b) and the Euler–Maclaurin formula we obtain, since $B(t)\mathbf{f}_m(t) \in PC_{2(M-m)+2}[a, b]$:

$$\sum_{j=0}^{J} C_j \mathbf{f}_m(t_j) = \int_a^b B(t)\mathbf{f}_m(t)\,dt + \sum_{j=1}^{J} \left\{ \frac{h}{2}[B(t_{j-1}+0)\mathbf{f}_m(t_{j-1}) + B(t_j - 0)\mathbf{f}_m(t_j)] - \int_{t_{j-1}}^{t_j} B(t)\mathbf{f}_m(t)\,dt \right\},$$

(5.13a)
$$= \int_a^b B(t)\mathbf{f}_m(t)\,dt - \sum_{k=1}^{M-m} \left(\frac{h}{2}\right)^{2k} S_k\{\mathbf{f}_m(t)\} + O(h^{2(M-m)+1}),$$

$$m = 1, 2, \cdots, M-1.$$

For $m = M$ we have $B(t)\mathbf{f}_M(t) \in PC_2[a, b]$ and so

(5.13b)
$$\sum_{j=0}^{J} C_j \mathbf{f}_M(t_j) = \int_a^b B(t)\mathbf{f}_M(t)\,dt + O(h).$$

Using the integral conditions from (5.9) in (5.13) yields, after rearranging some sums,

(5.13c)
$$\sum_{j=0}^{J} C_j \sum_{m=1}^{M} \left(\frac{h}{2}\right)^{2m} \mathbf{f}_m(t_j) = \sum_{m=1}^{M} \left(\frac{h}{2}\right)^{2m} S_m\{\mathbf{y}(t)\} + O(h^{2M+1}).$$

Thus (5.12) and the above yield

(5.14)
$$\sum_{j=0}^{J} C_j \left[\mathbf{u}_j - \mathbf{y}(t_j) - \sum_{m=1}^{M} \left(\frac{h}{2}\right)^{2m} \mathbf{f}_m(t_j) \right] = O(h^{2M+1}).$$

With $\boldsymbol{\phi}_j \equiv [\mathbf{u}_j - \mathbf{y}(t_j) - \sum_{m=1}^{M} (h/2)^{2m}\mathbf{f}_m(t_j)]$ in the stability result (5.6) we obtain (5.10) by means of (5.11) and (5.14). This completes the proof.

To actually employ this theorem for computing higher order accuracy approximations to the solution of (1.1), (5.1) we proceed as in (4.8)–(4.12) but use the difference problem (3.2), (5.5).

We point out that our problem (1.1), (5.1) could have been replaced by the following two-point boundary value problem:

(5.15)
$$\begin{aligned} \mathbf{y}'(t) &= A(t)\mathbf{y}(t) + \mathbf{g}(t), \\ \mathbf{z}'(t) &= B(t)\mathbf{y}(t), \quad \mathbf{z}(a) = 0, \quad \mathbf{z}(b) = \boldsymbol{\beta}. \end{aligned}$$

The difference method of § 3 applied to this problem involves systems of order $2n$, which are avoided in our direct approach. However the "midpoint" scheme applicable to (5.15) avoids the need for special care in treating the discontinuities of $B(t)$ as in (5.5b). This in fact can also be done in our direct treatment by using in (5.5a), in place of (5.5b), the coefficients:

(5.16)
$$C_0 \equiv \tfrac{1}{2}h_1 B(t_{1-1/2}), \quad C_J \equiv \tfrac{1}{2}h_J B(t_{J-1/2}),$$

$$C_j \equiv \tfrac{1}{2}[h_j B(t_{j-1/2}) + h_{j+1}B(t_{j+1/2})], \quad j = 1, 2, \cdots, J-1.$$

We have not employed these coefficients as the analysis becomes rather lengthy since the Euler–Maclaurin formula is no longer relevant. But the error estimates and $h \to 0$ extrapolation procedure remain valid.

Finally we observe that the procedures and analysis of § 3 and § 5 can easily be combined to treat (1.1) subject to constraints of the form

$$\sum_{\nu=1}^{N} B_\nu \mathbf{y}(\tau_\nu) + \int_a^b B(t)\mathbf{y}(t)\,dt = \boldsymbol{\beta}. \tag{5.17}$$

In fact the only change in the numerical method is simply to replace the C_j by the sum of those C_j defined in (5.5b) or (5.16) to those defined in (3.7a). Constraints of the form (5.17) can be represented by Stieltjes integrals with piecewise continuously differentiable integrators. The most general linear constraint applied to continuous $\mathbf{y}(t)$ (i.e., a linear functional on $C_0[a, b]$) is given by a Stieltjes integral with integrator of bounded variation. We cannot hope to treat this most general constraint by numerical methods of high order accuracy.

REFERENCES

[1] E. Coddington and N. Levinson, *Theory of Ordinary Differential Equations*, McGraw-Hill, New York, 1955.

[2] L. Fox, *The Numerical Solution of Two-Point Boundary-Value Problems in Ordinary Differential Equations*, Oxford University Press, London, 1957.

[3] W. B. Gragg, *On extrapolation algorithms for ordinary initial value problems*, this Journal, 2 (1965), pp. 384–403.

[4] P. Henrici, *Discrete Variable Methods in Ordinary Differential Equations*, John Wiley, New York, 1962.

[5] E. L. Ince, *Ordinary Differential Equations*, Dover, New York, 1944.

[6] E. Isaacson and H. B. Keller, *Analysis of Numerical Methods*, John Wiley, New York, 1966.

[7] H. B. Keller, *Numerical Methods for Two-Point Boundary-Value Problems*, Blaisdell, Waltham, Massachusetts, 1968.

[8] M. Lees, *Discrete methods for nonlinear two-point boundary value problems*, Numerical Solution of Partial Differential Equations, J. H. Bramble, ed., Academic Press, New York, 1966, pp. 59–72.

[9] V. L. Pereyra, *Highly accurate discrete methods for nonlinear problems*, MRC Rep. 749, Mathematics Research Center, U.S. Army, University of Wisconsin, Madison, 1967.

ACCURATE DIFFERENCE METHODS FOR NONLINEAR TWO-POINT BOUNDARY VALUE PROBLEMS*

HERBERT B. KELLER†

Abstract. We show that each *isolated* solution, $y(t)$, of the general nonlinear two-point boundary value problem (*): $y' = f(t, y)$, $a < t < b$, $g(y(a), y(b)) = 0$ can be approximated by the (box) difference scheme (**): $[u_j - u_{j-1}]/h_j = f(t_{j-1/2}, [u_j + u_{j-1}]/2)$, $1 \leqq j \leqq J$, $g(u_0, u_J) = 0$. For $h = \max_{1 \leqq j \leqq J} h_j$ sufficiently small, the difference equations (**) are shown to have a unique solution $\{u_j\}_0^J$ in some sphere about $\{y(t_j)\}_0^J$, and it can be computed by Newton's method which converges quadratically. If $y(t)$ is sufficiently smooth, then the error has an asymptotic expansion of the form $u_j - y(t_j) = \sum_{\nu=1}^{m} h^{2\nu} e_\nu(t_j) + O(h^{2m+2})$, so that Richardson extrapolation is justified.

The coefficient matrices of the linear systems to be solved in applying Newton's method are of order $n(J + 1)$ when $y(t) \in \mathbb{R}^n$. For separated endpoint boundary conditions: $g_1(y(a)) = 0$, $g_2(y(b)) = 0$ with $\dim g_1 = p$, $\dim g_2 = q$ and $p + q = n$, the coefficient matrices have the special block tridiagonal form $A \equiv [B_j, A_j, C_j]$ in which the $n \times n$ matrices $B_j(C_j)$ have their last q (first p) rows null. Block elimination and band elimination without destroying the zero pattern are shown to be valid. The numerical scheme is very efficient, as a worked out example illustrates.

1. Introduction. We study the application of the centered Euler or box scheme to very general nonlinear two-point boundary value problems of the form

$$\mathcal{N}y(t) \equiv y'(t) - f(t, y(t)) = 0, \qquad a \leqq t \leqq b; \tag{1.1a}$$

$$g(y(a), y(b)) = 0. \tag{1.1b}$$

Such problems may, of course, have nonunique solutions. But we show that *for each isolated solution* of (1.1), the difference scheme has, for a sufficiently fine net, a unique solution in some tube about the isolated solution; the numerical solution can be computed by Newton's method with quadratic convergence; there is an asymptotic error expansion proceeding in powers of h^2 so that Richardson extrapolation is valid and yields two orders of magnitude improvement per application. When the boundary conditions are of the separated endpoint type, then the Newton iterates can be computed by a block elimination procedure which is very efficient. The net employed can be nonuniform, while the solution and "coefficients" in the equation need only be piecewise smooth. With very little effort, our theory is extended to include multipoint boundary conditions of the form

$$g(y(\tau_1), y(\tau_2), \cdots, y(\tau_N)) = 0, \qquad a \leqq \tau_1 < \tau_2 < \cdots < \tau_N \leqq b.$$

These results generalize a previous study of accurate difference schemes for *linear* multipoint boundary value problems [3] to the nonlinear case. Indeed, the basic stability result from this earlier study is crucial in the present analysis. Somewhat similar results for nonlinear two-point boundary value problems have been given in [4, pp. 96–102], but only for linear boundary conditions under quite

---

* Received by the editors November 9, 1972, and in revised form March 1, 1973.

† Applied Mathematics, Firestone Laboratories, California Institute of Technology, Pasadena, California 91109. This work was supported by the Atomic Energy Commission under Contract AT(04-3)-767, Project Agreement no. 12.

unnatural restrictions. The present basic requirement that the solution of (1.1) be isolated is close to the minimal conditions to be expected.

The present theory has been extended to very general difference schemes which are defined in terms of one-step and multistep schemes for initial value problems. In fact any numerical scheme which is stable and consistent for initial value problems can be shown to yield a difference scheme to which our main theorem (trivially modified) applies. These results will be published in the future [10]. One particularly significant such extension due to R. Weiss [11] employs implicit Runge–Kutta schemes which with the appropriate m-point Lobatto quadrature points have $O(h^{2m-2})$ accuracy. Weiss further shows the equivalence of the difference schemes with appropriate collocation methods using piecewise polynomials. Recently H.-O. Kreiss [6] has developed a very general and quite complete theory of difference methods for linear boundary value problems. It is rather clear that his work can be extended to nonlinear boundary value problems by essentially the same techniques used in the present work.

A solution $y(t) \in \mathbb{R}^n$ *of* (1.1) *is said to be isolated if and only if the linearized problem*

$$L[y]\phi(t) \equiv \phi' - A(t)\phi(t) = 0, \qquad a \leqq t \leqq b; \tag{1.2a}$$

$$B_a\phi(a) + B_b\phi(b) = 0, \tag{1.2b}$$

with the $n \times n$ *matrices*

$$A(t) \equiv \partial f(t, y(t))/\partial y, \quad B_a \equiv \partial g(y(a), y(b))/\partial y(a), \quad B_b \equiv \partial g(y(a), y(b))/\partial y(b) \tag{1.2c}$$

has only the trivial solution $\phi(t) \equiv 0$. It is more or less well known that an isolated solution is "locally unique"; that is, no other solution exists in some tube about the isolated solution.

The difference scheme is employed on any (nonuniform) net $\{t_j\}_0^J$ with

$$\begin{aligned} t_0 &= a; \qquad t_j = t_{j-1} + h_j, \qquad 1 \leqq j \leqq J; \\ t_J &= b; \qquad h \equiv \max h_j \leqq \lambda \min h_j. \end{aligned} \tag{1.3a}$$

Here λ is a uniform bound on the ratio of maximum to minimum spacing for all families of nets we consider. Applied to (1.1), the box scheme (or centered Euler scheme) is

$$N_h u_j \equiv \frac{1}{h_j}[u_j - u_{j-1}] - f(t_{j-1/2}, \tfrac{1}{2}[u_j + u_{j-1}]) = 0, \qquad 1 \leqq j \leqq J, \tag{1.3b}$$

$$g(u_0, u_J) = 0. \tag{1.3c}$$

Our basic results may be stated as the following theorem.

MAIN THEOREM. *Let* (1.1) *have an isolated solution* $y(t) \in C_4[a, b]$. *Let* $f(t, z) \in C_3\{[a, b] \times S_\rho[y(t)]\}$, *where* $S_\rho[y(t)] \equiv \{z \mid z \in \mathbb{R}^n, \|z - y(t)\| \leqq \rho\} \subset \mathbb{R}^n$ *and* $g(v, w) \in C_2\{S_\rho[y(a)] \times S_\rho[y(b)]\}$. *Define the sphere* $S_\rho\{y(t_j)\} \equiv \{v_0, \cdots, v_J \mid v_j \in S_\rho \cdot [y(t_j)], 0 \leqq j \leqq J\} \subset \mathbb{R}^{n(J+1)}$. *Then for some* ρ *and* h_0 *sufficiently small, all nets* (1.3a) *with* $h \leqq h_0$ *are such that*:

(i) *The difference equations* (1.3) *have a unique solution* $\{u_j\} \in S_\rho\{y(t_j)\}$.

(ii) *The difference solution can be computed by Newton's method,* (3.5), *which converges quadratically for any initial iterate* $\{u_j^{(0)}\} \in S_{\rho_1}\{y(t_j)\}$, *provided* ρ_1 *and* ρ_1/h *are sufficiently small.*

(iii) $\|u_j - y(t_j)\| = O(h^2), 0 \leqq j \leqq J$.

(iv) *If* $f(t, z)$ *and* $y(t)$ *are sufficiently smooth, then there exist* $e_\nu(t)$ *such that*

$$u_j^{(k)} - y(t_{k,j}) = \sum_{\nu=1}^{m} \left(\frac{h_k}{2}\right)^{2\nu} e_\nu(t_{k,j}) + O(h_k^{2m+2}), \qquad 0 \leqq j \leqq J_k, \quad k = 0, 1, \cdots,$$

for all nets $\{t_{k,j}\}$ *satisfying* (4.10). *Here* $u_j^{(k)}$ *is the numerical solution on the k-th net.*

In § 2 we present some results from [3] which are valid for linear two-point boundary value problems; this is the basic stability theory we employ. Parts (i) and (ii) of the main theorem are proven in § 3 and parts (iii) and (iv) are proven in § 4. In § 5 we show that for separated endpoint boundary conditions, i.e., for $g_1(y(a)) = 0$, $g_2(y(b)) = 0$, the linearized equations which arise in Newton's method can be solved by very efficient block-elimination or band-elimination procedures. Some practical observations on the use of Newton's method in conjunction with Richardson extrapolation are given in § 6. A worked out example is also reported there. Our method has been used to solve nonlinear systems with as many as $n = 120$ equations, and this work is reported elsewhere [7]. In the Appendix we indicate how the basic results can be extended to apply to nonlinear multipoint boundary conditions.

It should be observed that all of our results apply equally well to the trapezoidal rule:

$$\frac{1}{h_j}[u_j - u_{j-1}] - \frac{1}{2}[f(t_{j-1}, u_{j-1}) + f(t_j, u_j)] = 0.$$

We have used centered Euler as in (1.3b) as it is frequently more efficient.

2. Summary of results for linear problems. We summarize here some of the results of [3] specialized for linear two-point boundary value problems of the form

$$Lv \equiv v' - A(t)v = g(t), \tag{2.1a}$$

$$B_a v(a) + B_b v(b) = \beta. \tag{2.1b}$$

It is well known and easily demonstrated (see [3]), that (2.1) has a unique solution for each $g(t) \in C[a, b]$ and each $\beta \in \mathbb{R}^n$ if and only if the $n \times n$ matrix

$$Q \equiv B_a Y(a, \tau) + B_b Y(b, \tau) \tag{2.2}$$

is nonsingular. Here $Y(t, \tau)$ is the fundamental solution matrix of (2.1a) defined for any $\tau \in [a, b]$ by

$$Y' = A(t)Y, \qquad a \leqq t \leqq b; \tag{2.3a}$$

$$Y(\tau) = I. \tag{2.3b}$$

In most applications we employ $\tau = a$ (or $\tau = b$).

The centered Euler or box scheme for (2.1) on the net (1.3a) is

$$L_h v_j \equiv h_j^{-1}(v_j - v_{j-1}) - A(t_{j-1/2})\tfrac{1}{2}(v_j + v_{j-1}) = g(t_{j-1/2}), \qquad 1 \leqq j \leqq J; \tag{2.4a}$$

$$B_a v_0 + B_b v_J = \beta. \tag{2.4b}$$

We combine Lemmas 3.1–3.3 of [3] to state the following.

LEMMA 2.1. *Let B_a, B_b and $A(t) \in C[a, b]$ be such that* (2.1) *has a unique solution. Then there exist constants h_0, K_1, K_2 such that for every net* (1.3a) *with $h \leqq h_0$ the linear difference equations* (2.4) *have a unique solution which is bounded by*

$$\|v_j\| \leqq K_1\|\beta\| + K_2 \max_{1 \leqq i \leqq J} \|g(t_{i-1/2})\|, \qquad 0 \leqq j \leqq J. \tag{2.5}$$

The linear system (2.4) can be written in the matrix-vector form

$$\mathbb{L}V = G \tag{2.6a}$$

by introducing the $n \times n$ matrices

$$L_j \equiv h_j^{-1}I + \tfrac{1}{2}A(t_{j-1/2}), \quad R_j \equiv h_j^{-1}I - \tfrac{1}{2}A(t_{j-1/2}), \tag{2.6b}$$

and the $n(J + 1)$ order matrix and vectors

$$\mathbb{L} \equiv \begin{pmatrix} B_a & & & B_b \\ -L_1 & R_1 & & \\ & \ddots & \ddots & \\ & & -L_J & R_J \end{pmatrix}, \quad V \equiv \begin{pmatrix} v_0 \\ v_1 \\ \vdots \\ v_J \end{pmatrix}, \quad G \equiv \begin{pmatrix} \beta \\ g(t_{1/2}) \\ \vdots \\ g(t_{J-1/2}) \end{pmatrix}. \tag{2.6c}$$

Lemma 2.1 clearly implies that $\mathbb{L}$ is nonsingular for all nets (1.3a) with $h \leqq h_0$. To get a bound on $\|\mathbb{L}^{-1}\|$, we use the operator norm induced by the vector norm on $V \in \mathbb{R}^{n(J+1)}$ given by

$$\|V\| \equiv \max_{0 \leqq j \leqq J} \|v_j\|, \tag{2.7}$$

where $\|v\|$ is any vector norm on $v \in \mathbb{R}^n$. Then (2.5) implies

$$\|\mathbb{L}^{-1}\| \leqq \max(K_1, K_2) \equiv K_0. \tag{2.8}$$

This furnishes a bound which is *uniform* on the family of matrices $\mathbb{L}$ defined for all nets (1.3a) with $h \leqq h_0$.

3. Solution of the difference equations. We first prove part (i) of the main theorem by means of contracting maps. Then we examine Newton's method. Write (1.3) in the vector form

$$\Phi(U) = 0, \tag{3.1a}$$

$$U \equiv \begin{pmatrix} u_0 \\ u_1 \\ \vdots \\ u_J \end{pmatrix}, \tag{3.1b}$$

$$\Phi(U) = \begin{pmatrix} g(u_0, u_J) \\ N_h u_1 \\ \vdots \\ N_h u_J \end{pmatrix}. \tag{3.1c}$$

Since $\mathbb{L}$ of (2.6) is nonsingular, employing (1.2c), we can write (3.1a) in the equivalent form

$$U = U - \mathbb{L}^{-1}\Phi(U) \equiv \Psi(U). \tag{3.2}$$

For any $V, W \in S_\rho\{y(t_j)\}$, we have

$$\begin{aligned} \Psi(V) - \Psi(W) &= [V - W] - \mathbb{L}^{-1}[\Phi(V) - \Phi(W)] \\ &= \mathbb{L}^{-1}[\mathbb{L} - \partial\tilde{\Phi}(V, W)/\partial U][V - W], \end{aligned} \tag{3.3a}$$

where

$$\frac{\partial\tilde{\Phi}(V, W)}{\partial U} = \int_0^1 \frac{\partial\Phi}{\partial U}(sV + [1 - s]W)\, ds. \tag{3.3b}$$

Here we have used the convexity of $S_\rho\{y(t_j)\}$ and the continuous differentiability of $f(t, z)$ and $g(v, w)$. From (1.2c) we recall that $A(t_{j-1/2}) = f_y(t_{j-1/2}, y(t_{j-1/2}))$. Now introduce $\hat{A}(t_{j-1/2}) \equiv f_y(t_{j-1/2}, \frac{1}{2}[y(t_j) + y(t_{j-1})])$ and denote by $\hat{\mathbb{L}}$ the matrix of (2.6b, c) with $A(t_{j-1/2})$ replaced by $\hat{A}(t_{j-1/2})$ for $1 \leqq j \leqq J$. Then

$$\begin{aligned} \left\| \hat{\mathbb{L}} - \frac{\partial\tilde{\Phi}}{\partial U} \right\| &\leqq \|\mathbb{L} - \hat{\mathbb{L}}\| + \left\| \hat{\mathbb{L}} - \frac{\partial\tilde{\Phi}}{\partial U} \right\|. \\ &\leqq K \max \|y(t_{j-1/2}) - \tfrac{1}{2}[y(t_j) + y(t_{j-1})]\| + K\rho, \end{aligned}$$

where K *is the maximum of the Lipschitz constants for* $f_y(t, z), g_v(v, w), g_w(v, w)$ *with respect to* z, v, w. If, as is implied by the hypothesis, $y(t) \in C_4[a, b]$, then clearly for some $M_0 > 0$,

$$\|\mathbb{L} - \partial\tilde{\Phi}/\partial U\| \leqq K(M_0 h^2 + \rho). \tag{3.3c}$$

Combined with (2.8), this result in (3.3a) yields

$$\|\Psi(V) - \Psi(W)\| \leqq \alpha\|V - W\|, \quad \alpha \equiv K_0 K(M_0 h^2 + \rho) < 1, \tag{3.4a}$$

for all $h \leqq h_0$, provided h_0 and ρ are sufficiently small.

The "center" of the sphere $S_\rho\{y(t_j)\}$ is denoted by $Y \equiv (y(t_0), \cdots, y(t_J))^T$. Then $\Phi(Y)$ can be estimated from (3.1c), (1.1) and (3.1b) by

$$\begin{aligned} \|\Phi(Y)\| &= \max_{1 \leqq j \leqq J} \|N_h y(t_j)\| \\ &= \max_{1 \leqq j \leqq J} \|N_h y(t_j) - N y(t_{j-1/2})\| \\ &\leqq M_1 h^2. \end{aligned}$$

This is just a bound on the local truncation errors which, with (3.2) and (2.8), now give

(3.4b) $$\|Y - \Psi(Y)\| \leqq K_0 M_1 h^2 \leqq (1 - \alpha)\rho,$$

for all $h \leqq h_0$, provided h_0 is sufficiently small.

From (3.4a, b) it follows that $\Psi(U)$ takes $S_\rho\{y(t_j)\}$ into itself and is contracting there. Thus (3.2) has a unique solution in $S_\rho\{y(t_j)\}$, and part (i) of the main theorem is established. We point out that the iteration scheme suggested by (3.2) is not practical since $\mathbb{L}$ is, of course, unknown. In many applications, Newton's method is extremely effective, so we present a theoretical justification for it showing the quadratic convergence.

Specifically, we define the sequence of Newton iterates U^ν by

(3.5a) $$U^0 \in S_{\rho_1}\{y(t_j)\};$$

(3.5b) $$\frac{\partial \Phi(U^\nu)}{\partial U}[U^{\nu+1} - U^\nu] = -\Phi(U^\nu), \qquad \nu = 0, 1, 2, \cdots.$$

Note, for any $U \in S_\rho\{y(t_j)\}$, that as in (3.3c), $\|\mathbb{L}^{-1}(\mathbb{L} - \partial\Phi(U)/\partial U)\| \leqq \alpha < 1$. Then since $\partial\Phi(U)/\partial U = \mathbb{L}[I - \mathbb{L}^{-1}(\mathbb{L} - \partial\Phi(U)/\partial U)]$, it follows from the Banach lemma that $\partial\Phi(U)/\partial U$ is nonsingular for all $U \in S_\rho\{y(t_j)\}$. Further, this lemma and (2.8) imply

(3.6a) $$\left\| \left(\frac{\partial \Phi(U)}{\partial U} \right)^{-1} \right\| \leqq \frac{K_0}{1 - \alpha}.$$

With the same Lipschitz constant K used in (3.3c), we have for all $U, V \in S_\rho\{y(t_j)\}$,

(3.6b) $$\|\partial\Phi(U)/\partial U - \partial\Phi(V)/\partial V\| \leqq K\|U - V\|.$$

The initial error in satisfying (3.1a) can be estimated as

(3.6c) $$\begin{aligned} \|\Phi(U^0)\| &\leqq \|\Phi(Y)\| + \|\Phi(U^0) - \Phi(Y)\| \\ &\leqq M_1 h^2 + \|\partial\tilde{\Phi}/\partial U\| \cdot \|U^0 - Y\| \\ &\leqq M_1 h^2 + 2(\lambda/h + C)\rho_1, \end{aligned}$$

where C is a bound on the norms of $g_v(v, w)$, $g_w(v, w)$ and $\frac{1}{2}f_u(t, u)$, and λ is defined in (1.3a). The quadratic convergence of Newton's method now follows from (3.6) in standard fashion provided h, ρ_1 and ρ_1/h are sufficiently small; see, for example, [5] or [8].

In practice, the basic problem is, as usual, to find any appropriate initial iterate U^0 which is within a distance ρ_1 from Y. We discuss some of the practical considerations in §§ 5 and 6. But it is important to note that as h is reduced, ρ_1 must also be reduced in order that $\|\Phi(U^0)\|$ be sufficiently small. The implication is that we must be able to guess at an $O(h)$ accurate solution to get our Newton scheme to converge. Once this is done for any fixed h, we shall see that we can then easily compute solutions accurate to $O(h^{2m})$ for some $m = 1, 2, \cdots$.

4. Error estimates. To establish part (iii) of the main theorem, let $v_j \equiv u_j - y(t_j)$, where $\{u_j\}$ is the solution of (1.3) and $y(t)$ is the solution of (1.1).

Then since $N_h u_j = N_h[y(t_j) + v_j] = 0$, the mean value theorem yields

(4.1a) $$\tilde{L}_h v_j = -N_h y(t_j) \equiv \tau_j[y], \qquad 1 \leqq j \leqq J.$$

Similarly, (1.1b) and (1.3c) imply

(4.1b) $$\tilde{B}_a v_0 + \tilde{B}_b v_J = 0.$$

Here $\tilde{L}_h$ is given by (2.4a) with $A(t_{j-1/2})$ replaced by

(4.2a) $$\tilde{A}_{j-1/2} \equiv \int_0^1 f_y\left(t_{j-1/2}, \frac{1}{2}[y(t_j) + y(t_{j-1})] + \frac{s}{2}[v_j + v_{j-1}]\right) ds$$

and

(4.2b) $$\tilde{B}_a \equiv \int_0^1 g_{y(a)}(y(a) + sv_0, y(b) + sv_J)\, ds,$$
$$\tilde{B}_b \equiv \int_b^1 g_{y(b)}(y(a) + sv_0, y(b) + sv_J)\, ds.$$

A standard Taylor expansion yields, with (1.1a),

(4.3) $$\|\tau_j[y]\| = \|Ny(t_{j-1/2}) - N_h y(t_j)\| \leqq M_1 h^2, \qquad 1 \leqq j \leqq J.$$

The system (4.1a, b) can now be written as

(4.4a) $$\tilde{\mathbb{L}} V = T,$$

(4.4b) $$V \equiv \begin{pmatrix} v_0 \\ v_1 \\ \vdots \\ v_J \end{pmatrix}, \quad T \equiv \begin{pmatrix} 0 \\ \tau_1[y] \\ \vdots \\ \tau_J[y] \end{pmatrix},$$

where $\tilde{\mathbb{L}}$ is defined by (2.6b, c) with $A_{j-1/2}$, B_a and B_b replaced by $\tilde{A}_{j-1/2}$, $\tilde{B}_a$ and $\tilde{B}_b$, respectively, of (4.2). It follows that

(4.5) $$\|\mathbb{L} - \tilde{\mathbb{L}}\| \leqq \frac{K}{2}\|V\|,$$

where K is the Lipschitz constant previously introduced. Since $\mathbb{L}$ has a uniformly bounded inverse, as in (2.8), write (4.4a) as $\mathbb{L}V = (\mathbb{L} - \tilde{\mathbb{L}})V + T$ and use (4.5) and (4.3) to deduce

(4.6) $$\|V\| \leqq \frac{K_0 K}{2}\|V\|^2 + K_0 M_1 h^2.$$

The scalar inequality $x \leqq \alpha x^2 + \beta$, with $\alpha > 0$ and $4\alpha\beta < 1$, implies that either $x \leqq x_-$ or $x \geqq x_+$ where $x_\pm = (1 \pm \sqrt{1 - 4\alpha\beta})/2\alpha$. To apply this result in (4.6), let us require h_0 to be so small that

(4.7a) $$2K_0^2 K M_1 h^2 < \tfrac{1}{2} \quad \text{for all } h \leqq h_0.$$

Then recalling $\|V\| = \|U - Y\| \leqq \rho$, we also require

$$\rho < (1 + \sqrt{1 - 2K_0^2 K M_1 h_0^2})/K_0 K \tag{4.7b}$$

and it follows that

$$\begin{aligned} \|V\| &\leqq (1 - \sqrt{1 - 2K_0^2 K M_1 h^2})/K_0 K \\ &\leqq \sqrt{2} K_0 M_1 h^2. \end{aligned} \tag{4.8}$$

Thus part (iii) is proven.

The asymptotic error expansions in part (iv) of the main theorem follow by showing that the coefficient functions, $e_\nu(t)$, can be defined recursively, with $e^0(t) \equiv y(t)$, by linear two-point boundary value problems of the form

$$\frac{d}{dt} e^\nu(t) - A(t)\, e^\nu(t) = \Theta_\nu(t), \qquad \nu = 1, 2, \cdots, \tag{4.9a}$$

$$B_a\, e^\nu(a) + B_b\, e^\nu(b) = \gamma_\nu, \qquad \nu = 1, 2, \cdots. \tag{4.9b}$$

In view of (1.2) and the fact that $y(t)$ is an isolated solution of (1.1), it follows that (4.9) has a unique solution for each, say continuous, $\Theta_\nu(t)$ and bounded γ_ν. We must, of course, specify how this inhomogeneous data is determined. To allow nonuniform spacing, we consider only sequences of nets $\{t_{k,j}\}$, $k = 0, 1, \cdots$, of the form

$$\begin{aligned} t_{k,0} &= a; \qquad t_{k,j} = t_{k,j-1} + h_{k,j}, \qquad 1 \leqq j \leqq J_k; \\ t_{k,J_k} &= b; \qquad h_k \equiv \max_j h_{k,j} \leqq \lambda \min_j h_{k,j}, \end{aligned} \tag{4.10a}$$

where for some piecewise C^∞-function $\phi(t)$, with jump discontinuities at most confined to the points $\{t_{0,j}\}_0^{J_0}$ of the initial net, the spacings are such that

$$h_{k,j} = h_k \phi(t_{k,j-1/2}) \equiv h_k \phi(\tfrac{1}{2}[t_{k,j} + t_{k,j-1}]); \qquad h_{k+1} < h_k. \tag{4.10b}$$

The proof and the "derivation" of the $\Theta_\nu(t)$ and γ_ν proceed essentially by induction and are by now standard. Indeed, the corresponding development for initial value problems as in [1] contains all the basic ideas and yields the form of the $\Theta_\nu(t)$. The modifications required for variable net spacing are contained in [3]. The boundary conditions are easily treated using expansions analogous to those for the differential equations. Thus the proof of the main theorem is concluded.

5. Separated endpoints and block elimination. Perhaps the most important practical observation in applying (1.3) with Newton's method (3.5) to solve the difference equations is the fact that many problems have *separated endpoint* boundary conditions. That is, $g(v, w)$ in (1.1b) can be written as

$$g(v, w) \equiv \begin{pmatrix} g_1(v) \\ g_2(w) \end{pmatrix}, \tag{5.1a}$$

where, say, $g_1(v)$ is a p-vector and $g_2(w)$ is a $q \equiv (n - p)$-vector. Then from (1.2c),

$$B_a \equiv \begin{pmatrix} M_a \\ 0_{q \times n} \end{pmatrix}, \quad B_b \equiv \begin{pmatrix} 0_{p \times n} \\ M_b \end{pmatrix}, \quad \begin{cases} M_a \equiv \dfrac{\partial g_1(y(a))}{\partial y} \text{ is } p \times n, \\[2ex] M_b \equiv \dfrac{\partial g_2(y(b))}{\partial y} \text{ is } q \times n. \end{cases} \tag{5.1b}$$

It is assumed that rank $M_a = p$ and rank $M_b = q$ for all arguments to be employed. This insures p independent constraints at $x = a$ and q independent constraints at $x = b$. If we write the first p-boundary conditions first and the last q-boundary conditions last, then $\Phi(U)$ is not as defined in (3.1c), but instead is

$$\Phi(U) \equiv \begin{pmatrix} g_1(u_0) \\ N_h u_1 \\ \vdots \\ N_h u_J \\ g_2(u_J) \end{pmatrix}. \tag{5.2}$$

The linear system for Newton's method applied to solve $\Phi(U) = 0$ now has the form

$$\mathbb{A}^\nu[U^{\nu+1} - U^\nu] = -\Phi(U^\nu), \tag{5.3a}$$

where

$$\mathbb{A}^\nu \equiv \begin{pmatrix} M_a^\nu & & & \\ -L_1^\nu & R_1^\nu & & 0 \\ 0 & & -L_J^\nu & R_J^\nu \\ & & & M_b^\nu \end{pmatrix},$$

(5.3b)

$$\begin{cases} L_j^\nu \equiv h_j^{-1} I + \dfrac{1}{2}\dfrac{\partial f}{\partial y}\left(t_{j-1/2}, \dfrac{1}{2}[u_j^\nu + u_{j-1}^\nu]\right), \\ R_j^\nu \equiv h_j^{-1} I - \dfrac{1}{2}\dfrac{\partial f}{\partial y}\left(t_{j-1/2}, \dfrac{1}{2}[u_j^\nu + u_{j-1}^\nu]\right), \\ M_a^\nu \equiv \dfrac{\partial g_1(u_0^\nu)}{\partial y}, \quad M_b^\nu \equiv \dfrac{\partial g_2(u_J^\nu)}{\partial y}. \end{cases}$$

The $\mathbb{A}^\nu$ are nonsingular when part (ii) of the main theorem holds, as they are obtained from row interchanges of the (nonsingular) $\partial\Phi(U^\nu)/\partial U$ introduced in § 3. Clearly each $\mathbb{A}^\nu$ is a band matrix of bandwidth at most $2n$ so that efficient band- or block-elimination methods can be used to solve (5.3a). In particular, efficient elimination schemes can be devised by writing the $\mathbb{A}^\nu$ in block tridiagonal form

$$\mathbb{A}^\nu \equiv [B_i^\nu, A_i^\nu, C_i^\nu], \qquad 0 \leqq i \leqq J, \tag{5.4}$$

where the A_i^ν, B_i^ν and C_i^ν are $n \times n$ matrices defined by equating the representations in (5.3b) and (5.4). For example, with $J = 3$ the old and new representations are schematically represented as Fig. 1. The solid squares are the $(-L_j^\nu)$ and (R_j^ν), the solid rectangles are M_a^ν and M_b^ν. The dotted lines indicate the rows of zeros that are adjoined and how the L_j^ν and R_j^ν are partitioned to form the A_i^ν, B_i^ν and C_i^ν. Thus each B_i^ν has its final q rows all zeros and each C_i^ν has its first p rows all zeros. It should be recalled that if the original problem (1.1) is linear with separated endpoint conditions, then our difference equations written in the form (5.2) already have a coefficient matrix of the same form as the $\mathbb{A}^\nu$ above.

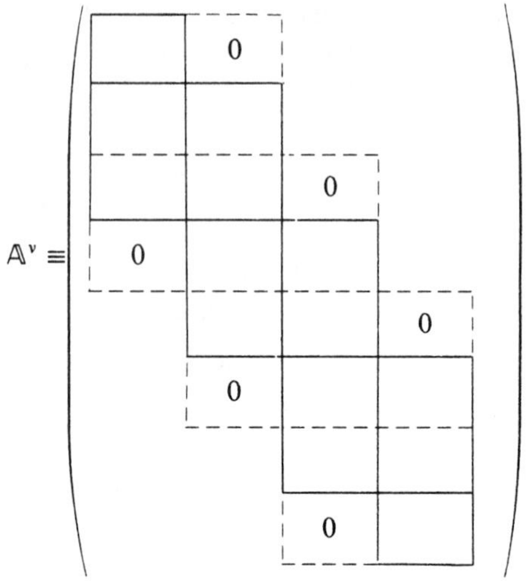

FIG. 1

We now examine methods for solving linear systems with coefficient matrices of the form (5.4) in which the submatrices have the zero elements indicated above. To simplify notation we drop the iteration index ν. Thus we consider nonsingular linear systems of order $n(J + 1)$ which have the form

$$\mathbb{A}\mathbf{X} = \mathbf{F}, \tag{5.4a}$$

where

$$\begin{aligned} &\mathbb{A} \equiv [B_j, A_j, C_j], \qquad 0 \leqq j \leqq J; \\ &\mathbf{X}^T \equiv (\mathbf{x}_0, \cdots, \mathbf{x}_J); \quad \mathbf{F}^T \equiv (\mathbf{f}_0, \cdots, \mathbf{f}_J), \end{aligned} \tag{5.4b}$$

and the $n \times n$ submatrices have the zero rows indicated in

$$B_j \equiv \begin{pmatrix} x & x & \cdots & x \\ \vdots & & & \vdots \\ x & x & \cdots & x \\ 0 & \text{———} & & 0 \\ \vdots & & & \vdots \\ 0 & \text{———} & & 0 \end{pmatrix} \begin{matrix} \left.\vphantom{\begin{matrix}x\\ \vdots\\ x\end{matrix}}\right\} p \\ \left.\vphantom{\begin{matrix}0\\ \vdots\\ 0\end{matrix}}\right\} q \end{matrix}, \qquad C_j \equiv \begin{pmatrix} 0 & \text{———} & & 0 \\ \vdots & & & \vdots \\ 0 & \text{———} & & 0 \\ x & x & \cdots & x \\ \vdots & & & \vdots \\ x & x & \cdots & x \end{pmatrix} \begin{matrix} \left.\vphantom{\begin{matrix}0\\ \vdots\\ 0\end{matrix}}\right\} p \\ \left.\vphantom{\begin{matrix}x\\ \vdots\\ x\end{matrix}}\right\} q \end{matrix}, \qquad p + q = n. \tag{5.4c}$$

To solve (5.4a), we seek factorizations of A into the form

$$\mathbb{A} = \mathbb{L}\mathbb{U}; \quad \mathbb{L} \equiv [\beta_j, \delta_j, \mathbf{0}], \quad \mathbb{U} \equiv [\mathbf{0}, \alpha_j, \gamma_j], \qquad 0 \leqq j \leqq J. \tag{5.5a}$$

Here $\alpha_j, \beta_j, \gamma_j$ and $\mathbf{0}$ are $n \times n$ matrices, the latter having all zero elements, so that $\mathbb{L}$ and $\mathbb{U}$ are also block-tridiagonal, in fact block-lower and -upper triangular as well. The solution of (5.4a) is then equivalent to solving the two systems

(5.5b)
$$\mathbb{L}\mathbf{Y} = \mathbf{F},$$

(5.5c)
$$\mathbb{U}\mathbf{X} = \mathbf{Y}.$$

The factorization in (5.5a) requires that

(5.6a)
$$\delta_0\alpha_0 = A_0;$$

(5.6b)
$$\delta_j\alpha_j = A_j - \beta_j\gamma_{j-1}, \qquad 1 \leqq j \leqq J;$$

(5.6c)
$$\beta_j\alpha_{j-1} = B_j, \qquad 1 \leqq j \leqq J;$$

(5.6d)
$$\delta_j\gamma_j = C_j, \qquad 0 \leqq j \leqq J-1.$$

The solution of (5.5b, c) is then obtained by solving the nth order systems

(5.6e)
$$\delta_0\mathbf{y}_0 = \mathbf{f}_0,$$
$$\delta_j\mathbf{y}_j = \mathbf{f}_j - \beta_j\mathbf{y}_{j-1}, \qquad 1 \leqq j \leqq J;$$

(5.6f)
$$\alpha_J\mathbf{x}_J = \mathbf{y}_J,$$
$$\alpha_j\mathbf{x}_j = \mathbf{y}_j - \gamma_j\mathbf{x}_{j+1}, \qquad J-1 \geqq j \geqq 0.$$

The factorization is not uniquely determined by (5.6a–d), if it indeed exists at all. We distinguish four common choices each of which uniquely defines the factorization (assuming existence); these are

(5.7)

case (i): $\delta_j \equiv I$;   *case* (iii): $\delta_j \equiv \begin{pmatrix} 1 & & \\ & \diagdown & \mathbf{0} \\ & & 1 \end{pmatrix}$, $\alpha_j \equiv \begin{pmatrix} x & & \\ \mathbf{0} & \diagdown & \\ & & x \end{pmatrix}$;

case (ii): $\alpha_j \equiv I$;   *case* (iv): $\delta_j \equiv \begin{pmatrix} x & & \\ & \diagdown & \mathbf{0} \\ & & x \end{pmatrix}$, $\alpha_j \equiv \begin{pmatrix} 1 & & \\ \mathbf{0} & \diagdown & \\ & & 1 \end{pmatrix}$.

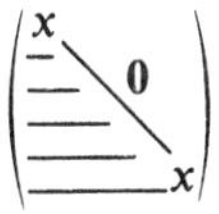

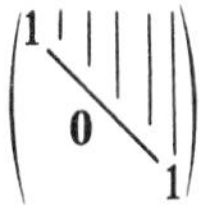

Cases (i) and (ii) are the standard block-tridiagonal factorization procedures, while cases (iii) and (iv) are standard band-factorization procedures or Gauss elimination accounting for zero elements. Assuming for the moment that they can all be carried out, it easily follows from (5.6c, d) and (5.4c) that *in all cases the β_j have all zeros in the last q rows*, while in cases (i), (iii) and (iv) the γ_j have all zeros in the first p rows. In case (ii), the γ_j may be full so that, as we shall see, this form of factorization should not be used. The operational counts for these schemes, accounting for all the null rows, are easily obtained by examining (5.6a–f). They are summarized in Table 1. In the totals and in (5.6e, f), we have dropped lower order terms not proportional to J. These counts for cases (i) and (iii) have also been given by Varah [9], who also observes that case (i) is preferable to case (iii) when $p/n < 0.38$ (based on cubic terms in the operational counts). Clearly, case (i) is always preferable to case (ii), while case (iii) is preferred to case (iv) when $p < q$.

The block-factorization cases (i) and (ii) can be performed with a restricted form of partial pivoting, which does not destroy the zero patterns in the B_j and C_j. This can be used to insure that these procedures are valid in our present applications. The restricted partial pivoting merely amounts to reordering the difference

TABLE 1

Operational counts for solving (5.4) using (5.6)

To solve	Case (i)	Case (ii)	Case (iii)	Case (iv)
(5.6a)	0	0	$n^3/3$	$n^3/3$
(5.6b)	$Jnpq$	Jn^2p	$J(npq + n^3/3 - n/3)$	$J(npq + n^3/3 - n/3)$
(5.6c)	$J(n^2p + n^3/3 - n/3)$	0	$Jp[n^2 + n]/2$	$Jp[n^2 - n]/2$
(5.6d)	0	$J(5n^3/6 + nq^2/2 + np/2 - n/3)$	$Jh[q^2 - q]/2$	$Jn[q^2 + q]/2$
(5.6e)	Jnp	$J(np + n^2)$	$J(pn + [n^2 - n]/2)$	$J(pn + [n^2 + n]/2)$
(5.6f)	$J(nq + n^2)$	Jn^2	$J(q - n + [n^2 + n]/2)$	$J(qn + [n^2 - n]/2)$
Totals (leading orders)	$J(4n^3/3 - nq^2 + 2n^2 - n/3)$	$J(4n^3/3 + np[p + 3]/2 + 2n^2 - n/3)$	$J(5n^3/6 + npq/2 + 2n^2 - n/3 + n/2[p - q])$	$J(5n^3/6 + npq/2 + 2n^2 - n/3 - n/2[p - q])$

equations for each mesh interval, but not interchanging equations for different intervals. The validity of this procedure was first proven by Mr. A. B. White [10], and we present a form of this basic result as the following theorem.

THEOREM. *Let the block tridiagonal matrix* $\mathbb{A}$ *of* (5.4b, c) *be nonsingular. Then with appropriate row interchanges within each of the n rows with indices k in* $jn + p < k \leqq (j + 1)n + p$ *for each* $j = 0, 1, \cdots, J - 1$, *the factorization* $\mathbb{A} = [\beta_j, I, 0][0, \alpha_j, C_j]$ *is valid.*

Proof. Suppose α_k has been computed for any $k = 0, 1, \cdots, J - 1$. Since $\alpha_0 = A_0$, we can start. Then we will show that:

(5.8a) the first p rows of α_k are linearly independent,

$$\text{rank}\begin{pmatrix} \alpha_k \\ B_{k+1} \end{pmatrix} = n. \tag{5.8b}$$

To do this, we define the square matrix of order $(J - k + 1)n$:

$$\mathbb{A}_k \equiv \begin{pmatrix} \alpha_k & C_k & & 0 \\ B_{k+1} & A_{k+1} & C_{k+1} & \\ & \ddots & \ddots & C_{J-1} \\ 0 & & B_J & A_J \end{pmatrix}. \tag{5.8c}$$

By the reduction procedure up to this stage, we must have

$$\det \mathbb{A} = \det \alpha_0 \cdots \det \alpha_{k-1} \det \mathbb{A}_k \neq 0. \tag{5.9}$$

If the first p rows of α_k are linearly dependent, then $\mathbb{A}_k$ is singular (since the first p rows of C_k are zeros). Thus (5.8a) holds. On the other hand, if rank $\begin{pmatrix} \alpha_k \\ B_{k+1} \end{pmatrix} < n$, then the first n columns of $\mathbb{A}_k$ are linearly independent and again $\mathbb{A}_k$ is singular, so (5.8b) holds.

To continue (or start if $k = 0$) the reduction, we can perform row interchanges among the rows of index $p + 1, \cdots, p + n$ in $\mathbb{A}_k$ to obtain a nonsingular matrix α_k in (5.8c). This follows from (5.8a, b) and the fact that the last $q = n - p$ rows of B_{k+1} are zeros. In these interchanges the zero patterns of C_k, C_{k+1} and B_{k+1} remain undisturbed, so the procedure can be continued.

Clearly in the final stage, $\alpha_J = \mathbb{A}_J$, and this must be nonsingular by (5.9). Thus the proof is complete.

It is of interest to note that while the first p rows of α_k are being processed, for the Gauss eliminations used in (5.6c, f), we can use column interchanges or row interchanges within the first p rows; for the final q rows, we switch over to row interchanges as in the theorem. With this mixed partial pivoting procedure, it is easily shown that the factorizations in cases (iii) and (iv) are valid under the hypothesis of the above theorem. Thus each of the factorizations indicated in (5.5) can be employed, with an appropriate restricted partial pivoting strategy, to solve the linear systems (5.3).

6. Practical considerations and an example. We point out that when part (iv) of the main theorem is applicable, Richardson extrapolation can be employed to get high accuracy with relatively few computations. Each extrapolation yields two orders of magnitude improvement in the numerical solution, provided that we can compute exact numerical solutions. Roundoff errors and iteration errors, of course, prevent this. However, if only $r - 1$ extrapolations are to be made, so that approximations accurate to $O(h^{2r})$ are sought, then we can tolerate roundoff and iteration errors of this magnitude. Now since Newton's method converges quadratically, by part (ii) of the main theorem, it follows that the νth iteration error will be $O(h^{2^\nu})$ if the initial iterate, U^0, is within $O(h)$ of the exact numerical solution. Then we need perform only $\nu \approx \ln(2r)/\ln(2)$ iterations to be consistent. For example, $\nu = 3$ iterations suffice for $O(h^8)$ accuracy. When an accurate approximate solution has been obtained on the crudest net, this then furnishes, say by interpolation, an at least $O(h)$ approximation to the solution on the next refined net. If this procedure is continued, the above indicated theory is applicable, and *we need only be concerned about the initial guess on the crudest net*. There is no general theory available to insure an adequate determination of such initial estimates. Likewise, roundoff control may require double precision for very high accuracy, but, of course, the machine word length is crucial here.

The techniques indicated above have been applied with the box scheme (1.3), Newton's method (3.5) and case (i) block-factorization in (5.6) to solve many problems of scientific and technical interest. However, we show here the application to a simple example that has been treated by Jerome and Varga [2] using variational techniques with splines. The problem is:

$$y'' = e^y, \tag{6.1a}$$

$$y(0) = y(1) = 0. \tag{6.1b}$$

This has the unique solution

$$y(t) = 2\ln\{C\sec[(C/2)(t - \tfrac{1}{2})]\}, \qquad C = 1.3360557. \tag{6.1c}$$

Formulated as a first order system, (6.1a, b) becomes

$$\begin{cases} y' = z \\ z' = e^y, \end{cases} \quad (1,0)\begin{pmatrix} y(0) \\ z(0) \end{pmatrix} = 0, \quad (1,0)\begin{pmatrix} y(1) \\ z(1) \end{pmatrix} = 0. \tag{6.2}$$

We first employ the scheme (1.3) with Newton's method (3.5) on four nets with the uniform spacings $h = 1/3, 1/6, 1/12, 1/24$. Then since $y(t)$ and $z(t)$ are analytic (from (6.1c)), part (iv) of the main theorem holds and three extrapolations can be made on the initial net. The results of these extrapolations are indicated in Table 2, where the errors after the indicated extrapolations are shown (in standard floating-decimal notation). Since the solution is symmetric about $x = 1/2$, we only show the results for $y(1/3)$, $z(1/3)$ and $z(0)$.

TABLE 2

Errors in computed solutions and in extrapolations

h	Number of extrapolations: 0	1	2	3	Error in
1/3	.161 (−02)				
		.727 (−05)			
1/6	.397 (−03)		.125 (−07)		
		.443 (−06)		.401 (−11)	$y(1/3)$
1/12	.990 (−04)		.192 (−09)		
		.275 (−07)			
1/24	.147 (−04)				
1/3	.100 (−02)				
		.487 (−05)			
1/6	.247 (−03)		.503 (−08)		
		.300 (−06)		.255 (−11)	$z(1/3)$
1/12	.613 (−04)		.761 (−10)		
		.187 (−07)			
1/24	.153 (−04)				
1/3	.335 (−02)				
		.176 (−04)			
1/6	.825 (−03)		.197 (−07)		
		.108 (−05)		.109 (−10)	$z(0)$
1/12	.205 (−03)		.297 (−09)		
		.673 (−07)			
1/24	.314 (−04)				

The convergence of the Newton iterates was very similar in all four cases. The *worst* case, for $h = 1/12$, is reported in Table 3. The data for all initial iterates was

TABLE 3

Convergence of Newton iterates

ν	$\max_j \lvert y_j^\nu - y_j^{\nu-1}\rvert$	$\max_j \lvert z_j^\nu - z_j^{\nu-1}\rvert$
1	.126 (+01)	.121 (+01)
2	.786 (−02)	.572 (−02)
3	.195 (−06)	.132 (−06)
4	.206 (−15)	.241 (−15)

chosen as $y_j^0 = (t_j - \frac{1}{2})^2 - \frac{1}{4}$ and $z_j^0 = 2t_j - 1$. No doubt fewer iterates would have sufficed if we had bothered to use more accurate initial guesses (determined as previously indicated). All computations were done in double precision on the Caltech IBM 370/155. The program was most ably designed and employed by Mr. S. A. Bolasna.

Note the high accuracy that can be obtained using relatively few net points; a total of 49 points was used on all the nets combined. From the first three nets, with a total of 24 points, we get absolute accuracy to within 2×10^{-8}, as is shown in Table 2. Also note that both the solution, $y(t)$, and its derivative, $z(t)$, are approximated to the same accuracy. These computations were repeated on other sets of nets, including several sets with nonuniform spacing, and very similar results were obtained. Jerome and Varga [2, Table II] get errors in $y(t)$ of at most 2.5×10^{-7} using natural cubic splines with only 10 nodes, that is, with spacing $h = 1/9$. We cannot make further comparisons here, as the details of their solution of this problem are unknown to us.

Appendix. Multipoint boundary conditions. If the two-point boundary conditions (1.1b) are replaced by the more general multipoint conditions:

$$\text{(A.1)} \qquad g(y(\tau_1), y(\tau_2), \cdots, y(\tau_N)) = 0, \qquad a \leqq \tau_1 < \tau_2 < \cdots < \tau_N \leqq b,$$

then for isolated solutions of (1.1a), (A.1), we must replace (1.2b) by

$$\text{(A.2a)} \qquad \sum_{\nu=1}^{N} B_\nu \phi(\tau_\nu) = 0,$$

$$\text{(A.2b)} \qquad B_\nu \equiv \frac{\partial}{\partial y(\tau_\nu)} g(y(\tau_1), \cdots, y(\tau_N)), \qquad 1 \leqq \nu \leqq N.$$

The net (1.3a) is required to be such that the boundary constraint points $\{\tau_\nu\}_0^N$ are contained in the net $\{t_j\}_0^J$, say,

$$\text{(A.3)} \qquad \tau_\nu = t_{j_\nu}, \qquad 1 \leqq \nu \leqq N.$$

The difference scheme is now (1.3b) and, replacing (1.3c), we see that

$$\text{(A.4)} \qquad g(u_{j_1}, u_{j_2}, \cdots, u_{j_N}) = 0.$$

The main theorem remains valid with obvious modifications. In particular, now: (1.1a), (A.1) is to have an isolated solution;

$$g(v_1, \cdots, v_N) \in C_2\{S_\rho[y(\tau_1)] \times \cdots \times S_\rho[y(\tau_N)]\};$$

the difference equations (1.3a), (A.4) have a unique solution.

The analysis of § 2 was originally carried out for the multipoint case in [2]. The changes required in (2.1b), (2.2), (2.4b) are obvious and lead to the replacement of $\mathbb{L}$ in (2.6c) by

$$\text{(A.5a)} \qquad \mathbb{L} \equiv \begin{pmatrix} C_0 & C_1 & \cdots & & C_J \\ -L_1 & R_1 & & & \\ & \ddots & \ddots & & \\ & & & -L_J & R_J \end{pmatrix},$$

$$\text{(A.5b)} \qquad C_k \equiv \begin{cases} 0 & \text{if } j \neq j_\nu, \\ B_\nu & \text{if } j = j_\nu, \qquad 1 \leqq \nu \leqq N. \end{cases}$$

Now, the analysis of § 3 remains valid after replacing $g(u_0, u_J)$ in (3.1c) by $g(u_{j_1}, u_{j_2}, \cdots, u_{j_N})$, employing $\mathbb{L}$ as above, and requiring that K be the maximum of the Lipschitz constants for $f_y(t, z)$, $\partial g(v_1, \cdots, v_N)/\partial v_\nu$ with respect to $z, v_\nu, \nu = 1, 2, \cdots, N$.

The error estimates through (4.8), and hence part (iii) of the modified main theorem, require only using

$$\text{(A.6a)} \qquad \tilde{B}_\nu = \int_0^1 \frac{\partial g}{\partial y(\tau_\nu)}(y(\tau_1) + sv_{t_1}, \cdots, y(\tau_N) + sv_{t_N})\,ds, \qquad 1 \leqq \nu \leqq N,$$

to replace (4.1b) by

$$\text{(A.6b)} \qquad \sum_{\nu=1}^{N} \tilde{B}_\nu v_{t_\nu} = 0.$$

The asymptotic error estimates require slightly more involved modifications. Of course, (4.9b) is replaced by

$$\text{(A.7)} \qquad \sum_{s=1}^{N} B_s\, e^\nu(\tau_s) = \gamma_\nu,$$

and the family of nets $\{t_{k,j}\}_0^{J_k}$, $k = 0, 1, 2, \cdots$, must each satisfy (A.3). The $\Theta_\nu(t)$ to be used in (4.9a) are unaltered and only the γ_ν for (A.7) need be changed. However, these are again easily obtained by formal Taylor expansions.

REFERENCES

[1] W. B. GRAGG, *On extrapolation algorithms for ordinary initial value problems*, this Journal, 2 (1965), pp. 384–403.

[2] J. W. JEROME AND R. S. VARGA, *Generalizations of spline functions and applications to nonlinear boundary value and eigenvalue problems*, Theory and Applications of Spline Functions, Academic Press, New York, 1969.

[3] H. B. KELLER, *Accurate difference methods for linear ordinary differential systems subject to linear constraints*, this Journal, 6 (1969), pp. 8–30.

[4] ———, *Numerical Methods for Two-Point Boundary-Value Problems*, Ginn–Blaisdell, Waltham, Mass., 1968.

[5] ———, *Newton's method under mild differentiability conditions*, J. Comput. System Sci., 4 (1970), pp. 15–28.

[6] H.-O. KREISS, *Difference approximations for boundary and eigenvalue problems for ordinary differential equations*, Math. Comp., 26 (1972), pp. 605–624.

[7] F. NIEUWSTADT AND H. B. KELLER, *Viscous flow past circular cylinders*, Computers and Fluids, 1 (1973), pp. 59–71.

[8] J. ORTEGA AND W. RHEINBOLDT, *Iterative Solution of Nonlinear Equations in Several Variables*, Academic Press, New York, 1970.

[9] J. M. VARAH, *On the solution of block tridiagonal systems arising from certain finite-difference equations*, Math. Comp., 26 (1972), pp. 859–868.

[10] A. B. WHITE, Ph.D. thesis, California Institute of Technology, Pasadena, Calif., in preparation.

[11] R. WEISS, *The application of implicit Runge–Kutta and collocation methods to boundary value problems*, to appear.

Reprinted with permission from *SIAM Journal On Numerical Analysis*, Vol. 12, No. 5 (1975), pp. 791–802.

DIFFERENCE METHODS FOR BOUNDARY VALUE PROBLEMS IN ORDINARY DIFFERENTIAL EQUATIONS*

H. B. KELLER† AND A. B. WHITE, JR.‡

Abstract. A general theory of difference methods for problems of the form

$$\mathcal{N}\mathbf{y} \equiv \mathbf{y}' - \mathbf{f}(t, \mathbf{y}) = \mathbf{0}, \qquad a \leqq t \leqq b, \quad \mathbf{g}(\mathbf{y}(a), \mathbf{y}(b)) = \mathbf{0},$$

is developed. On nonuniform nets, $t_0 = a$, $t_j = t_{j-1} + h_j$, $1 \leqq j \leqq J$, $t_J = b$, schemes of the form

$$\mathcal{N}_h\mathbf{u}_j \equiv \mathbf{G}_j(\mathbf{u}_0, \cdots, \mathbf{u}_J) = \mathbf{0}, \qquad 1 \leqq j \leqq J, \quad \mathbf{g}(\mathbf{u}_0, \mathbf{u}_J) = \mathbf{0}$$

are considered. For linear problems with unique solutions, it is shown that the difference scheme is stable and consistent for the boundary value problem *if and only if*, upon replacing the boundary conditions by an initial condition, the resulting scheme is stable and consistent for the initial value problem. For isolated solutions of the nonlinear problem, it is shown that the difference scheme has a unique solution converging to the exact solution if (i) the linearized difference equations are stable and consistent for the linearized initial value problem, (ii) the linearized difference operator is Lipschitz continuous, (iii) the nonlinear difference equations are consistent with the nonlinear differential equation. Newton's method is shown to be valid, with quadratic convergence, for computing the numerical solution.

1. Introduction. We present a new and rather comprehensive theory of general difference methods for approximating the solution of both linear and nonlinear boundary value problems for first order systems of ordinary differential equations. For linear problems with unique solutions our theory, in § 3, states essentially that a difference scheme is stable and consistent for the boundary value problem *if and only if* it is stable and consistent for the initial value problem. For isolated solutions of nonlinear problems our theory, in § 4, states that a difference scheme has a unique solution converging to the isolated solution if (i) the linearized difference equations are stable and consistent for the linearized initial value problem, (ii) the linearized difference operator is Lipschitz continuous, (iii) the nonlinear difference equations are consistent with the nonlinear problem. Newton's method is shown to be valid, with quadratic convergence, for computing the numerical solution.

The linear boundary value problems that we study include the general form

$$\mathcal{L}\mathbf{y} \equiv \mathbf{y}' - A(t)\mathbf{y} = \mathbf{f}(t), \qquad a \leqq t \leqq b, \tag{1.1a}$$

$$\mathcal{B}\mathbf{y} \equiv B_a\mathbf{y}(a) + B_b\mathbf{y}(b) = \boldsymbol{\beta}. \tag{1.1b}$$

Here $\mathbf{y}(t)$, $\mathbf{f}(t)$ and $\boldsymbol{\beta}$ are n-vectors, $A(t)$, B_a and B_b are $n \times n$ matrices; the elements of $A(t)$ and $\mathbf{f}(t)$ are in $C^N[a, b]$ while the solution $\mathbf{y}(t)$ is in $C^{N+1}[a, b]$. With little extra effort we may allow $A(t)$, $\mathbf{f}(t)$ and $\mathbf{y}(t)$ (and/or their derivatives) to have a finite number of jump discontinuities when two-point schemes are employed. The details of this device are contained in Keller [5] so we do not elaborate on it

* Received by the editors July 24, 1974.

† Applied Mathematics, California Institute of Technology 101-50, Pasadena, California 91109. This work was supported by the U.S. Atomic Energy Commission under Contract AT(04-3-767), Project Agreement 12.

‡ Computing Science Department, University of Texas, Austin, Texas 78712.

here. Multipoint boundary conditions of the form

$$\mathscr{B}\mathbf{y} \equiv \sum_{\nu=1}^{m} B_\nu \mathbf{y}(\tau_\nu) = \boldsymbol{\beta}, \qquad a \leqq \tau_1 < \tau_2 < \cdots < \tau_m \leqq b,$$

are also easily included.

The nonlinear problems are of the form

$$\text{(1.2a)} \qquad \mathscr{N}\mathbf{y} \equiv \mathbf{y}' - \mathbf{f}(t, \mathbf{y}) = \mathbf{0}, \qquad a \leqq t \leqq b,$$

$$\text{(1.2b)} \qquad \mathbf{g}(\mathbf{y}(a), \mathbf{y}(b)) = \mathbf{0},$$

where the n-vectors $\mathbf{f}(t, \mathbf{y})$ and $\mathbf{g}(\mathbf{v}, \mathbf{w})$ are assumed to have sufficient smoothness. We will be concerned only with isolated solutions of (1.2), that is, solutions $\mathbf{y} = \mathbf{y}(t)$ for which the linearized problem

$$\text{(1.3a)} \qquad \mathscr{L}[\mathbf{y}]\mathbf{z} = \mathbf{0}, \qquad a \leqq t \leqq b,$$

$$\text{(1.3b)} \qquad \mathscr{B}[\mathbf{y}]\mathbf{z} = \mathbf{0}$$

has *only* the trivial solution $\mathbf{z}(t) \equiv \mathbf{0}$. Here $\mathscr{L}[\mathbf{y}]$ and $\mathscr{B}[\mathbf{y}]$ are as defined in (1.1) but with the matrices

$$\text{(1.3c)} \quad A(t) \equiv A(t, \mathbf{y}(t)) \equiv \frac{\partial \mathbf{f}(t, \mathbf{y}(t))}{\partial \mathbf{y}}, \qquad B_x \equiv B_x[\mathbf{y}] \equiv \frac{\partial \mathbf{g}(\mathbf{y}(a), \mathbf{y}(b))}{\partial \mathbf{y}(x)}, \qquad x = a, b.$$

Again we can easily include more general boundary conditions, for example, the multipoint form

$$\mathbf{g}(\mathbf{y}(\tau_1), \mathbf{y}(\tau_2), \cdots, \mathbf{y}(\tau_m)) = \mathbf{0}.$$

The difference schemes employ arbitrary families of nets, say $\{t_j\}$, with

$$\text{(1.4a)} \qquad t_0 = a, \quad t_j = t_{j-1} + h_j, \quad 1 \leqq j \leqq J, \quad t_J = b,$$

and are subject only to the restriction that for some fixed $r > 0$,

$$\text{(1.4b)} \qquad h \equiv \max_j h_j \leqq r \min_k h_k.$$

If $\mathbf{u}^h \equiv (\mathbf{u}_0^T, \mathbf{u}_1^T, \cdots, \mathbf{u}_J^T)^T$ is to approximate $\mathbf{y}^h \equiv (\mathbf{y}^T(t_0), \mathbf{y}^T(t_1), \cdots, \mathbf{y}^T(t_J))^T$ for the linear problem (1.1), then our general difference schemes are formulated as

$$\text{(1.5a)} \qquad \mathscr{L}_h \mathbf{u}_j \equiv \sum_{k=0}^{J} C_{jk}(h)\mathbf{u}_k = \mathbf{F}_j(h; \mathbf{f}), \qquad 1 \leqq j \leqq J,$$

$$\text{(1.5b)} \qquad \mathscr{B}_h \mathbf{u}^h \equiv B_a \mathbf{u}_0 + B_b \mathbf{u}_J = \boldsymbol{\beta}.$$

For the nonlinear problem (1.2), our general difference schemes are formulated as

$$\text{(1.6a)} \qquad \mathscr{N}_h \mathbf{u}_j \equiv \mathbf{G}_j(\mathbf{u}^h) = \mathbf{0}, \qquad 1 \leqq j \leqq J,$$

$$\text{(1.6b)} \qquad \mathbf{g}(\mathbf{u}_0, \mathbf{u}_J) = \mathbf{0}.$$

The linearized difference equations obtained by linearizing (1.6) about $\mathbf{u}^h$ are

$$\text{(1.7a)} \qquad \mathscr{L}_h[\mathbf{u}^h]\mathbf{v}_j \equiv \sum_{k=0}^{J} C_{jk}(h, \mathbf{u}^h)\mathbf{v}_k = \mathbf{0}, \qquad 1 \leqq j \leqq J,$$

(1.7b) $$\mathscr{B}_h[\mathbf{u}^h]\mathbf{v}^h \equiv B_a[\mathbf{u}^h]\mathbf{v}_0 + B_b[\mathbf{u}^h]\mathbf{v}_J = 0,$$

where the $n \times n$ coefficient matrices are defined by

(1.7c) $$C_{jk}(h, \mathbf{u}^h) \equiv \frac{\partial \mathbf{G}_j(\mathbf{u}^h)}{\partial \mathbf{u}_k}, \qquad 0 \leqq j, k \leqq J,$$
$$B_a[\mathbf{u}^h] \equiv \frac{\partial \mathbf{g}(\mathbf{u}_0, \mathbf{u}_J)}{\partial \mathbf{u}_0}, \qquad B_b[\mathbf{u}^h] \equiv \frac{\partial \mathbf{g}(\mathbf{u}_0, \mathbf{u}_J)}{\partial \mathbf{u}_J}.$$

The general results of this paper are extensions of the work in [5] for linear problems and [7] for nonlinear problems. A form of these extensions is contained in the thesis of A. B. White [11]. An abstract form of the general technique used for the nonlinear case is given in [8].

2. Linear boundary value problems. For our basic theory, we need a result relating linear initial value problems and linear boundary value problems. However, it is simpler and more elegant to present the corresponding result for pairs of boundary value problems. Thus we consider first the pair of linear two-point problems $BV(\nu)$ for $\nu = 0, 1$:

(2.1a) $$\mathscr{L}\mathbf{y}^{(\nu)}(t) \equiv \frac{d\mathbf{y}^{(\nu)}_{(t)}}{dt} - A(t)\mathbf{y}^{(\nu)}(t) = \mathbf{f}(t), \qquad a < t < b,$$

$$\nu = 0, 1.$$

(2.1b) $$\mathscr{B}^{(\nu)}\mathbf{y}^{(\nu)} \equiv B_a^{(\nu)}\mathbf{y}^{(\nu)}(a) + B_b^{(\nu)}\mathbf{y}^{(\nu)}(b) = \boldsymbol{\beta},$$

These problems differ only in the matrices $B_a^{(\nu)}$ and $B_b^{(\nu)}$ that occur in the boundary conditions. Note that for all of our analysis ν could just as well be a continuous parameter, say in $0 \leqq \nu \leqq 1$, and thus our results apply to families of boundary value problems. We also define the corresponding pair of fundamental solutions, $Y^{(\nu)}(t)$, as the $n \times n$ matrix solutions of

(2.2a) $$\mathscr{L}Y^{(\nu)}(t) = 0, \qquad a < t < b,$$

(2.2b) $$\mathscr{B}^{(\nu)}Y^{(\nu)} = I, \qquad \nu = 0, 1.$$

An interesting equivalence theorem relating these problems is the following theorem.

THEOREM 2.3. *Let $BV(0)$ have a unique solution. Then $BV(1)$ has a unique solution if and only if $\mathscr{B}^{(1)}Y^{(0)}$ is nonsingular.*

Proof. Clearly we need only show that the homogeneous boundary value problem

(2.4) $$\mathscr{L}\mathbf{y}(t) = \mathbf{0}, \quad a < t < b, \quad \mathscr{B}^{(1)}\mathbf{y} = \mathbf{0}$$

has only the trivial solution if and only if $\mathscr{B}^{(1)}Y^{(0)}$ is nonsingular. However, every solution $\mathbf{y}(t)$ of $\mathscr{L}\mathbf{y} = \mathbf{0}$ has a unique representation of the form

(2.5) $$\mathbf{y}(t) = Y^{(0)}(t)\boldsymbol{\xi}$$

for some $\boldsymbol{\xi} \in E^n$. Indeed if $\mathscr{L}\mathbf{y} = \mathbf{0}$, then $\mathbf{y}(t) = \mathbf{y}^{(0)}(t)$ is the solution of $BV(0)$ with $\mathbf{f}(t) \equiv \mathbf{0}$ and $\boldsymbol{\beta} \equiv \mathscr{B}^{(0)}\mathbf{y}$. By hypothesis, this is the only such solution of $BV(0)$. But $Y^{(0)}(t)\boldsymbol{\beta}$ is also a solution of this problem, and so the unique representation

(2.5) is established. Now $\mathbf{y}(t)$ in (2.5) is a solution of (2.4) if and only if

$$\mathscr{B}^{(1)}\mathbf{y} = \mathscr{B}^{(1)}Y^{(0)}\boldsymbol{\xi} = \mathbf{0}.$$

Our result now follows since $\boldsymbol{\xi} = \mathbf{0}$ is the only possibility if and only if $\mathscr{B}^{(1)}Y^{(0)}$ is nonsingular. □

The result we actually need is a simple consequence of Theorem 2.3 and the uniqueness theorem for initial value problems, namely, Corollary 2.6.

COROLLARY 2.6. *Let $BV(0)$ have a unique solution. Then $Y^{(0)}(a)$ is nonsingular.*

Proof. With the choice $B_a^{(1)} \equiv I$, $B_b^{(1)} \equiv 0$, we see that $BV(1)$ becomes the initial value problem, the uniqueness of whose solutions is well known. Now apply Theorem 2.3. □

Finally, we point out that in all the above results we have not used the explicit form of the boundary conditions but merely the linearity of the boundary operators $\mathscr{B}^{(\nu)}$. Thus our results apply to any linear constraints which take $\mathbf{y}(t)$, $a \leqq t \leqq b$, into E^n. Obviously, this includes multipoint conditions of the form

$$\mathscr{B}^{(\nu)}\mathbf{y} \equiv \sum_{i=1}^{N} B_i^{(\nu)}\mathbf{y}(\tau_i), \qquad a \leqq \tau_1 < \tau_2 < \cdots < \tau_N = b, \tag{2.7}$$

where the $B_i^{(\nu)}$ are $n \times n$ matrices.

3. Difference methods for linear boundary value problems. The standard notions of truncation errors, consistency and stability for the scheme (1.5) applied to (1.1) can be defined as follows.

DEFINITION 3.1. (a) The *truncation errors* in scheme (1.5) applied to (1.1) are

$$\boldsymbol{\tau}_j\{\mathbf{y}\} \equiv \mathscr{L}_h\mathbf{y}(t_j) - \mathbf{F}_j(h, \mathbf{f}), \qquad 1 \leqq j \leqq J,$$

$$\boldsymbol{\tau}_0\{\mathbf{y}\} \equiv \mathscr{B}_h\mathbf{y} - \boldsymbol{\beta},$$

where $\mathbf{y}(t)$ is any solution of (1.1).

(b) The scheme (1.5) is consistent (accurate) of order p with (for) (1.1) provided there exist constants $K_0 > 0$ and $h_0 > 0$ such that

$$\|\boldsymbol{\tau}_j\{\mathbf{y}\}\| \leqq K_0 h^p, \qquad 0 \leqq j \leqq J,$$

for all nets (1.4) with $h \leqq h_0$ and for all solutions $\mathbf{y}(t)$ of (1.1).

(c) The scheme (1.5) is stable provided there exist positive constants K_1, K_2 and h_0 such that for any net function $\mathbf{v}^h$ defined on (1.4) and for all $h \leqq h_0$

$$\|\mathbf{v}_j\| \leqq K_1\|\mathscr{B}_h\mathbf{v}^h\| + K_2 \max_{1 \leqq i \leqq J} \|\mathscr{L}_h\mathbf{v}_i\|, \qquad 0 \leqq j \leqq J.$$

From these definitions we easily obtain the following well-known convergence theorem.

THEOREM 3.2. *If (1.5) is stable and consistent of order p for (1.1), then for all nets (1.4) with $h \leqq h_0$,*

$$\|\mathbf{y}(t_j) - \mathbf{u}_j\| \leqq K_0K_2h^p;$$

that is, the scheme (1.5) is convergent of order p for (1.1). Here $\mathbf{y}(t)$ is a solution of (1.1) and $\mathbf{u}^h$ is the solution of (1.5).

Proof. Let $\mathbf{v}_j \equiv \mathbf{y}(t_j) - \mathbf{u}_j$ and use the linearity of $\mathscr{L}_h$ to get that

$$\mathscr{L}_h\mathbf{v}_j = \mathscr{L}_h\mathbf{y}(t_j) - \mathscr{L}_h\mathbf{u}_j = \mathscr{L}_h\mathbf{y}(t_j) - \mathbf{F}_j(h, \mathbf{f}) = \boldsymbol{\tau}_j\{\mathbf{y}\}, \qquad 1 \leqq j \leqq J.$$

Similarly $\mathscr{B}_h\mathbf{v} = 0$, and the result now follows from stability. □

We introduce the matrix $\mathbb{A}_h$ and vectors $\mathbf{U}$, $\mathbf{F}$ all of order $nJ + n$ as

$$(3.3) \qquad \mathbb{A}_h \equiv \begin{pmatrix} B_a & 0 & \cdots & B_b \\ C_{10} & C_{11} & \cdots & C_{1J} \\ \vdots & \vdots & & \vdots \\ C_{J0} & C_{J1} & \cdots & C_{JJ} \end{pmatrix}, \quad \mathbf{U} \equiv \begin{pmatrix} \mathbf{u}_0 \\ \mathbf{u}_1 \\ \vdots \\ \mathbf{u}_J \end{pmatrix}, \quad \mathbf{F} \equiv \begin{pmatrix} \boldsymbol{\beta} \\ \mathbf{F}_1(h, \mathbf{f}) \\ \vdots \\ \mathbf{F}_J(h, \mathbf{f}) \end{pmatrix}.$$

Then the scheme (1.5) is simply

$$\mathbb{A}_h\mathbf{U} = \mathbf{F}.$$

Now we have an equivalent definition of stability as in the next lemma.

LEMMA 3.4. *The scheme* (1.5) *is stable if and only if there exist positive constants K and h_0 such that for all nets* (1.4) *with $h \leqq h_0$ the family of matrices $\mathbb{A}_h$ are nonsingular with uniformly bounded inverses, i.e.,*

$$(3.5) \qquad \|\mathbb{A}_h^{-1}\| \leqq K.$$

Proof. If $\|\cdot\|_n$ is the norm on E^n used in (3.1c), then we use as the norm on E^{nJ+n}: $\|\mathbf{X}\|_{nJ+n} = \max_{0 \leqq j \leqq J} \|\mathbf{x}_j\|_n$ where $\mathbf{X} = (\mathbf{x}_0^T, \cdots, \mathbf{x}_J^T)^T$. Using this vector norm, the induced norm on any matrix $B \equiv (B_{ij})$ of order $nJ + n$ with the B_{ij} of order n is given by: $\|B\|_{nJ+h} = \max_{0 \leqq i \leqq J} \sum_{j=0}^{J} \|B_{ij}\|_n$. Here of course $\|B_{ij}\|_n$ is the norm induced by $\|\cdot\|_n$. We now drop all subscripts on norms as their arguments suffice to identify the appropriate space.

To demonstrate Lemma (3.4) suppose (3.5) holds. Then for any $\mathbf{V} \equiv (\mathbf{v}_0^T, \cdots, \mathbf{v}_J^T)^T$ we have

$$\|\mathbf{v}_j\| \leqq \|\mathbf{V}\| = \|\mathbb{A}_h^{-1}\mathbb{A}_h\mathbf{V}\| \leqq K\|\mathbb{A}_h\mathbf{V}\|$$

$$\leqq K \max\left\{\|\mathscr{B}_h\mathbf{v}\|, \max_{1 \leqq i \leqq J} \|\mathscr{L}_h\mathbf{v}_i\|\right\}.$$

Thus (3.1c) follows with, say, $K_0 = K_1 = K$.

Now assume (3.1c) holds. It immediately follows that $\mathbb{A}_h$ is nonsingular since the homogeneous system $\mathbb{A}_h\mathbf{U} = 0$ has only the trivial solution. Then each vector $\mathbf{W} \in E^{nJ+n}$ can be represented in the form $\mathbf{W} = \mathbb{A}_h\mathbf{V}$ for some unique $\mathbf{V} \in E^{nJ+n}$. However, since (3.1c) implies, for all vectors $\mathbf{V}$, that

$$\|\mathbf{V}\| \leqq 2\max(K_1, K_2)\|\mathbb{A}_h\mathbf{V}\|,$$

it immediately follows that, for all $\mathbf{W} \neq \mathbf{0}$,

$$(3.6) \qquad \|\mathbb{A}_h^{-1}\mathbf{W}\|/\|\mathbf{W}\| \leqq 2\max(K_1, K_2).$$

Thus (3.5) holds with some $K \leqq 2\max(K_1, K_2)$. □

We present the basic stability result for difference schemes applied to the general pair of boundary value problems $BV(v)$ in (2.1). That is, we consider the two difference problems $BV_h(v)$:

$$\text{(3.7a)} \qquad \mathscr{L}_h u_j^{(\nu)} \equiv \sum_{k=0}^{J} C_{jk}(h)\mathbf{u}_k^{(\nu)} = \mathbf{F}_j(h,\mathbf{f}), \qquad 1 \leqq j \leqq J,$$

$$\text{(3.7b)} \qquad \mathscr{B}_h^{(\nu)}\mathbf{u}^{(\nu)} \equiv B_a^{(\nu)}\mathbf{u}_0^{(\nu)} + B_b^{(\nu)}u_J^{(\nu)} = \boldsymbol{\beta}, \qquad \nu = 0, 1.$$

Note that they differ only in the boundary conditions.

THEOREM 3.8. *Let each boundary value problem $BV(\nu)$, $\nu = 0, 1$, have a unique solution. Then the difference scheme $BV_h(0)$ is stable and consistent for $BV(0)$ if and only if $BV_h(1)$ is stable and consistent for $BV(1)$.*

Proof. The equivalence of the consistency for the two schemes is trivial since the schemes are identical when applied to any $\mathbf{u}_j$ with $1 \leqq j \leqq J$ and the boundary conditions in each case are exact.

To demonstrate the equivalence of stability we introduce

$$\text{(3.9)} \qquad \mathbb{A}_h^{(\nu)} \equiv \begin{pmatrix} B_a^{(\nu)} & 0 & \cdots & 0 & B_b^{(\nu)} \\ C_{10} & C_{11} & \cdots & & C_{1J} \\ \vdots & \vdots & & & \vdots \\ C_{J0} & C_{J1} & \cdots & & C_{JJ} \end{pmatrix}, \qquad \nu = 0, 1.$$

Suppose $BV_h(0)$ is stable. Then $\mathbb{A}_h^{(0)}$ is nonsingular for all $h \leqq h_0$ and for some $K > 0$,

$$\|(\mathbb{A}_h^{(0)})^{-1}\| \leqq K.$$

We introduce $\mathbb{D}_h$ as

$$\text{(3.10a)} \qquad \mathbb{D}_h \equiv \mathbb{A}_h^{(1)} - \mathbb{A}_h^{(0)} = \begin{pmatrix} (B_a^{(1)} - B_a^{(0)}) & 0 \cdots & (B_b^{(1)} - B_b^{(0)}) \\ 0 & 0 \cdots & 0 \\ \vdots & & \\ 0 & 0 \cdots & 0 \end{pmatrix}$$

Then by the assumed stability of $BV_h^{(0)}$, we can write

$$\text{(3.10b)} \qquad \mathbb{A}_h^{(1)} = [\mathbb{I} + \mathbb{D}_h(\mathbb{A}_h^{(0)})^{-1}]\mathbb{A}_h^{(0)}.$$

Now denote the block structure of $(\mathbb{A}_h^{(0)})^{-1}$ by means of

$$\text{(3.11a)} \qquad (\mathbb{A}_h^{(0)})^{-1} = (Z_{ij}^{(0)}),$$

where the $Z_{ij}^{(0)}$ are $n \times n$ matrices and $0 \leqq i, j \leqq J$. From the jth "column" of blocks we obtain, since $\mathbb{A}_h^{(0)}\mathbb{A}_h^{(0)-1} = \mathbb{I}$,

$$\text{(3.11b)} \qquad \mathbb{A}_h^{(0)} \begin{pmatrix} Z_{0j}^{(0)} \\ \vdots \\ Z_{Jj}^{(0)} \end{pmatrix} = \mathbb{I}_j \equiv \begin{pmatrix} 0 \\ \vdots \\ 0 \\ I \\ 0 \\ \vdots \\ 0 \end{pmatrix} \leftarrow j\text{th block}, \qquad 0 \leqq j \leqq J.$$

Using (3.11) and (3.10), we find that

$$\mathbb{A}_h^{(1)} = \begin{pmatrix} Q_{h0} & Q_{h1} & \cdots & Q_{hJ} \\ 0 & I & & \\ & & \ddots & \\ & & & I \end{pmatrix} \mathbb{A}_h^{(0)}, \tag{3.12a}$$

where

$$Q_{hj} \equiv B_a^{(1)} Z_{0j}^{(0)} + B_b^{(1)} Z_{Jj}^{(0)}, \qquad 0 \leqq j \leqq J. \tag{3.12b}$$

It follows from (3.12a) that $\mathbb{A}_h^{(1)}$ is nonsingular if and only if Q_{h0} is nonsingular. However, a glance at (3.7), (3.9) and (3.11) reveals that the $n \times n$ matrix $(Z_{i0}^{(0)})$ is just the difference approximation, using scheme $BV_h(0)$, to $Y^{(0)}(t_i)$, the fundamental solution for $BV(0)$ defined in (2.2). Since $BV_h(0)$ is stable and consistent (say, of order p), it follows from Theorem 3.2 that

$$\| Y^{(0)}(t_j) - Z_{j0}^{(0)} \| = O(h^p).$$

Then clearly

$$\| \mathscr{B}^{(1)} Y^{(0)} - Q_{h0} \| = O(h^p).$$

By Theorem 2.3 we have that $\mathscr{B}^{(1)} Y^{(0)}$ is nonsingular and hence the Banach lemma now implies, for h_0 sufficiently small, that Q_{h0} is nonsingular and, in fact, $\|Q_{h0}^{-1}\| \leqq C$ for all $h \leq h_0$ and some constant C independent of h.

Thus $\mathbb{A}_h^{(1)}$ is nonsingular and its inverse is

$$(\mathbb{A}_h^{(1)})^{-1} = \mathbb{A}_h^{(0)-1} \begin{pmatrix} Q_{h0}^{-1} & -Q_{h0}^{-1} Q_{h1} & \cdots & -Q_{h0}^{-1} Q_{hJ} \\ & I & & \\ & & \ddots & \\ & & & I \end{pmatrix}.$$

Using $\|Q_{hj}\| \leqq \|B_a^{(1)}\| \cdot \|Z_{0j}^{(0)}\| + \|B_b^{(1)}\| \cdot \|Z_{Jj}^{(0)}\|$, we obtain

$$\sum_{j=1}^{J} \|Q_{hj}\| \leqq (\|B_a^{(1)}\| + \|B_b^{(1)}\|)K,$$

and so

$$\|(\mathbb{A}_h^{(1)})^{-1}\| \leqq K \max \{\|I\|, C[\|I\| + K(\|B_a^{(1)}\| + \|B_b^{(1)}\|)]\}.$$

Thus the stability of $BV_h(1)$ follows from that of $BV_h(0)$.

The converse is proven by merely interchanging the superscripts $\nu = 0$ and $\nu = 1$ in the above arguments. ☐

Now the relevant application of Theorem 3.8 to the scheme (1.5) applied to (1.1) is simply Corollary 3.13.

COROLLARY 3.13. *Let* (1.1) *have a unique solution. Then the difference scheme* (1.5) *is stable and consistent for* (1.1) *if and only if the scheme*

$$\mathscr{L}_h \mathbf{v}_j = \mathbf{F}_j(h, \mathbf{f}), \qquad 1 \leqq j \leqq J, \tag{3.14a}$$

$$\mathbf{v}_0 = \boldsymbol{\alpha} \tag{3.14b}$$

is stable and consistent for the initial value problem

(3.15a) $$\mathscr{L}\mathbf{y} = \mathbf{f}(t), \qquad a < t < b,$$

(3.15b) $$\mathbf{y}(a) = \boldsymbol{\alpha}.$$

Proof. We need simply identify (1.1) with $BV(1)$ and (3.15) with $BV(0)$. The latter clearly has a unique solution as it is just an initial value problem. Then $BV_h(1)$ is taken as (1.5) and $BV_h(0)$ is taken as (3.14). Our result follows by applying Theorem 3.8. □

The schemes allowed in (1.5) are extremely general. In fact our theory now enables us to use the very well developed initial value theory of Dahlquist [1] and Henrici [2] to determine stable difference methods for linear boundary value problems. For one-step schemes, the results are particularly simple. Thus if (1.1) has a unique solution and in (1.5), we take for $j = 1, 2, \cdots, J$

(3.16a) $$\begin{gathered} C_{jk}(h) \equiv 0 \quad \text{for } k \neq j-1, j, \\ C_{j,j-1}(h) \equiv -\frac{1}{h_j} I + \tilde{C}_{j,j-1}(h), \qquad C_{j,j}(h) \equiv \frac{1}{h_j} I + \tilde{C}_{j,j}(h), \end{gathered}$$

(3.16b) $$\|\tilde{C}_{j,j-1}(h)\| \leqq M, \|\tilde{C}_{j,j}(h)\| \leqq M \quad \text{for all } h \leqq h_0,$$

then (1.5) *is stable and convergent for* (1.1) *if* (1.5) *is consistent with* (1.1). This result easily follows from Corollary 3.13 and Theorem 1 of Isaacson and Keller [4, p. 396], which implies the stability of (1.5) with coefficients satisfying (3.16). We point out that one-step schemes for first order systems are "compact as possible" in the terminology of Kreiss [9], and thus we obtain his results for such systems and extend them to nonuniform nets.

There are of course many schemes for initial value problems that are not treated in the above cited works. For example, the midpoint rule for initial value problems is stable, given appropriate starting data; but by altering the scheme at only one point (while not affecting consistency), it can be made unstable. Conversely some schemes which are unstable for initial value problems become stable when some of the initial data are replaced by conditions at the end of the interval. These examples, pointed out by H.-O. Kreiss, serve to stress the form in which the initial and boundary conditions are required to enter in the present theory.

Finally we note that asymptotic error expansions are easily obtained when the corresponding truncation error expansions are known by simply using the stability result. This is done in some detail for special one-step schemes in [5]. For more general schemes devised from initial value methods, we can readily employ the expansions given by Gragg [13] for one-step methods and by Henrici [2] and Engquist [12] for multistep methods.

4. Difference methods for nonlinear boundary value problems. The definitions of truncation errors, consistency and stability for the scheme (1.6) applied to (1.2) are as follows.

DEFINITION 4.1. (a) The *truncation errors* in scheme (1.6) applied to (1.2) are

$$\tau_0\{\mathbf{y}\} \equiv \mathbf{g}(\mathbf{y}(a), \mathbf{y}(b)), \qquad \tau_j\{\mathbf{y}\} \equiv \mathcal{N}_h\mathbf{y}(t_j), \qquad 1 \leqq j \leqq J,$$

where $\mathbf{y}(t)$ is any solution of (1.2).

(b) The scheme (1.6) *is consistent (accurate) of order p* for the solution $\mathbf{y}(t)$ of (1.2) provided there exist constants $K_0 > 0$ and $h_0 > 0$ such that

$$\|\boldsymbol{\tau}_j\{\mathbf{y}\}\| \leqq K_0 h^p, \qquad 0 \leqq j \leqq J,$$

for all nets (1.4) with $h \leqq h_0$.

(c) The scheme (1.6) *is stable for* $\mathbf{y}^h$ provided there exist positive constants K_ρ, ρ and h_0 such that for all net functions $\mathbf{v}^h, \mathbf{w}^h \in S_\rho(\mathbf{y}^h) \equiv \{\mathbf{u}^h : \|\mathbf{u}_j - \mathbf{y}_j\| \leqq \rho,\ 0 \leqq j \leqq J\}$ and all nets (1.4) with $h \leqq h_0$,

$$\|\mathbf{v}_j - \mathbf{w}_j\| \leqq K_\rho \max\{\|\mathcal{N}_h \mathbf{v}_k - \mathcal{N}_h \mathbf{w}_k\|,\ 1 \leqq k \leqq J;\ \|\mathbf{g}(\mathbf{v}_0, \mathbf{v}_J) - \mathbf{g}(\mathbf{w}_0, \mathbf{w}_J)\|\}.$$

In analogy with Theorem 3.2, we now have the well-known Theorem 4.2.

THEOREM 4.2. *Let* $\mathbf{y}(t)$ *be a solution of* (1.2) *and for all nets* (1.4) *with* $h \leqq h_0$ *let* $\mathbf{u}^h$ *be a solution of* (1.6) *in* $S_\rho(\mathbf{y}^h)$ *where* $\mathbf{y}_j \equiv \mathbf{y}(t_j)$. *If* (1.6) *is accurate of order p for* $\mathbf{y}(t)$ *and stable for* $\mathbf{y}^h$, *then on all nets* (1.4) *with* $h \leqq h_0$

$$\|\mathbf{y}(t_j) - \mathbf{u}_j\| \leqq K_0 K_\rho h^p.$$

Proof. By Definition 4.1(c) with $\mathbf{v}_j \equiv \mathbf{y}(t_j)$ and $\mathbf{w}_j \equiv \mathbf{u}_j$,

$$\|\mathbf{y}(t_j) - \mathbf{u}_j\| \leqq K_\rho \max\{\|\mathcal{N}_h \mathbf{y}(t_k) - \mathcal{N}_h \mathbf{u}_k\|,\ 1 \leqq k \leqq J;\ \|\mathbf{g}(\mathbf{y}(a), \mathbf{y}(b)) - \mathbf{g}(\mathbf{u}_0, \mathbf{u}_J)\|\}.$$

Using (1.2b) and (1.6a, b), we get the result upon recalling Definition 4.1 (a, b). □

The basic problems are of course to insure that (1.6) has a solution in $S_\rho(\mathbf{y}^h)$ for all $h \leqq h_0$ and to verify stability. We could apply the general theory developed in [8] to get these results, but in the interest of completeness we indicate the details. For stability we have the following lemma.

LEMMA 4.3. *Let* $\mathbf{y}(t)$ *be an isolated solution of* (1.2) *and assume*

(i) *the linear difference scheme*

$$\mathscr{L}_h[\mathbf{y}^h]\mathbf{v}_j = 0, \qquad 1 \leqq j \leqq J, \quad \mathbf{v}_0 = \mathbf{z}_0, \tag{4.3a}$$

defined in (1.7) *is stable and consistent for the initial value problem*

$$\mathscr{L}[\mathbf{y}]\mathbf{z} = 0, \quad a \leqq t \leqq b, \quad \mathbf{z}(a) = \mathbf{z}_0; \tag{4.3b}$$

(ii) *for some* $\rho > 0$, $K_L > 0$, $h_0 > 0$, *all* $\mathbf{w}^h \in S_\rho(\mathbf{y}^h)$ *and for all* $h \leqq h_0$,

$$\|\mathscr{L}_h[\mathbf{y}^h] - \mathscr{L}_h[\mathbf{w}^h]\| \leqq K_L \|\mathbf{y}^h - \mathbf{w}^h\|. \tag{4.4a}$$

$$\|B_x[\mathbf{y}^h] - B_x[\mathbf{w}^h]\| \leqq K_L \max\{\|\mathbf{y}(a) - \mathbf{w}_0\|, \|\mathbf{y}(b) - \mathbf{w}_J\|\}, \qquad x = a, b. \tag{4.4b}$$

Then the scheme (1.6) *is stable for* $\mathbf{y}^h$ *provided* ρ *is sufficiently small.*

Proof. Define $\mathbb{A}_h[\mathbf{w}^h]$ for any $\mathbf{w}^h \in S_\rho(\mathbf{y}^h)$ by using (1.7c) with $\mathbf{u}^h$ replaced by $\mathbf{w}^h$ in (3.3). We claim that $\mathbb{A}_h[\mathbf{y}^h]$ is nonsingular and for some constant $K > 0$,

$$\|\mathbb{A}_h^{-1}[\mathbf{y}^h]\| \leqq K \quad \text{for all } h \leqq 0. \tag{4.5}$$

This follows from Corollary 3.13, Lemma 3.4 and 4.3(a, b), since $\mathbf{y}(t)$ is assumed an isolated solution, and thus (1.3) has a unique solution.

Let us write the nonlinear difference operators of (1.6) in the vector form

$$\mathbf{\Phi}(\mathbf{u}^h) \equiv \begin{pmatrix} \mathbf{g}(\mathbf{u}_0, \mathbf{u}_J) \\ \mathcal{N}_h \mathbf{u}_1 \\ \vdots \\ \mathcal{N}_h \mathbf{u}_J \end{pmatrix}. \tag{4.6}$$

Then by the assumed differentiability of the $\mathbf{G}_j(\cdot)$ and $\mathbf{g}(\cdot, \cdot)$,

$$\mathbf{\Phi}(\mathbf{v}^h) - \mathbf{\Phi}(\mathbf{w}^h) = \hat{\mathbb{A}}_h[\mathbf{v}^h, \mathbf{w}^h](\mathbf{v}^h - \mathbf{w}^h), \tag{4.7a}$$

where

$$\hat{\mathbb{A}}_h[\mathbf{v}^h, \mathbf{w}^h] \equiv \int_0^1 \mathbb{A}_h[s\mathbf{v}^h + (1 - s)\mathbf{w}^h]\, ds. \tag{4.7b}$$

It follows from (4.4) that for all $\mathbf{v}^h, \mathbf{w}^h \in S_\rho(\mathbf{y}^h)$

$$\|\hat{\mathbb{A}}[\mathbf{v}^h, \mathbf{w}^h] - \mathbb{A}_h[\mathbf{y}^h]\| \leqq \rho K_L.$$

Thus if ρ is so small that $\rho K_L K < 1$, the Banach lemma implies $\hat{\mathbb{A}}_h[\cdot, \cdot]$ nonsingular and in fact

$$\|\hat{\mathbb{A}}_h^{-1}[\mathbf{v}^h, \mathbf{w}^h]\| \leqq \frac{K}{1 - \rho K_L K}.$$

Stability as in Definition 4.1(c) is simply, using (4.6),

$$\|\mathbf{v}^h - \mathbf{w}^h\| \leqq K_\rho \|\mathbf{\Phi}(\mathbf{v}^h) - \mathbf{\Phi}(\mathbf{w}^h)\|$$

and it clearly follows with $K_\rho = K/(1 - \rho K_L K)$. □

The existence of a unique solution in $S_\rho(\mathbf{y}^h)$ of the difference equations (1.6) for each $h \leqq h_0$ is established by contraction mappings applied to

$$\mathbf{u}^h = \mathbf{u}^h - \mathbb{A}_h^{-1}[\mathbf{y}^h]\mathbf{\Phi}(\mathbf{u}^h).$$

The proof assumes consistency as in Definition 4.1(a, b) and the hypothesis of Lemma 4.3. The details are contained in [8, Thm. 3.6] and are similar to part of the argument in [7, § 3]. Combining these results with those of Theorem 4.2 and Lemma 4.3, we have the following basic theorem.

THEOREM 4.8. *Let* $\mathbf{y}(t)$ *be an isolated solution of* (1.2). *Let the difference scheme* (1.6) *be accurate of order p for* $\mathbf{y}(t)$ *and satisfy the hypothesis* (i) *and* (ii) *of Lemma* 4.3. *Then for* $\rho > 0$ *and* $h_0 > 0$, *both sufficiently small, the difference equations* (1.6) *have for each* $h \leqq h_0$ *a unique solution* $\mathbf{u}^h \in S_\rho(\mathbf{y}^h)$ *with*

$$\|\mathbf{y}(t_j) - \mathbf{u}_j\| \leqq M h^p,$$

for some constant $M > 0$.

To actually compute the numerical solution, we employ Newton's method in the form

$$\mathbf{u}_0^h \in S_{\rho_0}(\mathbf{y}^h), \tag{4.9a}$$

$$\mathbb{A}_h[\mathbf{u}_\nu^h](\mathbf{u}_{\nu+1}^h - \mathbf{u}_\nu^h) = -\mathbf{\Phi}(\mathbf{u}_\nu^h), \qquad \nu = 0, 1, 2, \cdots. \tag{4.9b}$$

The quadratic convergence is easily established under the hypothesis of Theorem 4.8 with some $\rho_0 \leqq \rho$. For any $\mathbf{v}^h \in S_{\rho_0}(\mathbf{y}^h)$, we have the identity

$$\mathbb{A}_h[\mathbf{v}^h] = \mathbb{A}_h[\mathbf{y}^h]\{\mathbb{I} - \mathbb{A}_h^{-1}[\mathbf{y}^h](\mathbb{A}_h[\mathbf{y}^h] - \mathbb{A}_h[\mathbf{v}^h])\}.$$

Using (4.4a, b), (4.5) and the Banach lemma, we have that $A_h[\mathbf{v}^h]$ is nonsingular with

$$\|\mathbb{A}_h^{-1}[\mathbf{v}^h]\| \leqq K_{\rho_0} = \frac{K}{1 - \rho_0 K_L K}. \tag{4.10}$$

From (4.9b) with $\nu = 0$ we obtain, using (4.7),

$$\mathbf{u}_1^h - \mathbf{u}_0^h = -\mathbb{A}_h^{-1}[\mathbf{u}_0^h]\mathbf{\Phi}(\mathbf{y}^h) + \mathbb{A}_h^{-1}[\mathbf{u}_0^h]\hat{\mathbb{A}}_h[\mathbf{u}_0^h, \mathbf{y}^h](\mathbf{y}^h - \mathbf{u}_0^h).$$

Now note that

$$\mathbb{A}_h^{-1}[\mathbf{u}_0^h]\hat{\mathbb{A}}_h[\mathbf{u}_0^h, \mathbf{y}^h] \equiv \mathbb{I} + \mathbb{A}_h^{-1}[\mathbf{u}_0^h](\hat{\mathbb{A}}_h[\mathbf{u}_0^h, \mathbf{y}^h] - \mathbb{A}_h[\mathbf{u}_0^h]).$$

Using (4.10) and the Lipschitz continuity (4.4), we find that for any $\rho_0 \leqq \rho$ there exists some $C > 0$ such that

$$\|\mathbb{A}_h^{-1}[\mathbf{u}_0^h]\hat{\mathbb{A}}_h[\mathbf{u}_0^h, \mathbf{y}^h]\| \leqq C.$$

Thus we finally get, recalling (4.1a, b), that

$$\|\mathbf{u}_1^h - \mathbf{u}_0^h\| \leqq K_{\rho_0} K_0 h^p + C\rho_0. \tag{4.11}$$

From (4.10) and (4.11), the quadratic convergence of Newton's method follows in standard fashion (see [6] or [10]). The convergence proof in [7] is unnecessarily restrictive, as has been observed by F. de Hoog [3], since sharp estimates of $\|\mathbf{\Phi}(\mathbf{u}_0^h)\|$ were sought rather than of $\|\mathbb{A}_h^{-1}[\mathbf{u}_0^h]\mathbf{\Phi}(\mathbf{u}_0^h)\|$ as we do above. In particular, we stress that it is not necessary that ρ_0 be reduced with h, and thus a much larger sphere is shown to be in the domain of attraction for the root in question.

In closing, we note that the complete nonlinear theory goes over for the general nonlinear multipoint boundary conditions. The details are quite similar to those contained in the Appendix to [7] and so we do not repeat them here.

REFERENCES

[1] G. Dahlquist, *Convergence and stability in the numerical integration of ordinary differential equations*, Math. Scand., 4 (1956), pp. 33–53.

[2] P. Henrici, *Discrete Variable Methods in Ordinary Differential Equations*, John Wiley, New York, 1962.

[3] F. de Hoog, Private communication, 1974.

[4] E. Isaacson and H. B. Keller, *Analysis of Numerical Methods*, John Wiley, New York, 1966.

[5] H. B. Keller, *Accurate difference methods for linear ordinary differential systems subject to linear constraints*, this Journal, 6 (1969), pp. 8–30.

[6] ———, *Newton's method under mild differentiability conditions*, J. Comput. System Sci., 4 (1970), pp. 15–28.

[7] ———, *Accurate difference methods for nonlinear two-point boundary value problems*, this Journal, 11 (1974), pp. 305–320.

[8] ———, *Approximation methods for nonlinear problems with application to two-point boundary value problems*, Math. Comp., to appear (1975).

[9] H.-O. Kreiss, *Difference approximations for boundary and eigenvalue problems for ordinary differential equations*, Math. Comp., 26 (1972), pp. 605–624.

[10] J. ORTEGA AND W. RHEINBOLDT, *Iterative Solution of Nonlinear Equations in Several Variables*, Academic Press, New York, 1970.

[11] A. B. WHITE, *Numerical solution of two-point boundary value problems*, Thesis, Calif. Inst. of Tech., Pasadena, Calif., 1974.

[12] B. ENGQUIST, *Asymptotic error expansions for multistep methods*, Research Rep., Computer Science Dept., Uppsala University, Uppsala, Sweden, 1969.

[13] W. B. GRAGG, *On extrapolation algorithms for ordinary initial value problems*, this Journal, 2 (1965), pp. 384–403.

Reprinted with permission from *SIAM Journal On Numerical Analysis,* Vol. 14, No. 5 (1977), pp. 779–791.

A NUMERICAL METHOD FOR SINGULAR TWO POINT BOUNDARY VALUE PROBLEMS*

Dedicated to R. D. Richtmyer in Honor of his 65th Birthday

D. C. BRABSTON† AND H. B. KELLER‡

Abstract. The numerical solution of boundary value problems for linear systems of first order equations with a regular singular point at one endpoint is considered. The standard procedure of expanding about the singularity to get a nonsingular problem over a reduced interval is justified in some detail. Quite general boundary conditions are included which permit unbounded solutions. Error estimates are given and some numerical calculations are presented to check the theory.

1. Introduction. We consider the numerical solution of linear two point boundary value problems with a regular singularity at one endpoint. In particular the class of problems is formulated as:

(1.1a) $$\mathscr{L}\mathbf{y}(t) \equiv \mathbf{y}'(t) - A(t)\mathbf{y}(t) = \mathbf{f}(t), \qquad 0 < t \leqq 1;$$

(1.1b) $$\mathscr{B}\mathbf{y}(t) \equiv \lim_{t \downarrow 0} [B_0(t)\mathbf{y}(t) + B_1\mathbf{y}(1) - \mathbf{b}(t)] = \mathbf{0};$$

(1.1c) $$A(t) \equiv t^{-1}R + A_0(t).$$

Here $\mathbf{y}(t)$, $\mathbf{f}(t)$, $\mathbf{b}(t)$ are n-vectors while R, $A_0(t)$, $B_0(t)$, B_1 are $n \times n$ matrices. We assume $A_0(t)$ is, say, analytic on $(0, \delta_0]$ and sufficiently smooth on $(0, 1]$. Similarly $B_0(t)$, $\mathbf{f}(t)$ and $\mathbf{b}(t)$ are smooth on $(0, 1]$ but may be singular at $t = 0$. The solution $\mathbf{y}(t) \in C^1(0, 1]$ but need not even be bounded as $t \to 0$. The form of the boundary conditions (1.1b) includes the typical constraints of wave propagation problems (i.e. incoming or outgoing waves) and those due to symmetry in singular coordinate systems. With $B_0(x) = \tilde{B}_0 t^{-R}$ conditions can be imposed on the regular part of the solution, as is done by Natterer [10]. That seems to be the first work to study numerical methods for fairly general linear systems of the form (1.1a, c).

We examine the more or less standard procedure of expanding about the singular point, solving a regular boundary value problem over a reduced interval excluding the singular point and matching this solution to the expansion. To study this process we require an existence and uniqueness theory for (1.1) which is developed in § 2. Then in § 3 we introduce a regular boundary value problem on some interval $[\delta, 1]$ and show that it is equivalent to (1.1). The regular problem cannot be determined explicitly so in § 4 we study its replacement by a truncated regular problem which can be obtained explicitly. In § 5 we examine the error between the exact solution of (1.1) and a numerical solution of the truncated regular problem. Finally numerical examples are presented in § 6.

The extension of our methods to problems with two regular singular endpoints or even interior regular singularities is easily carried out, as shown in Brabston [1]. The study of unbounded intervals can be included if the point at

* Received by the editors March 23, 1976, and in revised form July 23, 1976. This work was supported in part by ERDA under Grant AT-04-3-767.

† Data Systems Department, TRW Systems Group, Redondo Beach, California 90278.

‡ Applied Mathematics, California Institute of Technology, Pasadena, California 91125.

infinity is a regular singular point. This expansion technique or procedures very close to it have been used for many years. However, a study of the scalar case has only recently been made by Gustaffson [3]. Other studies of difference methods for special singular second order problems have been carried out by Jamet [5], Natterer [11], and Russell and Shampine [12]. A study of the trapezoidal and centered Euler difference schemes applied to bounded solutions of (1.1) and more general equations is in progress by de Hoog and Weiss [4].

2. Existence and uniqueness theory. A fundamental solution matrix, $Y(t)$, for (1.1a) can be defined, using some nonsingular $n\times n$ matrix Y_0, by:

(2.1a) $$\mathscr{L}Y(t)=0, \qquad 0<t\leqq 1;$$

(2.1b) $$Y(\delta_0)=Y_0.$$

It is well known (see Coddington and Levinson [2, pp. 118–122]) that when R in (1.1c) has no eigenvalues separated by a positive integer:

(2.2a) $$Y(t)=P(t)t^R, \qquad 0<t\leqq\delta_0;$$

(2.2b) $$P(t)=\sum_{k=0}^{\infty}P_kt^k, \qquad P_0=I,\cdots;$$

where $P(t)$ is analytic and nonsingular on $0\leqq t\leqq\delta_0$. A simple transformation of (1.1a, c) insures the eigenvalue condition so we merely assume it to hold in all that follows (see, however, the example in § 2.1). Note that Y_0 cannot be chosen arbitrarily in (2.1b) if we use the representation (2.2).

Every solution of (1.1a) can be written as

(2.3a) $$\mathbf{y}(t)=Y(t)\mathbf{c}+\mathbf{y}_p(t), \qquad 0<t\leqq 1;$$

where the particular solution, $\mathbf{y}_p(t)$, satisfies:

(2.3b) $$\mathscr{L}\mathbf{y}_p(t)=\mathbf{f}(t), \qquad 0<t\leqq 1, \quad \mathbf{y}_p(\delta_0)=\mathbf{y}_0.$$

In fact we have that

(2.3c) $$\mathbf{y}_p(t)=Y(t)\left\{\int_{\delta_0}^{t}Y^{-1}(\tau)\mathbf{f}(\tau)\,d\tau+Y^{-1}(\delta_0)\mathbf{y}_0\right\}.$$

Thus $\mathbf{y}(t)$, given by (2.3a), will satisfy the boundary conditions (1.1b) if and only if:

(2.4a) $$\lim_{t\downarrow 0}\{[B(t)+B_1Y(1)]\mathbf{c}-[\mathbf{g}(t)-B_1\mathbf{y}_p(1)]\}=\mathbf{0}.$$

Here:

(2.4b) $$B(t)\equiv B_0(t)Y(t);$$

(2.4c) $$\mathbf{g}(t)\equiv\mathbf{b}(t)-B_0(t)\mathbf{y}_p(t).$$

So our existence and uniqueness theory is reduced to a study of the existence and uniqueness of a solution $\mathbf{c}$, of (2.4a).

Suppose, for the moment, that the possible singularities of $B_0(t)$, $\mathbf{b}(t)$ and $\mathbf{f}(t)$ are of the forms:

(2.5) $$B_0(t) \equiv \hat{B}_0(t)t^{R_0}, \quad \mathbf{b}(t) \equiv t^{R_1}\hat{\mathbf{b}}(t), \quad \mathbf{f}(t) \equiv t^{R_2}\hat{\mathbf{f}}(t),$$

where $\hat{B}_0(t)$, $\hat{\mathbf{b}}(t)$ and $\hat{\mathbf{f}}(t)$ are analytic on $[0, \delta_0]$. Then using the Jordan forms of the $n \times n$ matrices R, R_0, R_1 and R_2 with (2.2) and (2.5) in (2.4b, c) it easily follows that $B(t)$ and $\mathbf{g}(t)$ have a finite number of distinct singularities. More precisely we find, for some positive integers q and K that there exists a set of q scalar functions $\varphi_\nu(t)$, $\nu = 1, 2, \cdots, q$, satisfying:

(2.6a) $$\lim_{t \downarrow 0} |\varphi_\nu(t)| = \infty,$$

(2.6b) $$\lim_{t \downarrow 0} |\varphi_\nu(t)/\varphi_{\nu+1}(t)| = 0,$$

(2.6c) $$\lim_{t \downarrow 0} t^K \varphi_q(t) = 0;$$

such that:

(2.6d) $$B(t) = M_0(t) + \sum_{\nu=1}^{q} \varphi_\nu(t) M_\nu;$$

(2.6e) $$\mathbf{g}(t) = \mathbf{g}_0(t) + \sum_{\nu=1}^{q} \varphi_\nu(t)\mathbf{g}_\nu.$$

Here $M_0(t)$ and $\mathbf{g}_0(t)$ are analytic on $[0, \delta_0]$ while the M_ν and $\mathbf{g}_\nu$ are constant matrices and vectors. If $B_0(t)$, $\mathbf{b}(t)$ and $\mathbf{f}(t)$ do not have the forms in (2.5) we require that their singularities be such that expansions of the form (2.6) are valid for some set $\{\varphi_\nu(t)\}_1^q$. Obviously finite sums of terms of the form used in (2.5) are easily included.

Now we have the basic existence and uniqueness result.

THEOREM 2.7. *Let the expansions* (2.6) *hold. Define the* $n(q+1)$ *order matrix* M *and vector* $\mathbf{g}$ *by*:

(2.7a) $$M \equiv \begin{pmatrix} M_0(0) + B_1 Y(1) \\ M_1 \\ \vdots \\ M_q \end{pmatrix};$$

(2.7b) $$\mathbf{g} \equiv \begin{pmatrix} \mathbf{g}_0(0) - B_1 \mathbf{y}_p(1) \\ \mathbf{g}_1 \\ \vdots \\ \mathbf{g}_q \end{pmatrix}.$$

Then (1.1a, b, c) *has a solution if and only if* rank M = rank $(M, \mathbf{g})$. *The solution is unique if and only if* rank $M = n$.

Proof. We have shown above that our result is equivalent to the solvability of (2.4a). Using (2.6a, b, d, e) in (2.4a) we find that the coefficients of each singularity function, $\varphi_\nu(t)$, must vanish. This yields, in brief, the system

$$Mc = g. \tag{2.7c}$$

Now the theorem follows from the elementary rank condition for linear systems. □

2.1. An example. It is easy to determine the singularities, $\varphi_\nu(t)$, and the constant quantities M_ν and $\mathbf{g}_\nu$, $1 \leqq \nu \leqq q$, of (2.6) explicitly when the forms (2.5) hold. In general, however, we cannot evaluate $Y(1)$ or $\mathbf{y}_p(1)$ so that M and $\mathbf{g}$ may not be fully evaluated. An interesting exception is afforded by the inhomogeneous axisymmetric potential equation:

$$v'' + \frac{\sigma}{t} v' = (\sigma - 2)t^{-\sigma} \sin t - t^{(1-\sigma)} \cos t, \qquad 0 < t \leqq 1, \quad \sigma > 0. \tag{2.8a}$$

This or related forms are used by several authors to study numerical methods for singular points [3], [4], [5], [11], [12]. We write the equivalent first order system:

$$\mathbf{y}' - \begin{pmatrix} 0 & 1 \\ 0 & -\sigma/t \end{pmatrix} \mathbf{y} = \begin{pmatrix} 0 \\ (\sigma-2)t^{-\sigma} \sin t - t^{1-\sigma} \cos t \end{pmatrix}, \qquad \mathbf{y} \equiv \begin{pmatrix} v \\ v' \end{pmatrix}, \tag{2.8b}$$

and consider boundary conditions of the general form (1.1b). Since $\sigma = 1$ yields eigenvalues of R which differ by an integer it is a special case. However we easily find that the fundamental solution matrix for (2.8b) is:

$$Y(t) = \begin{pmatrix} 1 & s_\sigma(t) \\ 0 & t^{-\sigma} \end{pmatrix}, \qquad \text{where } s_\sigma(t) \equiv \begin{cases} \ln t, & \sigma = 1; \\ t^{1-\sigma}(1-\sigma)^{-1}, & \sigma \neq 1. \end{cases} \tag{2.9a}$$

A particular solution of (2.8b) is, for all σ:

$$\mathbf{y}_p(t) = \begin{pmatrix} t^{1-\sigma} \cos t \\ (1-\sigma)t^{-\sigma} \cos t - t^{1-\sigma} \sin t \end{pmatrix}. \tag{2.9b}$$

For constant $B_0(t) \equiv (B^0_{ij})$ and $\mathbf{b}(t) \equiv (b_1, b_2)^T$ we use (2.9) in (2.4b, c) to get the $\{\phi_\nu(t)\}_1^q$ of (2.6) as:

$$\begin{aligned} &0 < \sigma < 1 : q = 1, && \phi_1(t) = t^{-\sigma}; \\ &\sigma = 1 : q = 2, && \phi_1(t) = \ln t, \quad \phi_2(t) = t^{-1}; \\ &\sigma > 1 : q = [\sigma] + 1, && \phi_1(t) = t^{[\sigma]-\sigma}, \cdots, \phi_{q-1}(t) = t^{1-\sigma}, \quad \phi_q(t) = t^{-\sigma}. \end{aligned} \tag{2.9c}$$

Here $[\sigma]$ is the largest integer less than σ. With $B_1 \equiv (B^1_{ij})$ we easily obtain the matrices M_ν and vectors $\mathbf{g}_\nu$ that enter into (2.7). Rather than list these quantities we state some of the conclusions that they imply; more details can be found in Brabston [1].

A. For $0 < \sigma < 1$ a unique solution exists for every constant $\mathbf{b}(t) \equiv \mathbf{b}$ provided:

$$B_0 = \begin{pmatrix} B^0_{11} & 0 \\ B^0_{21} & 0 \end{pmatrix} \quad \text{and} \quad M_0(0) + B_1 Y(1) = \begin{pmatrix} B^0_{11} + B^1_{11} & B^1_{11}(1-\sigma)^{-1} + B^1_{12} \\ B^0_{21} + B^1_{21} & B^1_{21}(1-\sigma)^{-1} + B^1_{22} \end{pmatrix}$$

is nonsingular.

In terms of the scalar formulation (2.8a) this special form of B_0 implies that $v'(t)$ cannot enter into the boundary conditions at $t=0$.

B. For $\sigma \geqq 1$ a unique solution exists for every constant $\mathbf{b}(t) \equiv \mathbf{b}$ provided: $B_0 \equiv 0$ and B_1 is nonsingular, that is an initial value problem must be posed with initial point $t=1$.

Of course many special solutions exist for B_0 and B_1 which violate the above general conditions. For example

$$B_0 \equiv \begin{pmatrix} 1 & 0 \\ 0 & 1 \end{pmatrix}, \quad B_1 \equiv \begin{pmatrix} 0 & 0 \\ 0 & 1 \end{pmatrix}, \quad \mathbf{b} = \begin{pmatrix} 0 \\ (\sigma-1)(1-\cos 1)-\sin 1 \end{pmatrix}$$

correspond to the boundary conditions

$$v(0)=1, \quad \lim_{t \downarrow 0} v'(t) + v'(1) = (\sigma-1)(1-\cos 1) - \sin 1.$$

For $0<\sigma<1$ this yields the unique solution

$$v(t) = 1 - t^{1-\sigma}(1-\cos t).$$

3. Reduction to a regular problem. We show how to replace the singular problem (1.1) by a regular problem on a reduced interval $[\delta, 1]$. Suppose (1.1) has a solution under the hypothesis of Theorem 2.7. This solution must have the form (2.3a) for some fixed $\mathbf{c}$. Then for any point $\delta \in (0, \delta_0]$ since $Y(\delta)$ is nonsingular, the vector $\mathbf{c}$ is given by:

$$\mathbf{c} = Y^{-1}(\delta)[\mathbf{y}(\delta) - \mathbf{y}_p(\delta)]. \tag{3.1}$$

But this vector $\mathbf{c}$ must satisfy (2.7c) and thus we get that

$$B_{0\delta}\mathbf{y}(\delta) + B_{1\delta}\mathbf{y}(1) = \mathbf{b}_\delta \tag{3.2a}$$

where we have used $\mathbf{y}(1) = Y(1)\mathbf{c} + \mathbf{y}_p(1)$ and introduced:

$$B_{0\delta} \equiv \begin{pmatrix} M_0(0) \\ M_1 \\ \vdots \\ M_q \end{pmatrix} Y^{-1}(\delta), \quad B_{1\delta} \equiv \begin{pmatrix} B_1 \\ 0 \\ \vdots \\ 0 \end{pmatrix}, \quad \mathbf{b}_\delta \equiv \begin{pmatrix} \mathbf{g}_0(0) \\ \mathbf{g}_1 \\ \vdots \\ \mathbf{g}_q \end{pmatrix} + B_{0\delta}\mathbf{y}_p(\delta). \tag{3.2b}$$

It follows that the regular two point problem consisting of

$$\mathscr{L}\mathbf{u}(t) \equiv \mathbf{u}(t) - A(t)\mathbf{u}(t) = \mathbf{f}(t), \qquad \delta \leqq t \leqq 1 \tag{3.3}$$

subject to (3.2) has a solution when (1.1) has a solution given by (2.3a). Indeed we have even more as follows.

THEOREM 3.4. *Let* $\mathbf{y}_p(t)$ *be a particular solution of* (1.1a), *say as defined in* (2.3b). *Then the singular problem* (1.1) *has a unique solution if and only if the regular problem* (3.2)–(3.3) *has a unique solution. Further, the unique solutions of both problems are identical on* $[\delta, 1]$.

Proof. If either problem has a solution it can be represented by (2.3a) with possibly differing constant vectors **c**. Using this form in (3.2a) yields the linear system

$$[B_{0\delta}Y(\delta)+B_{1\delta}Y(1)]\mathbf{c}=\mathbf{b}_\delta-[B_{0\delta}\mathbf{y}_p(\delta)+B_{1\delta}\mathbf{y}_p(1)]. \tag{3.5}$$

However, this system is precisely (2.7c). Thus both problems lead to the identical linear system and our results follow. □

4. The truncated regular problem. To determine the matrix $B_{0\delta}$ and inhomogeneous data $\mathbf{b}_\delta$ of (3.2) for the reduced regular problem we require $Y(\delta)$ and $\mathbf{y}_p(\delta)$. These are not generally known so we approximate them. Taking δ sufficiently small we truncate the power series representation of $P(t)$ in (2.2) to get, say:

$$Y^N(t)\equiv P^N(t)t^R,\qquad 0<t\leqq\delta; \tag{4.1a}$$

$$P^N(t)\equiv\sum_{k=0}^{N}P_kt^k. \tag{4.1b}$$

Using $Y^N(t)$ in place of $Y(t)$ in (2.3c) we also define the truncated particular solution,

$$\mathbf{y}_p^N(t)\equiv Y^N(t)\left\{\int_{\delta_0}^{t}[Y^N(\tau)]^{-1}\mathbf{f}(\tau)\,d\tau+Y^N(\delta_0)^{-1}\mathbf{y}_0\right\}. \tag{4.1c}$$

Of course, as we shall see, any other approximations to $Y(t)$ and $\mathbf{y}_p(t)$ would do provided they are sufficiently accurate at $t=\delta$. In the indicated expansion technique we must take N sufficiently large so that all singularities are included; that is so that $Y(t)-Y^N(t)$ and $\mathbf{y}_p(t)-\mathbf{y}_p^N(t)$ are regular at $t=0$ and in fact vanish as $t\downarrow 0$. This can be assured, if the highest order singularity occurs in t^R, by taking $N\geqq\max\operatorname{Re}[-\lambda_k(R)]+1$. We shall in fact assume that

$$\max_{0\leqq t\leqq\delta}(\|Y(t)-Y^N(t)\|,\ \|\mathbf{y}_p(t)-\mathbf{y}_p^N(t)\|)\leqq\Delta(N,\delta) \tag{4.2a}$$

where $\Delta(N,\delta)\downarrow 0$ as $N\uparrow\infty$ for any fixed $\delta\in(0,\delta_0]$. We require N to be so large that

$$\|Y^{-1}(\delta)\|\Delta(N,\delta)<1. \tag{4.2b}$$

Then $Y^N(\delta)$ is nonsingular, by the Banach lemma, and

$$\|Y^N(\delta)^{-1}\|\leqq\frac{\|Y^{-1}(\delta)\|}{1-\|Y^{-1}(\delta)\|\Delta(N,\delta)}. \tag{4.2c}$$

We have already used this assumption in writing (4.1c).

In analogy with (2.4b, c) we define the truncated quantities:

$$B^N(t)\equiv B_0(t)Y^N(t), \tag{4.3a}$$

$$\mathbf{g}^N(t)\equiv\mathbf{b}(t)-B_0(t)\mathbf{y}_p^N(t). \tag{4.3b}$$

If $B_0(t)$ and/or $\mathbf{f}(t)$ are singular at $t=0$ we require that N is taken sufficiently large so that $B(t)-B^N(t)$ and $\mathbf{g}(t)-\mathbf{g}^N(t)$ vanish as $t\downarrow 0$. This is always possible when the assumptions (2.6) are valid. Indeed then $N\geqq K$ and the equality holds if $B_0(t)$ and

$\mathbf{f}(t)$ are regular. Under the above assumptions it follows that

(4.3c) $$B^N(t) = M_0^N(t) + \sum_{\nu=1}^{q} \varphi_\nu(t) M_\nu, \qquad M_0^N(0) = M_0(0);$$

(4.3d) $$\mathbf{g}^N(t) = \mathbf{g}_0^N(t) + \sum_{\nu=1}^{q} \varphi_\nu(t) \mathbf{g}_\nu, \qquad \mathbf{g}_0^N(0) = \mathbf{g}_0(0).$$

Here $M_0^N(t)$ and $\mathbf{g}_0^N(t)$ are "truncations" of $M_0(t)$ and $\mathbf{g}_0(t)$, respectively, in (2.6d, e). Note that the M_ν, $\mathbf{g}_\nu$ and $\varphi_\nu(t)$ are retained exactly while the value of $M_0^N(t)$ and $\mathbf{g}_0^N(t)$ at $t=0$ are exactly the values of $M_0(0)$ and $\mathbf{g}_0(0)$, respectively.

A truncated regular boundary value problem is now defined as:

(4.4a) $$\mathscr{L}\mathbf{y}^N(t) = \mathbf{f}(t), \qquad \delta \leqq t \leqq 1,$$

(4.4b) $$B_{0\delta}^N \mathbf{y}^N(\delta) + B_{1\delta}\mathbf{y}^N(1) = \mathbf{b}_\delta^N.$$

Here we recall (3.2b) and that $Y^N(\delta)$ is nonsingular to introduce:

(4.4c) $$B_{0\delta}^N \equiv B_{0\delta} Y(\delta) Y^N(\delta)^{-1}; \qquad \mathbf{b}_\delta^N \equiv \mathbf{b}_\delta + B_{0\delta}^N \mathbf{y}_p^N(\delta) - B_{0\delta}\mathbf{y}_p(\delta).$$

We note that the truncated regular problem (4.4) does not employ $Y(\delta)$ or $\mathbf{y}_p(\delta)$ in its specification of (4.4c) but only $Y^N(\delta)$ and $\mathbf{y}_p^N(\delta)$. Unfortunately even for very large N the truncated and untruncated problems need not be equivalent when $q \geqq 1$. Indeed since any solution of (4.4a) can be represented as

(4.5) $$\mathbf{y}^N(t) = Y(t)\mathbf{c}^N + \mathbf{y}_p(t), \qquad \delta \leqq t \leqq 1,$$

it will be a solution of (4.4b) provided the constant $\mathbf{c}^N$ satisfies:

(4.6a) $$M^N \mathbf{c}^N = \mathbf{g}^N$$

where, if we recall (2.7a, b), (3.2b) and (4.4c):

(4.6b) $$\begin{aligned} M^N &= B_{0\delta}^N Y(\delta) + B_{1\delta} Y(1) \\ &= M Y^N(\delta)^{-1} Y(\delta) + B_{1\delta} Y(1)[I - Y^N(\delta)^{-1} Y(\delta)] \end{aligned}$$

(4.6c) $$\mathbf{g}^N = \mathbf{g} + B_{0\delta}^N[\mathbf{y}_p^N(\delta) - \mathbf{y}_p(\delta)].$$

Thus (4.4) has a solution if and only if rank M^N = rank $(M^N, \mathbf{g}^N)$ and a solution is unique if rank $M^N = n$. We recall that (3.2)–(3.3) has a unique solution if and only if (2.7c) has a unique solution, that is rank M = rank $(M, \mathbf{g}) = n$. If $\|Y^N(\delta) - Y(\delta)\|$ is small but nonzero while $\mathbf{y}_p^N(\delta) = \mathbf{y}_p(\delta)$ the linear systems (4.6a) and (2.7c) are still not in general equivalent. They would be in this case if $B_{1\delta}Y(1)$ $[I - Y^N(\delta)^{-1} Y(\delta)] \equiv 0$.

However when (3.2)–(3.3) has a unique solution we can be assured that a subset of the boundary conditions in (4.4b) does yield, for N sufficiently large, a uniquely solvable problem for (4.4a). More precisely we state this as follows.

THEOREM 4.7. *Let* (1.1) *have a unique solution and the expansions* (2.6) *hold. Then there exists some* $n \times n(q+1)$ *projection matrix* P *such that* PM *is nonsingular. For any such* P *the problem*:

(4.7a) $$\mathscr{L}\mathbf{y}^N(t) = \mathbf{f}(t), \qquad \delta \leqq t \leqq 1;$$

(4.7b) $$P[B_{0\delta}^N \mathbf{y}^N(\delta) + B_{1\delta}\mathbf{y}^N(1) - \mathbf{b}_\delta^N] = \mathbf{0}$$

has a unique solution for all N sufficiently large.

Proof. Since (1.1) has a unique solution, then by Theorem 2.7 the matrix M has rank n. So some $n \times n$ submatrix of M, say PM, must be nonsingular; this yields the existence of P.

A solution of (4.7a) must have the form (4.5). It easily follows using (4.6) that (4.7b) will be satisfied if and only if

$$PM^N \mathbf{c}^N = P\mathbf{g}^N. \tag{4.8a}$$

However, recalling (4.2), we have

$$\|PM - PM^N\| \leqq K_0 \frac{\|Y^{-1}(\delta)\|\Delta(N, \delta)}{1 - \|Y^{-1}(\delta)\|\Delta(N, \delta)} \tag{4.8b}$$

where $K_0 = \|(M_0^T(0), M_1^T, \cdots, M_q^T)^T\|$. Then PM^N is nonsingular for $\Delta(N, \delta)$ sufficiently small. □

By applying P to (2.7c) we easily conclude, under the above hypothesis and using (4.8), (4.4c), (4.2) and (3.2b), that for $\Delta(N, \delta)$ sufficiently small:

$$\|\mathbf{c}^N - \mathbf{c}\| \leqq K_1(N, \delta)\Delta(N, \delta) \tag{4.9a}$$

where

$$K_1(N, \delta) \equiv \frac{\|(PM)^{-1}\|(1 + \|(PM)^{-1}\| \cdot \|P\mathbf{g}\|)K_0\|Y^{-1}(\delta)\|}{1 - [K_0\|(PM^{-1})\| + 1]\|Y^{-1}(\delta)\|\Delta(N, \delta)}. \tag{4.9b}$$

We recall from (4.5), (2.3a) and Theorem 3.4 that $\mathbf{y}(t)$ the unique solution of (1.1) and $\mathbf{y}^N(t)$ the unique solution of the truncated problem (4.7) satisfy

$$\mathbf{y}^N(t) - \mathbf{y}(t) = Y(t)[\mathbf{c}^N - \mathbf{c}], \qquad \delta \leqq t \leqq 1. \tag{4.10}$$

In actual applications we do not know M but only M^N. Thus we pick some P to assure PM^N is nonsingular for all $N \geqq N_0$, say. In our experience it has always been $Q = M_0(0) + B_1 Y(1)$, the first $n \times n$ submatrix of M, that is nonsingular.

5. The numerical solution and error estimates. When the singular problem (1.1) has a unique solution and (2.6) holds then we can determine a truncated regular two point problem (4.7) which also has a unique solution. There are many ways in which the solution, $\mathbf{y}^N(t)$, of this regular problem can be approximated. In particular we assume a stable accurate of order r difference scheme is used on a quasi-uniform net $\{t_j\}_0^J$ with $h = \max_j h_j$:

$$t_0 = \delta; \qquad t_j = t_{j-1} + h_j, \quad 1 \leqq j \leqq J; \qquad t_J = 1. \tag{5.1}$$

(With fine net spacing near t_0 we may also introduce netpoints in $0 < t < \delta$ and on $t > 1$ to employ deferred corrections in an efficient manner, if high order accuracy is required; see [8].) Denoting the numerical solution of (4.7) on the net (5.1) by $\mathbf{u}_j$ we have that:

$$\|\mathbf{y}^N(t_j) - \mathbf{u}_j\| \leqq K_2(N, \delta)h^r, \qquad 0 \leqq j \leqq J. \tag{5.2}$$

The theory which insures (5.2) is thoroughly developed; see for instance [9] for quite general difference schemes or [6] for the Box scheme employed in our calculations.

From (4.9), (4.10) and (5.2) we have that

(5.3) $\|\mathbf{y}(t_j)-\mathbf{u}_j\|\leqq\|Y(t_j)\|K_1(N,\delta)\Delta(N,\delta)+K_2(N,\delta)h^r, \qquad 0\leqq j\leqq J, \quad \delta\leqq t_j\leqq 1.$

The constant $K_1(N, \delta)$ is given in (4.9b) while $K_2(N, \delta)$ from (5.2) depends upon the magnitude of higher derivatives of $\mathbf{y}^N(t)$ as well as the stability constant of the difference scheme over $\delta\leqq t\leqq 1$ (see [6], [9] for more details). For *fixed* δ both $K_1(N, \delta)$ and $K_2(N, \delta)$ are uniformly bounded for all $N\geqq N_0(\delta)$, some sufficiently large integer. In most singular problems we can estimate $\Delta(N, \delta)$ by

(5.4) $$\Delta(N,\delta)=O(\delta^{N-q})$$

for some integer q. This can easily be done when the only singularities are those in $Y(t)$, see Brabston [1]. When (5.4) holds we need only take $N-q=r$ to be compatible with any fixed difference scheme of accuracy h^r when h and δ are comparable. However, for δ too small $K_1(N, \delta)$ of (4.9b) may degrade the accuracy if $\|Y^{-1}(\delta)\|\Delta(N, \delta)$ is too large. Also for $\delta=O(h)$ the stability constants of the difference scheme may cause $K_2(N, \delta)$ to become unbounded. So this is not always a sure way to estimate the least value to use for N. But in general it is reasonable to take $N>q+r$ in actual calculations. We see some of these effects in our calculations of § 6.

An approximation to the solution $\mathbf{y}(t)$ of (1.1) on $0<t<\delta$ can be obtained from the finite difference solution by defining:

(5.5a) $$\mathbf{c}_h^N\equiv[Y^N(\delta)]^{-1}(\mathbf{u}_0-\mathbf{y}_p^N(\delta)),$$

(5.5b) $$\mathbf{u}(t)\equiv Y^N(t)\mathbf{c}_h^N+\mathbf{y}_p^N(t), \qquad 0<t\leqq\delta.$$

Here $\mathbf{c}_h^N$ is defined in analogy with $\mathbf{c}$ of (3.1) from the exact solution (2.3a). Thus we get, recalling (4.2) and (5.2) with $t_0=\delta$:

(5.6a) $$\|\mathbf{y}(t)-\mathbf{u}(t)\|\equiv\|\mathbf{e}(t)\|\leqq\|Y(t)(\mathbf{c}-\mathbf{c}_h^N)\|+K_3(N,\delta)\Delta(N,\delta),$$

(5.6b) $$\|\mathbf{c}-\mathbf{c}_h^N\|\leqq K_4(N,\delta)\Delta(N,\delta)+K_5(N,\delta)h^r; \qquad 0<t\leqq\delta.$$

If any row of $Y(t)$ remains bounded as $t\to 0$ the corresponding component of the error can be made arbitrarily small at $t=0$. Otherwise our error bounds blow up as $t\to 0$. This should not be unexpected, however, as in our formulation we have allowed the exact solution to have singularities at $t=0$. The relative error, say as measured by $\|\mathbf{y}(t)-\mathbf{u}(t)\|/\|\mathbf{y}(t)\|$ as $t\to 0$, does remain bounded and can be made arbitrarily small on $[0, \delta]$ as $N\to\infty$ and $h\to 0$.

6. Numerical examples. We have computed approximations to the solutions of (2.8a) with σ values and boundary conditions given by:

(6.1a) $$\sigma=\tfrac{1}{2}, \quad \lim_{t\downarrow 0} v(t)=1, \quad v(1)=2;$$

(6.1b) $$\sigma=\tfrac{3}{2}, \quad \lim_{t\downarrow 0} v(t)=1, \quad v(1)=\cos 1.$$

By our analysis in § 2 it follows that a unique solution exists in each case. These are easily found to be:

(6.2a) $$v(t)=1+(1-\cos 1)t^{1/2}+t^{1/2}\cos t,$$

(6.2b) $$v(t)=1-t^{-1/2}+t^{-1/2}\cos t.$$

Note that the derivatives of the solutions in both cases are singular at $t=0$.

With the use of the formulation (2.8b) our boundary conditions become:

$$\lim_{t\downarrow 0}\begin{pmatrix}1 & 0\\ 0 & 0\end{pmatrix}\mathbf{y}(t)+\begin{pmatrix}0 & 0\\ 1 & 0\end{pmatrix}\mathbf{y}(1)=\mathbf{b}$$

where: $\mathbf{b}^T=(1,2)$ for $\sigma=\frac{1}{2}$ and $\mathbf{b}^T=(1,\cos 1)$ for $\sigma=\frac{3}{2}$.

Trivial Taylor expansions in (2.9) yield $\mathbf{y}_p^N(t)$ and $Y^N(t)$ and we find that

$$\Delta(N,\delta)=\begin{cases}\dfrac{1}{1-\sigma}\delta^{1-\sigma}, & N=0,\\[2ex] \dfrac{2N+5-\sigma}{(2N+4)!}\delta^{2N+3-\sigma}, & N\geqq 1.\end{cases}\tag{6.3}$$

For $\sigma>1$ we must take $N\geqq(\sigma-3)/2$ and $N\geqq 1$ in order to include all the singularities as required to insure (4.2). Proceeding as in § 4 we obtain the truncated reduced boundary conditions as in (4.4b, c). For $\sigma=\frac{1}{2}$ these become:

$$B_{0\delta}^N\equiv\begin{pmatrix}1 & -2\delta\\ 0 & 0\end{pmatrix},\quad B_{1\delta}\equiv\begin{pmatrix}0 & 0\\ 1 & 0\end{pmatrix},$$
$$\mathbf{b}_\delta^N\equiv\begin{pmatrix}1-2\sum_{k=0}^{N}\dfrac{(-1)^k}{(2k+1)!}\delta^{2k+5/2}\\ 2\end{pmatrix}\tag{6.4a}$$

and for $\sigma=\frac{3}{2}$ they yield,.upon dropping a pair of redundant conditions:

$$B_{0\delta}^N\equiv\begin{pmatrix}1 & 0\\ 0 & 0\end{pmatrix},\quad B_{1\delta}\equiv\begin{pmatrix}0 & 0\\ 1 & 0\end{pmatrix},$$
$$\mathbf{b}_\delta^N\equiv\begin{pmatrix}1-\sum_{k=0}^{N}\dfrac{(-1)^k}{(2k+2)!}\delta^{2k+3/2}\\ \cos 1\end{pmatrix}.\tag{6.4b}$$

Note that the $B_{0\delta}^N$ are independent of N and hence turn out to be *exact* in this example; the $\mathbf{b}_\delta^N$ are not exact, however.

We use the Box scheme (or centered Euler) as presented in [6] to solve the system (2.8b) subject to (4.4b) using the quantities in (6.4). The net was chosen to be uniform with spacing $h=(1-\delta)/J$. The difference equations were solved on an IBM 370/158 in double precision using a stable block elimination procedure (not the one implied in [6] but rather case i) of eq. (5.7) in [7]). It takes approximately 476 milliseconds to solve one problem with $J=80$ intervals and the computing time is linear in J. The results thus computed are $O(h^2)$ accurate approximations. We also used one Richardson extrapolation to get $O(h^4)$ accurate approximations.

A particularly important and sensitive measure of the accuracy is $\|\mathbf{c}-\mathbf{c}_h^N\|$. In Table 1 this error is tabulated for both problems, for a sequence of refined nets both for the Box scheme (I) and for one extrapolation (II). We used $\delta=0{\cdot}1$ and $N=6$ in this series of calculations. The theoretical estimate (5.6b) is in part borne out by these results. In particular since here $\Delta(N,\delta)\approx O(10^{-14})$ the dominant

TABLE 1

The errors, $\|\mathbf{c}-\mathbf{c}_h^N\|$, for varying nets with $\delta = 0{\cdot}1$ and $N = 6$

	$\sigma = \frac{1}{2}$		$\sigma = \frac{3}{2}$	
J	I(h^2)	II(h^4)	I(h^2)	II(h^4)
10	.813 (.2)	.218 (−3)	—	—
20	.220 (−2)	.173 (−4)	.929 (−4)	.785 (−6)
40	.562 (−3)	.117 (−5)	.238 (−4)	.511 (−7)
60	.251 (−3)	.235 (−6)	.106 (−4)	.102 (−7)
80	.141 (−3)	.748 (−7)	.599 (−5)	.323 (−8)

term should be $O(h^r)$ for $r = 2$ or 4. Taking ratios as indicated we obtain the results in Table 2. The theoretically correct limiting ratios are 4 and 16. The unextrapolated calculations are relatively closer to the theory. This is probably caused by the $\Delta(N, \delta)$ term whose influence is greater for the more accurate $O(h^4)$ results.

TABLE 2

Ratios of errors $\|\mathbf{c}-\mathbf{c}_h^N\|$ for successively bisected nets with J_1 and J_2 intervals each

J_1/J_2	$\sigma = \frac{1}{2}$		$\sigma = \frac{3}{2}$	
10/20	3.70	12.60	—	—
20/40	3.91	14.79	3.90	15.36
40/80	3.99	15.64	3.97	15.82

To assess this term we vary N while keeping h fixed with $J = 80$ and δ fixed at 10^{-1} for the $\sigma = \frac{1}{2}$ case. The results using one Richardson extrapolation are as follows:

N:	0	1	2	3
$\|\mathbf{c}-\mathbf{c}_h^N\|$:	.134 (−4)	.812 (−7)	.748 (−7)	.748 (−7)

If the $O(h^r)$ term in (5.6b) is negligible, with $r = 4$ since we use extrapolation these values should vary as $\Delta(N, \delta)$. From (6.3) we see that this does occur in going from $N = 0$, $\Delta(0, \delta) = O(\delta^{1/2})$ to $N = 1$, $\Delta(1, \delta) = O(\delta^{9/2})$. After that the increase in N, with $\Delta(N, \delta) = O(\delta^{(4N+5)/2})$ is apparently masked by the truncation error which is $O(10^{-8})$ with $J = 80$.

The variation with δ is more difficult to check since the variation of the coefficients, $K_\nu(N, \delta)$, in (5.6b) is not known and may be quite sensitive. Using the Box scheme, with $O(h^2)$ accuracy, $\sigma = \frac{1}{2}$, $N = 6$ and $J = 80$ we find the following

δ:	5×10^{-2}	10^{-1}	2×10^{-1}	4×10^{-1}
$\|\mathbf{c}-\mathbf{c}_h^N\|$:	.636 (−3)	.141 (−3)	.258 (−4)	.279 (−5)

If, as is likely, the $\Delta(N, \delta)$ term is negligible compared to the $O(h^2)$ term then ratios of successive terms should be proportional to $[(1-\delta_\nu)/(1-\delta_{\nu+1})]^2$ since

$h_\nu = (1-\delta_\nu)/80$ in each case. This assumes that $K_4(N, \delta)$ does not vary much with δ. The ratio of errors gives:

$$4.59, \quad 5.47, \quad 9.25$$

while the squares of the ratio of net spacings gives:

$$1.114, \quad 1.266, \quad 1.778.$$

Thus there is no satisfactory agreement here and we assume that the coefficient variations with δ cannot be neglected.

TABLE 3
The errors in $v(t)$ and $v'(t)$ over $0<t<\delta=10^{-1}$

	$\sigma=\frac{1}{2}$		$\sigma=\frac{3}{2}$	
t	$\|e_1(t)\|$	$\|e_2(t)\|$	$\|e_1(t)\|$	$\|e_2(t)\|$
0.100	.894 (−4)	.447 (−3)	.139 (−15)	.300 (−4)
0.075	.775 (−4)	.515 (−3)	.915 (−6)	.459 (−4)
0.051	.635 (−4)	.629 (−3)	.244 (−5)	.835 (−4)
0.016	.356(−4)	.112(−2)	.906(−5)	.475(−3)
0.006	.218 (−4)	.183 (−2)	.186 (−4)	.206 (−2)
0.001	.894 (−5)	.447 (−2)	.539 (−4)	.300 (−1)

Finally we examine the error over $0<t\leqq\delta$ when (5.5) is used. The results are shown for both problems with $\delta=10^{-1}$, $N=6$ and $J=80$ in Table 3. Here $|e_1(t)|$ and $|e_2(t)|$ are the absolute errors in approximating $v(t)$ and $v'(t)$, respectively. From (2.9a) we see that the first row of $Y(t)$ is nonsingular for $\sigma=\frac{1}{2}$ but it is singular for $\sigma=\frac{3}{2}$. The second row is singular in both cases. Then the error estimate in (5.6a) indicates that $|e_1(t)|$ should remain bounded as $t\to 0$ in the $\sigma=\frac{1}{2}$ case and may become unbounded for $\sigma=\frac{3}{2}$. The error $|e_2(t)|$ may become unbounded in both cases. Specifically as $t\downarrow 0$, $|e_1(t)|$ decays like $t^{1/2}$ and $|e_2(t)|$ grows like $t^{-1/2}$ for $\sigma=\frac{1}{2}$. For $\sigma=\frac{3}{2}$, $|e_1(t)|$ grows like $t^{-1/2}$ (but with some specious values) and $|e_2(t)|$ grows like $t^{-3/2}$. This is consistent with (5.6a) and (2.9a).

REFERENCES

[1] D. BRABSTON, *Numerical solution of singular endpoint boundary value problems*, part II of Ph.D. thesis, Calif. Inst. of Tech., Pasadena, 1974.

[2] E. CODDINGTON AND N. LEVINSON, *Ordinary Differential Equations*, McGraw-Hill, New York, 1955.

[3] B. GUSTAFSSON, *A numerical method for solving singular boundary value problems*, Numer. Math., 21 (1973), pp. 328–344.

[4] F. DE HOOG AND R. WEISS, *Difference methods for boundary value problems with a singularity of the first kind*, MRC Tech. Summary Rep. 1536, Math. Res. Center, Univ. of Wisc., Madison, 1975.

[5] P. JAMET, *On the convergence of finite difference approximation to one-dimensional singular boundary-value problems*, Numer. Math., 14 (1970), pp. 355–378.

[6] H. B. KELLER, *Accurate difference methods for linear ordinary differential systems subject to linear constraints*, this Journal, 6 (1969), pp. 8–30.

[7] ———, *Accurate difference methods for nonlinear two-point boundary value problems*, this Journal, 11 (1974), pp. 305–320.

[8] H. B. KELLER AND V. PEREYRA, *Deferred corrections for two-point boundary value problems*, in preparation.

[9] H. B. KELLER AND A. WHITE, *Difference methods for boundary value problems in ordinary differential equations*, this Journal, 12 (1975), pp. 791–802.

[10] F. NATTERER, *A generalized spline method for singular boundary value problems of ordinary differential equations*, Linear Algebra Appl., 7 (1973), pp. 189–216.

[11] ———, *Das Differenzenverfahren für singuläre Rand-Eigenwertaufgaben gewöhnlicher Differentialgleichungen*, Numer. Math., 23 (1975), pp. 387–409.

[12] R. D. RUSSELL AND L. F. SHAMPINE, *Numerical methods for singular boundary value problems*, this Journal, 12 (1975), pp. 13–36.

Numerical Solution of Bifurcation and Nonlinear Eigenvalue Problems

Herbert B. Keller

1. Introduction.

We show in this paper how a large class of bifurcation and nonlinear eigenvalue problems can be solved and the numerical methods justified. Only equilibrium problems are treated here, say in the general form

$$G(u, \lambda) = 0 \,, \tag{1.1}$$

where $G \;:\; \mathbf{B} \times \mathbf{R} \to \mathbf{B}$ for some Banach space, $\mathbf{B}$. We find it most instructive to refer occasionally to the canonical example of the matrix eigenvalue problem:

$$Au - \lambda u = 0 \,, \tag{1.2}$$

where A is an n×n real matrix with real eigenvalues.

By a smooth branch or arc of solutions

$$\Gamma_{ab} : [u(s), \lambda(s)] \quad , \qquad s_a \leq s \leq s_b \,, \tag{1.3}$$

we mean a one-parameter family of solutions of (1.1), $u(s) \in \mathbf{B}$, $\lambda(s) \in \mathbf{R}$ depending twice continuously differentiably on some parameter $s \in [s_a, s_b]$. Of course, the parameter, s, is quite arbitrary on each such branch and this fact is crucial in our study. In figures 1a and 1b we sketch, respectively, typical solution branches for problems (1.1) and (1.2). The "points" at λ_A and λ_B represent "simple" bifurcation, while λ_D is a multiple bifurcation point. The "arcs" Γ_j in figure 1b for $j \neq 0$ represent the eigenspaces for A belonging to the eigenvalues λ_j while Γ_0 is the trivial solution. For simple eigenvalues the bifurcation is simple (*i.e.*, a one parameter family of solutions branches off). At $\lambda = \lambda_C$ on branch Γ_1 in figure 1a we have what is called a "limit point" in the applied mechanics literature. These present no difficulties analytically or computationally in our current theory, although they seem to have been quite troublesome in the past [1,12,20]. Note that all the branches $\Gamma_1, \Gamma_2, \ldots$ in the eigenvalue problem are composed entirely of these limit points.

We assume that the basic problem is to compute large segments of solution branches of (1.1) including all branches bifurcating from each segment.

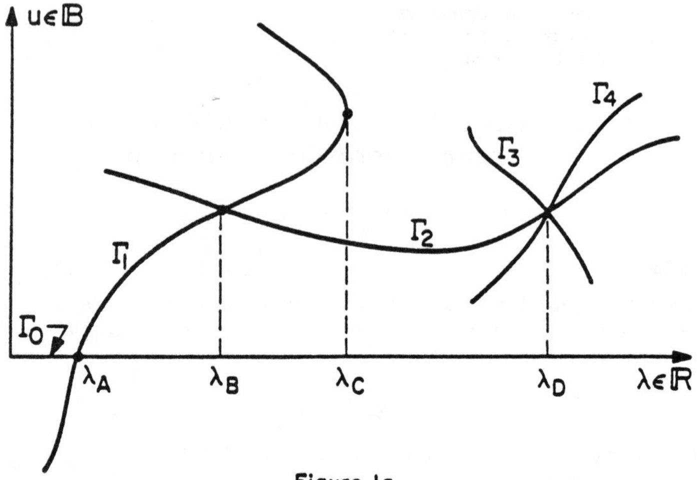

Figure 1a.

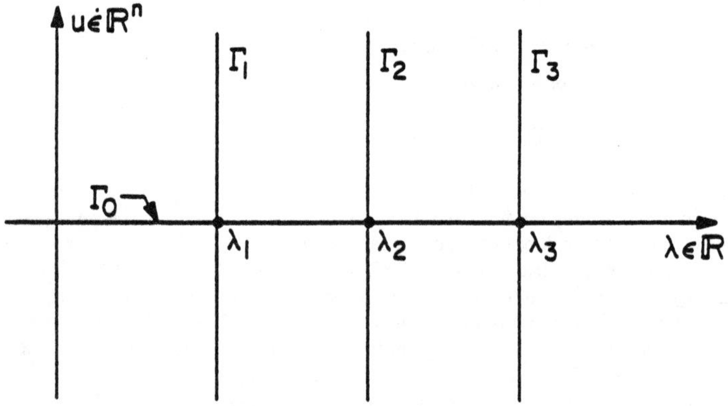

Figure 1b.

2. Parametrization and Continuation of Solution Arcs.

The standard approach is almost invariably to use λ, one of the naturally occurring parameters of the problem, as the parameter defining solution arcs, $u(\lambda)$. Indeed, if for some $\lambda = \lambda_0$ a solution $u = u_0$ of (1.1) is isolated, that is

$$G_u^0 \equiv G_u(u_0, \lambda_0) \tag{2.1}$$

is nonsingular, and if $G(u, \lambda)$ is C_1 in some ρ_0-sphere about $[u_o, \lambda_0]$, then the implicit function theorem insures the existence of a unique smooth arc of solutions $u = u(\lambda)$ for $|\lambda - \lambda_0| < \rho_1$, say. Furthermore, with our assumed smoothness, it follows that $du(\lambda)/d\lambda$ exists and satisfies:

$$G_u(u(\lambda), \lambda)\frac{du}{d\lambda} = -G_\lambda(u(\lambda), \lambda)\,. \tag{2.2}$$

Many procedures are now available for extending or approximating the solution branch through $[u_0, \lambda_0]$. The implicit function theorem suggests contraction mapping techniques while (2.2) suggests predictor-corrector continuation. A nice survey of the latter ideas (for finite dimensional problems) is given by Rheinboldt [18]. In particular, we have frequently used [3] one step of Euler's method in (2.2) as a predictor:

$$u^0(\lambda + \delta\lambda) = u(\lambda) + \delta\lambda\frac{du(\lambda)}{d\lambda} \tag{2.3a}$$

to supply the initial iterate for Newton's method to solve (1.1) at $\lambda + \delta\lambda$:

$$G_u^\nu \delta u^\nu(\lambda + \delta\lambda) = -G^\nu\ ,\ \nu = 0, 1, \dots\ ; \tag{2.3b}$$

where:

$$\begin{cases} G_u^\nu \equiv G_u(u^\nu(\lambda + \delta\lambda), \lambda + \delta\lambda)\ ,\ G^\nu \equiv G(u^\nu(\lambda + \delta\lambda),\ \lambda + \delta\lambda)\ , \\ u^{\nu+1}(\lambda + \delta\lambda) = u^\nu(\lambda + \delta\lambda) + \delta u^n u(\lambda + \delta\lambda)\,. \end{cases} \tag{2.3c}$$

It is not difficult to base existence proofs on such techniques provided the arc consists of isolated solutions.

All of the indicated continuation procedures may fail or encounter difficulties as a nonisolated solution is approached; that is a point $[u_0, \lambda_0]$ where G_u^0 is singular. We also call these singular points. As we shall see in §3, the above indicated continuation procedures could skip over some singular points but not over limit points as at $\lambda = \lambda_C$ in figure 1a. Also at bifurcation points some special procedures are required to switch from one branch to another. A simple analysis shows that the scheme (2.3) is incapable of tracing out any nontrivial solutions of the eigenvalue problem (1.2)!

To circumvent these difficulties we recall that the parametrization of solution arcs is at our disposal. Thus we are free to impose some additional constraint or normalization on the solution and we do this, quite generally, by replacing (1.1) by:

$$\text{(2.4)} \qquad a)\ G(u,\lambda)=0\ , \qquad b)\ N(u,\lambda,s)=0\ .$$

Here $N : \mathbf{B}\times\mathbf{R}^2 \to \mathbf{R}$ and $s \in \mathbf{R}$ is the independent parameter on the solution arc. We shall show several choices for N which make s an approximation to "arclength" on the solution branch. Then, as we shall see, limit points essentially disappear, it is easy to jump over singular points or to compute them, relatively large steps in s can be taken and it is easy to switch branches at bifurcation points. Note that we are not simply changing the parameter in the problem as was first done in [12] or in another way in [1].

By introducing $x \in \mathbf{X} \equiv \mathbf{B}\times\mathbf{R}$ and $P : \mathbf{X}\times\mathbf{R} \to \mathbf{X}$ as

$$\text{(2.5a)} \qquad x \equiv [u,\lambda]\ ,\ P(x,s) \equiv \begin{pmatrix} G(u,\lambda) \\ N(u,\lambda,s) \end{pmatrix}$$

a solution arc of (1.1) or (2.4) is $x(s) \equiv [u(s),\lambda(s)]$ and it satisfies

$$\text{(2.5b)} \qquad P(x(s),s)=0\ .$$

For fixed s a solution $x(s)$ is isolated if

$$\text{(2.6)} \qquad \mathcal{A}(s) \equiv P_x(x(s),s) = \begin{pmatrix} G_u(u(s),\lambda(s)) & G_\lambda(u(s),\lambda(s)) \\ N_u(u(s),\lambda(s),s) & N_\lambda(u(s),\lambda(s),s) \end{pmatrix}$$

is nonsingular. Furthermore, on a smooth arc $\dot{x}(s) \equiv dx(s)/ds$ satisfies

$$\text{(2.7)} \qquad \mathcal{A}(s)\dot{x}(s) = -\begin{pmatrix} 0 \\ N_s(u(s),\lambda(s),s) \end{pmatrix}\ .$$

Now continuation in s could proceed in exact analogy with our prior continuation in λ. However, the possible advantages of our reformulations are clarified by the fact that P_x can be nonsingular while G_u is singular. Indeed, this is but a special case of the basic:

Lemma 2.8. *Let* $\mathbf{B}$ *be a Banach space and consider the linear operator* $\mathcal{A} : \mathbf{B} \times \mathbf{R}^\nu \to \mathbf{B} \times \mathbf{R}^\nu$ *of the form:*

$$\mathcal{A} \equiv \begin{pmatrix} A & B \\ C^* & D \end{pmatrix} \qquad \textit{where} \qquad \begin{cases} A : \mathbf{B} \to \mathbf{B} \,, \ B : \mathbf{R}^\nu \to \mathbf{B} \,; \\ C^* : \mathbf{B} \to \mathbf{R}^\nu \,, \ D : \mathbf{R}^\nu \to \mathbf{R}^\nu \,. \end{cases}$$

i) If A *is nonsingular, then* $\mathcal{A}$ *is nonsingular iff:*

$$D - C^* A^{-1} B \qquad \textit{is nonsingular} \,. \tag{2.8a}$$

ii) If A *is singular and*

$$\dim \mathcal{N}(A) = \operatorname{codim} \mathcal{R}(A) = \nu \,, \tag{2.8b}$$

then $\mathcal{A}$ *is nonsingular iff:*

$$\begin{aligned} &c_0) \ \dim \mathcal{R}(B) = \nu \quad , \quad c_1) \ \mathcal{R}(B) \cap \mathcal{R}(A) = 0 \,, \\ &c_2) \ \dim \mathcal{R}(C^*) = \nu \quad , \quad c_3) \ \mathcal{N}(A) \cap \mathcal{N}(C^*) = 0 \,. \end{aligned} \tag{2.8}$$

iii) If A *is singular and* $\dim \mathcal{N}(A) > \nu$*, then* $\mathcal{A}$ *is singular.*

Proof. Not difficult to work out; it will appear elsewhere [10]. ♣

Using Lemma 2.8 and our reformulation of (1.1) it is easy to prove a variety of bifurcation theorems which we do not report here. We only use in the present work the special case of Lemma 2.8 with $\nu = 1$. Then, given (2.8b), the conditions in (2.8c) simply reduce to

$$B \notin \mathcal{R}(A) \quad \text{and} \quad C^* \notin \mathcal{R}(A^*) \,. \tag{2.9}$$

As a simple example of the use of our procedure, consider the eigenvalue problem (1.2) subject to either of the normalizations

$$a) \ \ N_0(u, \lambda, s) \equiv \lambda - s \,; \qquad b) \ \ N_1(u, \lambda, s) \equiv \|u\|_2^2 - s^2 \,. \tag{2.10}$$

Using $N_0 = 0$ the predictor-corrector continuation scheme generates the trivial solution branch, $u = 0$, $\lambda =$ arbitrary, starting from any point on this branch. Alternatively, using $N_1 = 0$, our scheme traces out the one-dimensional eigenspace belonging to any simple eigenvalue $\lambda - \lambda_0$. Switching over from one of these normalizations to the other at a simple bifurcation point $(u, \lambda) = (0, \lambda_0)$ allows us to trace out the two branches through this point. Indeed, this is the key to our method for switching branches at bifurcation points in the general case, see §5.

A somewhat natural parametrization of a solution branch $[u(s), \lambda(s)]$ is to use for s a form of arclength. That is, for some $\theta \in (0,1)$:

$$N_\theta(u,\lambda,s) \equiv \theta||\dot{u}(s)||^2 + (1-\theta)|\dot{\lambda}(s)|^2 - 1 = 0 . \tag{2.10c}$$

This form is not the most practical one to use, even for merely proving existence theorems and so we use approximations to it. Assuming a solution of (1.1) known, say, $[u,\lambda] = [u_0,\lambda_0]$, we set $[u_0,\lambda_0] = [u(s_0),\lambda(s_0)]$ and define over $s_0 \leq s < s_1$:

$$\begin{aligned} N_2(u,\lambda,s) \equiv \theta||u(s) - u(s_0)||^2 + (1-\theta)|\lambda(s) - \lambda(s_0)|^2 \\ -(s-s_0)^2 = 0 . \end{aligned} \tag{2.10d}$$

Alternatively, if in addition to $[u_0,\lambda_0]$ we know $[\dot{u}_0,\dot{\lambda}_0]$ satisfying (2.10c) at $s = s_0$, then we can use on $s_0 \leq s < s_1$:

$$N_3(u,\lambda,s) \equiv \theta\dot{u}^*(s_0)[u(s) - u(s_0)] + (1-\theta)\dot{\lambda}(s_0)[\lambda(s) - \lambda(s_0)] - (s-s_0) . \tag{2.10e}$$

Here $\dot{u}^*(s_0) \in \mathbf{B}^*$ is the dual element to $\dot{u}(s_0)$ such that $\dot{u}^*(s_0)\dot{u}(s_0) = ||\dot{u}(s_0)||^2$. (The existence of such an element is assured by the Hahn-Banach theorem.) We call N_2 or N_3 pseudo-arclength normalizations and examine some of their properties in §3. Previous attempts to use arclength as a parameter in solving nonlinear algebraic systems have been made in [13,14].

3. Continuation About Regular and Limit Points.

We shall justify continuation procedures using the normalization N_3 on solution arcs composed of "regular points" or "normal limit points". Specifically, let $[u_0,\lambda_0]$ be a solution of (1.1) and let $[\dot{u}_0,\dot{\lambda}_0]$ satisfy

$$a)\quad G_u^0\dot{u}_0 + G_\lambda^0\dot{\lambda}_0 = 0 , \qquad b)\quad ||\dot{u}_0||^2 + |\dot{\lambda}_0|^2 > 0 . \tag{3.0}$$

Then we say that $[u_0,\lambda_0]$ is a **regular solution (point)** if in addition:

$$G_u^0 \equiv G_u(u_0,\lambda_0) \qquad \text{is nonsingular} . \tag{3.1}$$

We call $[u_0,\lambda_0]$ a **normal limit solution (point)** if (3.0) holds but in place of (3.1) we have:

$$a)\quad \dim\mathcal{N}(G_u^0) = \operatorname{codim}\mathcal{R}(G_u^0) = 1 , \qquad b)\quad G_\lambda^0 \notin \mathcal{R}(G_u^0) . \tag{3.2}$$

Theorem 3.3. *Let $[u_0,\lambda_0]$ be either a regular solution or a normal limit solution. Let $G(u,\lambda)$ have two continuous derivatives in some sphere about $[u_0,\lambda_0]$. Then with*

$[u(s_0), \lambda(s_0)] \equiv [u_0, \lambda_0]$, $[\dot{u}(s_0), \dot{\lambda}(s_0)] \equiv [\dot{u}_0, \dot{\lambda}_0]$ and $\dot{u}^*(s_0)$ as defined after (2.10e) there exists a unique smooth arc of solutions $[u(s), \lambda(s)]$ of (2.3) using $N \equiv N_3$ on $|s - s_0| \le \rho$ for some sufficiently small $\rho > 0$. On this solution arc the Fréchet derivative $\mathcal{A}(s)$ of (2.6) is nonsingular.

Proof. All of these results will follow from the implicit function theorem applied to (2.4) at $[u, \lambda, s] = [u_0, \lambda_0, s_0]$ if $\mathcal{A}(s_0)$ is nonsingular. We first consider $[u_0, \lambda_0]$ to be a normal limit solution. By (3.2b) in (3.0a) it then follows that $\dot{\lambda}_0 = 0$ and hence $\dot{u}_0 \in \mathcal{N}(G_u^0)$. Now (3.2a) implies since $\dot{u}_0^* \dot{u}_0 \neq 0$ that $\dot{u}_0^* \notin \mathcal{R}(G_u^{0*})$. This result and (3.2) now yield with $\nu = 1$ in Part ii) of Lemma 2.8 that

$$\mathcal{A}(s_0) = \begin{pmatrix} G_u^0 & G_\lambda^0 \\ \theta \dot{u}_0^* & (1-\theta)\dot{\lambda}_0 \end{pmatrix}$$

is nonsingular [use the form in (2.9)].

Next, let $[u_0, \lambda_0]$ be a regular solution. If $\dot{\lambda}_0 = 0$, then by (3.0a) and (3.1), $\dot{u}_0 = 0$. This contradicts (3.0b) so $\dot{\lambda}_0 \neq 0$ at a regular point. Then (3.0a) and (3.1) imply

$$\dot{u}_0 / \dot{\lambda}_0 = -(G_u^0)^{-1} G_\lambda^0 \,.$$

Now, by Part i) of Lemma 2.8, $\mathcal{A}(s_0)$ above is nonsingular iff

$$(1-\theta)\dot{\lambda}_0 - \theta \dot{u}_0^* (G_u^0)^{-1} G_\lambda^0 \neq 0 \,.$$

That is, using the above, iff

$$[\theta \dot{u}_0^* \dot{u}_0 + (1-\theta)\dot{\lambda}_0^2]/\dot{\lambda}_0 = [\theta ||\dot{u}_0||^2 + (1-\theta)|\dot{\lambda}_0|^2]/\dot{\lambda}_0 \neq 0 \,.$$

Since$\dot{\lambda}_0 \neq 0$ and $||\dot{u}_0|| \neq 0$, it follows that $\mathcal{A}(s_0)$ is nonsingular. ♣

Clearly any smooth branch of solutions composed of regular points or normal limit points can be determined using, say, Euler-Newton continuation on (2.4) with the normalization $N \equiv N_3$. We could easily justify the normalization $N = N_2$ for smooth arcs, since on them:

$$u(s) - u(s_0) = \dot{u}(s_0)(s - s_0) + \mathcal{O}(|s - s_0|^2) \,,$$
$$\lambda(s) - \lambda(s_0) = \dot{\lambda}(s_0)(s - s_0) + \mathcal{O}(|s - s_0|^2) \,.$$

When using (2.10e) over a sequence of intervals $[s_0, s_1], [s_1, s_2], \ldots$, it is a good policy to impose the arclength condition

$$\theta ||\dot{u}(s)||^2 + (1-\theta)|\dot{\lambda}(s)|^2 = 1$$

occasionally, say at each joint, $s = s_k$. The resulting arc is then only piecewise smooth in s, with jump discontinuities in the length of the tangent vector $[\dot{u}(s), \dot{\lambda}(s)]$ at $s = s_k$.

Specifically, if $[\dot{u}(s_{\bar{k}}), \dot{\lambda}(s_{\bar{k}})]$ is the limit as $s \uparrow s_k$, then we use on $[s_k, s_{k+1}]$

$$[\dot{u}(s_k), \dot{\lambda}(s_k)] = c[\dot{u}(s_{\bar{k}}), \dot{\lambda}(s_{\bar{k}})] ,$$
$$c^{-2} = \theta||\dot{u}(s_{\bar{k}})||^2 + (1-\theta)|\dot{\lambda}(s_{\bar{k}})|^2 .$$

This renormalization allows more uniform steps in s to be taken during the continuation process.

4. Continuation Past Singular Points.

A solution $x(s) = [u(s), \lambda(s)]$ of (2.4) is said to be **singular** or a **singular point** if $\mathcal{A}(s)$ in (2.6) is singular. We will consider smooth arcs of solutions $x(s)$ for $s_a \le s \le s_b$ on which only $x(s_0)$ for some $s_0 \in (s_a, s_b)$ is singular. Under mild smoothness conditions, various continuation procedures can "jump" over the singular point in going from s_a to s_b. A simple way to do this is by using Euler-chord or Euler-Newton continuation, as we proceed to show.

We assume, as stated above, that

$$\text{(4.0)} \qquad \mathcal{A}(s) \equiv P_x(x(s), s) \quad \text{is nonsingular for} \quad s \in [s_a, s_b] - \{s_0\} .$$

Then the tangent, $\dot{x}(s_a)$, is uniquely defined by

$$\text{(4.1}a\text{)} \qquad \mathcal{A}(s_a)\dot{x}(s_a) = -P_s(x(s_a), s_a) ,$$

and an approximation to $x(s)$ is

$$\text{(4.1}b\text{)} \qquad x^0(s) \equiv x(s_a) + [s - s_a]\dot{x}(s_a) .$$

Using this approximation we consider the chord (or special Newton) method:

$$\text{(4.2)} \quad \begin{array}{ll} a) & \mathcal{A}^0(s) \equiv P_x(x^0(s), s) \\ b) & \mathcal{A}^0(s)\left[x^{\nu+1}(s) - x^\nu(s)\right] = -P(x^\nu(s), s) , \ \nu = 0, 1, \dots . \end{array}$$

We could also try Newton's method, with $y^0(s) \equiv x^0(s)$:

$$\text{(4.3)} \quad \left.\begin{array}{ll} a) & \mathcal{A}^\nu(s) \equiv P_x(y^\nu(s), s) \\ b) & \mathcal{A}^\nu(s)\left[y^{\nu+1}(s) - y^\nu(s)\right] = -P(y^\nu(s), s) \end{array}\right\} \ \nu = 0, 1, \dots .$$

To get convergence we need only show that $x^0(s)$ is in the appropriate domain of attraction about $x(s)$ and that the $\mathcal{A}^\nu(s)$ are all nonsingular. We do this in

Theorem 4.4. *Let $x(s)$ be a twice continuously differentiable arc of solutions of (2.5) on $[s_a, s_b]$. Let (4.0) hold and for some positive functions $K(s)$, $\kappa(s)$ and $\rho(s)$ defined on $[s_a, s_b]$:*

$$(4.4) \quad \begin{aligned} &a) \quad \|P_x(y,s) - P_x(x(s),s)\| \leq K(s)\|y - x(s)\| \quad \forall\, y \text{ in} \|y - x(s)\| \leq \rho(s)\,; \\ &b) \quad \max_{s_a \leq t \leq s}\|\ddot{x}(t)\| \leq \kappa(s)\,. \end{aligned}$$

For $s \in [s_a, s_b] - \{s_0\}$ define

$$(4.4c) \qquad M(s) \equiv \|\mathcal{A}^{-1}(s)\|\,,$$

and for some positive $\theta(s) < 1/3$ define

$$(4.4d) \qquad r(s) \equiv \min\left[\rho(s), \frac{\theta(s)}{M(s)K(s)}\right]\,.$$

Then if

$$(4.4e) \qquad |s - s_a|^2\kappa(s) \leq 2r(s)\,,\ s \neq s_0\,,$$

the chord iterates $x^\nu(s) \to x(s)$ with geometric convergence factor

$$\frac{2\theta(s)}{1 - \theta(s)}\,.$$

Proof. It easily follows from (4.1) and (4.4) that

$$\|x^0(s) - x(s)\| \leq \tfrac{1}{2}|s - s_a|^2\kappa(s)\,.$$

Then (4.4d,e) imply that $\|x^0(s) - x(s)\| \leq r(s)$. Now (4.4a,c) yield

$$\|\mathcal{A}^{-1}(s)\left[\mathcal{A}^0(s) - \mathcal{A}(s)\right]\| \leq M(s)K(s)r(s) \leq \theta(s)\,.$$

Thus the Banach Lemma insures that $\mathcal{A}^0(s)$ is nonsingular with

$$\|\mathcal{A}^0(s)^{-1}\| \leq \frac{M(s)}{1 - \theta(s)}\,.$$

We can now define

$$H(y,s) \equiv y - \mathcal{A}^0(s)^{-1}P(y,s)$$

and for all y, z in $\|y - x(s)\| \leq r(s)$:

$$\begin{aligned}\|H(y,s) - H(z,s)\| &= \|\mathcal{A}^0(s)^{-1}\left[\mathcal{A}^0(s)(y-z) - (P(y,s) - P(z,s))\right] , \\ &\leq \frac{M(s)}{1-\theta(s)} K(s) 2r(s)\|y-z\| , \\ &\leq \frac{2\theta(s)}{1-\theta(s)}\|y-z\| .\end{aligned}$$

Since $\theta(s) < 1/3$, we get that $H(y,s)$ is contracting on $\|y - x(s)\| \leq r(s)$ and the theorem follows. ♣

Note that $M(s)$ must become unbounded and thus $r(s) \to 0$ as $s \to s_0$. Our result thus uses the fact (see figure 2) that there is a cone about $x(s)$ with vertex at $x(s_0)$ and interior to this cone the chord method converges. To jump over a singular point, the tangent to $x(s)$ at $x(s_a)$ need only penetrate the cone for some $s > s_0$ for which (4.4e) holds. Clearly, if the curvature of the solution arc is not too great over $[s_a, s_b]$ this can be achieved.

Newton's method can also be shown to converge in the same cone. However, Newton's method may even converge in a much larger region, including a cylindrical tube about $x(s)$. Thus the singular solution $x(s_0)$ can be determined directly in such cases not only by bisection. Unfortunately, the status of the convergence of Newton's method at singular points is not completely clear at the present time. The main idea and, no doubt, the behavior to be expected in many cases is explained in the basic paper of Rall [16]. But the details of a proof with reasonable sufficient conditions seem to be lacking. Progress in this direction has recently been made by Reddien [17].

5. Switching Branches at Bifurcation Points.

Bifurcation points are solutions at which two or more smooth branches of solutions of (1.1) have non-tangential intersections. In particular, they are singular points, say $[u_0, \lambda_0]$, at which:

$$\text{(5.0)} \qquad \begin{aligned} a) \qquad & \dim \mathcal{N}(G_u^0) = \operatorname{codim} \mathcal{R}(G_u^0) = m , \\ b) \qquad & G_\lambda^0 \in \mathcal{R}(G_u^0) . \end{aligned}$$

From (5.0a) we have the existence of elements $\phi_j \in \mathbf{B}$ and $\psi_j^* \in \mathbf{B}^*$ such that:

$$\text{(5.1)} \qquad \left.\begin{aligned} \mathcal{N}(G_u^0) &= \text{span}\{\phi_1, \phi_2, \ldots, \phi_m\} \\ \mathcal{N}(G_u^{0*}) &= \text{span}\{\psi_1^*, \psi_2^*, \ldots, \psi_m^*\} \end{aligned}\right\} \psi_i^*\phi_j = \delta_{ij} \,;\, i,j = 1,2,\ldots,m .$$

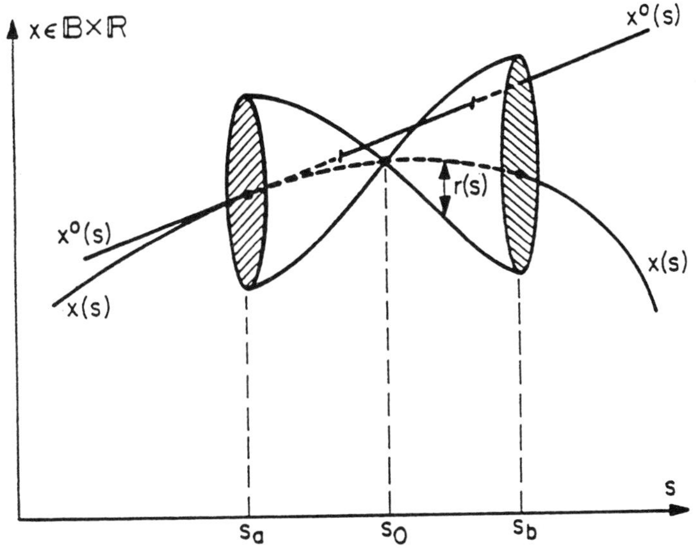

Figure 2

In addition, (5.0b) implies the existence of a unique element $\phi_0 \in \mathbf{B}$ such that:

(5.2) $$G_u^0\phi_0 + G_\lambda^0 = 0\ ;\ \psi_j^*\phi_0 = 0\ ,\ 1 \le j \le m\ .$$

Let $[u(s), \lambda(s)]$ be any smooth branch of (1.1) through the bifurcation point, say with $u(s_0) = u_0$, $\lambda(s_0) = \lambda_0$. Then, since

(5.3a) $$G_u^0\dot{u}(s_0) + G_\lambda^0\dot{\lambda}(s_0) = 0$$

it follows from (5.1)-(5.2) that

(5.3b) $$\dot{u}(s_0) = \sum_{j=0}^{m} \alpha_j\phi_j$$

where

(5.3c) $$\alpha_0 = \dot{\lambda}(s_0)\ ;\ \alpha_j = \psi_j^*\dot{u}(s_0)\ ,\ 1 \le j \le m\ .$$

We get on differentiation of $G(u(s), \lambda(s)) = 0$ at $s = s_0$:

(5.4) $$G_u^0\ddot{u}_0 = -\left[G_{uu}^0\dot{u}(s_0)\dot{u}(s_0) + 2G_{u\lambda}^0\dot{u}(s_0)\dot{\lambda}(s_0) + G_{\lambda\lambda}^0\dot{\lambda}(s_0)\dot{\lambda}(s_0)\right] - G_\lambda^0\ddot{\lambda}(s_0)\ .$$

Since $G_u^0\ddot{u}(s_0) \in \mathcal{R}(G_u^0)$ and $G_\lambda^0 \in \mathcal{R}(G_u^0)$, the bracketed term on the right side of (5.4) is also in $\mathcal{R}(G_u^0)$. Then $\psi_i^*[\ldots] = 0$ must hold for this term and so using (5.3) it follows that the $m+1$ scalars $\alpha_0, \ldots, \alpha_m$ must satisfy the quadratic system:

(5.5a) $$\sum_{j=1}^{m}\sum_{k=1}^{m} a_{ijk}\alpha_j\alpha_k + 2\sum_{j=1}^{m} b_{ij}\alpha_j\alpha_0 + c_1\alpha_0^2\ ,\ 1 \le i \le m\ ;$$

where

(5.5b) $$a_{ijk} \equiv \psi_i^* G_{uu}^0\phi_j\phi_k\ ,\ b_{ij} \equiv \psi_i^*[G_{uu}^0\phi_0 + G_{u\lambda}^0]\phi_j\ ,\ 1 \le j,k \le m\ ;$$
$$c_i \equiv \psi_i^*(G_{uu}^0\phi_0\phi_0 + 2G_{u\lambda}^0\phi_0 + G_{\lambda\lambda}^0)\ .$$

Thus the tangent, $[\dot{u}(s_0), \dot{\lambda}(s_0)]$, to every smooth branch through the bifurcation point $[u_0, \lambda_0]$ must have the form (5.3b,c) and satisfy (5.5). Conversely, if at a solution $[u_0, \lambda_0]$

of (1.1) conditions (5.0) hold and (5.5) has $r \geq 2$ distinct nontrivial roots,[†] then $[u_0, \lambda_0]$ is a bifurcation point with at lease r smooth solution branches of (1.1) intersecting there. Essentially, this result is proven in [11]. In the important special case $m = 1$, the algebraic bifurcation equations (5.5) reduce to the single quadratic

$$a_{111}\alpha_1^2 + 2b_{11}\alpha_1\alpha_0 + c_1\alpha_0^2 = 0 \ . \tag{5.6a}$$

If $[\alpha_0, \alpha_1]$ is one nontrivial root of this quadratic, then the other root is distinct, provided

$$\begin{cases} a_{111}\alpha_1 + b_{11}\alpha_0 \neq 0 & \text{when} \quad \alpha_0 \neq 0 \ ; \text{ or} \\ b_{11} \neq 0 & \text{when} \quad \alpha_0 = 0 \ . \end{cases} \tag{5.6b}$$

If $[u_1(s), \lambda_1(s)]$ is a smooth branch of solutions through the bifurcation point with tangent $[\dot{u}_1(s_0), \dot{\lambda}_1(s_0)]$ determined by $[\alpha_0, \alpha_1]$ in (5.3b,c) with $m = 1$, then the condition (5.6b) can be written as:

$$\begin{cases} \psi_1^*[G_{uu}^0\dot{u}_1(s_0) + G_{u\lambda}^0\dot{\lambda}_1(s_0)]\phi_1 \neq 0 & \text{for} \quad \alpha_0 \neq 0 \ , \text{ or} \\ \psi_1^*[G_{uu}^0\phi_0 + G_{u\lambda}^0]\phi_1 \neq 0 & \text{for} \quad \alpha_0 = 0 \ . \end{cases} \tag{5.6c}$$

This is essentially the form of the bifurcation condition given by Crandall and Rabinowitz [5]. However, it is seldom pointed out that this condition insures the existence of two distinct roots of a quadratic;[††] Rheinboldt [18] makes this quite explicit for the class of problems he treats.

Method I. An obvious way to determine branches bifurcating at $[u_0, \lambda_0]$ is to determine several distinct roots of (5.5), use them in (5.3b,c) to construct several distinct tangent vectors $[\dot{u}_k(s_0), \dot{\lambda}_k(s_0)]$, $k = 1, 2, \ldots$, then use each of these tangents in $N_3 = 0$ of (2.10e) and proceed as previously indicated. For $m = 1$, simple bifurcation, similar devices have been used and suggested in analytical and perturbation studies [7,18,21]. Rheinboldt [18] uses an approximation to the second root in some of his numerical methods to predict a point on the "second" branch. This idea can be used quite generally to obtain approximations to the coefficients $\{a_{ijk}, b_{ij}, c_i\}$ in (5.5) if the ϕ_j and ψ_j^* are known (or sufficiently well approximated). Thus, we define:

† Since (5.5a) is homogeneous, a root is determined only to within a multiplicative constant. Roots are said to be distinct if they are not scalar multiples of each other.

†† In bifurcation from the trivial solution $[u_1(s), \lambda_1(s)] = [0, s]$, it follows that $\phi_0 \equiv 0$, since $G_\lambda^0 = 0$. Hence, $c_1 \equiv 0$ and the quadratic (5.6a), in the case $m = 1$ of simple bifurcation, has the two solutions: $\alpha_1 = 0$, $\alpha_1/\alpha_0 = -2b_{11}/a_{111}$. Clearly, $b_{11} \neq 0$ if (5.6c) holds, since $\dot{u}_1(s_0) = 0$.

$$
(5.7)\quad
\begin{aligned}
&a) && a_{ijk}(\epsilon) \equiv \psi_i^* \tfrac{1}{\epsilon}[G_u(u_0 + \epsilon\phi_j, \lambda_0) - G_u^0]\phi_k \,, \\
&b) && b_{ij}(\epsilon) \equiv \psi_i^* \tfrac{1}{\epsilon}\{[G_u(u_0 + \epsilon\phi_j, \lambda_0) - G_u^0]\phi_0 \\
& && \qquad + G_\lambda(u_0 + \epsilon\phi_j, \lambda_0) - G_\lambda^0]\} \\
&c) && c_i(\epsilon) \equiv \psi_1^* \tfrac{1}{\epsilon}\{[G_u(u_0 + \epsilon\phi_0, \lambda_0) - G_u^0]\phi_0 \\
& && \qquad + 2[G_\lambda(u_0 + \epsilon\phi_o, \lambda_0) - G_\lambda^0] \\
& && \qquad\qquad + [G_\lambda(u_0, \lambda_0 + \epsilon) - G_\lambda^0]\} \,.
\end{aligned}
$$

Clearly, $\{a_{ijk}(\epsilon), b_{ij}(\epsilon), c_i(\epsilon)\} \to \{a_{ijk}, b_{ij}, c_i\}$ as $\epsilon \to 0$. The above scheme avoids the need for determining second Fréchet derivatives.

Method II. There are other devices which avoid the need to evaluate the coefficients or roots of (5.5). This assumes that one branch through the bifurcation point has been determined. Then the tangent $[\dot{u}(s_0), \dot{\lambda}(s_0)]$ can also be assumed known on this branch. The idea is simply to seek solutions on some subset "parallel" to the tangent but displaced from the bifurcation point in some direction "normal" to the tangent.

For example, in the case $m = 1$ of simple bifurcation, the known solution branch $[u_1(s), \lambda_1(s)]$ has the tangent at the bifurcation, $s = s_0$ given by (5.3b,c). An "orthogonal" to this tangent in the plane spanned by $[\phi_1, 0]$ and $[\phi_0, 1]$ is also given by (5.3b,c) but with α_1 and α_0 replaced by:

$$
(5.8a)\qquad \hat{\alpha}_1 = \alpha_0(1 + ||\phi_0||^2) \,, \; \hat{\alpha}_0 = -\alpha_1||\phi_1||^2 \,.
$$

Then we seek solutions in the form:

$$
(5.8b)\qquad
\begin{aligned}
u_2 &= u_1(s_0) + \epsilon[\hat{\alpha}_0\phi_0 + \hat{\alpha}_1\phi_1] + v \,, \\
\lambda_2 &= \lambda_1(s_0) + \epsilon\hat{\alpha}_0 + \eta \,.
\end{aligned}
$$

These are to satisfy:

$$
(5.9)\qquad
\begin{aligned}
G(u_2, \lambda_2) &= 0 \,, \\
N(u_2, \lambda_2) \equiv (\hat{\alpha}_0\psi_0^* + \hat{\alpha}_1\psi_1^*)v + \hat{\alpha}_1\eta &= 0 \,.
\end{aligned}
$$

We use Newton's method to solve (5.9) for $v \in \mathbf{B}$ and $\eta \in \mathbf{R}$ with the initial estimate $(v^0, \eta^0) = (0, 0)$. Here ϵ must be taken sufficiently large so that the scheme does not return to $(u_1(s_0), \lambda_1(s_0))$ as the solution.

Method III. Another way to determine a branch bifurcating from a known branch $[u(s), \lambda(s)]$ at $s = s_0$ is to apply a constructive existence theory, using iterations, say as in [8,11]. To sketch the basic idea, we consider simple bifurcation and seek the bifurcated branch in the form

$$(5.10a)\qquad \left\{\begin{array}{l} \left.\begin{cases} u = u_1(\sigma) + \epsilon[\phi_1 + v] \\ \lambda = \lambda_1(\sigma) \end{cases}\right\} \text{ if } \dot\lambda_1^0 \neq 0\,, \\ \left.\begin{cases} u = u_1(\sigma) + \epsilon[\phi_0 + v] \\ \lambda = \lambda_1(\sigma) + \epsilon \end{cases}\right\} \text{ if } \dot\lambda_1^0 = 0\,, \end{array}\right. \quad ;\ \psi_1^* v = 0\,.$$

Then (1.1) is written, using (5.10a), as

$$(5.10b)\qquad G_u^0 v = G_u^0 v - \begin{cases} \epsilon^{-1} G(u_1(\sigma) + \epsilon[\phi_1 + v], \lambda_1(\sigma))\,, & \text{if } \dot\lambda_1^0 \neq 0 \\ \epsilon^{-1} G(u_1(\sigma) + \epsilon[\phi_0 + v], \lambda_1(\sigma) + \epsilon)\,, & \text{if } \dot\lambda_1^0 = 0 \end{cases} \quad ;\ \psi_1^* v = 0\,.$$

To insure that the right-hand side is in $\mathcal{R}(G_u^0)$, we try to pick $\sigma = s$ such that $h(s; \epsilon, v) = 0$ where

$$(5.10c)\qquad h(s;\epsilon,v) \equiv \begin{cases} \psi_1^* \left[G_u^0 v - \frac{1}{\epsilon} G\left(u_1(s) + \epsilon \begin{Bmatrix} \phi_1 \\ \phi_0 \end{Bmatrix} + \epsilon v, \lambda_1(s) + \epsilon \right) \right] \text{ if } \dot\lambda_1^0 \begin{cases} \neq 0 \\ = 0 \end{cases}, \epsilon \neq 0\,; \\ \psi_1^* \left[G_u^0 v - G_u(u_1(s), \lambda_1(s)) \left(\begin{Bmatrix} \phi_1 \\ \phi_0 \end{Bmatrix} + v \right) - G_\lambda(u_1(s), \lambda_1(s)) \begin{Bmatrix} 0 \\ 1 \end{Bmatrix} \right] \text{ if } \dot\lambda_1^0 \begin{cases} \neq 0 \\ = 0 \end{cases}, \epsilon = 0\,. \end{cases}$$

It easily follows that $h(s_0; 0, 0) = 0$ and

$$(5.10d)\qquad h_s^0 \equiv h_s(s_0; 0, 0) = \begin{cases} -\psi_1^* \left[G_{uu}^0 \dot u_1(s_0) + G_{u\lambda}^0 \dot\lambda_1(s_0) \right] \phi_1 & \text{if } \dot\lambda_1^0 \neq 0\,, \\ -\psi_1^* \left[G_{uu}^0 \phi_0 + G_{u\lambda}^0 \right] \dot u_1(s_0) & \text{if } \dot\lambda_1^0 = 0\,. \end{cases}$$

Thus, as in (5.6), $h_s^0 \equiv h_s(s_0; 0, 0) \neq 0$ and so the implicit function theorem yields a root, $s = \sigma(\epsilon, v)$, of $h(s; \epsilon, v) = 0$. We use this root in (5.10b) and then, by contraction maps, it is easily shown that (5.10b) has a unique solution $v = v(\epsilon)$ for $|\epsilon|$ sufficiently small [10].

The main difficulty in applying the above procedure is in solving $h(s; \epsilon, v) = 0$ for s at each iterate $v = v^\nu$, say. Of course, if λ occurs linearly in the problem and it is used as the parameter, s, then this is trivial. But when λ occurs nonlinearly as it must for secondary bifurcation, modifications must be introduced. Several have been proposed in [4,6,15,19].

For example, given the νth iterate, (σ^ν, v^ν), we can use the chord method to define $\sigma^{\nu+1}$ as in

$$m^0\sigma^{\nu+1} = m^0\sigma^\nu - h(\sigma^\nu; \epsilon, v^\nu)\ , \quad m^0 \equiv \psi_1^* B\ ; \tag{5.11a}$$

and then $v^{\nu+1}$ is obtained from

$$\begin{aligned} G_u^0 v^{\nu+1} &= G_u^0 v^\nu - \tfrac{1}{\epsilon} G\left(u_1(\sigma^\nu) + \epsilon\left[\begin{Bmatrix}\phi_1\\ \phi_0\end{Bmatrix} + v^\nu\right], \lambda_1(\sigma^\nu) + \begin{Bmatrix}0\\ \epsilon\end{Bmatrix}\right) - B[\sigma^{\nu+1} - \sigma^\nu] \\ &\qquad \text{if } \dot\lambda_1^0 \begin{Bmatrix}\neq 0\\ = 0\end{Bmatrix}; \\ \psi_1^* v^{\nu+1} &= 0\ . \end{aligned} \tag{5.11b}$$

Applying ψ_1^* to the right-hand side above we see that it is in $\mathcal{R}(G_u^0)$. Furthermore, with the choice

$$B = B^0 \equiv \begin{cases} \left[G_{uu}^0 \dot u_1(s_0) + G_{u\lambda}^0 \dot\lambda_1(s_0)\right]\phi_1 & \text{if } \dot\lambda_1(s_0) \neq 0\ , \\ \left[G_{uu}^0 \phi_0 + G_{u\lambda}^0\right] \dot u_1(s_0) & \text{if } \dot\lambda_1(s_0) = 0\ . \end{cases} \tag{5.12a}$$

It follows from (5.10b) that $m^0 = h_s^0 \neq 0$. There is no difficulty in showing convergence of the above scheme. To avoid the evaluation of second derivatives we can use the trick in (5.7) and take:

$$\begin{aligned} B = B(\epsilon) &\equiv \tfrac{1}{\epsilon}\left[G_u\left(u_1(s_0) + \epsilon\begin{Bmatrix}\phi_1\\ \phi_0\end{Bmatrix}, \lambda_1(s_0) + \epsilon\begin{Bmatrix}0\\ 1\end{Bmatrix}\right) - G_u\big(u_1(s_0), \lambda_1(s_0)\big)\right]\dot u_1(s_0) \\ &\quad + \tfrac{1}{\epsilon}\left[G_\lambda\big(u_1(s_0) + \epsilon \dot u_1(s_0), \lambda_1(s_0)\big) - G_\lambda\big(u_1(s_0), \lambda_1(s_0)\big)\right]\dot\lambda_1(s_0) \\ &\qquad \text{if } \dot\lambda_1(s_0)\begin{Bmatrix}\neq 0\ ,\\ = 0\ .\end{Bmatrix} \end{aligned} \tag{5.12b}$$

Clearly, $B(\epsilon) = B^0 + \mathcal{O}(\epsilon)$ so that $\psi_1^* B(\epsilon) = h_s^0 + \mathcal{O}(\epsilon)$ and the proofs proceed as before. This modification is due to Rheinboldt [19] where it is justified for finite dimensional problems. He also shows that if the solution branch $[u_1(s), \lambda_1(s)]$ in (5.11) is replaced by the "parabolic" approximation:‡

$$\begin{aligned} \hat u_1(s) &= u_1(s_0) + (s - s_0)\dot u_1(s_0) + \tfrac{1}{2}(s - s_0)^2 \ddot u_1(s_0) \\ \hat\lambda_1(s) &= \lambda_1(s_0) + (s - s_0)\dot\lambda_1(s_0) + \tfrac{1}{2}(s - s_0)^2 \ddot\lambda_1(s_0) \end{aligned} \tag{5.13}$$

the procedure still converges to a bifurcated solution.

‡ In [19] the parameter $s \equiv \lambda$ is employed but the above indicated generalization can be shown to work with no additional difficulties.

Method IV. A final method for determining the bifurcating branch at a simple eigenvalue is to use a technique based on a modification [10] of the Crandall and Rabinowitz [5] proof of bifurcation. Thus we again seek solutions of the form (5.10a) and define

$$(5.14)\qquad \begin{aligned} &a) \quad g(v,s;\epsilon) \equiv \begin{cases} \frac{1}{\epsilon} G\left(u_1(s) + \epsilon\left[\left\{\begin{smallmatrix}\phi_1\\ \phi_0\end{smallmatrix}\right\} + v\right], \lambda_1(s) + \epsilon\left\{\begin{smallmatrix}0\\ 1\end{smallmatrix}\right\}\right) & if \quad \dot\lambda_1^0 \left\{\begin{smallmatrix}\neq 0\\ =0\end{smallmatrix}\right., \epsilon \neq 0\,; \\ G_u\left(u_1(s), \lambda_1(s)\right)\left[\left\{\begin{smallmatrix}\phi_1\\ \phi_0\end{smallmatrix}\right\} + v\right] + G_\lambda(u_1(s))\left\{\begin{smallmatrix}0\\ 1\end{smallmatrix}\right\} & if \quad \dot\lambda_1^0 \left\{\begin{smallmatrix}\neq 0\\ =0\end{smallmatrix}\right., \epsilon = 0\,. \end{cases} \\ &b) \quad N(v,s;\epsilon) \equiv \psi_1^* v\,. \end{aligned}$$

Now we note that

$$(5.15a)\qquad g(0,s_0;0) = 0\,,\quad N(0,s_0;0) = 0\,,$$

and the Fréchet derivative at $(v,s;\epsilon) = (0,s_0;0)$

$$(5.15b)\qquad \mathcal{A}^0 = \left.\frac{\partial(g,N)}{\partial(v,s)}\right|_{(0,s_0;0)} = \begin{pmatrix} G_u^0 & B^0 \\ \psi_1^* & 0 \end{pmatrix},$$

where B^0 is given in (5.12a). If (5.6c) holds, then by our Lemma 2.8 it follows that $\mathcal{A}^0$ is nonsingular. Now the usual implicit function theorem shows that

$$(5.15c)\qquad \begin{cases} g(v,s;\epsilon) = 0\,, \\ N(v,s;\epsilon) = 0\,, \end{cases}$$

has a smooth solution $(v(\epsilon), \sigma(\epsilon))$ for each $|\epsilon| \le \epsilon_0$ and using this solution in (5.10a) yields the bifurcating branch of solutions.

In solving (5.15c), we never use $\epsilon = 0$ so that even when employing Newton's method, second derivatives need not be computed. Further, we can use the device of Rheinboldt and replace $[u_1(s), \lambda_1(s)]$ in (5.10a) and (5.14a) by $[\hat u_1(s), \hat\lambda_1(s)]$ of (5.13). The implicit function theorem still holds as above and now we get a solution $(\hat v(\epsilon), \hat\sigma(\epsilon))$ of the modified (5.15c) to use in the modified form of (5.10a). The rigorous justification of this procedure is straightforward and obviously the method can be made constructive, [10].

6. Numerical Methods.

To apply the previously indicated procedures we must assume that stable, convergent numerical methods are known for approximating the solutions of the linearized problems which arise in Newton's method or in the chord method. Indeed, even more is required to rigorously justify the numerical methods used in switching branches at a bifurcation point.

We must be assured that bases for $\mathcal{N}(G_u^0)$ and $\mathcal{N}(G_u^{0*})$ can be accurately determined. For the case of simple bifurcation this is not very difficult since the theory of numerical methods for computing eigenfunctions belonging to simple eigenvalues for broad classes of linear operators is well developed. Indeed, the only works thus far to justify numerical methods at bifurcation points consider simple bifurcation from a trivial branch, [2,22,23]. We shall not present a general convergence theory here but rather indicate the practical aspects in actually carrying out our procedures.

Basically, the problem (2.4) is discretized in some form which we indicate by:

$$G_h(u_h, \lambda_h) = 0 \ , \ N_h(u_h, \lambda_h, s) = 0 \ . \tag{6.1}$$

Here (u_h, λ_h) represents the approximation to (u, λ) on the net or family of nets which is parameterized by h. If (2.4) has a smooth isolated solution, then under modest assumptions on the consistency of $[G_h, N_h]$ with $[G, N]$ and on the stability and Lipschitz continuity of the linearized difference operators, say $\mathcal{A}_h$, the general theory in [9] assures us that (6.1) has a unique solution which can be computed by Newton's method. The bulk of the computations occur in this case and we write the difference operators linearized about (u_h, λ_h) say in the form:

$$\mathcal{A}_h \equiv \begin{pmatrix} A_h & b_h \\ c_h^* & d_h \end{pmatrix} . \tag{6.2}$$

From the comparison with $\mathcal{A}$ in (2.6) we see that A_h is a form of difference approximation to G_u. Indeed, A_h is in general a matrix of (large) order h^{-m} when the basic problem is formulated in $\mathbf{R}^m$. Similarly, b_h and c_h^* are column and row vectors approximating G_λ and N_u, respectively, while d_h is a scalar approximating N_λ.

The basic computational problem is to solve linear systems in the form

$$\mathcal{A}_h \begin{pmatrix} \delta u_h \\ \delta \lambda_h \end{pmatrix} = \begin{pmatrix} r_h \\ \rho_h \end{pmatrix} . \tag{6.3}$$

To do this we need only determine y_h and z_h satisfying

$$a) \quad A_h y_h = b_h \quad , \qquad b) \quad A_h z_h = r_h \quad , \tag{6.4}$$

and then:

$$c) \quad \delta\lambda_h = (-c_h^* z_h + \rho_h)/(d_h - c_h^* y_h) \quad , \quad d) \quad \delta u_h = z_h - \delta\lambda_h y_h \quad . \tag{6.4}$$

Of course, we solve (6.4a,b) by some Gaussian elimination procedure with some form of pivoting to get (neglecting the permutations for clarity of presentation)

(6.4) $\qquad e) \quad A_h = L_h U_h \,.$

Since the bulk of the computations occur in determining the factorization (6.4e), it follows that our normalization procedure costs very little extra effort. Now we sketch an algorithm for generating "all" the solution branches of (1.1) using the indicated techniques and assuming that only simple bifurcation occurs.

ALGORITHM

i) Using Euler-Newton continuation, generate an approximate solution arc: $x_{1h}(s) \equiv (u_{1h}(s), \lambda_{1h}(s))$, skipping over any "singular" points that are encountered.

ii) Return to the neighborhood of each singular point and locate it accurately (*i.e.*, use false position or bisection to determine the zero, s_{0h}, of det $A_h(s)$). In particular, simple roots or more generally odd order roots are easily sensed by the sign change in det A_h. However, one must remember to account for the row or column interchanges in the LU-decomposition.

iii) Test for limit point or bifurcation at each singularity. To do this we need an approximation ψ^*_{1h} to ψ^*_1, the null vector of G^{0*}_h. We do this as we also compute ϕ_{1h} an approximation to ϕ_1 the null vector of G^0_h. This is easily and efficiently done by inverse iteration. In fact, if we are really close to a singular point, then it suffices to use, say:

$$(6.5) \qquad \begin{aligned} &a) \qquad A_h\phi_h = \phi^0_h \;, \\ &b) \qquad A^*_h\psi_h = U^*_h L^*_h \psi_h = \phi_h \;; \end{aligned}$$

where $\phi^0_h = \delta u_h$ is the last correction in the Newton scheme used to compute $u_h(s_{0h})$. As we already have the LU-decomposition in (6.4), these calculations are not costly.

The test is in the form:

$$(6.6) \qquad \psi^*_{1h} b_h \begin{cases} \doteq 0\,, & \text{seek bifurcation ;} \\ \neq 0\,, & \text{a limit point .} \end{cases}$$

iv) To switch over to a bifurcating branch, we must also compute $[\dot{u}_{1h}(s), \dot{\lambda}_{1h}(s)]$, an approximation to the tangent to the solution branch, $x_{1h}(s)$, at the point s_{0h} best approximating the bifurcation point. However, this will have been computed in step i) or ii) if, as we assume, the normalization N_3 of (2.10e) has been employed. Then we can easily determine ϕ_{0h} an approximation to ϕ_0 of (5.2) as follows. We set

$$(6.7) \qquad a) \quad \alpha_{0h} = \dot{\lambda}_{1h}(s_{0h}) \quad , \qquad b) \quad \alpha_{1h} = \psi^*_{1h}\dot{u}_{1h}(s_{0h})/\psi^*_{1h}\phi_{1h} \quad ,$$

and then compute

$$(6.7) \qquad c) \quad \phi_{0h} = \frac{1}{\alpha_{0h}}[\dot{u}_{1h}(s_{0h}) - \alpha_{1h}\phi_{1h}] \quad .$$

[Of course, if $\alpha_{0h} = 0$, we cannot use this simple procedure. In its place we must use a solution of (6.4a) at $s = s_{0h}$ and subtract a multiple of ϕ_{1h} so that $\psi^*_{1h}\phi_{0h} = 0$.]

Now, to use Method I, we approximate $a_{111}(\delta)$ and $b_{11}(\delta)$ of (5.7b) by:

$$\begin{aligned}
&a) \quad a_{111h}(\delta) = \psi^*_{1h}\tfrac{1}{\delta}[A_h(u_{1h}(s_0) + \delta\phi_{1h}, \lambda_{1h}(s_{0h})) - A_h(u_{1h}(s_{0h}), \lambda_{1h}(s_{0h}))]\phi_{1h} \\
(6.8) \quad &b) \quad b_{11h}(\delta) = \psi^*_{1h}\tfrac{1}{\delta}[A_h(u_{1h}(s_{0h}) + \delta\phi_{1h}, \lambda_{1h}(s_{0h})) - A_h(u_{1h}(s_{0h}), \lambda_{1h}(s_{0h}))]\phi_{0h} \\
&\qquad + \psi^*_{1h}\tfrac{1}{\delta}[b_h(u_{1h}(s_{0h}) + \delta\phi_{1h}, \lambda_{1h}(s_{0h})) - b_h(u_{1h}(s_{0h}), \lambda(s_{0h}))] \, .
\end{aligned}$$

Then we approximate the other root of (5.6a) by $[\bar{\alpha}_{0h}, \bar{\alpha}_{1h}]$, where

$$(6.8) \qquad c) \qquad \bar{\alpha}_{1h}/\bar{\alpha}_{0h} = -\left(\frac{\alpha_{1h}}{\alpha_{0h}} + \frac{2b_{11h}(\delta)}{a_{111h}(\delta)}\right) \, .$$

The tangent to the bifurcated branch is approximated by:

$$(6.9) \qquad \dot{u}_{2h}(s_{0h}) = \bar{\alpha}_{1h}\phi_{1h} + \bar{\alpha}_{0h}\phi_{0h} \, , \dot{\lambda}_{2h}(s_{0h}) = \bar{\alpha}_{0h} \, .$$

Using (6.9) in the normalization N_3, we simply return to step i) and generate the bifurcating branch.

To use Method II we proceed as before but do not bother with (6.8). Rather, we approximate $[\hat{\alpha}_0, \hat{\alpha}_1]$ of (5.8a) by:

$$(6.10a) \qquad \hat{\alpha}_{0h} = -\alpha_{1h}||\phi_{1h}||^2 \, , \, \hat{\alpha}_{1h} = \alpha_{0h}(1 + ||\phi_{0h}||^2) \, .$$

Then we seek a solution of (6.1) in the form:

$$(6.10b) \qquad \begin{aligned} u_h &= u_{1h}(s_{0h}) + \epsilon[\hat{\alpha}_{0h}\phi_{0h} + \hat{\alpha}_{1h}\phi_{1h}] + v_h \, , \\ \lambda_h &= \lambda_{1h}(s_{0h}) + \epsilon\hat{\alpha}_{0h} + \eta_h \, ; \end{aligned}$$

where $N_h(\cdot)$ is taken as:

$$(6.10c) \qquad N_h(u_h, \lambda_h) \equiv (\hat{\alpha}_{0h}\phi^*_{0h} + \hat{\alpha}_{1h}\phi^*_{1h})v_h + \hat{\alpha}_{0h}\eta_h \, .$$

Once a solution $[v_h, \eta_h]$ is obtained, a new tangent vector is computed and we return to step i) using the normalization N_3. Indeed, the above indicated computations involve but minor modifications from the procedure of step i).

Method III has been discussed in more detail by Rheinboldt [19]. It has also been used for bifurcation from the trivial solution in [15,22,23].

To our knowledge, the new Method IV has not yet been used in actual calculations. However, it is in the process of being tested at the present time.

7. A Simple Example.

We have used the procedures of §6 on several examples of the form:

$$u_{xx} + f(x, u; \lambda) = 0 \ , \ u(0) = u(1) = 0 \ , \tag{7.1a}$$

where

$$f(x, u; \lambda) \equiv 2q(\lambda) + \pi^2 \lambda p(u - q(\lambda)x(1 - x)) \ . \tag{7.1b}$$

If $p(0) = 0$, then a solution of (7.1) is given by

$$u_1(x, \lambda) = q(\lambda)x(1 - x) \ , \ \lambda = \text{arb} \ . \tag{7.2}$$

The linearized problem about $u_1(x)$ is

$$\phi_{xx} + \pi^2 \lambda p_u(0)\phi = 0 \ , \ \phi(0) = \phi(1) = 0 \ . \tag{7.3}$$

Thus, if $p_u(0) = 1$, the eigenvalues of (7.3) are

$$\lambda_k = k^2 \ , \ k = 1, 2, \ldots \ . \tag{7.4}$$

We have used several choices for $q(\lambda)$ and $p(z)$ but we show here, in figure 3, the computed results for the choice

$$q(\lambda) \equiv \lambda^2 e^{-\lambda/2} \ , \ p(z) \equiv z - z^2 - z^3 \cdots - z^8 - 2z^9 \ . \tag{7.5}$$

The difference scheme used was the Collatz Mehrstellenverfaren, which is $\mathcal{O}(h^4)$ accurate and the net spacing was $h = 1/20$. The procedures of §6 were applied using the pseudoarclength normalizations N_2 and N_3, and Method II was used to switch branches at bifurcation points. (The norm, $||x||_2$, was allowed to go "negative" if it went to zero along a branch; thus Γ_1 is a smooth curve on figure 2.) The only initial guess required was $[u, \lambda] \equiv [0, 0]$ employed near $\lambda = 0$ to start the Euler-Newton continuation on the branch Γ_0,

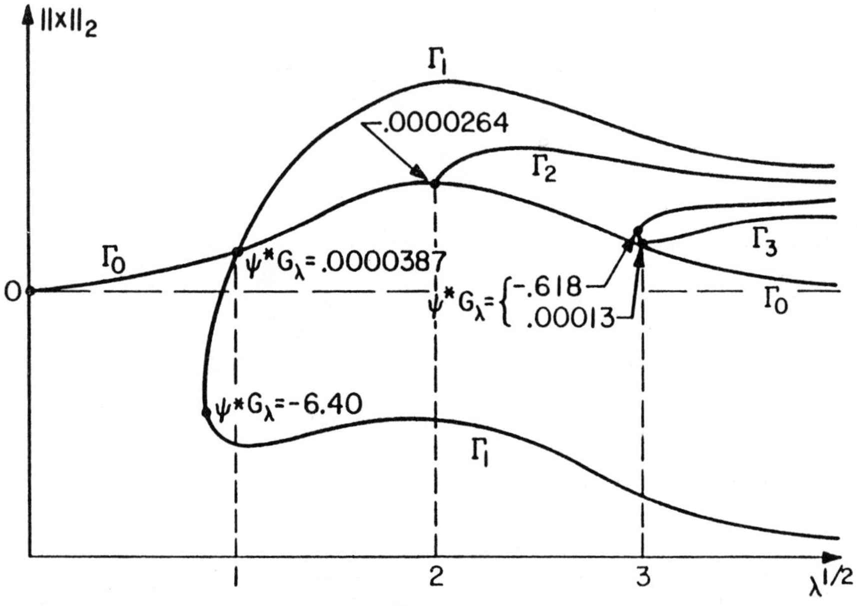
||x||$_2$
Γ$_1$
.0000264
Γ$_2$
Γ$_0$
ψ*G$_λ$=.0000387
O
ψ*G$_λ$={ -.618 .00013
Γ$_3$
Γ$_0$
ψ*G$_λ$=-6.40
Γ$_1$
1
2
3
λ$^{1/2}$

Figure 3

which is the basic solution $u_1(x,\lambda)$. This branch was easily computed and sign changes in $\det A_h(\lambda)$ were noted near the first three eigenvalues. Upon refining the location of these potential bifurcation or limit points, using false position, the eigenvector approximations ϕ_h and ψ_h were computed. The test $\psi^* G_\lambda = 0$ indicated bifurcation at each point and Method II easily switched to the branch Γ_1, Γ_2 or Γ_3 bifurcating from the corresponding point $\lambda^{1/2} = 1, 2$ or 3. These branches were extended in either direction by simply changing the sign of the ϵ used in defining the normal vector in Method II. Then continuation generated the branch. On two of these extensions, Γ_1 and Γ_3, new zeros of $\det A_h(\lambda)$ were found, but they failed the necessary test for bifurcation. Indeed, we see in figure 3 that they are simple limit points. The branch Γ_2 as shown in figure 3 is actually covered twice, but it does not show on the figure due to symmetry. There is a fundamental difference between bifurcations from odd and even "eigenvalues" but our scheme for computing has no difficulties with either case.

We also started our procedure at a remote point on the bifurcated branch Γ_1. It, of course, located the basic solution Γ_0 as a bifurcation from this branch and then proceeded to find the remainder of the branches in figure 3. To completely automate our procedure, we would have to devise step control techniques to allow optimum steps in the arclength parameter, s. Also net selection, variable order (via deferred corrections or Richardson extrapolation) and accuracy tests should be included. However, further testing with Methods I-IV should be carried out before general purpose codes are seriously contemplated. Furthermore, since the bulk of the computations occur in the continuation process, the choice of Euler-Newton must be reconsidered. Rheinboldt [18] has initiated serious studies of this question.

References

1. Anselone, P. M. and Moore, R. H., "An extension of the Newton-Kantorovič method for solving nonlinear equations with an application to elasticity", *J. Math. Anal. Appl.* **13** (1966) 476-501.

2. Atkinson, K. E., "The numerical solution to a bifurcation problem", *SIAM J. Num. Anal.*

3. Bauer, L., Keller, H. B. and Reiss, E. L., "Axisymmetric buckling of hollow spheres and hemispheres", *Comm. Pure. Appl. Math.* **23** (1970) 529-568.

4. Chen, Y. M. and Christiansen, P. L., "Application of a modified Newton's iteration method to construct solutions of eigenvalue problems of nonlinear partial differential equations", *SIAM J. Num. Anal.* **18** (1970) 335-345.

5. Crandall, M. G. and Rabinowitz, P. H., "Bifurcation from simple eigenvalues", *J. Funct. Anal.* **8** (1971) 321-340.

6. Decmoulin, Y. M. and Chen, Y. M., "An iteration method for solving nonlinear eigenvalue problems" *SIAM J. Appl. Math.* **28** (1975) 588-595.

7. Gallagher, R. H., "Perturbation procedures in nonlinear finite element structural analysis", Lecture Notes in Mathematics, No. 461, *Comp. Mech.*, Springer-Verlag, N.Y. (1975) 75-89.

8. Keller, H. B., "Nonlinear Bifurcation", *J. Diff. Eqs.* **7** (1970) 417-434.

9. Keller, H. B., "Approximation methods for nonlinear problems with application to two-point boundary value problems", *Math. of Comp.* **29** (1975) 464-474.

10. Keller, H. B., *Bifurcation Theory and Nonlinear Eigenvalue Problems*, unpublished Lecture Notes, Caltech (1976).

11. Keller, H. B. and Langford, W. F., "Iterations, perturbations and multiplicities for nonlinear bifurcation problems", *Arch. Rat. Mech. Anal.* **48** (1972) 83-108.

12. Keller, H. B. and Wolfe, A., "On the nonunique equilibrium states and buckling mechanism of spherical shells", *SIAM J. Appl. Math.* **13** (1965) 674-705.

13. Klopfenstein, R. W., "Zeros of nonlinear functions", *J. ACM* **8** (1961) 366-373.

14. Kubiček, M., Algorithm 502, "Dependence of solution of nonlinear systems on a parameter", *ACM-TOMS* **2** (1976) 98-107.

15. Langford, W. F., "Numerical solution of bifurcation problems for ordinary differential equations", preprint, McGill University, Montreal (1976).

16. Rall, L., "Convergence of the Newton process to multiple solutions", *Numer. Math.* **9** (1966) 23-37.

17. Reddien, G. W., "On Newton's method for singular problems", to be submitted for publication.

18. Rheinboldt, W. C., "Numerical continuation methods for finite element applications", in *Proc. U.S.-German Symp. on Formulation and Computational Algorithms in Finite Element Analysis*, MIT Press, 1976.

19. Rheinboldt, W. C., "Numerical methods for a class of finite dimensional bifurcation problems", to appear in *SIAM J. Num. Anal.*

20. SImpson, R. B., "A method for the numerical determination of bifurcation states of nonlinear systems of equations", *SIAM J. Num. Anal.* **12** (1975) 439-451.

21. Thurston, G. A., "Continuation of Newton's method through bifurcation points", *J. Appl. Mech.* (1969) 425-430.

22. Varol, Y. L. and Westreich, D., "Applications of Galerkins' method to bifurcation and two point boundary value problems", Math. Tech. Report-153, Ben Gurion University of the Negev, Israel (1976).

23. Weiss, R. K., "Bifurcation in difference approximations to two-point boundary value problems", *Math. of Comp.* **29** (1975) 746-760.

Acknowledgements

All the numerical work of §7 and much of the development of the test examples was carried out by Dr. E. Doedel. This work is supported by ERDA under contract AT(04-3)-767 Project Agreement No. 12 and by the ARO under contract DAAG29-75-C-0009.

Applied Mathematics 217-50
California Institute of Technology
Pasadena, California 91125

Shooting and Parallel Shooting Methods for Solving the Falkner-Skan Boundary-Layer Equation*

TUNCER CEBECI† AND HERBERT B. KELLER‡

Douglas Aircraft Company and California Institute of Technology

Received July 9, 1970

We present three accurate and efficient numerical schemes for solving the Falkner-Skan equation with positive or negative wall shear. Newton's method is employed, with the aid of the variational equations, in all the schemes and yields quadratic convergence. First, ordinary shooting is used to solve for the case of positive wall shear. Then a nonlinear eigenvalue technique is introduced to solve the inverse problem in which the wall shear is prescribed and the pressure distribution is to be determined. With this approach the reverse flow solutions (i.e., negative wall shear) are obtained. Finally, a parallel shooting method is employed to reduce the sensitivity of the convergence of the iterations to the initial estimates.

1. INTRODUCTION

Laminar boundary layers exhibiting similarity have long been the subject of numerous studies since they play an important role in illustrating the main physical features of boundary-layer phenomena. They also provide a basis for approximate methods of calculating more complex, nonsimilar flows. In the case of two-dimensional flows, when the external velocity at the edge of the boundary layer, u_e, is of the form $u_e \sim x^{(\beta/2-\beta)}$, the equations of the incompressible laminar boundary layer become similar and can be reduced to

$$f''' + ff'' + \beta[1 - (f')^2] = 0. \tag{1}$$

* This work was partially supported by the U. S. Army Research Office, Duram, under Contract DAHC 04-68-C-0006.

† California State College, Long Beach, Calif., and Douglas Aircraft Company, Long Beach, Calif.

‡ Applied Mathematics, Firestone Laboratories, California Institute of Technology, Pasadena, Calif. 91109.

The usual boundary conditions are

$$f(0) = f'(0) = 0, \tag{2a}$$

$$\lim_{\eta_\infty \to \infty} f'(\eta_\infty) = 1. \tag{2b}$$

Equation (1) is the well-known Falkner–Skan equation, which has provided many fruitful sources of information about the behavior of incompressible boundary layers. Its solutions have been extensively studied and reported in the literature for various values of β. Most of these studies have concentrated on accelerating ($\beta > 0$), constant ($\beta = 0$), and decelerating ($\beta < 0$) flows ahead of the separation point (i.e., the point of zero wall shear). For all flows ahead of the separation point the wall shear, which is proportional to $f''(0)$, is greater than zero. However, physically relevant solutions exist only for values of β in the range of $-0.19884 \leqslant \beta \leqslant 2$. Zero wall shear corresponds to $\beta = -0.19884$. Flows for which the wall shear is less than zero are called reverse flows and correspond to flows beyond the separation point. They were first obtained by Stewartson [1]. These solutions exist only for β in the range : $-0.19884 < \beta < 0$. Thus there are two physically relevant solutions of (1), (2), in this latter β-range.

Equations (1) and (2) form a third-order nonlinear two-point boundary value problem for which no closed-form solutions are known. Thus numerical methods are usually employed and of these the most popular is the shooting method. This consists in solving an initial-value problem for (1) in which one keeps $f(0)$ and $f'(0)$ fixed at their proper values (zero in this case) and tries various values of $f''(0)$ in order to satisfy (2b). The systematic method by which new values of $f''(0)$ are determined is one of the main features of this paper and it is found to be far better than the usual "cut-and-try" methods that have been applied [2]. In Fig. 1 solutions of the initial value problems are illustrated for values of $f''(0)$ in the neighborhood of the "exact" value required to satisfy $f'(\infty) = 1$. For β negative, it has been observed that solutions for all values of $f''(0)$, sufficiently near the correct one, meet the proper boundary conditions at infinity. However, the desired solution, as described in Ref. [2] and [3], is the one that approaches $f' = 1$ most rapidly from below, as indicated in Fig. 1a. The other solutions for negative values of β have no apparent physical meaning. For β positive, solutions for all values of $f''(0)$ but the correct value diverge as shown in Fig. 1b.

The present paper utilizes the shooting methods as developed and described in [4]. For the simple shooting method the "cut-and-try" searching technique is replaced by Newton's method. This generally provides quadratic convergence of the iterations and decreases the computation time. In Section 2 we describe this application to the Falkner–Skan equation for positive wall shear values. It is found that Newton's method as we employ it automatically determines the physically relevant solution for negative β values.

For reverse flows, with negative wall shear, we proceed in another way; solving what may be termed a nonlinear eigenvalue problem [4]. That is, we fix $f''(0)$ to be the desired negative wall shear value and determine the appropriate value for β by iterations. This procedure is described in Section 3 and again Newton's method is used in conjunction with the shooting method. Quadratic convergence was usually obtained. It should be noted that this type of approach also becomes important in problems (usually in nonsimilar flows) in which pressure distribution for a prescribed wall shear is to be found; i.e., so-called inverse problems.

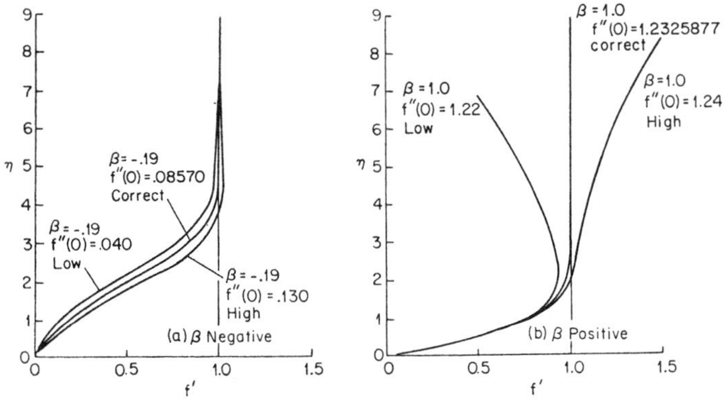

FIG. 1. Typical solutions of the Falkner–Skan boundary-layer equation.

One difficulty in both of the above applications is that the initial estimate of $f''(0)$ or of β must occasionally be very close to the exact value in order for the method to converge. These difficulties can be largely eliminated by employing the parallel shooting method [4]. We illustrate this in Section 4 only for flows with $f''(0) > 0$. Newton's method is now found to be extremely insensitive to the accuracy of the initial estimates and quadratic convergence is always observed.

2. Simple Shooting with Newton's Method. Positive Wall Shear

We first replace (1) by a system of three first-order ordinary differential equations. If the unknowns f, f', and f'' are denoted by f, u, and v, respectively, the system of three first-order equations can be written as

$$f' = u, \tag{3}$$

$$u' = v, \tag{4}$$

$$v' = -fv - \beta(1 - u^2). \tag{5}$$

In vector form this system can be written as

$$\mathbf{y}' = \mathbf{g}(\mathbf{y}), \tag{6}$$

where

$$\mathbf{y} = \begin{pmatrix} f \\ u \\ v \end{pmatrix} \qquad \mathbf{g}(\mathbf{y}) \equiv \begin{pmatrix} u \\ v \\ -fv - \beta(1 - u^2) \end{pmatrix}. \tag{7}$$

The boundary conditions given in (2) are replaced by

$$f(0) = 0, \qquad u(0) = 0 \tag{8a}$$

$$u(\eta_\infty) = 1. \tag{8b}$$

Here η_∞ is some "sufficiently large" value which is easily determined in the calculations. It varies with β but this aspect of the problem will not be discussed further. We denote the value of wall shear, $v(0)$, by

$$v(0) = s. \tag{8c}$$

The problem is to find s such that the solution of the initial value problem (6) and (8a, c) satisfies the outer boundary condition (8b). That is, if we indicate the solution of this initial value problem by $[f(\eta, s), u(\eta, s), v(\eta, s)]$ then we seek s such that

$$u(\eta_\infty, s) - 1 \equiv \varphi(s) = 0. \tag{9a}$$

To solve (9a) we employ Newton's method [5]. For some initial estimate s^0 of the root this yields the iterates s^ν defined by

$$s^{\nu+1} = s^\nu - \frac{\varphi(s^\nu)}{[d\varphi(s^\nu)/ds]} \equiv s^\nu - \frac{u(\eta_\infty, s^\nu) - 1}{[\partial u(\eta_\infty, s^\nu)/\partial s]} \quad \nu = 0, 1, 2, \ldots. \tag{9b}$$

In order to obtain the derivative of u with respect to s, we take the derivatives of (6), (8a), and (8c) with respect to s. This leads to the following linear differential equations, known as the variational equations for (3), (4) and (5):

$$F' = U, \tag{10}$$

$$U' = V, \tag{11}$$

$$V' = -fV - vF + 2u\beta U; \tag{12}$$

and to the initial conditions

$$F(0) = 0, \qquad U(0) = 0, \qquad V(0) = 1. \tag{13}$$

Here

$$F \equiv \frac{\partial}{\partial s}, \qquad U \equiv \frac{\partial u}{\partial s}, \qquad V \equiv \frac{\partial v}{\partial s}. \tag{14}$$

Once the initial-value problem given by (6), (8a), and (10) through (13) are solved, $u(\eta_\infty, s^\nu)$ and $U(\eta_\infty, s^\nu) = \partial u(\eta_\infty, s^\nu)/\partial s$ are known and consequently the next approximation to $v(0)$, namely, $s^{\nu+1}$ can be computed from (9b). We use a fourth-order Runge–Kutta method to solve the initial value problems. (More efficient schemes should be employed if possible but in practice we are at the mercy of our programmers).

The above procedure was used to study flows for which wall shear was greater than or equal to zero. Calculations[1] for various values of β with various initial estimates of s^0 showed that the method is quite effective.

The convergence properties of the iterations depend upon the value of $\beta > 0$, the initial estimates s^0 must be more accurate as β increases and conversely for the decelerating flows, $\beta < 0$. For example, at $\beta = 0.5$ the converged wall shear is $v(0) = 0.92768$. When $s^0 = 0.1$, 0.2, 0.3, and 0.4 the iterations diverge while for $s^0 = 0.5$ they converge. Similarly at $\beta = 1.0$ the correct wall shear is $v(0) = 1.23259$ and the values $s^0 < 0.8$ lead to divergence while $s^0 = 0.9$ yields convergence. Some details of the convergence rates are displayed in Table I. Obviously a good initial guess, s^0, for some value of β is obtained by employing the converged value, $v(0)$, for a close value of β. In this way we easily determined the "exact" solution given in [3] never using more than three iterations.

TABLE I

Some Iterations for Accelerating and Decelerating Flows

Iteration No. ν	$\beta = 0.5$ s^ν	$\beta = 1$ s^ν	$\beta = -0.05$ s^ν	$\beta = -0.10$ s^ν
0	0.50000	0.9	0.10000	0.10000
1	0.539332	1.697839	0.396211	0.355755
2	0.623504	1.351750	0.400320	0.319287
3	0.822382	1.244502	0.4003238	0.3192733
4	0.923408	1.232734	—	—
5	0.927675	1.232590	—	—
6	0.927680	—	—	—

[1] All calculations were single-precision on an IMB 360/65. The convergence test was: $|s^{\nu+1} - s^\nu| < 10^{-6}$.

3. Nonlinear Eigenvalue Problem. Negative Wall Shear

To obtain the reverse flow solutions, we solve the system (6), (8a), (8b), and (8c) as a "nonlinear eigenvalue" problem with β as the unknown parameter. That is, by (8c) the value $v(0)$ of the wall shear is specified and we seek the appropriate value of β, the pressure gradient parameter. To do this we again employ shooting techniques and Newton's method. Specifically, we solve the initial value problem (6), (8a), and (8c) with $v(0) = s$ fixed and seek to vary β so that (8b) is satisfied. That is, if we denote the solution of this initial value problem by $[f(\eta, \beta), u(\eta, \beta), v(\eta, \beta)]$, then we seek β such that

$$u(\eta_\infty, \beta) - 1 \equiv \Psi(\beta) = 0 \tag{15}$$

Newton's method applied to this equation yields the iterates β^ν defined by

$$\beta^{\nu+1} = \beta^\nu - \frac{\Psi(\beta^\nu)}{[d\Psi(\beta^\nu)/d\beta]} \equiv \beta^\nu - \frac{u(\eta_\infty, \beta^\nu) - 1}{[\partial u(\eta_\infty, \beta^\nu)/\partial\beta]}, \qquad \nu = 0, 1, \ldots. \tag{16}$$

The derivative $\partial u/\partial\beta$ is now obtained from the solution of the variational equations

$$p' = r, \tag{17}$$

$$r' = q, \tag{18}$$

$$q' = -fq - vp - (1 - u^2) + 2\beta ur, \tag{19}$$

subject to the initial conditions

$$p(0) = 0, \qquad q(0) = 0, \qquad r(0) = 0, \tag{20}$$

$$p = \frac{\partial f}{\partial\beta}, \qquad r = \frac{\partial u}{\partial\beta}, \qquad q = \frac{\partial v}{\partial\beta}, \tag{21}$$

and the system (17)–(20) is obtained by differentiating (3)–(5), (8a) and (8c) with respect to β.

Again, the fourth-order Runge–Kutta method is used to obtain the solution of the systems given by (6), (8a), (8c), and (17)–(20). The Newton iterates are then evaluated as in (16). This procedure has been applied with equal success to both nonseparating flows, $v(0) > 0$, and reverse flows, $v(0) < 0$. However, we discuss here only the results for reverse flows. As in the previous approach, convergence was obtained when the initial guess, β^0, was "reasonably" close to the correct value, β^*. Table II shows a comparison of reverse-flow solutions obtained by the present method and those obtained by Stewartson [1].

TABLE II

Comparison of Reverse-Flow Solutions

Number of Iterations ν	Fixed Parameter $v(0)$	Initial Estimate β^0	Converged Value β^*	Stewartson [1]
5	0	−0.26	−0.198851	—
4	−0.001	−0.26	−0.198826	—
5	−0.04	−0.26	−0.196348	—
5	−0.097	−0.26	−0.180552	—
3	−0.097	−0.18	−0.180553	−0.18
3	−0.132	−0.16	−0.152118	−0.15
7	−0.132	−0.26	−0.079596	—
3	−0.141	−0.1	−0.101763	−0.1
2	−0.108	−0.05	−0.049745	−0.05
5	−0.097	−0.05	−0.040286	—
2	−0.074	−0.025	−0.024789	−0.025
5	−0.04	−0.01	−0.009162	—

Figure 2 presents a plot of the nondimensional velocity profiles for the reverse flows. The results show that as the singular point $\beta = 0$, is approached, the magnitude of the reverse flow velocity decreases and the boundary-layer thickness increases. In the region very close to $\beta = 0$, it is difficult to obtain solutions.

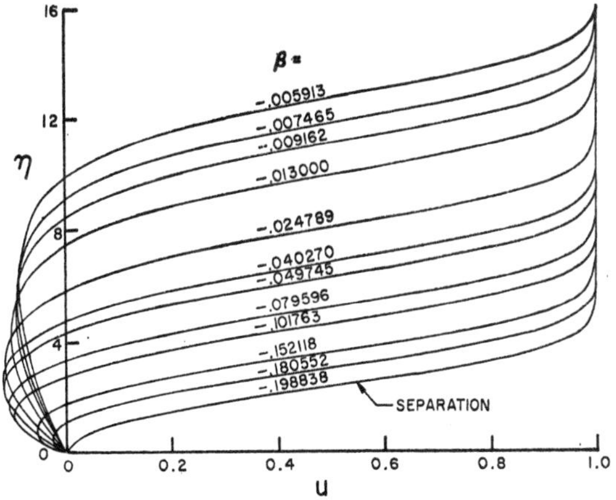

FIG. 2. Dimensionless velocity profiles for reverse flows.

4. Parallel Shooting, Positive Wall Shear

The sensitivity to the initial guess of the simple shooting methods, described in Sections 2 and 3 can be reduced by using the parallel shooting method [4]. According to this method, the total interval $[0, \eta_\infty]$ is divided into a number of subintervals, the appropriate initial-value problems are integrated over each subinterval, and then all of the "initial" data are adjusted *simultaneously* in order to satisfy the boundary conditions and appropriate continuity conditions at the subdivision points.

In the present study, we have arbitrarily divided the total transformed boundary-layer thickness η_∞ into three[2] subintervals: $[0, \eta^I]$, $[\eta^I, \eta^{II}]$, $[\eta^{II}, \eta_\infty]$. Over each subinterval the system (6) is solved subject to the initial conditions

$$\text{(I)}\ \mathbf{y}(0) = \begin{pmatrix} 0 \\ 0 \\ s \end{pmatrix}, \quad \text{(II)}\ \mathbf{y}(\eta^I) = \begin{pmatrix} a_1 \\ b_1 \\ c_1 \end{pmatrix}, \quad \text{(III)}\ \mathbf{y}(\eta^{II}) = \begin{pmatrix} a_2 \\ b_2 \\ c_2 \end{pmatrix}. \tag{22}$$

We denote the solutions of (6) over the subintervals I, II, III by $\mathbf{y}^I(\eta, s)$, $\mathbf{y}^{II}(\eta, a_1, b_1, c_1)$, and $\mathbf{y}^{II},(\eta, a_2, b_2, c_2)$, respectively. Then we impose the continuity conditions

$$\mathbf{y}^I(\eta^I, s) = \begin{pmatrix} a_1 \\ b_1 \\ c_1 \end{pmatrix} = \mathbf{y}^{II}(\eta^I, a_1, b_1, c_1), \tag{23a}$$

$$\mathbf{y}^{II}(\eta^{II}, a_1, b_1, c_1) = \begin{pmatrix} a_2 \\ b_2 \\ c_2 \end{pmatrix} = \mathbf{y}^{III}(\eta^{II}, a_2, b_2, c_2), \tag{23b}$$

and the boundary condition (8b) which becomes

$$(\mathbf{y}^{III})_2 \equiv u^{III}(\eta_\infty, a_2, b_2, c_2) = 1. \tag{23c}$$

This system of equations can also be written in vector form as

$$\boldsymbol{\varphi}(\mathbf{s}) \equiv \begin{bmatrix} \varphi_1 \\ \varphi_2 \\ \varphi_3 \\ \varphi_4 \\ \varphi_5 \\ \varphi_6 \\ \varphi_7 \end{bmatrix} = \begin{bmatrix} f^I(\eta^I, s) - a_1 \\ u^I(\eta^I, s) - b_1 \\ v^I(\eta^I, s) - c_1 \\ f^{II}(\eta^{II}, a_1, b_1, c_1) - a_2 \\ u^{II}(\eta^{II}, a_1, b_1, c_1) - b_2 \\ v^{II}(\eta^{II}, a_1, b_1, c_1) - c_2 \\ u^{III}(\eta_\infty, a_2, b_2, c_2) - 1 \end{bmatrix} = 0 \tag{24a}$$

[2] The choice turned out to be quite satisfactory for the cases studied in this paper.

where, in transposed form,

$$\mathbf{s}^T = (s, a_1, b_1, c_1, a_2, b_2, c_2) \tag{24b}$$

The system (24a) has seven equations and seven unknowns. We solve this system by Newton's method which now yields the iterates $\mathbf{s}^\nu$ defined by:

$$\mathbf{s}^{\nu+1} = \mathbf{s}^\nu - \left[\frac{\partial \boldsymbol{\varphi}(\mathbf{s}^\nu)}{\partial \mathbf{s}}\right]^{-1} \boldsymbol{\varphi}(\mathbf{s}^\nu), \qquad \nu = 0, 1, \ldots. \tag{25}$$

To find the Jacobian matrix $[(\partial \boldsymbol{\varphi}/\partial \mathbf{s})(\mathbf{s}^\nu)]$, we solve the following variational systems:

$$\begin{array}{lll}
\text{(I)} & 0 \leqslant \eta \leqslant \eta^I & \\
& F' = U & \Big\| \; F^1(0) = 0, \\
& U' = V & \Big\| \; U^1(0) = 0, \\
& V' = -f^I V - V^I F + 2\beta u^I U & \Big\| \; V^1(0) = 1; \\
\text{(II)} & \eta^I \leqslant \eta \leqslant \eta^{II} & \\
& F' = U & \Big\| \; F^2(\eta^I) = 1, \quad F^3(\eta^I) = 0, \quad F^4(\eta^I) = 0; \\
& U' = V & \Big\| \; U^2(\eta^I) = 0, \quad U^3(\eta^I) = 1, \quad U^4(\eta^I) = 0; \\
& V' = -f^{II} V - V^{II} F + 2\beta u^{II} U & \Big\| \; V^2(\eta^I) = 0, \quad V^3(\eta^I) = 0, \quad V^4(\eta^I) = 1; \\
\text{(III)} & \eta^{II} \leqslant \eta \leqslant \eta_\infty & \\
& F' = U & \Big\| \; F^5(\eta^{II}) = 1, \quad F^6(\eta^{II}) = 0, \quad F^7(\eta^{II}) = 0; \\
& U' = V & \Big\| \; U^5(\eta^{II}) = 0, \quad U^6(\eta^{II}) = 1, \quad U^7(\eta^{II}) = 0; \\
& V' = -f^{III} V - V^{III} F + 2\beta u^{III} U & \Big\| \; V^5(\eta^{II}) = 0, \quad V^6(\eta^{II}) = 0, \quad V^7(\eta^{II}) = 1.
\end{array} \tag{26}$$

For example, in the first subinterval we solve the system of equations by using the given initial conditions in that subinterval and obtain the solution $F^1(\eta)$, $U^1(\eta)$, and $V^1(\eta)$. In the second subinterval, we solve the system of equations three times using the three sets of initial conditions with superscripts 2, 3, and 4. We denote these solutions by $F^j(\eta)$, $U^j(\eta)$, and $V^j(\eta)$ with $J = 2, 3$, and 4. Similarly, by solving the system of equations in the third subinterval using the three sets of initial conditions, 5, 6, and 7 we obtain the solutions $F^j(\eta)$, $U^j(\eta)$, and $V^j(\eta)$ with $j = 5, 6$, and 7. Using these solutions, the Jacobian matrix can be shown to be [4]

$$A^{(\nu)} \equiv \frac{\partial \boldsymbol{\varphi}(s^{(\nu)})}{\partial \mathbf{s}} \equiv \begin{bmatrix}
F^1(\eta^I) & -1 & 0 & 0 & 0 & 0 & 0 \\
U^1(\eta^I) & 0 & -1 & 0 & 0 & 0 & 0 \\
V^1(\eta^I) & 0 & 0 & -1 & 0 & 0 & 0 \\
0 & F^2(\eta^{II}) & F^3(\eta^{II}) & F^4(\eta^{II}) & -1 & 0 & 0 \\
0 & U^2(\eta^{II}) & U^3(\eta^{II}) & U^4(\eta^{II}) & 0 & -1 & 0 \\
0 & V^2(\eta^{II}) & V^3(\eta^{II}) & V^4(\eta^{II}) & 0 & 0 & -1 \\
0 & 0 & 0 & 0 & U^5(\eta_\infty) & U^6(\eta_\infty) & U^7(\eta_\infty)
\end{bmatrix} \tag{27}$$

In terms of this matrix the Newton iterates (25) are determined by solving the simple linear system:

$$A^{(\nu)}(\mathbf{s}^{\nu+1} - \mathbf{s}^{\nu}) = -\boldsymbol{\varphi}(\mathbf{s}^{\nu}). \tag{28}$$

In summary, our parallel shooting procedure proceeds by solving three initial value problems for (6) with initial data $\mathbf{s}^{\nu}$ as in (22). Along with these we solve the seven linear variational problems (26) to evaluate $A^{(\nu)}$ from (27). Finally, $\mathbf{s}^{\nu+1}$ is determined from (28) and one iteration cycle is completed.

Parallel shooting procedures were developed [4] to eliminate a shortcoming of simple shooting methods, frequently called "instability." This phenomenon is caused by the fact that rapidly growing solutions of the initial value problems magnify various errors (truncation as well as roundoff). Incorrect guesses at the appropriate unknown initial data are effectively truncation errors. Thus it may be expected that parallel shooting will reduce the sensitivity of the convergence of iteration procedures (like Newton's method) to the magnitude of the initial errors. This speculation was indeed borne out in the present calculations. Of course, the initial guess for parallel shooting is more complicated, in the present case requiring seven values rather than one. However, this additional complexity caused no difficulty and in the calculations presented below, we simply assumed a linear velocity profile from which all of the initial values (i.e., $\mathbf{s}^0$) were obtained by an integration and a differentiation. Tables III, IV, and V present the results obtained

TABLE III

Comparison of Calculated Results with Those of Reference [3]

	$v(0)$	
β	Parallel Shooting	Ref. [3]
−0.195	0.55177	0.055172
−0.19	0.085702	0.085700
−0.10	0.319278	0.319270
−0.05	0.400330	0.400323
0	0.469603	0.469600
0.10	0.587037	0.587035
0.20	0.686711	0.686708
0.40	0.854423	0.854421
0.60	1.120269	1.120268
1.00	1.232561	1.232588
1.20	1.335724	1.335772
1.60	1.521516	1.521514

TABLE IV

Convergence of $v(0)$; Parallel Shooting

Iteration No. ν	$\beta = 0.5$ $v(0)$	$\beta = 1.0$ $v(0)$	$\beta = -0.05$ $v(0)$
0	0.167	0.167	0.167
1	0.9445	2.23939	0.44785
2	0.927396	1.422649	0.399251
3	0.927683	1.243555	0.400330
4	0.927682	1.232634	—
5	—	1.232591	—

TABLE V

Convergence of **s** values for $\beta = 0.5$

Iteration No.	$v(0)$	a_1	b_1	c_1	a_2	b_2	c_2
0	0.167	0.25	0.5	0.167	0.5	1.0	0.167
1	0.9445	0.38951	0.698044	0.461947	2.302359	1.091020	0.083129
2	0.927396	0.380950	0.680833	0.444003	2.195862	0.994259	0.014278
3	0.927683	0.381092	0.681117	0.444287	2.197072	0.994964	0.014322
4	0.927682	0.381092	0.681116	0.444286	2.197067	0.994862	0.014324

in this manner. Table III gives a comparison of $v(0)$-values calculated by the parallel shooting method with those given in [3]. The agreement is very good and the disagreement which is in the sixth decimal place, is probably due to the round-off error since single precision arithmetic was used in the present calculation. In all these calculations the subintervals were arbitrarily taken as $0 \leqslant \eta \leqslant 1$, $1 \leqslant \eta \leqslant 3$ and $3 \leqslant \eta \leqslant 6$. The choice turned out to be satisfactory. When the subintervals were changed, almost identical results were obtained.

Table IV indicates the convergence of the iterations for three β values. These results, as for all other β values studied by parallel shooting, show excellent convergence properties. A comparison of Tables I and IV shows that parallel shooting has much less sensitivity to the initial guess.

A study was also made of the convergence of the components of $\mathbf{s}^\nu$ other than $s = v(0)$. Table V presents the results for $\beta = 0.5$. It is seen that the values of **s** other than $v(0)$ all converge much faster than $v(0)$ and this was typical. Note that for practical purposes convergence has occurred in the second iteration (i.e. three significant digits).

References

1. K. Stewartson, Further solutions of the Falkner–Skan equation, *Phil. Mag.* **12** (1931), 865–896.
2. D. R. Hartree, On an equation occuring in Falkner and Skan's approximate treatment of the equations of the boundary layer, *Proc. Cambridge Phil. Soc.* **33** (1937), 223–239.
3. A. M. O. Smith, "Improved Solutions of the Falkner and Skan Boundary-Layer Equation," Fund Paper *J. of Aero. Sci.* Sherman M. Fairchild, 1954.
4. H. B. Keller, "Numerical Methods for Two-Point Boundary-Value Problems," pp. 54–68, Ginn-Blaisdell Pub. Co., Waltham, Mass., 1968.
5. E. Isaacson and H. B. Keller, "Analysis of Numerical Methods," p. 54, John Wiley and Sons, New York, 1966.

Reprinted with permission from *SIAM Journal On Applied Mathematics,* Vol. 38, No. 1 (1980), pp. 52–64.

THE VON KARMAN SWIRLING FLOWS*

MARIANELA LENTINI† AND HERBERT B. KELLER†

Abstract. Computations of the Karman swirling flows are reported. They show four such flows and clearly indicate that an infinite sequence exists.

1. Introduction. In 1921 Karman showed that for the steady axi-symmetric flow of a viscous incompressible fluid above an infinite plane rotating disk, the Navier–Stokes equations reduce to a system of ordinary differential equations which are also the boundary layer equations for this flow. Subsequently there have been numerous studies, both theoretical and numerical, of this and closely related flows. The most recent calculations of Zanbergen and Dijkstra (1977) show that two distinct families of solutions exist, parametrized by the ratio $\gamma \equiv \Omega_\infty/\Omega_0$, of angular velocity at ∞ to that of the disk. We find that there are at least four such families and our results clearly suggest an infinite number of such flows. In particular flows with $\gamma = 0$, having zero angular velocity at infinity, would exist in each family. A mechanism to establish such flows is suggested by a study of the results and the way in which they have been computed. Recent asymptotic expansions support our conjecture and supply quantitative agreement.

The success of our calculations is due to three important features. First we establish rather carefully the proper asymptotic boundary conditions that must be satisfied in order to have bounded solutions. These are highly nonlinear conditions and they enable us to compute on shorter finite intervals than would otherwise be possible. Secondly we employ a technique for generating or continuing families of solutions in which the "free" parameter of the family need not be single valued. This enables solution branches to "fold back" and easily yields nonunique solutions. Finally the code used to solve the families of two point boundary value problems is rather sophisticated and automatically adjusts the net and order of accuracy to yield efficient solutions. Indeed all of the results reported here were obtained in single precision on a P.D.P. 10 using less than 200 net-points (in fact less than 120 points for most solutions).

In § 2 we formulate the equations and boundary conditions. In § 3 we describe the numerical methods and arclength continuation procedure. In § 4 we present the results of our calculations and discuss some of their implications.

2. Formulation and asymptotic boundary conditions. Cylindrical coordinates (r, ϕ, z) are taken such that the disk rotating with angular velocity Ω_0, is in the plane $z = 0$. Then with the dimensionless axial variable

$$\zeta = (\Omega_0/\nu)^{1/2} z,$$

the von Karman representation of the velocity field is

$$u = r\Omega_0 f'(\zeta), \qquad v = r\Omega_0 g(\zeta), \qquad w = -2(\nu\Omega_0)^{1/2} f(\zeta). \tag{1}$$

The Navier–Stokes equations reduce in this case to

$$\text{(2)} \qquad \begin{aligned} &\text{a)} \quad f'''(\zeta) + 2f(\zeta)f''(\zeta) = [f'(\zeta)]^2 + \gamma^2 - g^2(\zeta), \\ &\text{b)} \quad g''(\zeta) + 2f(\zeta)g'(\zeta) = 2f'(\zeta)g(\zeta). \end{aligned}$$

* Received by the editors December 12, 1978.

† California Institute of Technology, Pasadena, California 91125. This research was supported by CONICIT (Caracas, Venezuela), by DOE under contract No. EY-76-S-03-0767 Project Agreement No. 12 and by the U.S. Army Research Office under contract No. DAAG29-78-C-0011.

If there is no suction or injection of fluid at the disk the appropriate boundary conditions are

$$f(0)=0, \qquad f'(0)=0, \qquad g(0)=1. \tag{3}$$

The constant of integration, γ, in (2a) is determined by the angular velocity of the fluid at infinity which we write as

$$\lim_{\zeta\to\infty} \Omega_0 g(\zeta)=\Omega_\infty \equiv \gamma\Omega_0. \tag{4}$$

The asymptotic behavior of the flow at infinity has been studied by McLeod (1969). He finds that the decay is exponential; monotone for $\gamma=0$ and oscillatory for $\gamma\neq 0$. These results justify the integration leading to (2a) and assure that:

$$\lim_{\zeta\to\infty} [f(\zeta), f'(\zeta), f''(\zeta), g(\zeta), g'(\zeta)]=[f_\infty, 0, 0, \gamma, 0]. \tag{5}$$

Here f_∞ represents the unknown inflow ($f_\infty>0$) or outflow ($f_\infty<0$) at infinity. To obtain appropriate boundary conditions we must examine the linearized problem about the flow (5) at infinity. To do this we introduce the new dependent variables

$$\begin{aligned} \mathbf{y}(\zeta) &\equiv [y_1(\zeta), y_2(\zeta), y_3(\zeta), y_4(\zeta), y_5(\zeta)]^T \\ &=[f(\zeta)-f_\infty, f'(\zeta), f''(\zeta), g(\zeta)-\gamma, g'(\zeta)]^T. \end{aligned} \tag{6}$$

Then the reduced Navier–Stokes equations (2) can be written as the first order system:

$$\mathbf{y}'(\zeta)=A(f_\infty, \gamma)\mathbf{y}(\zeta)+Q(\mathbf{y}) \tag{7}$$

where:

$$A(f_\infty, \gamma)\equiv\begin{pmatrix} 0 & 1 & 0 & 0 & 0 \\ 0 & 0 & 1 & 0 & 0 \\ 0 & 1 & -2f_\infty & -2\gamma & 0 \\ 0 & 0 & 0 & 0 & 1 \\ 0 & 2\gamma & 0 & 0 & -2f_\infty \end{pmatrix},$$

(8)

$$Q(\mathbf{y})\equiv\begin{pmatrix} 0 \\ 0 \\ y_2^2(\zeta)-2y_1(\zeta)y_3(\zeta)-y_4^2(\zeta) \\ 0 \\ 2y_2(\zeta)y_4(\zeta)-2y_1(\zeta)y_5(\zeta) \end{pmatrix}.$$

A general theory for nonlinear systems of the form (7) on semi-infinite intervals has been developed by Lentini and Keller (1978). For present purposes the main result of that study is that bounded solutions can be insured by a condition which insures bounded solutions of the corresponding linear problem (i.e. (7) with $Q(\mathbf{y})\equiv 0$). The condition on the linear problem is the vanishing of the projection of the solution into the subspace of $\mathbb{R}^5$ which is the union of: i) the invariant subspace of $A(f_\infty, \gamma)$ belonging to eigenvalues with real part positive; and ii) the complement of the eigenspace in any invariant subspace of $A(f_\infty, \gamma)$ belonging to eigenvalues with real part zero. These subspaces are the only ones in which solutions of the linear problem can grow unboundedly. Thus our condition is simply to require that components in these subspaces vanish at infinity. In addition the components in the eigenspaces belonging to eigenvalues with real part zero can be specified at either end of the interval.

To represent the above indicated conditions analytically we must determine the eigenvalues and eigenvectors of $A(f_\infty, \gamma)$. Since f_∞ is not known in advance the resulting conditions are highly nonlinear. Specifically the eigenvalues of $A(f_\infty, \gamma)$ are

$$\begin{aligned} \lambda_1 = \overline{\lambda_2} &= -f_\infty + a(f_\infty, \gamma) + ib(f_\infty, \gamma), \\ \lambda_3 &= 0, \\ \lambda_4 = \overline{\lambda_5} &= -f_\infty - a(f_\infty, \gamma) + ib(f_\infty, \gamma), \end{aligned} \tag{9}$$

where

$$a(f_\infty, \gamma) \equiv \frac{1}{\sqrt{2}}[(f_\infty^4 + 4\gamma^2)^{1/2} + f_\infty^2]^{1/2}, \qquad b(f_\infty, \gamma) \equiv \frac{1}{\sqrt{2}}[(f_\infty^4 + 4\gamma^2)^{1/2} - f_\infty^2]^{1/2}.$$

For $\gamma \neq 0$ we have $\operatorname{Re} \lambda_1 = \operatorname{Re} \lambda_2 > 0$ and $\operatorname{Re} \lambda_4 = \operatorname{Re} \lambda_5 < 0$ since $a(f_\infty, \gamma) > |f_\infty|$. Thus for $\gamma \neq 0$ we need only the two dimensional projection into the invariant subspace spanned by the eigenvectors belonging to λ_1 and λ_2. A laborous calculation of P and P^{-1}, for which $P^{-1}AP = \operatorname{diag}(\lambda_1, \lambda_2, \lambda_3, \lambda_4, \lambda_5)$, yields, from the first two rows $\mathbf{q}_\nu^T$, $\nu = 1, 2$ of P^{-1}, the required projection conditions: $\lim_{\zeta\to\infty} \mathbf{q}_\nu^T \mathbf{y}(\zeta) = 0$, $\nu = 1, 2$; or explicitly:

$$\begin{aligned} &\lim_{\zeta\to\infty} \left[(f_\infty + a) y_2(\zeta) + y_3(\zeta) - \frac{\gamma}{a} y_4(\zeta) \right] = 0, \\ &\lim_{\zeta\to\infty} \left[\frac{b^2}{\gamma} a y_2(\zeta) + (f_\infty + a) y_4(\zeta) + y_5(\zeta) \right] = 0. \end{aligned} \tag{10}$$

When $\gamma = 0$ then $b = 0$, $a = |f_\infty|$ and from (9): $\lambda_1 = \lambda_2 = 0$ if $f_\infty > 0$ or $\lambda_4 = \lambda_5 = 0$ if $f_\infty < 0$. In either case it can be shown that zero is an eigenvalue of algebraic multiplicity three but only of geometric multiplicity two. So the above indicated theory only requires one projection condition in this case. But we can use as the asymptotic boundary conditions the limiting form of (10) as $\gamma \to 0$ provided $f_\infty > 0$. This gives the required projection condition and as an extra condition, a projection onto one of the null vectors. (We recall that such a condition can be imposed at either end.) Also McLeod (1969) has shown that $f_\infty > 0$ for $\gamma = 0$ so the limiting form of (10) is a correct condition is this case.

3. Numerical methods and continuation procedures. The calculations employ the variables

$$\begin{aligned} \mathbf{v}(\zeta) &\equiv [v_1(\zeta), v_2(\zeta), v_3(\zeta), v_4(\zeta), v_5(\zeta)]^T, \\ &= [f(\zeta), f'(\zeta), f''(\zeta), g(\zeta) - \gamma, g'(\zeta)]^T. \end{aligned}$$

Then the Navier–Stokes equations (2) become

$$\mathbf{v}'(\zeta) = A(\gamma)\mathbf{v}(\zeta) + Q(\mathbf{v}(\zeta)). \tag{11}$$

Here $Q(\mathbf{v})$ is as defined in (8) and $A(\gamma) \equiv A(0, \gamma)$, also using (8). The boundary conditions are obtained from (3) and the asymptotic conditions (10) evaluated at some finite point ζ_∞:

$$\begin{aligned} &v_1(0) = 0, \qquad v_2(0) = 0, \qquad v_4(0) = 1 - \gamma, \\ &h_1(\mathbf{v}, \gamma) \equiv [f_\infty + a(f_\infty, \gamma)] v_2(\zeta_\infty) + v_3(\zeta_\infty) - \frac{\gamma}{a(f_\infty, \gamma)} v_4(\zeta_\infty) = 0, \\ &h_2(\mathbf{v}, \gamma) \equiv \frac{b^2(f_\infty, \gamma)}{\gamma} a(f_\infty, \gamma) v_2(\zeta_\infty) + [f_\infty + a(f_\infty, \gamma)] v_4(\zeta_\infty) + v_5(\zeta_\infty) = 0. \end{aligned} \tag{12}$$

We recall that $v_1(\zeta_\infty) = f_\infty$ in this formulation so that the conditions $h_\nu(\mathbf{v}, \gamma) = 0$, $\nu = 1, 2$ in (12) are nonlinear.

The choice of ζ_∞ is varied throughout the calculations. One check that an adequate value has been used is, following Zanbergen and Dijkstra (1977), to test the identities:

$$(13)\qquad \begin{aligned} &[f''(0)]^2 - [g'(0)]^2 = 4\int_0^\infty f(\zeta)([f''(\zeta)]^2 - [g'(\zeta)]^2)\,d\zeta,\\ &f''(0)g'(0) + \frac{2}{3}\gamma^3 - \gamma^2 + \frac{1}{3} = 4\int_0^\infty f(\zeta)f''(\zeta)g'(\zeta)\,d\zeta. \end{aligned}$$

In addition ζ_∞ was varied until the computed results stabilize to at least four significant digits.

The basic numerical method used to approximate solutions of (11), (12) is a two point box or trapezoidal rule on a nonuniform mesh $\{\zeta_j\}_0^J$ placed on $[0, \zeta_\infty]$ with $\zeta_0 = 0$ and $\zeta_J = \zeta_\infty$. This scheme is second order accurate; that is the error is $\mathcal{O}(h^2)$ where h is the maximum step width. The resulting nonlinear difference equations and boundary conditions are represented as

$$(14)\qquad G(V; \gamma) = 0,$$

where V is a vector of the $5(J+1)$ unknowns, $\mathbf{v}(\zeta_j)$, $j = 0, 1, \cdots, J$. To solve these equations Newton's method is used; that is

$$(15)\qquad \begin{aligned} &\text{a)}\quad V^0(\gamma) \equiv \text{initial estimate},\\ &\text{b)}\quad \frac{\partial G}{\partial V}(V^\nu; \gamma)\delta V^\nu = -G(V^\nu; \gamma),\\ &\text{c)}\quad V^{\nu+1}(\gamma) = V^\nu + \delta V^\nu(\gamma),\\ &\text{d)}\quad \text{STOP IF } \|\delta V^\nu(\gamma)\| < \varepsilon. \end{aligned}$$

The $\mathcal{O}(h^2)$ accurate solution obtained in this way is used to readjust the net, possibly adding netpoints, so that the local truncation error is approximately equidistributed on the new net. Then higher order accurate solutions, $[\mathcal{O}(h^4)$ or $\mathcal{O}(h^6)]$ are computed by applying deferred corrections. A final net readjustment and refinement may also be made. All of these details are done automatically by the code which is a modification of PASVA 3 developed by Lentini and Pereyra (1977).

For $\gamma = 1$ the solution is given by a rigid body rotation and so the calculations are easily started there. When a solution $V(\gamma)$ has been computed we then use as the initial estimate of the solution at $\gamma + \delta\gamma$:

$$(16)\qquad V^0(\gamma + \delta\gamma) = V(\gamma) + \delta\gamma V_\gamma(\gamma).$$

Here $V_\gamma(\gamma)$ is the solution of the variational problem about $V(\gamma)$:

$$(17)\qquad \frac{\partial G}{\partial V}(V; \gamma)V_\gamma = -\frac{\partial G}{\partial \gamma}(V; \gamma).$$

Of course the Jacobian matrix occurring in (17) is essentially known in LU-factored form from the last Newton iterate in (15) and so V_γ can be computed with almost no additional effort. This continuation procedure allows relatively large steps to be taken in γ and is a natural one to use in conjunction with Newton's method. We use it until $\partial G/\partial V$ shows signs of becoming singular (easily detected during the continuation in γ

as we keep track of $\Delta(\gamma) \equiv \det [\partial G(V(\gamma), \gamma)/\partial V])$. Then we switch over to a pseudo-arclength continuation procedure devised by Keller (1977).

In this new procedure we introduce a parameter, s, and seek $V = V(s)$ *and* $\gamma = \gamma(s)$ to satisfy

$$\text{(18)} \qquad \begin{aligned} &G(V, \gamma) = 0, \\ &N(V, \gamma, s) \equiv \dot{V}_0^T[V(s) - V_0] + \dot{\gamma}_0[\gamma(s) - \gamma_0] - [s - s_0] = 0. \end{aligned}$$

Here it is assumed that at $s = s_0$, say, we know $V_0, \gamma_0, \dot{V}_0, \dot{\gamma}_0$ such that

$$\text{(19)} \qquad \begin{aligned} &\text{a)} \quad G(V_0; \gamma_0) = 0, \\ &\text{b)} \quad \frac{\partial G}{\partial V}(V_0; \gamma_0)\dot{V}_0 + \frac{\partial G}{\partial \gamma}(V_0; \gamma_0)\dot{\gamma}_0 = 0, \\ &\text{c)} \quad \|\dot{V}_0\|^2 + \dot{\gamma}_0^2 = 1. \end{aligned}$$

We solve (18) by Newton's method and even though $\partial G(V_0; \gamma_0)/\partial V$ is singular, the Jacobian of the "inflated" system may be nonsingular. Indeed it is shown that this is the case for "normal limit points" which occur where the family of solutions $V(\gamma)$ become double valued. After $[V(s), \gamma(s)]$ have been computed we easily compute $[\dot{V}(s), \dot{\gamma}(s)]$ satisfying (19b, c) (with s_0 replaced by s) and the sign is determined such that

$$\text{(20)} \qquad \dot{V}_0^T\dot{V}(s) + \dot{\gamma}_0\dot{\gamma}(s) > 0.$$

There is no difficulty in computing the double valued solution branches in this way and the amount of computation is but very slightly more than that for the direct continuation in γ. Away from the normal limit points we return to the γ-continuation procedure (16), (17) if $|\delta\gamma|$ can be larger than $|\gamma(s) - \gamma(s_0)|$ in (18).

The pseudo-arclength continuation procedure is but part of a more general scheme in which bifurcation can also be detected; see Keller (1977). However no bifurcations were found in our calculations.

4. Results and discussion. Starting at $\gamma = 1$ with $\zeta_\infty = 15$ we compute a branch of solutions over $1 \geqq \gamma \geqq 0$. Then $\zeta_\infty = 25$ was used to continue this branch over $0 \geqq \gamma \geqq \gamma_1$. The limit point value $\gamma_1 = -.16057$ was accurately determined by a bisection procedure, described in Lentini (1978), which locates the zero of $\dot{\gamma}(s)$ during the pseudo-arclength continuation. This value agrees with the value $\gamma^* = -.16057$ of Zanbergen and Dijkstra (1977) and of A. White (1978). The solutions over $1 \geqq \gamma \geqq \gamma_1$ are shown in Figs. 1a, b, c. For $1 \geqq \gamma \geqq 0$ they have been obtained by Rogers and Lance (1960). For $\gamma = 0$, the results agree with the original Karman solution. On the entire branch $1 \geqq \gamma \geqq \gamma_1$ our results agree as well as we can check them with those of Zanbergen and Dijkstra (1977) and of White (1978).

Apparently the steady solutions of the form (1) which have been computed by Rogers and Lance (1960) for $\gamma < \gamma_1$ do not connect to the sheet we have determined. A linearized stability analysis for time dependent solutions of the similarity form (1) easily shows that our solutions at γ_1 have *neutral* stability.

The arclength-continuation procedure smoothly goes over to yield a second sheet or branch of solutions over $\gamma_1 \leqq \gamma \leqq \gamma_2 = 0.07452$. Again γ_2 is accurately determined by bisection. This branch has also been computed by Zanbergen and Dijkstra (1977) and A. White (1978). They all had difficulties in determining solutions up to γ_2. Our solutions on this branch are shown in Figs. 2a, b, c. Since $\gamma_2 > 0$ we have determined, on this branch, another solution of the original Karman problem, $\gamma = 0$. During the

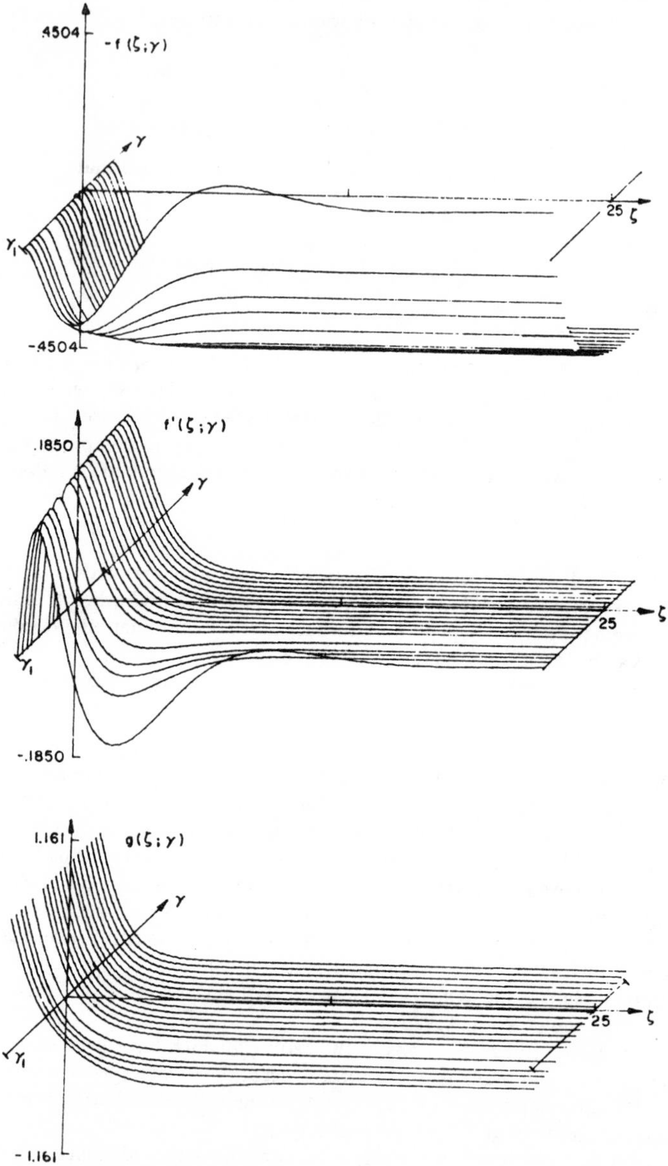

FIG. 1. *Velocity profiles on part of the first branch*: $-0.16057 = \gamma_1 \leqq \gamma \leqq .08$ a) *Axial velocity*: $-f$. b) *Radial velocity*: f'. c) *Angular velocity*: g.

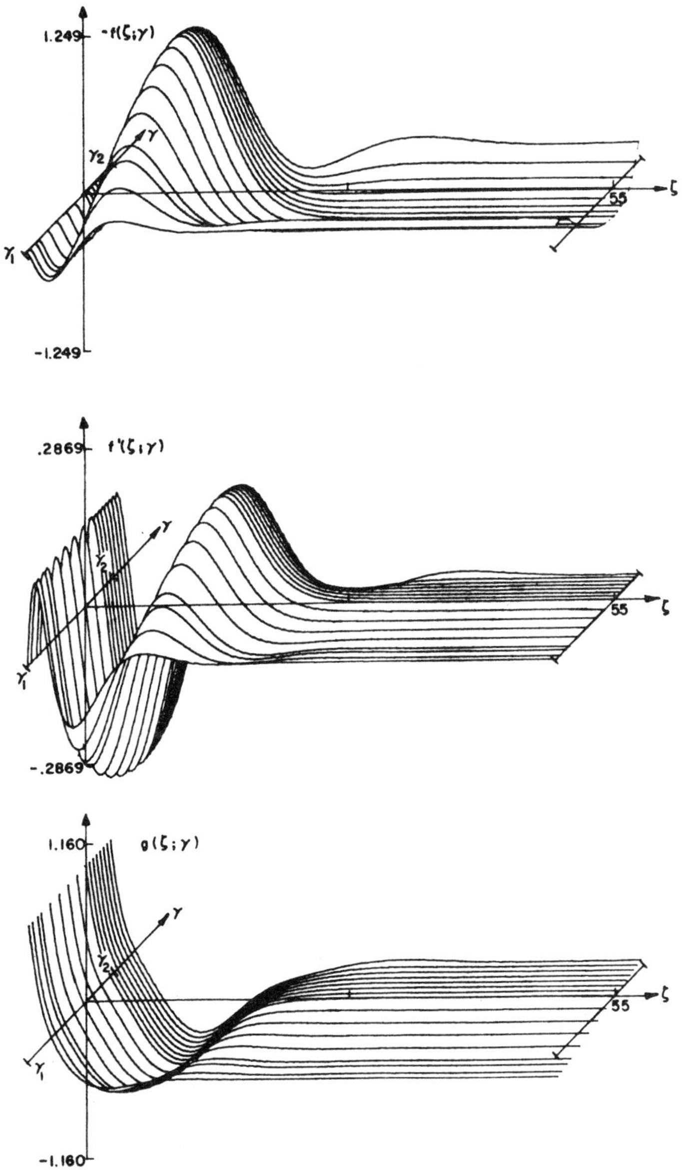

FIG. 2. *Velocity profiles on the second branch*: $-0.16057 = \gamma_1 \leqq \gamma \leqq \gamma_2 = 0.07452$. a) *Axial velocity*: $-f$. b) *Radial velocity*: f'. c) *Angular velocity*: g.

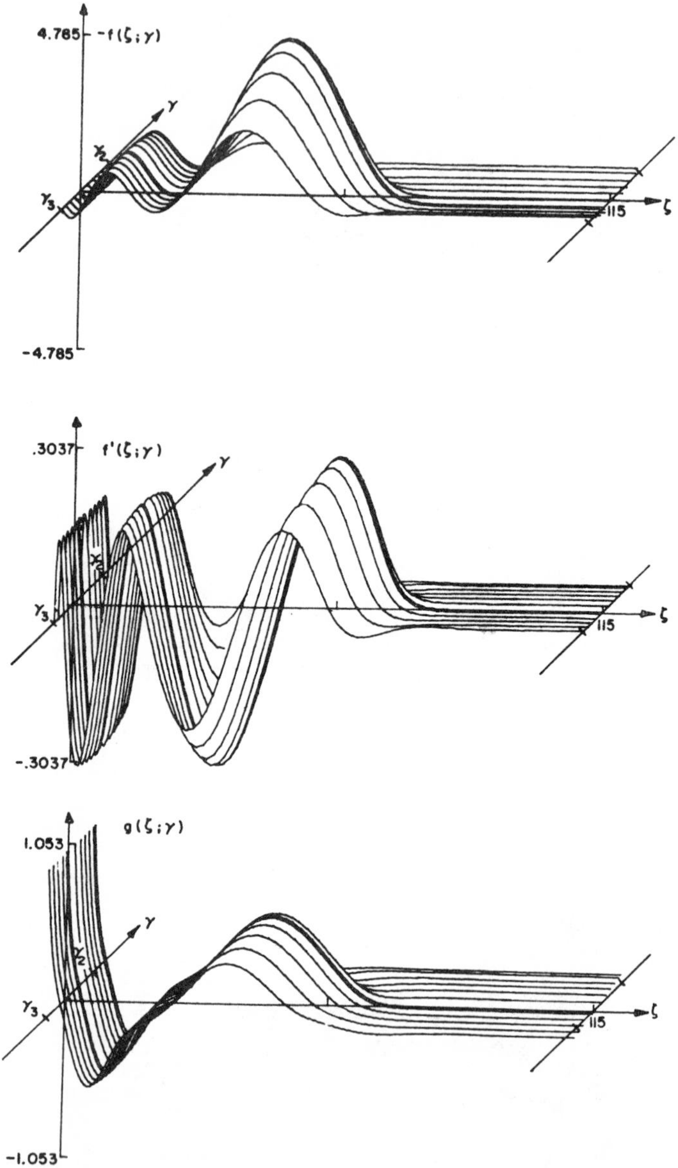

FIG. 3. *Velocity profiles on the third branch*: $-0.0574 = \gamma_3 \leqq \gamma \leqq \gamma_2 = 0.07452$. a) *Axial velocity*: $-f$. b) *Radial velocity*: f'. c) *Angular velocity*: g.

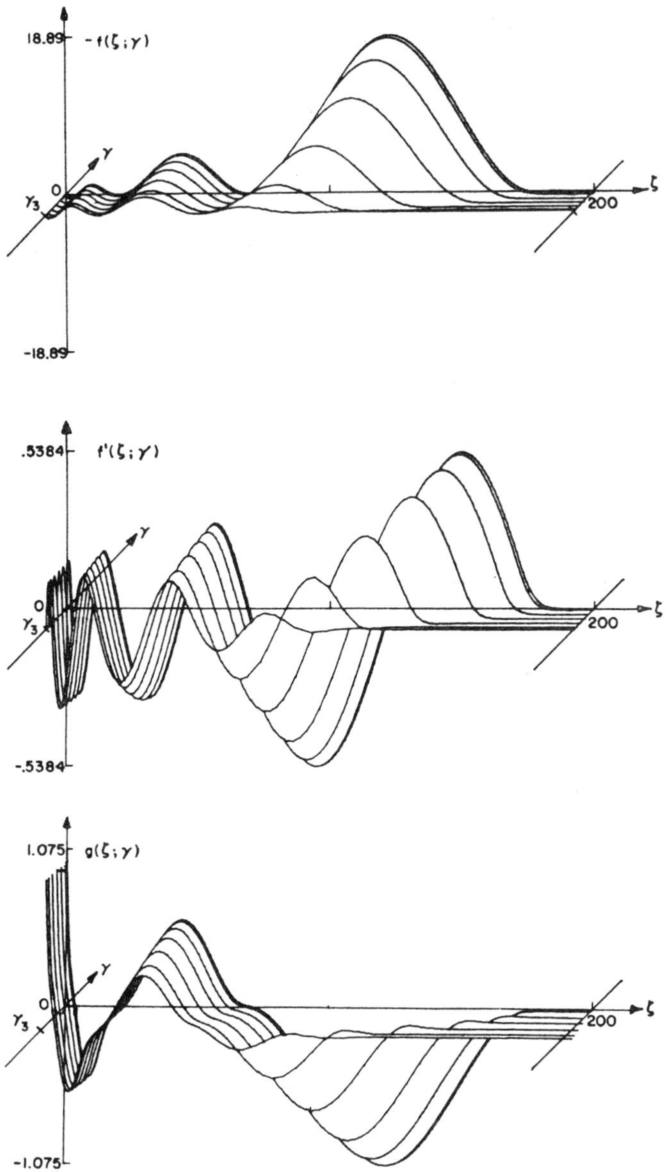

FIG. 4. *Velocity profiles on part of the fourth branch*: $-0.0574 = \gamma_3 \leqq \gamma \leqq 0$. a) *Axial velocity*: $-f$. b) *Radial velocity*: f'. c) *Angular velocity*: g.

generation of this second branch we increase ζ_∞ to 45. Since γ increases "into" the page in Figs. 1, 2, 3, 4, the outermost curves, for $\gamma = \gamma_1$, on Figs. 1a, b, c are supposed to be identical with the corresponding curves on Figs. 2a, b, c. Scale changes account for the different appearance. Note in particular that the axial velocity in Fig. 2a reaches a maximum near $\gamma = 0$, the second Karman swirling flow.

Again our procedures smoothly reverse at γ_2 and give a new third sheet or branch of solutions over $\gamma_2 \geqq \gamma \geqq \gamma_3 = -0.0574$. They are shown in Figs. 3a, b, c. Note that another Karman swirling flow is obtained and again the maximum axial flow occurs near this solution, $\gamma = 0$ in Fig. 3a. Now ζ_∞ had to be increased to 95. But at $\gamma = \gamma_3$ our computations go over into a fourth branch $\gamma_3 \leqq \gamma \leqq \gamma_4$ and we have not bothered to determine γ_4. Indeed we have only continued this branch up to $\gamma = 0$ to obtain a fourth Karman swirling flow. The solutions on this branch used ζ_∞ up to $\zeta_\infty = 200$ and they are shown in Figs. 4a, b, c.

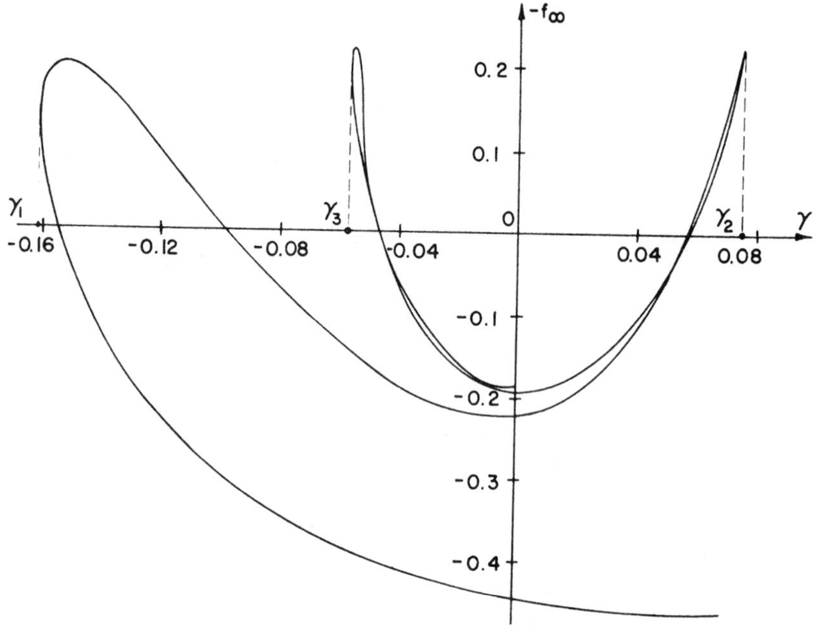

FIG. 5. *The axial flow at infinity, $-f(\infty, \gamma)$, as a function of $\gamma \equiv \Omega_\infty/\Omega_0$.*

In Fig. 5 we plot the axial flow at infinity, $-f_\infty$, vs. the ratio $\gamma = \Omega_\infty/\Omega_0$ of angular velocity at infinity to that of the disk. This graph clearly suggests how the sheets of solutions fold over at the critical ratios, γ_n, $n = 1, 2, 3$ to form new branches of solutions. It seems safe to conjecture that an infinite sequence of critical ratios $\{\gamma_n\}$ exists, that these ratios alternate about $\gamma = 0$ and converge to zero as a limit, and thus that there are infinitely many solutions of the Karman swirling flow problem. To focus attention on these flows we show in Figs. 6a, b, c the four such solutions we have computed. These are also included in Figs. 1, 2, 3, 4 for the parameter value $\gamma = 0$. It is rather remarkable and an impressive independent check on the calculations that these various solutions

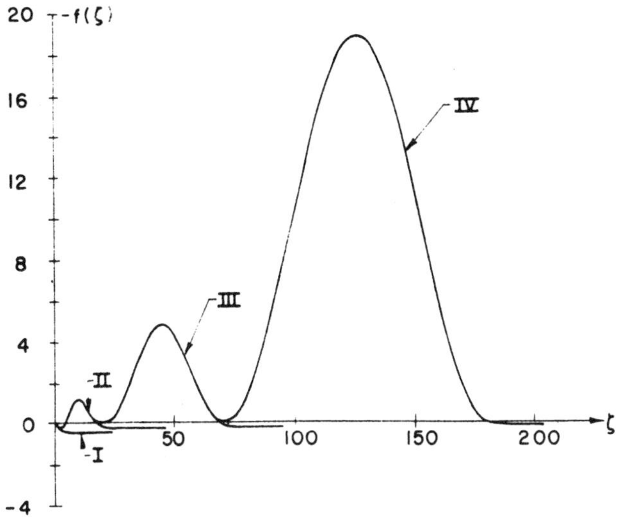

FIG. 6a. *The axial velocity, $-f(\zeta)$, in the first four Karman swirling flows, $\gamma = 0$.*

coincide to graphical accuracy (and better) over large intervals. Recall that they were generated by continuation procedures involving rather large departures from all of the swirling flow solutions.

The general qualitative rule for the formation of the successive higher order Karman swirling flows is apparent from a study of Fig. 6. In fact R. Rosales (1978) has

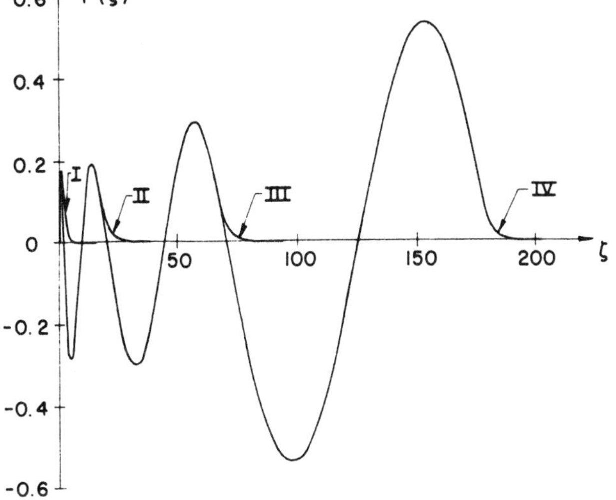

FIG. 6b. *The radial velocity, $f'(\zeta)$, in the first four Karman swirling flows, $\gamma = 0$.*

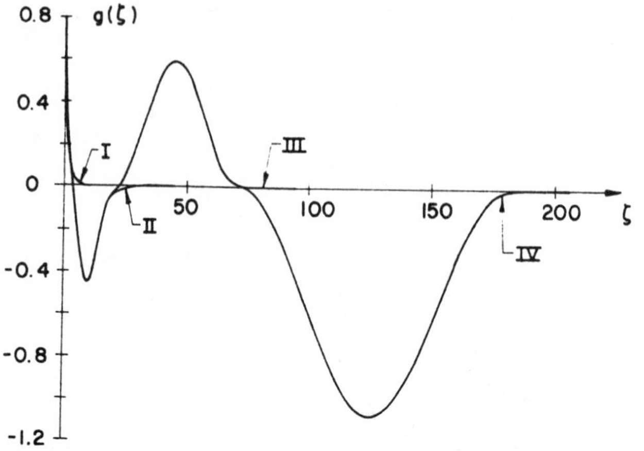

FIG. 6c. *The angular velocity, $g(\zeta)$, in the first four Karman swirling flows, $\gamma = 0$.*

been able to make it quantitative by matching more carefully the expansions for large blowing by Kuiken (1971) with those for small suction by Ockendon (1972). Specifically we denote the nth swirling flow by $[f_n(\zeta), g_n(\zeta)]$, $n = 1, 2, 3, \cdots$ and let ζ_n be the location of the positive maximum of $(-f_n(\zeta))$ on $\zeta \geqq 0$. Then $-f_n(\zeta_n)$ is the largest outward velocity from the disk (i.e. largest "blowing" rate) in the nth solution. (Note that the classical Karman swirling flow, $n = 1$, has no blowing.) To leading order in the expansion for large blowing Rosales obtains:

$$\text{(21)}\quad \begin{aligned} &\text{a)} \quad f_n(\zeta_n)/f_{n-1}(\zeta_{n-1}) \approx 4, && n = 3, 4, \cdots, \\ &\text{b)} \quad (\zeta_n/\zeta_{n-1})\left(\frac{2^{n-1}-1}{2^n-1}\right) \approx 1, && n = 3, 4, \cdots, \\ &\text{c)} \quad f_n(\infty)\left(\frac{16|f_n(\zeta_n)|}{g_n^2(\zeta_n)}\right)^{1/3} \approx 1.2096\cdots, && n = 2, 3, 4, \cdots. \end{aligned}$$

In Table I we list the critical values we have computed for the quantities entering the formulae in (20). Then in Table II we tabulate the quantities listed in (20) using the data from Table I. Additional quantities have been checked with their expansions and give equally good agreement.

TABLE I
Some critical values for the swirling flows.

	$n = 1$	$n = 2$	$n = 3$	$n = 4$
ζ_n	—	10.3033	45.332	126.269
$-f_n(\zeta_n)$	—	1.24838	4.78435	18.8929
$g_n(\zeta_n)$	—	.464972	.598110	1.07453
$f_n(\infty)$	.442237	.223859	.193794	.187078

TABLE II
Comparison of computed and asymptotic values of some quantities for the swirling flows.

	$n=2$	$n=3$	$n=4$	$n \to \infty$
$f_n(\zeta_n)/f_{n-1}(\zeta_{n-1})$	—	3.832	3.949	4
$(2^{n-1}-1)\zeta_n/(2^n-1)\zeta_{n-1}$	—	1.886	1.300	1
$f_n(\infty)[16\lvert f_n(\zeta_n)\rvert/g_n^2(\zeta_n)]^{1/3}$	1.0120	1.1591	1.1962	1.2096 · · ·

Acknowledgment. The three dimensional graphs were plotted using a program of Dr. R. K.-H. Szeto to whom we are grateful.

REFERENCES

T. VON KARMAN (1921), *Uber laminare und turbulente Reibung.* Z. Angew. Math. Mech., 1, pp. 232–252.

H. B. KELLER (1977), *Numerical solution of bifurcation and nonlinear eigenvalue problems,* Applications of Bifurcation Theory, Academic Press, New York, pp. 359–384.

H. K. KUIKEN (1971), *The effect of normal blowing on the flow near a rotating disk of infinite extent,* J. Fluid Mech., 47, pp. 789–798.

M. LENTINI AND V. PEREYRA (1977), *An adaptive finite-difference solver for non-linear two-point boundary value problems,* SIAM J. Numer. Anal., 14, pp. 91–111.

M. LENTINI (1978), *Boundary value problems over semi-finite intervals,* Ph.D. Thesis, California Institute of Technology, Pasadena, CA 91125.

M. LENTINI AND H. B. KELLER (1978), *Boundary value problems over semi-infinite intervals,* in preparation.

J. B. MCLEOD (1969), *The asymptotic form of solutions of von Karman's swirling flow problem,* Quart. J. Math. Oxford, (2) 20, pp. 483–496.

H. OCKENDON (1972), *An asymptotic solution for steady flow above an infinite rotating disk with suction,* Quart. J. Mech. and Applied Math., XXV, pp. 291–301.

M. H. ROGERS AND G. N. LANCE (1960), *The rotationally symmetric flow of a viscous fluid in the presence of an infinite rotating disk,* J. Fluid Mech., 7, pp. 617–631.

R. R. ROSALES (1978), *Asymptotic expansions of Karman swirling flows,* in preparation.

A. B. WHITE (1978), *Multiple solutions for rotationally symmetric incompressible, viscous flow.* Center for Numer. Anal., Report CNA-132, University of Texas at Austin.

P. J. ZANDBERGEN AND D. DIJKSTRA (1977), *Non-unique solutions of the Navier–Stokes equations for the Karman swirling flow,* J. Engineering Mathematics 11, pp. 167–188.

A CATALOG OF SELECTED

DOVER BOOKS

IN SCIENCE AND MATHEMATICS